GRASSES AND LEGUMES
IN
BRITISH AGRICULTURE

GRASSES AND LEGUMES IN BRITISH AGRICULTURE

Edited by

C. R. W. Spedding and E. C. Diekmahns

Grassland Research Institute, Hurley

Bulletin 49

Commonwealth Bureau of Pastures and Field Crops

C·A·B

Commonwealth Agricultural Bureaux

First published in 1972 by the Commonwealth Agricultural Bureaux
Farnham Royal, Bucks, England
SBN 85198 016 3

Printed in Great Britain by
Alden & Mowbray Ltd
at the Alden Press, Oxford

Foreword

GRASSLAND in all its varying forms accounts for about 73% of the mainland of Britain, and although no accurate records exist, it is estimated that of the total acreage of grassland in England and Wales some 60–70% has at one time or another been cultivated and sown. Grass and forage legumes are, therefore, by far the most important crops of this country, and are likely to remain so because of the ease with which they grow in our climate.

Much has been written about the grass crop in the general context of land use and animal production systems, but this book is the first attempt in this country to describe in precise biological terms the agronomic characteristics of the main crop grasses and legumes grown in Britain, and to assess their value for animal production. It is intended to be a record of our present knowledge and thus to serve as a handbook for research workers, teachers and students at universities and agricultural colleges, agricultural advisory officers and progressive farmers.

This book is a joint production of the Grassland Research Institute and the Commonwealth Bureau of Pastures and Field Crops. Over the years members of the staff of this research station have accumulated a great deal of information and experience that cannot easily be communicated in the form of scientific papers and the more formal reports and pronouncements of research workers. In order to ensure that this accumulated wisdom was made available, the then Director of the Bureau, Mr. C. L. Skidmore, and I decided in November 1966 to try to ensure that it was written down and set in the context of progress and development elsewhere in the world. At that time we had not envisaged a volume of this size, but it was always our desire to make it as comprehensive and informative as possible. It soon became apparent that to achieve this we should have to rely on contributions from other centres in addition to those from Hurley. At this stage, Professor C. R. W. Spedding, as Editor, and Mr. E. C. Diekmahns, as Assistant Editor, took on the detailed planning of the book and sought the collaboration of the various chapter authors. These were drawn from present and past members of the staff of the Institute; from a few of our colleagues at the Welsh Plant Breeding Station, Aberystwyth and the Plant Breeding Institute in Cambridge; from the National Institute of Agricultural Botany in Cambridge, as well as Mr. R. Boyce of the Commonwealth Bureau of Pastures and Field Crops, Hurley, and Mr. L. P. Smith of the Meteorological Office in Bracknell. To all of these we are most grateful for the time and trouble they have taken in helping with this enterprise.

The editorial task of trying to weld the individual styles of so many contributors into a readable and coherent whole can be imagined. The Editor and his Assistant, and the staff of the Commonwealth Bureau of Pastures and Field Crops are to be congratulated on the satisfactory way in which this has been accomplished. Much of the painstaking work of editing, checking and processing the manuscripts was undertaken by Mr. E. C. Diekmahns, Head of the Information Section in the Institute. Unfortunately he died suddenly of a heart attack in July 1971 just as his contribution to this project was nearing completion. Although he did not live to see the finished work, this book will serve as one of the more permanent records of his time at the Institute.

I am sure that this book will become a reference for all who are concerned with the science of grass and leguminous forage crops, in this and other countries, for many years to come.

E. K. Woodford
Director,
Grassland Research Institute,
Hurley

December 1971

Contents

Plates

Contributors

(and their specific chapters)

D. T. A. Aldrich, 15–22, 33–6
R. C. Anslow, 15–22, 33–6, 40
M. Borrill, 6
R. Boyce, 15–22, 29, 33–6
A. H. Charles, 39
C. R. Clement, 2
J. P. Cooper, 4
A. J. Corrall, 15–22
D. W. Cowling, 14–22, 33–6
J. W. Dent, 15–22, 33–6
E. C. Diekmahns, 25
W. Ellis-Davies, 27
E. A. Garwood, 14–22, 32–6
J. O. Green, 15–22, 33–6, 40
C. E. Harris, 15–22, 33–6
A. J. Heard, 12, 30
M. J. Hopper, 15–22, 32
A. A. Idle, 7, 15–22, 33–6
O. R. Jewiss, 9, 11
J. G. W. Jones, 44
L. Jones, 8, 26
D. D. Kydd, 15–22, 33–6
D. A. Lambert, 15–22, 33–6
J. Morrison, 7, 15–22, 33–6
D. F. Osbourn, 5, 15–22, 41
C. E. Powell, 14–22, 32–6
W. F. Raymond, 5, 41
M. J. Robson, 14–22, 32–6
H. H. Rogers, 38
J. Rossiter, 15–22, 33–6
G. J. A. Ryle, 10
A. Smith, 23, 37, 43
L. P. Smith, 3
C. R. W. Spedding, 1, 46
W. Stiles, 14–22, 32–6
J. C. Tayler, 45
R. Thaine, 15–22, 33–6
P. T. Thomas, 6, 24
D. J. Thomson, 15–22, 33–6
C. G. Waldock, Index
D. C. Whitehead, 13, 15–22, 31, 33–6
R. J. Wilkins, 42
R. D. Williams, 28

Referees

The following served as referees for various chapters

G. Alderman, B.Sc.
D. T. A. Aldrich, B.Sc.(Agric.), Dip. Agric.
R. A. Arnott, B.Sc.
R. D. Baker, B.Sc., M.Sc., C.D.F.M.
E. W. Bean, B.Sc., Ph.D.
N. R. Brockington, B.Sc., Dip. Agric. Sci.
C. R. Clement, B.Sc., Ph.D.
J. P. Cooper, M.Sc., Ph.D., D.Sc., F.I.Biol.
A. J. Corrall
J. W. Dent, B.Sc., F.R.I.C.
J. S. W. Dickens, B.Sc., Ph.D.
W. Ellis-Davies, B.Sc.
R. Gair, B.Sc.
J. O. Green, B.Sc., M.Sc.
D. J. Griffiths, M.Sc., Ph.D.
J. Hodgson, B.Sc., Ph.D.
L. Jones, B.Sc.
L. H. P. Jones, B.Agr.Sci., Ph.D., A.R.A.C.I., F.I.Biol.
E. L. Leafe, B.Sc., Ph.D.
D. Wynn Parry, M.Sc.
E. T. Roberts, B.Sc., A.R.C.S., M.I.Biol.
M. J. Robson, B.Sc., Ph.D., M.I.Biol.
H. H. Rogers, B.Sc., Dip. Agric. Sci.
G. J. A. Ryle, B.Sc., Ph.D., M.I.Biol.
J. L. Stoddart, B.Sc., Ph.D.
T. T. Treacher, B.Sc., Ph.D.
R. J. Wilkins, B.Sc., Ph.D.
R. D. Williams, B.Sc., Ph.D., M.I.Biol.
T. E. Williams, B.Sc.
A. Zaleski, M.A.

Acknowledgments

THANKS are due to the Agricultural Research Council, and to the Director and Governing Body of the Grassland Research Institute, for facilities granted in the compilation of this book.

It is a pleasure to record our gratitude to the contributors, to the Directors of contributing Institutes for their co-operation, and to all those who served as referees.

We wish to acknowledge the following for permission to reproduce figures or photographs.

Messrs Faber & Faber Ltd, and Professor J. T. Coppock for Figs 2.1a, 2.1b and 2.2.

Dr Yvonne Aitken and the Editor of the *Aust. J. agric. Res.* for Plates 16A, B and C.

The remaining figures were prepared by Mr B. D. Hudson, the photographs (Chapters 11, 15, 16, 19–22, 27, and 33–6) were taken by Mr. A. E. King and the index was prepared by Lt. Col. C. G. Waldock.

We wish especially to thank Mr. I. A. N. Wilson for considerable assistance in the preparation of the whole manuscript, and Dr R. D. Williams for his help in seeing the book through the press.

C. R. W. Spedding

E. C. Diekmahns

Preface

THE object of this book is to set down clearly the findings of research concerning the most important crop grasses and legumes in British agriculture, with particular reference to agronomic characteristics such as growth habit, establishment, yield and quality.

The intended readership includes advisory officers, technical representatives, teachers and students at universities and agricultural colleges, progressive farmers and research workers.

The idea of such a book was first mooted in November 1966, and it has taken longer than was originally hoped to bring it to publication. It would probably never have been completed if a less ambitious time-table had been adopted at the outset, but the time actually taken is not unreasonable in relation to the immense amount of work that has gone into it.

Forty-three people have contributed to the text and each chapter has been independently refereed. Since the chapters on each of the main grass and legume species are the result of editorial combination of contributions from twenty-two different authors, the difficulties of checking successive drafts may be imagined.

The book represents a recognition that research workers have a responsibility to make their findings available, not merely in the form of papers (scientific and popular) embodying main results and conclusions, but also in terms of the vast accumulation of descriptive information. This is partly because it may be useful to others and partly because it is necessary to be clear about what is and what is not already known.

Thus, although the book has been made as readable as possible, it has many of the attributes of a work of reference.

In all this, a great deal of help has been received from the publishers and it is a pleasure to record the appreciation of the editors for the efforts of the Director and staff of the Commonwealth Agricultural Bureau of Pastures and Field Crops.

Mr E. C. Diekmahns contributed greatly to the preparation of the book and worked very hard on it until his untimely death in July 1971. Having been actively associated with the enterprise from the beginning, it is sad that he did not see it completed.

I should like to record a special appreciation for all his efforts and for those of Dr R. D. Williams, who took over at a critical time and completed Mr Diekmahns' task.

C. R. W. Spedding

PART I

INTRODUCTION

CHAPTER 1

Grasses and Legumes as Crops

A GREAT variety of grass and legume species occurs in the natural grassland of the world and has formed the basis of much animal production. Characteristically, such grassland has been regarded as a natural resource to be exploited with the minimum of further investment, whether in terms of fertilizer, cultivation or even enclosure. In Britain, Sir George Stapledon and Dr. William Davies (1948) drew attention to the untapped potential of grassland and Davies (1960) stressed the need to regard grass as a crop. Nevertheless, much of the thinking about grassland, both in practice and in research, has been greatly coloured by this background of association with low-cost exploitation, low output of a highly seasonal nature, low quality and unpredictable quantity.

Questions that would be regarded as important both to ask and to answer for other crops, have been less confidently posed and less clearly answered for the grass and legume constituents of grassland. If these plants are to play their full part in feeding the world, however, it is vital to treat them as crops and to describe their attributes as crop plants. It is true that large areas of the world's grassland must continue for some time to be exploited with minimal inputs, for mainly economic reasons, but it is also true that many more of these areas are amenable to growing grass as a crop than is usually assumed. In the British Isles, probably a high proportion of the acreage on which grass is ever grown comes into this category. Nor need we assume that any particular cultivations are necessarily involved. It is quite certain that a vast area is capable of growing herbage crops, though the factors which decide whether they ought to be grown are so many and various that meaningful generalization is impossible.

Pasture grasses have many advantages, however: they are capable of producing a high yield of high-quality food suitable for ruminants in a variety of physiological states; they can be grown *as crops* where few others can grow; many are perennials and can remain productive for many years if well managed; they form plant populations capable of photosynthesis and growth during a large part of the year, whenever the temperature permits. In addition the legumes often associated with grass are independent of fertilizer nitrogen.

These and other attributes have yet to be fully exploited in practice or, indeed, explored in research and, as already mentioned, decisions as to whether grass or legume crops constitute the best use of land must take other factors into account.

This book is chiefly concerned with herbage crops in the situation where these issues have already been decided in favour of a grass or a legume, or both. In these circumstances, there remain a great many species and varieties to choose from.

Once the purpose for which the crop is to be grown has been decided, what is required is a description, in terms relevant to the choice that has to be made, of the main crop plants available. New cultivars, however, are constantly being bred, selected, tested and evaluated by many organizations in this country, notably the Welsh Plant Breeding Station, Aberystwyth, and the Plant Breeding Institute and the National Institute of Agricultural Botany, both at Cambridge. The best *cultivar* available for any particular purpose must therefore be expected to change rapidly, even from one year to the next.

This book has therefore concentrated on a clear description and assessment of the main grassland species, selected for their importance as cultivated crop plants when sown as single species or in simple mixtures, within British agriculture. The species chosen have, however, much wider relevance, being common constituents of sown temperate grassland throughout the world. The information available for each species has for the most part been obtained from studies on named varieties, and where it has been considered necessary to do so, the variety has been specified. Most of the information has been derived from research in the UK, but work done in, for example, New Zealand, the USA, Australia and the Netherlands has also been drawn on where relevant.

As with so many agricultural investigations that have to be conducted in the field, research concerned with the evaluation of grasses and legumes has largely been carried out on a comparative basis. This gives rise to some difficulty in describing each species, but an attempt has been made to avoid a mass of comparative statements and to set down, as clearly as possible, what is known and what is not known about each grass and legume. This information has been assembled for 8 grasses in Part II and 4 legumes in Part III and a chapter has been devoted to each.

The choice of these species is not quite as simple as might be supposed. Certainly they are those most commonly sown and in this sense are therefore important. However, many other species have not received anything like the same attention, and it remains possible that they may become important and, if subjected to intensive selection, might prove more useful than some of the species currently sown.

After the 8 chapters on individual grasses, therefore, and similarly after the 4 legume chapters, there is an additional chapter on the species omitted from the main list, with some indication as to why they are regarded as less important or useful. In some cases, it is difficult to give a reason for the apparent neglect of a particular species.

Preceding these sections on particular grasses and legumes are explanatory chapters designed to assist in interpreting the descriptions of species. Definitions of terms and the reasons for using them are given in some of these preliminary chapters, while others are concerned with morphology and with simple descriptions of growth and reproduction in grasses and legumes. Although all the data about one species have generally been placed in the chapter devoted to it, where the bulk of the information available, such as that on chemical composition, pests and diseases, or response to light or management, is common to many grasses or many legumes, it has been brought together in one of the explanatory chapters and only the exceptions noted briefly in the chapters dealing with individual species.

In assembling what is known about the species detailed in the chapters in Parts II and III, particular attention has been paid to the identification of gaps in the information available. In many cases, it has not been possible to arrive at clear-cut statements, and reliance has to be placed on recorded experience, not always consistent. This is especially so in relation to such attributes as longevity and persistency, which are, in any event, not easy to distinguish. Longevity refers to a potential but has to be expressed in one environment or another; persistency refers primarily to how long a species lasts in defined agricultural systems.

Parts II and III constitute the major part of this book; in the remainder of Part I two major aspects of herbage crop productivity are considered—first (in Chapters 2, 3 and 4) the environment in which crop production is to take place, and secondly (in Chapter 5) the definition of the product. The environment is dealt with in 3 parts—the soil, the climate generally and solar radiation in particular. These three topics cover between them the physical environment in which crop production takes place.

Since the product derived from herbage crops can take many forms, depending chiefly on the species grown and the way the herbage is processed, it is essential to be clear about the factors that determine quality and their relationship to the use to be made of the crop. Chapter 5 deals with those aspects of quality in herbage that are frequently referred to in later chapters.

In the final section (Part IV), species are assessed, comparatively, from the point of view of establishment, herbage yield and herbage quality. An attempt is also made to assess the species described in terms of their value for specific purposes (e.g. cutting or grazing) or for specific animal production processes. This assessment entails some consideration of whether species should be sown singly or in mixtures; this is a complicated subject and a separate chapter (Chapter 38) has therefore been included to set out the principles involved.

References

DAVIES, W. 1960. The grass crop, its development, use and maintenance. London: E. and F.N. Spon Ltd, 2nd edn, pp. 363.

STAPLEDON, R. G.; DAVIES, W. 1948. Ley farming. London: Faber and Faber Ltd, 2nd edn, pp. 182.

CHAPTER 2

The Soil Environment

IN THE Atlantic period (5 000–3 000 BC) almost the whole of Britain below 2 000 ft (610 m) was covered with deciduous forest and there were probably no extensive areas of grassland. The brown forest soils developed during this period have been modified since Neolithic times to grow the crops required by man and, today, almost all the brown forest type of soil has been converted to the agricultural type (Jacks, 1954).

Evidence from lacustrine sediments and from pollen analysis (Dimbleby, 1961) suggests that when forest was cleared in prehistoric times, food crops were grown until the supply of nutrients was largely used up, after which grasses and legumes became the dominant vegetation. When the herbage was not grazed or cut, the land frequently, though not always, reverted to forest. In later ages, as the result of increased demand for crops and grass, natural reversion to forest was largely eliminated.

Swards developed in this way arose as a direct consequence of man's activities and cannot be properly called natural grassland. Both the way in which the grass was used—as meadows or as pastures—and the nature of the soil determined the species and ecotypes present in the sward. Extreme conditions of soil and climate resulted in the cultivated forest soils changing into peaty gleys or podzols. The reclamation of the resultant moors and heaths by drainage, application of lime or marl, subsoiling, fertilizers and irrigation, both for the production of crops and of grass, has proceeded sporadically for several centuries whenever the return for the considerable investment involved has been attractive enough. Under less extreme conditions, more productive unsown swards were maintained, with a species composition determined by soil conditions, climate and management. Production of herbage from these better swards, always seriously limited by a lack of N, was increased by practices which encouraged the growth of legumes, such as liming and the use of crushed bones and phosphatic fertilizers.

The growing of grass, grass/legume mixtures or legumes on arable land as part of a systematic rotation of crops, was recognized during the 17th century as a valuable system of farming. Blith (1652) emphasized the importance of designing a rotation such that the land would be returned to grass before soil fertility had become too severely depleted by arable cropping. The first use of

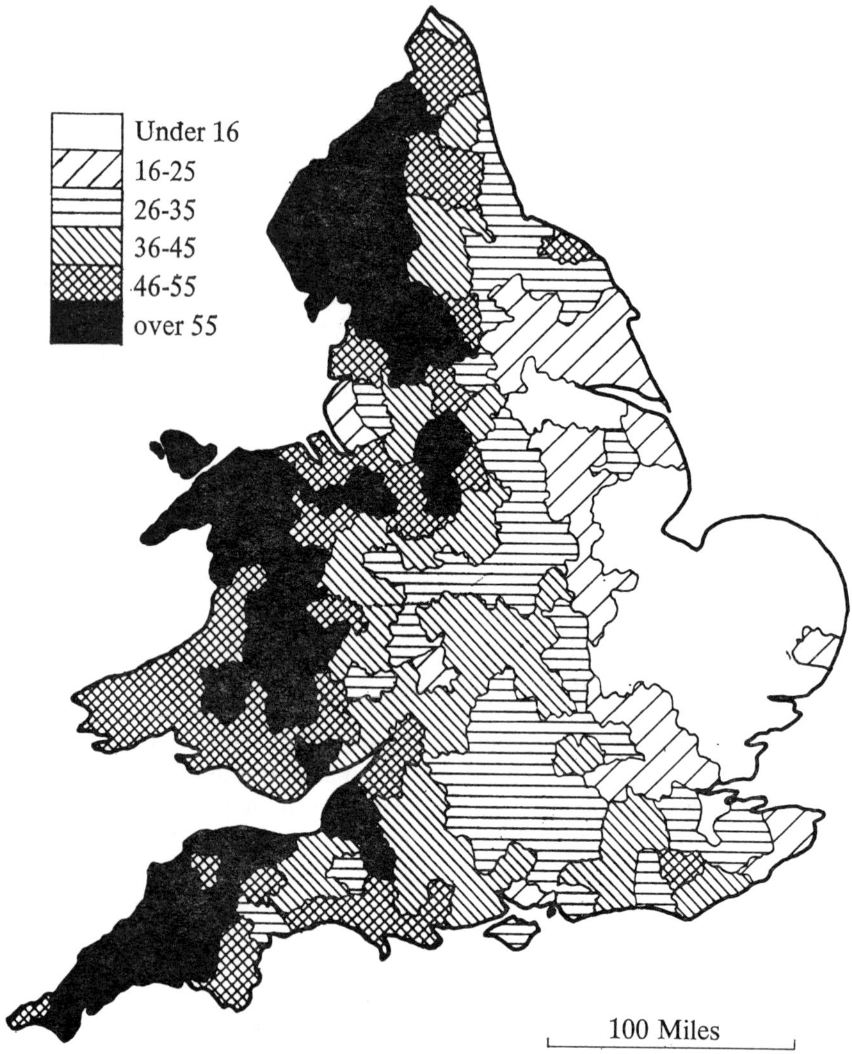

Fig. 2.1a. Percentage of arable land under temporary grass. (From Coppock, 1964)

cultivated species of grasses and legumes was for the establishment of such temporary swards on land which also grew arable crops. These cultivated species are still the ones most commonly used today for the same purpose. The proportion of arable land at present under temporary grass increases from east to west, as is shown in Fig. 2.1a. Conversely, the proportion of one-year leys is greatest in the east (Fig. 2.1b). These patterns of distribution of sown grass are probably related to the relative profitability of grass and

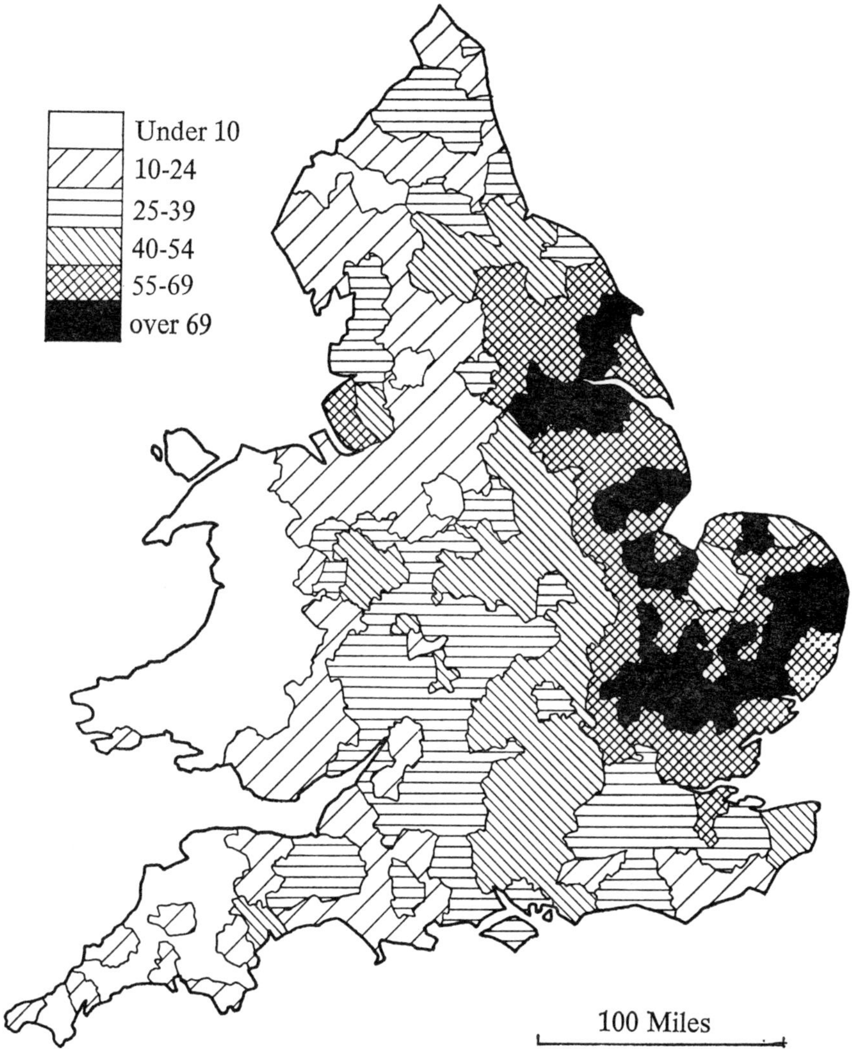

Fig. 2.1b. One-year leys as a percentage of leys sown in 1957. (From Coppock, 1964)

arable crops in the different localities as determined mainly by climate and in part by differences in topography and soils.

There are no detailed soil maps covering the whole country. The Soil Survey of England and Wales has published a number of soil maps with accompanying memoirs or bulletins giving an account of geography, geology, climate, vegetation and land use of the district surveyed (Soil Survey of Great Britain. Report No. 18, 1965), but most of England and Wales yet remains to be surveyed. In the absence of soil maps, the one-inch geological map can provide information on factors influencing soils and relief. Drift maps should be used where available, since drift deposits are not always derived from the underlying rock. General characteristics of the soil, such as drainage, texture, stoniness and depth, can often be derived. However, neither a general-purpose soil survey nor a geological survey can provide all the information required to evaluate soils in relation to the potential production of a specific crop such as grass. Gibbons (1961) emphasized that only special surveys can provide this information and that the criteria to be used will vary with the crop and the climate. Nevertheless, some attempt to classify land on a national scale according to its agricultural potential has been made in Coppock's *Agricultural Atlas of England and Wales* (1964), which gives the following table based on that first constructed by Stamp in 1930. Coppock points out that although the map may be out of date, it has not been superseded.

Table 2.1. Land suitable for cropping with cultivated grasses and legumes

	Percentage of the total area of England and Wales
1st class	5·3
1st class (heavy)	3·2
Good	26·1
Good (heavy)	13·3
Medium (light)	7·0
Medium (other)	25·0
Total	79·9

From Coppock (1964)

With the use of N and other fertilizers, the potential yield of herbage is likely to be similar on all the classes of land included in Table 2.1 and shown unshaded in Fig. 2.2. However, this potential would only be reached where growth was not limited by lack of water.

Sandy heaths such as the Breckland and Bagshot Sands are shown lightly shaded in Fig. 2.2. Although classified as poor land by Coppock, there is evidence that with adequate fertilizers plus lime where necessary, high yields of herbage are technically possible on such soils (Baker & Chard, 1964).

Most of the land shown cross-hatched in Fig. 2.2 is over 1 000 ft. (305 m) in altitude and the yield of DM will often be limited by a combination of exposure, thin stony soils and steep topography which not only directly affects plant growth but also makes cultivation, fertilizer application and grassland management difficult. The use of cultivated species of grasses and

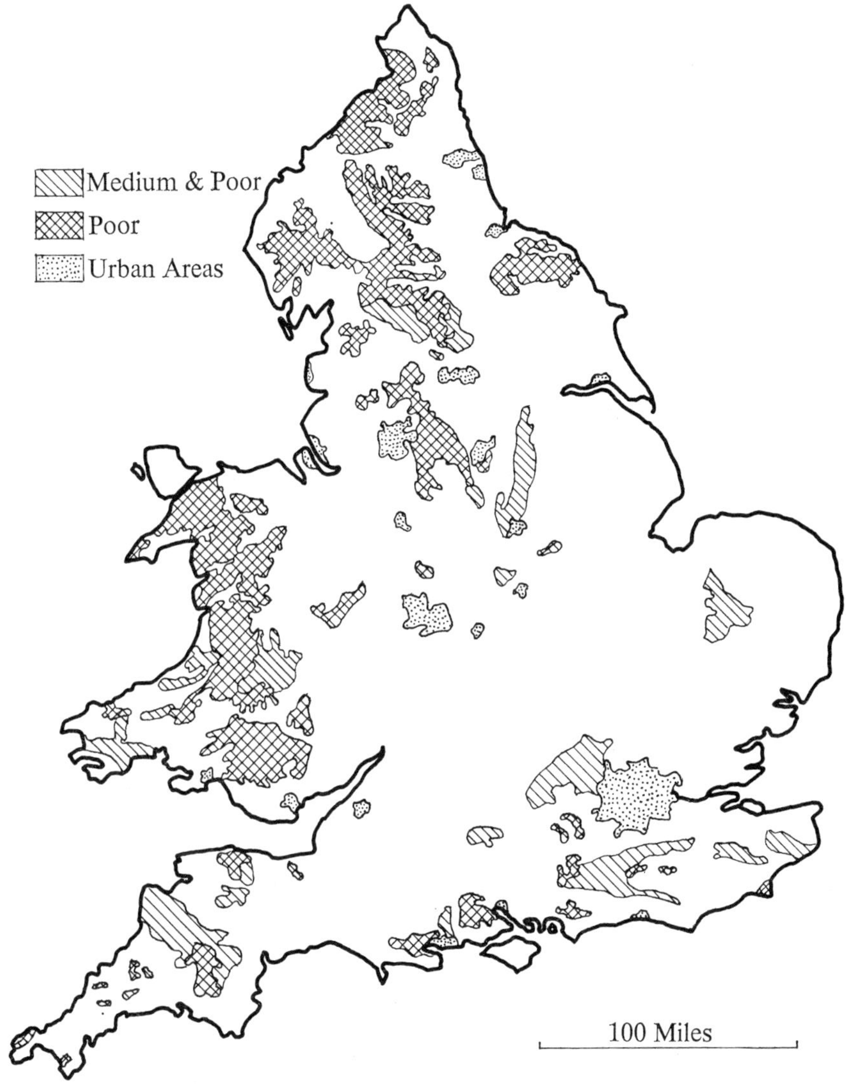

Fig. 2.2. Distribution of soil potential for grassland in England and Wales. (From Coppock, 1964)

legumes may be of considerable importance in these regions but will probably be confined to the small proportion of the total area where conditions for growth and herbage production are comparable to those of lowland situations.

Within the classes of land ranging from first class to medium, both the nutrient status and physical condition of the soils will vary widely. Agriculture is concerned with modifying such conditions to favour higher crop yields and quality.

From an examination of the results of experiments made available mainly during the last decade, Cooke (1967) concluded that we now know in some detail how to use fertilizers to supply sufficient nutrients from the soil to meet the full needs of grasses and legumes grown for forage.

Those attributes of soils which limit their suitability as physical media for the production and utilization of herbage crops are consequently assuming increased importance. Avery (1964) states that with our present knowledge of how to meet the nutrient needs of plants, "the ability of land to produce crops is largely determined by the physical characteristics of the site and soil".

Physical conditions determine the level of production from grass both by directly affecting plant growth and by imposing limitations on the establishment, management and utilization of sown swards. In this latter context, topography is obviously important. Curtis, Doornkamp and Gregory (1965) have suggested that slope boundaries should be drawn at $2\frac{1}{2}°$, $8\frac{1}{2}°$, 13°, 17° and 31°. All normal operations associated with managing and utilizing grassland are possible up to the $8\frac{1}{2}°$ boundary, but become difficult, especially the harvesting of green grass for silage, drying or feeding fresh, on slopes steeper than this. However, ploughing, fertilizer application, sowing and haymaking can be done without special equipment on slopes up to 13°. Ploughing and sowing are possible on slopes not exceeding 17°, but mechanical harvesting and carting become increasingly difficult and swards on such areas would be used mainly for grazing.

A high content of water in soils due to inadequate drainage or a low rate of infiltration can reduce herbage production in several ways:

1. Wet soil is slow to warm and growth may be delayed in spring.
2. Surface water can kill grasses and legumes and encourage rushes and other weeds.
3. The compressive strength of soils is greatly reduced as water content increases. Damage to the structure of the soil and to the plant can result from treading and wheel-slip when the soil cannot support the load applied. Puddling the soil surface further reduces the rate of infiltration, resulting in more water being held at the surface, with cumulative damage to both crop and soil.

The direct effect of too much or too little water on herbage yield varies from one species to another; the effect of a lack of water is considered elsewhere. In Table 2.2, a low soil moisture suction (high water content) is seen to reduce the yield of lucerne, to have little effect on that of timothy, and to increase the yield of brome grass. Transport of O_2 through roots varies between species and this may cause differences in their growth response to high soil water content and poor aeration. For example, Arikardo (1964) found that Italian ryegrass had a better ventilating system in its roots than oats. Aspects of soil structure, such as pore size and bulk density, may affect penetration by the roots of some species more than others.

Waddington and Baker (1965) found that although soil penetration by *Poa pratensis* roots was reduced when O_2 diffusion rates were less than $5–9 \times 10^{-8}$ g cm^{-2} min^{-1}, the yield of herbage was not reduced. This is a lower rate of diffusion than that which has reduced yields of other crops (Kristensen

Table 2.2. Yields of grasses and legumes as affected by different soil moisture levels (g/pot)

	Field capacity	Water suction 0 cm	Water suction 40 cm
Brome grass	2·30	3·98	2·58
Timothy	5·78	5·55	5·03
Lucerne	8·75	1·40	5·30
Ladino clover	8·21	7·10	7·38

From Finn *et al.* (1961)

& Lemon, 1964) and supports the view that apart from circumstances at the time of sowing, many grasses tolerate a wide range of physical soil conditions.

Grass and grass/legume swards can bring about changes in the physical conditions of soils. Jacks (1954) states "after man himself, grass is the most important agent in producing typical farm-land soil".*

From Table 2.3 (p. 13), which brings together results from a number of experiments, some on clay and others on sandy loam, it will be seen that the aggregate stability of soil was, in general, greater under ryegrass mixtures than under mixtures with other grasses. More organic matter also accumulated in soils under ryegrass than in soils under cocksfoot swards. The following increases were recorded over a period of 3 years (Clement & Williams, 1964):

	Increase in organic carbon % in dry soil to 15-cm depth
S23 ryegrass and white clover	0·21
S37 cocksfoot and white clover	0·16
S.E. ±	0·009

* The concept is not a new one: Worlidge (1668) wrote, "Ray-grass by which they improve any cold, sour, clay, weeping lands, for which it is best, but good also for drier upland grounds, especially stony, light, or sandy lands, which is unfit for sainfoin, hath the precedence of all other species".

There is, however, no clear evidence that such differences between cultivated species of grass affect the yields of crops grown in rotation.

The quantity and rate of supply of nutrients available to grasses and legumes varies widely between soils, depending on parent material, degree of weathering, previous crops and fertilizer treatments. Analysis of the soil or of the herbage (the latter may have the greater diagnostic value) must be used to determine the soil nutrient supply of any particular field. The following classification represents a very broad generalization:

The supply of nutrients to herbage crops from the soil

	Sands	Loams	Clays
N	Very deficient	Inadequate	Inadequate
P	Often adequate	Often adequate	Often adequate
K	Inadequate	Adequate	Adequate

Such procedures serve to emphasize that no agricultural soil can be expected by itself to meet the N requirement of grass swards. For example, cultivated soils low in organic matter may supply only 30–40 kg N/ha per year. Conversely, many soils of different textures may supply sufficient P for grasses and legumes, particularly when phosphatic fertilizers have been regularly applied to arable crops in the rotation. Grasses have been placed in the category of plants able to obtain P from soils with low levels of availability (Bingham, 1962). Sandy (and also calcareous) soils are unlikely to meet unaided the K requirement of cut swards for more than one year. When swards are used only for grazing, soil K supplies may be adequate on loams and heavier land. Some clay soils appear to be able to supply enough K to maintain high yields of cut grass for many years (Arnold & Close, 1961).

Plant nutrients are commonly concentrated in the uppermost horizons of soils under grass. Fertilizers, animal excreta and leaf litter fall on the surface of the soil and are transported to deeper horizons in roots only by the activity of animals or in percolating water. This contrasts with tilled soils in which fertilizer and crop residues are redistributed each year to the depth of cultivation.

Insoluble materials, organic matter and nutrients strongly adsorbed on soil particles are moved down the profile by the soil fauna. Those species of earthworms which produce casts may also bury superficial deposits with soil brought up from depth. Such processes are slow and, over a period of 3–4 years under grass, probably bring about only a minor redistribution of nutrients. It has been shown by Clement and Williams (1967) that after 3 years under grass, the accumulation of organic nitrogen is still largely confined to the top 2–3 cm of soil.

Soluble substances, and to some extent small particles held in suspension,

Table 2.3. The influence of sown grasses on the development of water-stable aggregates (WSA) under leys

	Lias clay				Reading Beds Sandy loam			
Experiment No.	E77	E127	E114	E51	H9	H16	H8	H15
Date sown	1947	1949	1948	1946	1950–1	1951–2	1951	1950
Date sampled	1951	1953	1952	1952	1953–4	1954–5	1955	1953
Swards	% WSA> 2·0 mm				% WSA> 0·5 mm			
Perennial ryegrass:								
S24	...	...	...	...	...	8·3	...	...
S24 + white clover S100	13·1	...	20·4	...	...	7·3	7·5	6·4
S23 + white clover S100	...	5·8	15·5	...	8·6	...	6·2	...
Cocksfoot:								
S37 + white clover S100	9·5	...	...	13·6	7·1	5·4	5·4	...
S143 + white clover S100	...	3·4	15·0	...	...	...	...	4·0
Timothy:								
S51 + white clover S100	...	4·6	...	...	...	...	...	...
S48 + white clover S100	8·5	...	...	14·6	...	...	7·0	...
S48 + red clover S123	...	...	...	...	...	5·8	...	...
Meadow fescue:								
S53 + white clover S100	...	3·9	...	15·6	...	...	...	...
S53, S215 + white clover S100	...	...	...	...	...	...	5·4	...
Red fescue:								
S59 + white clover S100	...	...	...	...	...	...	6·4	...
White clover:								
S100 + S184	...	...	...	...	...	6·8	7·7	...
S100	...	...	...	...	...	...	...	5·8
Standard error	1·25	1·02	Paddock*	Paddock*	0·43	0·47	1·5	Paddock*
Error degrees of freedom (n)	4	6	...	...	15	12	6	...

* Data taken from paddocks in the same field but with no replication.

From Clement and Williams (1958)

can be washed and leached down the profile. The amount and extent of movement will depend on the proportion of the material, the soil, the crop and the climate. Nitrate N is one of the most mobile nutrients. Burns and Dean (1964) examined some of the factors which regulate the movement and leaching of nitrate from fertilizers, and from these and other studies, Allison (1966) concluded that there is little loss of nitrate by leaching between May and October. He emphasized the importance of crops, such as grass, which not only absorb nutrients as these become available but also transpire whenever meteorological conditions permit and so reduce to a minimum the amount of water moving down the profile. However, when N fertilizers were applied to grass during the summer and autumn in amounts greater than the crop could take up, large quantities of mineral N—mainly nitrate—were detected in late winter at a depth of 45–60 cm in his experiments (Table 2.4).

Table 2.4. The influence of fertilizer N applied to a grass sward on the distribution of mineral N within the soil profile

Fertilizer to ley during previous year kg N/ha per month	Sample depth cm	NH_4-N+NO_3-N in soil when sampled from field ppm.
20	0–15	17·8
	15–30	24·2
	30–45	8·1
	45–60	4·8
100	0–15	49·1
	15–30	59·7
	30–45	82·2
	45–60	73·9

After Clement (1965)

British soils range from acid (pH 3·5) to moderately alkaline (pH 8·2). Most cultivated soils have pH values above 4·5. Acid soils not only have a low pH, but may also be deficient in Ca and other bases and contain phytotoxic levels of soluble Al and Mn. In the moderately acid range (pH 4·5–5·5), therefore, there is not necessarily any close connection between the pH of a soil and its suitability for a given crop (Russell, 1961). Many herbage crops can tolerate acid soils, more particularly so in areas of high rainfall. Table 2.5 opposite shows that only red clover gave a reduced yield at pH 4·5–5·3. Unfortunately in this experiment the grasses were N-deficient.

For maximum yields of cultivated grasses and legumes in the eastern counties of England, it is generally desirable to maintain the pH of the soil above pH 6; in the western counties and in Wales, pH 5·5 is probably adequate (Small, 1946). Periodic liming, most commonly with ground $CaCO_3$, enables a wide range of herbage crops to be grown on more acid soils. When,

Table 2.5. Average yields of grasses and clovers (g DM)

pH at sowing	7·2	6·75	5·95	4·5	3·6
pH after 12 months	6·8	6·4	6·0	5·3	4·4
Lolium perenne	8·9	8·9	8·7	9·2	8·8
L. multiflorum	19·1	22·7	22·8	23·9	18·9
Dactylis glomerata	6·9	5·9	7·9	7·2	7·3
Trifolium pratense	84·5	122·0	123·0	89·0	29·3
T. repens	43·8	65·1	87·5	89·2	24·8

After McNeur (1953)

however, a soil is very acid throughout the profile, a single application of lime to the cultivated horizon may not provide an environment suitable for crops which are very intolerant of acidity, such as lucerne. Table 2.6 is taken

Table 2.6. Effect of lime applications on soil pH

		Ground limestone	
	Nil	5 tons/ac cultivated into seedbed	10 tons/ac cultivated into seedbed
Depth (in.)		pH values	
0–3	4·4	7·0	7·2
3–6	4·0	4·2	4·7
6–9	4·4	4·4	4·3
9–12	4·3	4·2	4·4
12–15	4·3	4·3	4·5
15–18	4·3	4·0	4·9

The pH values were determined a year after the lime was applied. Lucerne roots failed to penetrate the acid subsoil and the crop was therefore susceptible to drought.

From Baker and Chard (1964)

from a report by Baker and Chard (1964), on herbage production on the very acid Bagshot Sands.

References

ALLISON, F. E. 1966. The fate of nitrogen applied to soils. *Adv. Agron.* **18**, 219–58.

ARIKARDO, H. 1964. Studies on the development of the ventilating system in relation to the tolerance against excess-moisture injury in various crops. 13. Difference between Italian ryegrass and oats in tolerance against excess-moisture injury. *Proc. Crop Sci. Soc. Japan* **32**, 353–7.

ARNOLD, P. W.; CLOSE, B. M. 1961. Release of non-exchangeable potassium from some British soils cropped in the glasshouse. *J. agric. Sci., Camb.* **57**, 295–304.

AVERY, B. W. 1964. The soils and land use of the district around Aylesbury and Hemel Hempstead. London: H.M.S.O.

BAKER, H. K.; CHARD, J. R. A. 1964. Investigations into the establishment, growth and use of grasses and legumes on Bagshot sands. *N.A.A.S. q. Rev.* **66**, 62–8.

Bingham, F. T. 1962. Chemical soil tests for available phosphorus. *Soil Sci.* **94,** 87–95.
Blith, W. 1652. The English improver improved. London: J. Wright.
Burns, G. R.; Dean, L. A. 1964. The movement of water and nitrate around bands of sodium nitrate in soils and glass beads. *Proc. Soil Sci. Soc. Am.* **28,** 470–4.
Clement, C. R. 1965. Accumulation of soil organic matter and grassland productivity. *Tech. Rep. Grassld Res. Inst., Hurley* pp. 35–7.
Clement, C. R.; Williams, T. E. 1958. An examination of the method of aggregate analysis by wet sieving in relation to the influence of diverse leys on arable soils. *J. Soil Sci.* **9,** 252–66.
Clement, C. R.; Williams, T. E. 1964. Leys and soil organic matter. 1. The accumulation of organic carbon in soils under different leys. *J. agric. Sci., Camb.* **63,** 377–83.
Clement, C. R.; Williams, T. E. 1967. Leys and soil organic matter. 2. The accumulation of nitrogen in soils under different leys. *J. agric. Sci., Camb.* **69,** 133–8.
Cooke, G. W. 1967. The control of soil fertility. London: Crosby Lockwood and Son Ltd, pp. 526.
Coppock, J. T. 1964. An agricultural atlas of England and Wales. London: Faber and Faber Ltd.
Curtis, B. F.; Doornkamp, J. C.; Gregory, K. J. 1965. The description of relief in field studies of soil. *J. Soil Sci.* **16,** 16–30.
Dimbleby, G. W. 1961. Soil pollen analysis. *J. Soil Sci.* **12,** 1–11.
Finn, B. J.; Bourget, S. J.; Nielsen, K. F.; Dow, B. K. 1961. Effect of different soil moisture tensions on grass and legume species. *Can. J. Soil Sci.* **41,** 16–23.
Gibbons, F. R. 1961. Some misconceptions about what soil surveys can do. *J. Soil Sci.* **12,** 96–100.
Jacks, G. V. 1954. Soil. Edinburgh: Thomas Nelson.
Kristensen, K. J.; Lemon, E. R. 1964. Soil aeration and plant root relations. 3. Physical aspects of oxygen diffusion in the liquid phase of the soil. *Agron. J.* **56,** 295–301.
McNeur, A. J. 1953. Effect of soil acidity on growth of pasture plants. 1. Pot experiments with grasses and clovers in acidified soils. *N.Z. Jl Sci. Technol. Sect. A* **35,** 311–26.
Russell, E. W. 1961. Soil conditions and plant growth. London: Longmans, 9th edn.
Small, J. 1946. pH and plants. London: Baillière, Tindall and Cox.
Waddington, D. V.; Baker, J. H. 1965. Influence of soil aeration on the growth and chemical composition of three grass species. *Agron. J.* **57,** 253–8.
Worlidge, J. 1668. Systema agriculturae being the mystery of husbandry discovered and layd open. London: Samuel Speed.

CHAPTER 3

The Climatic Environment

THAT grass is the most widely grown crop in Britain and that the land appears green most of the year is a matter of common observation; nevertheless the suitability of its climate for grass growth varies considerably from place to place and from year to year.

Outstanding features of the British climate are the moderate range of temperatures and the well distributed rainfall throughout the growing season. The main physical influences on the weather are the Atlantic Ocean to the west and the continent of Europe to the south-east. The deep ocean tends to cause mild winters and cool summers and a continuity of light to moderate rainfall; the adjacent land mass is the source of very cold winter weather and of occasional hot, dry weather in summer.

Climatic data are usually presented in the form of averages, but in practice, we find, there is a wide variation above and below the mean. It is the frequency of given meteorological events which ultimately decides the suitability of an area for a given agricultural activity. A frequency of more than 3 years out of 10 for any seriously adverse weather can be regarded as a critical level; any greater frequency considerably increases the risk of encountering two adverse years in succession, with consequent profound economic effects on a farming enterprise.

Average winter temperatures in Britain are, of course, lower in the north than in the south, but in very cold winters the lowest temperatures are generally encountered in the south-east. The run of average isotherms from north-west to south-east combines these two features, thus tending to obscure their agricultural implications.

Spring arrives first in the south-west, advancing northwards fairly rapidly in the west, but more slowly in the east. Summer warmth is greatest in the Home Counties around London, and least in the coastal areas of the north-west. Autumn approaches from the north-west and spreads southward, reaching the south-west last.

Superimposed on this general seasonal timing is the effect of altitude and to a lesser extent the aspect and exposure of a site. Mean temperature decreases and cloudiness increases with altitude, although this latter effect is modified on the lee side of a hill range, so that an increase of sunshine can be expected in areas such as the Welsh Marches. Rainfall also increases with height and hill ranges again tend to create a 'rain-shadow' in their lee, as for example in the Evesham area.

The effect of temperature in regulating the onset and the end of the grass-growing season has never been defined with precision, but there is much evidence to suggest that a mean temperature of about 42 or 43°F (6·1°C) is the threshold above which growth generally begins. This threshold temperature is that measured in a screen at a standard height of 4 ft (122 cm) above the crop; in the present context, however, it is worth noting that there is little difference between the temperature at a soil depth of 1 ft (30 cm) and that at 4 ft above soil level in spring and autumn.

Using 43°F (6·1°C) as a threshold, and soil temperatures as the criterion, the altitude correction to the grass-growing season amounts to a reduction of 5–6 days per 100 ft (30·5 m.) in England. Using 42°F (5·6°C) and air temperatures, the correction varies from 4 days per 100 ft in Scotland, to 6 days in northern England and 9 days in southern England. These are average values; in practice, the growth retarding effect of altitude tends to be greatest in early springs and least in late springs. It is worthy of note that in Europe the height effect is smaller, being probably only 2 days per 100 ft in Switzerland.

Using a mean air temperature of 42°F(5·6°C) as the criterion, meteorological records suggest that the south coast from the Isle of Wight westwards, lowland Cornwall, the peninsulas of Pembroke and Caernarvon, and all Anglesey have mean growing seasons of 365 days, which is clearly an overstatement; the choice of threshold may, of course, be incorrect, but the fact also remains that the use of average temperatures will often exclude much of the effect of periods of cold during winter.

It is therefore more practical to use 43°F (6·1°C) as the threshold, and, taking account of the correction for altitude, it is possible to draw, in general terms, a map showing the length of the grass-growing season as determined by temperature alone (Fig. 3.1). From this map, it will be seen that the lowland areas of England appear to have an average growing season of some 250 days, in Scotland the figure for the lowlands is about 20–30 days less and in inland Ireland up to 50 days more. The coastal areas of the south-west have averages in excess of 300 days, and the only parts of the British Isles which appear to have no close season for grass are the coastal areas of south-west Ireland. The growing season in hill areas is difficult to represent on a small-scale map, but averages of less than 250 in Wales, less than 225 in the Pennines and less than 200 in Scotland are indicated. Experience suggests that these figures are reasonably accurate, but it must be remembered that in years with late springs and early autumns, the season may be shortened by up to 45 days.

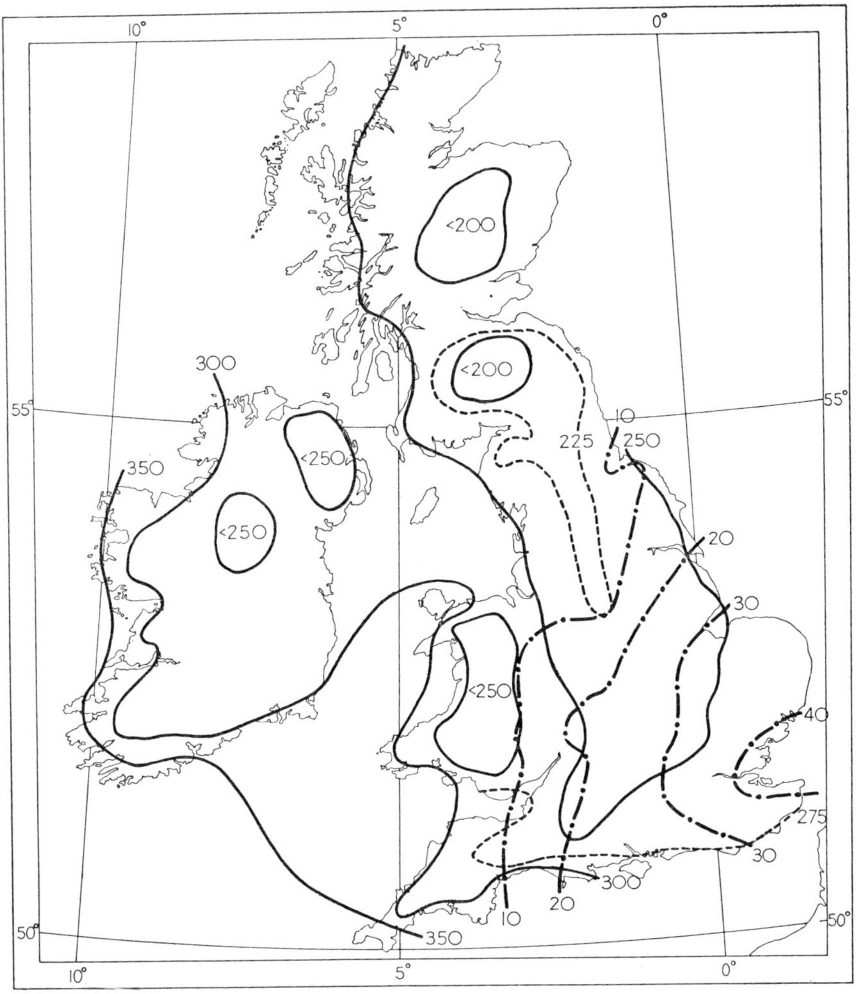

Fig. 3.1. Average number of days with mean air temperature above 43°F (6·1°C) (—— and ----). Average number of days when growth is severely restricted by partial drought (-·-·-).

A minimum temperature, however, is not the sole criterion for grass growth; an adequate supply of soil moisture is also essential. Investigations have shown that if the soil moisture deficit within the grass root range exceeds 2 in. (5 cm) on good average soils, growth is seriously retarded. If the number of days of partial drought is calculated in this manner using average figures for rainfall and potential transpiration, a figure can be derived which should be subtracted from growing season as determined from temperature alone. These averages, shown also in Fig. 3.1, are probably a slight underestimate

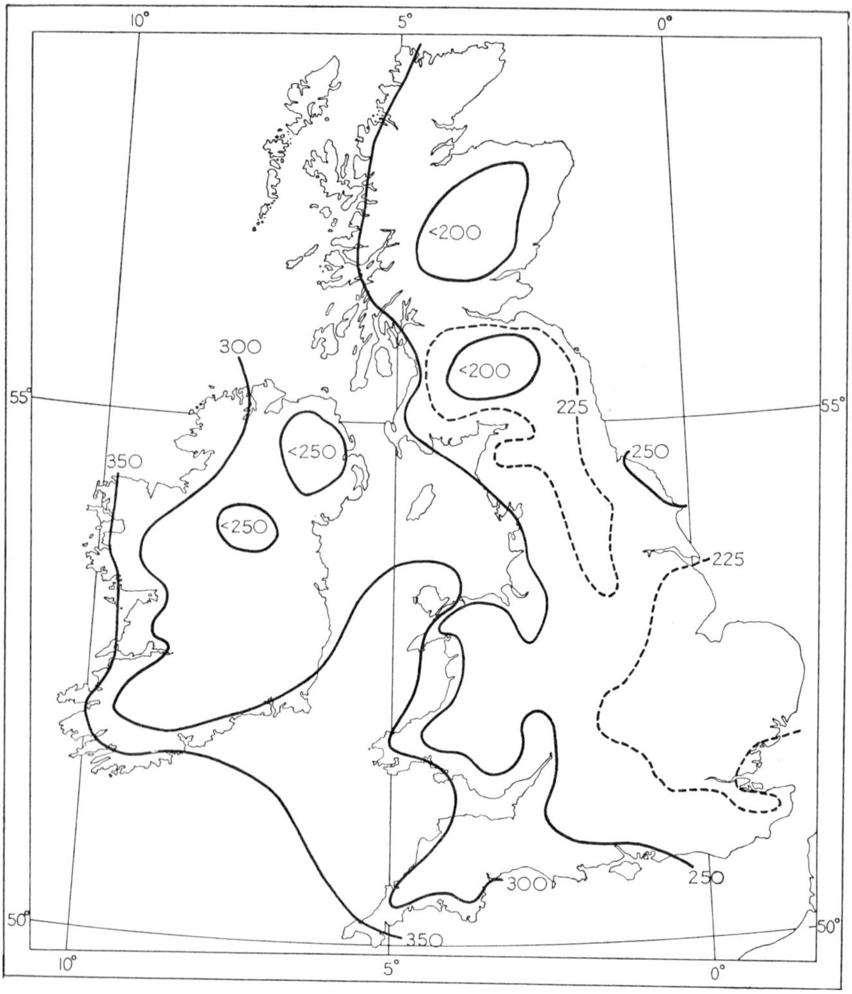

Fig. 3.2. Average length of the growing season for grasses (days).

as they do not take into account years such as 1959 when grass over most of Britain suffered from water shortage.

If the two sets of lines in Fig. 3.1 are combined, Fig. 3.2 is obtained, which gives an estimate of the average growing season for grass on the basis of both temperature and moisture. This implies that 250–300 days is the figure appropriate to most lowland grassland areas in Britain. The average figure for the predominantly arable areas of England is 225 days.

The length of season is not in itself sufficient to indicate the amount of growth; transpiration must also be taken into account. An attempt may be made to do this by reference to a parameter called effective transpiration,

Table 3.1. Effective transpiration and grassland farming

Percentage of grass in farmland	Average effective transpiration
>75% grass (24 counties)	15·8 in. (39·5 cm)
60–74% grass (15 counties)	14·7 in. (36·8 cm)
35–59% grass (11 counties)	14·2 in. (35·5 cm)
<35% grass (10 counties)	13·45 in. (33·6 cm)

From Smith, L. P. 1967. Meteorology and the pattern of British grassland farming. *Agric. Met.* **4**, 321–38.

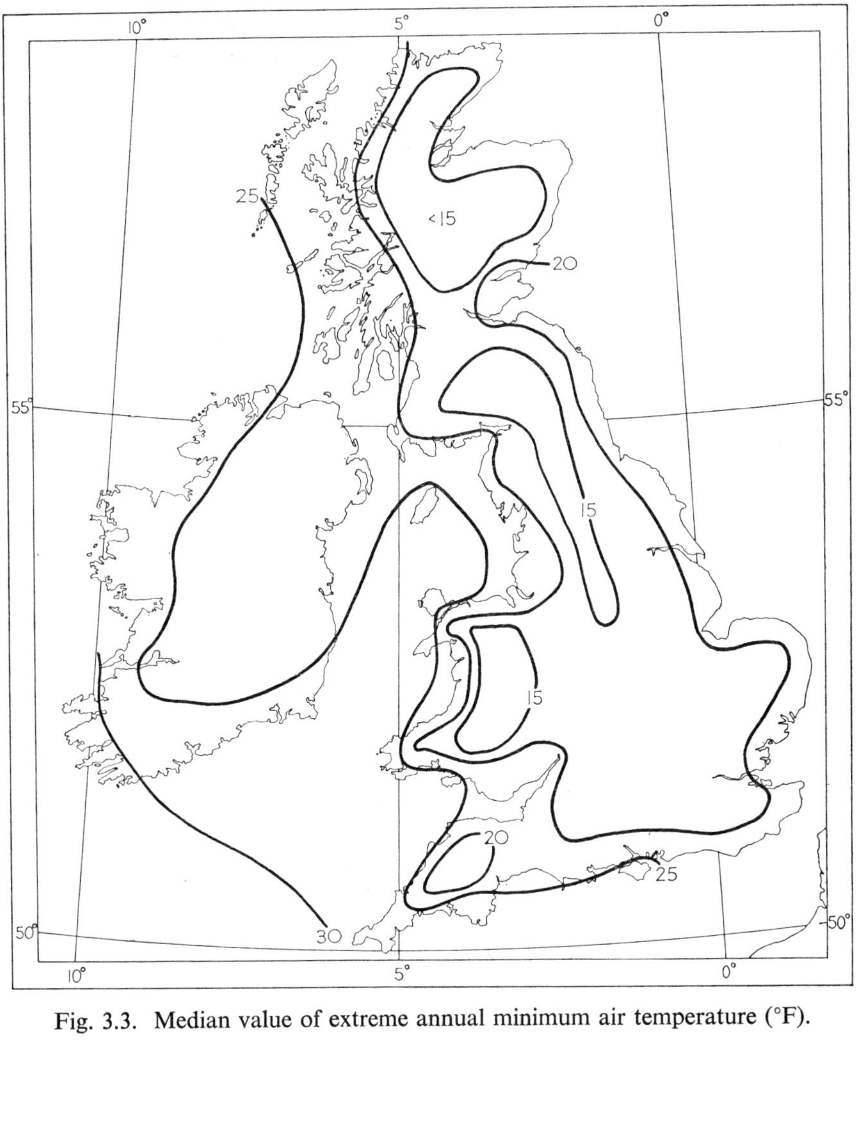

Fig. 3.3. Median value of extreme annual minimum air temperature (°F).

which is the sum of the potential transpiration of grass during the growing season when the soil moisture deficit does not exceed 2 in. (5 cm). Figures based on this parameter are shown in Table 3.1 for England and Wales.

Two adverse winter factors must be considered, firstly the leaching caused by excessive winter rainfall and secondly the liability to severe frosts.

Use of rainfall and transpiration averages can determine the relative liability to excess winter rain with the following results:

Low risk. All areas lying to the east of a line Flamborough Head–Worcester–Dungeness, with the lowest liability in Huntingdonshire, Cambridgeshire and Essex.

Medium risk. All remaining lowland England except Devon and Cornwall and the east coastal areas of Scotland; central and north-eastern Ireland.

High risk. All remaining areas, most of which have an average September-March rainfall of 30 in. (762 mm) or more.

Assessment of the severe frost risk can be obtained from examination of the annual minimum temperature, the median values of which are shown in Fig. 3.3. The outstanding feature of the distribution is the ameliorating effect of proximity to the coast, especially the Atlantic coast. This coastal effect is naturally greatest in the immediate vicinity of the sea, but temperatures at as little as 5 or 10 miles (8–17 km) inland may not differ greatly from the inland figures proper. Over most of the inland grasslands of England, Scotland and Wales, at least 10 degrees of frost (°F) (−5·6°C) are to be expected in 5 years out of 10 and, strangely enough, Ireland fares little better.

It could be argued that Britain could grow more grass if it had slightly more summer rain in the drier areas and slightly more sunshine, but these requirements are meteorologically incompatible. Due to the absence of severe winters in the first half of the twentieth century, the growing season has probably been at its longest for many centuries; long-term climatic analysis suggests that there have been many periods in the past when conditions were not quite so favourable.

CHAPTER 4

The Light Environment

LIGHT influences the growth and production of forage grasses and legumes in three ways: (1) it provides the energy supply for all growth and development through the process of photosynthesis; (2) it affects leaf structure and expansion of the leaf surface and hence the size of the photosynthetic system; (3) it controls the timing of floral development and seed production, and hence the distribution of dry matter in the plant.

Light energy input

Under British conditions (50–60°N.) the duration and intensity, and therefore the daily input, of light energy vary greatly through the year, from photoperiods of less than 9 h and an energy supply of about 50 cal total radiation/cm^2 per day in mid-winter, to photoperiods of over 18 h and 400–500 cal/cm^2 per day in midsummer (visible radiation usually comprises about 45% of the total radiation input). Regional variation is greatest in winter, mean December values ranging from about 20 cal in north Scotland to about 60 in south-west England, but varies relatively less with latitude in June, with a range of 400 to 450 cal (Taylor & Smith, 1961).

Energy conversion

The interception and conversion of light energy by the grass crop through the year can be considered at three levels of complexity: (i) the individual leaf or young seedling growing without mutual shading, (ii) the primary grass canopy, in which there is total interception of light, (iii) the established sward in the field under regular grazing or cutting.

THE INDIVIDUAL LEAF OR YOUNG SEEDLING

At low light intensities, the photosynthetic rate of the individual leaf increases rapidly as light intensity rises, but in most temperate forage species light saturation is reached at 20 000–30 000 lux and an increase above this intensity has no further effect on photosynthesis. Correspondingly, at low light intensities the efficiency of conversion of the incoming light energy is high (about 15%), but decreases to about 3% above 30 000 lux (Cooper, 1970). For much of the day in the British summer, therefore, the light intensity is well above the saturation level for the individual leaves of most temperate grasses and legumes.

This light saturation level and the associated maximum photosynthetic rate varies considerably between species and between varieties. In many subtropical grasses such as *Paspalum dilatatum* and *Cynodon dactylon* photosynthesis increases with light intensity up to 60 000 lux or more, compared to a saturation level of about 30 000 lux in such temperate grasses as ryegrass and cocksfoot (Brown *et al.*, 1966; de Jager, 1967). Red clover (Hesketh & Moss, 1963; Brown *et al.*, 1966) and white clover (Beinhart 1962; McCree & Troughton, 1966a) have saturation levels similar to cocksfoot, while lucerne shows rather higher values (Brown *et al.*, 1966).

Within perennial ryegrass, large differences between varieties and genotypes in maximum photosynthetic rate occur, associated with variation in such structural features as mesophyll thickness and number and size of mesophyll cells (Wilson & Cooper, 1967), while in red clover, Kendall and Taylor (1963) found significant differences in maximum photosynthesis between individual clones. The photosynthetic activity of a leaf also depends on the light and temperature conditions under which it develops. Light saturation is reached at lower intensities in the thinner leaves grown at low light intensities than in the thicker leaves developed at high intensities.

Similar conclusions can be drawn from the growth analysis of individual widely spaced seedlings (Black, 1956). In shading experiments involving many temperate grasses (cocksfoot, meadow fescue, Italian ryegrass, perennial ryegrass, timothy) and legumes (lucerne, alsike clover, red clover, white clover), the net assimilation rate (NAR) and the relative growth rate (RGR) increased with light intensity up to full summer daylight, although the leaf area ratio decreased (Blackman & Black, 1959). It was estimated that in the UK the maximum RGR would occur at or around full summer daylight for the grasses, and at somewhat above this value for the legumes. Similarly, in perennial and Italian ryegrass grown in controlled environments (Cooper, 1967), NAR increased with light intensity up to 20 000–30 000 lux but the leaf area ratio, which is a measure of the proportion of photosynthetic tissue, decreased, so that the RGR showed little change between 8 000 and 20 000 lux. Energy conversion was greatest at the lower light intensities, reaching 13–15% at 2 000 lux and declining to 3–4% at 20 000 lux. The same material showed a seasonal variation in NAR from a maximum of about 1·0 mg/cm^2 per day in the summer (2–3% conversion) to a minimum of below 0·3 mg/cm^2 per day in winter (>6% conversion), although low temperatures were also limiting in the winter.

Species and varietal differences in NAR have also been reported. Chatterjee

(1961) found marked differences between a North African and a British variety of tall fescue in winter, while MacColl and Cooper (1967) reported small but consistent differences between local varieties of tall fescue, cocksfoot and perennial ryegrass from northern and Mediterranean origins. In perennial ryegrass, Wilson and Cooper (1967) found marked ecotypic differences in NAR at both high and low light intensities, which could be related to variation in the photosynthetic activity of the individual leaf.

THE PRIMARY CANOPY

Energy conversion and dry matter (DM) production of the crop canopy are more complex than in individual leaves or plants (Donald & Black, 1958). After germination or defoliation, photosynthesis is initially limited by the amount of leaf surface exposed to light, but as the leaf area increases, the rate of photosynthesis is reduced by mutual shading, although the crop growth rate continues to increase. Eventually, sufficient leaf area is developed to intercept all the incoming light energy, and a steady crop growth rate is obtained. If the canopy is allowed to grow further, crop growth rate may decrease due to an increase in the amount of actively respiring tissue below the level of light penetration. Under adequate insolation, however, a steady crop growth rate can be maintained for a considerable time, as shown by Brougham (1956) for ryegrass, and McCree and Troughton (1966b) for white clover.

The critical or optimum leaf area index at which the sward intercepts all the available light, and hence shows the maximum crop growth rate, differs with the spatial distribution of the leaves in the canopy and also with the amount of light falling upon them. For instance, in a white clover sward with horizontal leaves arranged at right angles to the incoming light, 60% of the light may be intercepted by the first layer of leaf area, while in ryegrass, with more erect leaves, only 30% will be intercepted (Monteith, 1966). Similarly, the more widely dispersed leaves of lucerne intercept less light per unit layer of leaf area than do those of white clover (Brown *et al.*, 1966). As a result, white clover has a lower critical leaf area index of about 3–4, compared with 7–8 in ryegrass and 6–7 in timothy, with consequent differences in maximum crop growth rate (Brougham 1958, 1960). Similarly, the higher the light intensity, the greater the leaf area required for complete light interception, and the greater the maximum crop growth rate (Stern & Donald, 1961).

At low light intensities the efficiency of conversion by a ryegrass canopy is similar to that by the individual leaf (about 15%), but the dilution of light energy as it passes through the sward means that at higher intensities the canopy is more efficient than the individual leaf, giving values of 5–6% conversion (150–200 kg DM/ha per day) in the summer (Cooper, 1970). Similarly, in the Netherlands, de Wit (1959) has calculated potential rates of DM production of more than 200 kg/ha per day for a ryegrass canopy for the period May–August. In other words, light saturation of the crop canopy occurs at a higher intensity than for an individual leaf, though the estimates of 50 000–60 000 lux for canopies of lucerne, white clover and cocksfoot reported by Brown *et al.* (1966) are still less than those of bright sunlight in the summer.

PRODUCTION IN THE SWARD

Under sward conditions the effects of defoliation and regrowth are superimposed on the light interception by the primary canopy and the photosynthetic activity of the individual leaf. There are comparatively few records of the potential production and energy conversion by grass or legume swards in which water and nutrients were not limiting, and in which defoliation treatments were planned to maintain adequate light interception through the year. However, in small plots of a perennial ryegrass sward outdoors, grown without limitations of water and soil nutrients, an overall conversion of light energy of more than 3% through the year has been possible (Cooper, 1970), corresponding to over 20 000 kg DM/ha per annum. The seasonal distribution of DM production varied from < 10 kg/ha per day between October to March, to > 150 kg/ha per day in May–June, but the variation in energy conversion was rather less, from 2–3% in the winter period to over 4% in the summer; temperature limitations were obviously operating in the winter. These estimates of potential production, although high, are approached by a number of records from intensively managed grassland, using high fertilizer input (Armitage & Templeman, 1964).

Leaf growth and expansion of the leaf surface

In addition to its direct influence on the rate of photosynthesis, light also influences DM production through its effects on the rate of leaf and tiller expansion and hence the size of the photosynthetic system.

In most temperate grasses and legumes, an increase in the light intensity leads to more rapid leaf appearance (Anslow, 1966), although in perennial ryegrass there is little further effect at intensities above 30 000 lux (Cooper, 1967) and in white clover above 20 000 lux (Beinhart, 1963). The rate of tillering in grasses, and the degree of branching in white clover are also increased, partly because faster leaf production provides more leaf axil sites from which tillers can develop, and partly because more of the sites develop tillers (Langer, 1963; Beinhart, 1963). The effects of light intensity on leaf length and width are compensatory (Friend, 1966; Cooper, 1967). Increased intensity leads to a reduction in leaf length, but an increase in width and, consequently, leaf area is greatest at intermediate light intensities. A higher intensity also results in a thicker leaf, and a greater weight and increased chlorophyll content per unit area. These morphogenetic effects on leaf structure, particularly leaf thickness, may have important effects on photosynthetic activity (Wilson & Cooper, 1967). The distribution of DM between shoot and root is also strongly influenced by light; lower intensities reduce root growth more than shoot growth (Black, 1957).

The effect of photoperiod on leaf growth and tillering is less clear, since photoperiodic effects have often been confounded with those of increased energy input (Black, 1957). In the temperate grasses, including tall fescue, ryegrass, cocksfoot and meadow fescue, longer photoperiods increase leaf length, and hence individual leaf area, but slightly reduce the rate of leaf appearance (Ryle, 1966 b, c).

The extent of the response to photoperiod will differ with species and variety, and is often related to the climatic origin of the material. In ryegrass and cocksfoot, for instance, longer photoperiods increase leaf expansion far more in varieties of northern origin than in those from the Mediterranean region (Cooper, 1964).

Floral development and seed production

The development of the inflorescence has a marked effect on the seasonal production of the sward. It results in an increase in the size of existing shoots or tillers and is usually associated with reduced root growth and inhibition of new tillers (Langer, 1963). Stem elongation also increases the vertical spacing of the leaf canopy and so makes for more effective light interception. At the same time, flowering and seed production are usually associated with increased lignification and reduced digestibility.

Most temperate grasses and legumes are long-day plants, in the sense that they require photoperiods longer than a certain minimum for the initiation and subsequent development of the inflorescence (Cooper, 1960a; Evans, 1964). The light intensity needed for the induction of flowering is very low, 100 lux being adequate for ryegrass, cocksfoot and timothy. The exact photoperiodic requirements vary with the climatic and agronomic origin of the material (Cooper, 1963). In varieties of ryegrass and cocksfoot from the Mediterranean, where flowering must take place in early spring before the beginning of the dry season, floral development can occur in photoperiods of 9–10 h, but later-flowering material of the same species from northern Europe may require 13–14 h or more. Similar climatic variation in flowering requirements has been reported for timothy (Evans & Allard, 1934) and for red clover (Gorman, 1955; Schultze, 1957). Even within the same climatic region, the photoperiod requirement, and hence the time of flowering, differs with cultivar. In the British Isles, early-flowering ryegrass varieties such as Irish and S24 will initiate heads in mid-March at a photoperiod of 10–11 h, followed by ear emergence in early May. The late-flowering S23 and S101, on the other hand, do not form heads until early May at a photoperiod of 13–14 h, and ears emerge some 4–5 weeks later (Cooper, 1963).

In addition to this requirement for long days for floral development, many perennial forage grasses and legumes need exposure to winter conditions of cold and/or short days before they become able to produce inflorescences. In perennial ryegrass, this winter requirement can be satisfied either by cold or by short days of less than 9 h (Cooper & Calder, 1964), while in cocksfoot photoperiods of less than 17 h suffice (Calder 1964). Smooth-stalked meadow-grass seems to require both short days and cold (Lindsey & Peterson, 1964), but in tall fescue a 14-h photoperiod is sufficient, provided that temperatures are not high (Templeton *et al.*, 1961). In white clover also, some varieties or genotypes appear to require short days or low-temperature exposure before flowering (Thomas, 1961; Beatty & Gardner, 1961).

The degree of winter requirement is often related to the perenniality of the species or variety. In ryegrass, for instance, the annual Westerwolds ryegrass has no winter requirement, while perennial ryegrass has an obligatory requirement, and many plants of Italian ryegrass show a quantitative response only (Cooper, 1960b). This association does not apply to species with perennial organs and most varieties of lucerne, red clover and timothy, although perennial, have little or no winter requirement (Cooper, 1960a).

These flowering requirements for light and temperature can have marked effects if locally adapted varieties are transferred to contrasting climatic regions. In South Australia, for instance, with a Mediterranean climate and lower latitude, many northern European grass cultivars flower irregularly or fail to produce heads, because of insufficient exposure to low temperatures and/or too short a photoperiod in summer. Conversely, when transferred to northern Europe, certain Mediterranean cultivars flower very early in the spring, with an associated lack of persistency and decrease in nutritive value (Knight, 1963).

Even after initiation in an appropriate photoperiod, subsequent floral development and seed production are affected by light intensity and photoperiod. In many grasses, such as ryegrass, cocksfoot and meadow fescue, a reduction in light intensity reduces both the number of tillers which form ears and the fertility of those which do so, mainly by reducing the number of florets on the primary branches, rather than the number of primary branches themselves (Ryle, 1966a).

By contrast, in both ryegrass (Ryle, 1965) and timothy (Ryle & Langer, 1963) shortening the photoperiod results in a longer ear due to an increase in the number of axillary primordia and the development of more spikelets after initiation.

References

ANSLOW, R. C. 1966. The rate of appearance of leaves on tillers of the Gramineae. *Herb Abstr.* **36,** 149–55.

ARMITAGE, E. R.; TEMPLEMAN, W. G. 1964. Response of grassland to nitrogenous fertilizer in the west of England. *J. Br. Grassld Soc.* **19,** 291–7.

BEATTY, D. W.; GARDNER, F. P. 1961. Effect of photoperiod and temperature on flowering of white clover, *Trifolium repens* L. *Crop Sci.* **1,** 323–6.

BEINHART, G. 1962. Effects of temperature and light intensity on CO_2 uptake, respiration and growth of white clover. *Pl. Physiol., Lancaster* **37**, 709–15.

BEINHART, G. 1963. Effects of environment on meristematic development, leaf area and growth of white clover. *Crop Sci.* **3**, 209–13.

BLACK. J. N. 1957. The influence of varying light intensity on the growth of herbage plants. *Herb. Abstr.* **27**, 89–98.

BLACKMAN, G. E.; BLACK, J. N. 1959. Physiological and ecological studies in the analysis of plant environment. 11. A further assessment of the influence of shading on the growth of different species in the vegetative phase. *Ann. Bot.* **23**, 51–63.

BROUGHAM, R. W. 1956. Effect of intensity of defoliation on regrowth of pasture. *Aust. J. agric. Res.* **7**, 377–87.

BROUGHAM, R. W. 1958. Interception of light by the foliage of pure and mixed stands of pasture plants. *Aust. J. agric. Res.* **9**, 39–52.

BROUGHAM, R. W. 1960. The relationship between the critical leaf area, total chlorophyll content and maximum growth-rate of some pasture and crop plants. *Ann. Bot.* **24**, 463–74.

BROWN, R. H.; BLASER, R. E.; DUNTON, H. L. 1966. Leaf area index and apparent photosynthesis under various microclimates for different pasture species. *Proc. 10th int. Grassld Congr., Helsinki, 1966* pp. 108–13.

CALDER, D. M. 1964. Stage development and flowering in *Dactylis glomerata*. *Ann. Bot.* **28**, 187–206.

CHATTERJEE, B. N. 1961. Analysis of ecotypic differences in tall fescue (*Festuca arundinacea* Schreb.). *Ann. appl. Biol.* **49**, 560–2.

COOPER, J. P. 1960a. The use of controlled life cycles in the forage grasses and legumes. *Herb. Abstr.* **30**, 71–9.

COOPER, J. P. 1960b. Short day and low temperature induction in *Lolium*. *Ann. Bot.* **24**, 232–46.

COOPER, J. P. 1963. Species and population differences in climatic response. In: Evans, L. T. [Ed.] The environmental control of plant growth. New York and London: Academic Press, pp. 381–400.

COOPER, J. P. 1964. Climatic variation in forage grasses. 1. Leaf development in climatic races of *Lolium* and *Dactylis*. *Jnl appl. Ecol.* **1**, 45–62.

COOPER, J. P. 1967. Growth analysis of spaced seedlings; base populations of selection programme. *Rep. Welsh Pl. Breed. Stn 1966* pp. 15–18.

COOPER, J. P. 1970. Potential production and energy conversion in temperate and tropical grasses. *Herb. Abstr.* **40**, 1–15.

COOPER, J. P.; CALDER, D. M. 1964. The inductive requirements for flowering of some temperate grasses. *J. Br. Grassld* Soc. **19**, 6–14.

DONALD, C. M.; BLACK, J. N. 1958. The significance of leaf area in pasture growth. *Herb. Abstr.* **28**, 1–6.

EVANS, L. T. 1964. Reproduction. In: Barnard, C. [Ed.] Grasses and grasslands. London and Melbourne: Macmillan & Co. Ltd, pp. 126–53.

EVANS, M. W.; ALLARD, H. A. 1934. Relation of length of day to growth of timothy. *J. agric. Res.* **48**, 571–86.

FRIEND, D. J. C. 1966. The effects of light and temperature on the growth of cereals. In: Milthorpe, F. L.; Ivins, J. D. [Eds] The growth of cereals and grasses. Proceedings of the Twelfth Easter School in Agricultural Science, University of Nottingham, 1965. London: Butterworth and Co., pp. 181–99.

GORMAN, L. W. 1955. Effect of photoperiod on varieties of red clover (*Trifolium pratense* L.). *N.Z. Jl Sci. Technol. Sect. A* **37**, 40–54.

HESKETH, J. D.; MOSS, D. N. 1963. Variation in the response of photosynthesis to light. *Crop Sci.* **3**, 107–11.

JAGER, J. M. DE 1967. Measurement of photosynthesis: infra-red gas analysis. *Rep. Welsh Pl. Breed. Stn 1966* pp. 22–7.

KENDALL, W. A.; TAYLOR, N. L. 1963. Rates of respiration and photosynthesis in clones of red clover. *Crop Sci.* **3**, 146–51.

KNIGHT, R. 1963. A comparison of seasonal plant growth in a northern European and a Mediterranean climate. *Rep. Welsh Pl. Breed. Stn 1962* pp. 102–14.

LANGER, R. H. M. 1963. Tillering in herbage grasses. *Herb. Abstr.* **33,** 141–8.

LINDSEY, L. K.; PETERSON, M. L. 1964. Floral initiation and development in *Poa pratensis. Crop Sci.* **4,** 540–4.

MACCOLL, D.; COOPER, J. P. 1967. Climatic variation in forage grasses. 3. Seasonal changes in growth and assimilation in climatic races of *Lolium, Dactylis* and *Festuca. J. appl. Ecol.* **4,** 113–27.

MCCREE, K. J.; TROUGHTON, J. H. 1966a. Prediction of growth rate at different light levels from measured photosynthesis and respiration rates. *Pl. Physiol., Lancaster* **41,** 559–66.

MCCREE, K. J.; TROUGHTON, J. H. 1966b. Non-existence of an optimum leaf area index for the production rate of white clover grown under constant conditions. *Pl. Physiol., Lancaster* **41,** 1615–22.

MONTEITH, J. L. 1966. Analysis of microclimate in cereals and grasses. In: Milthorpe, F. L.; Ivins, J. D. [Eds] The growth of cereals and grasses. Proceedings of the Twelfth Easter School in Agricultural Science, University of Nottingham, 1965. London: Butterworth and Co., pp. 123–37.

RYLE, G. J. A. 1965. Effects of daylength and temperature on ear size in S24 perennial ryegrass. *Ann. appl. Biol.* **55,** 107–44.

RYLE, G. J. A. 1966a. Effects of photoperiod in the glasshouse on the growth of leaves and tillers of three perennial grasses. *Ann. appl. Biol.* **57,** 257–68.

RYLE, G. J. A. 1966b. Effects of photoperiod in growth cabinets on the growth of leaves and tillers in three perennial grasses. *Ann. appl. Biol.* **57,** 269–80.

RYLE, G. J. A. 1966c. Physiological aspects of seed yield in grasses. In: Milthorpe, F. L.; Ivins, J. D. [Eds]. The growth of cereals and grasses. Proceedings of the Twelfth Easter School in Agricultural Science, University of Nottingham, 1965. London: Butterworth and Co., pp. 106–18.

RYLE, G. J. A.; LANGER, R. H. M. 1963. Studies on the physiology of flowering in timothy (*Phleum pratense* L.). 2. Influence of daylength and temperature on the size of the inflorescence. *Ann. Bot.* **27,** 233–44.

SCHULTZE, E. 1957. Photoperiodic experiment on perennial fodder plants. 1. *Z. Acker- u. PflBau* **103,** 198–226.

STERN, W. R.; DONALD, C. M. 1961. Relationship of radiation, leaf area index and crop growth rate. *Nature, Lond.* **189,** 597–8.

TAYLOR, S. M.; SMITH, L. P. 1961. Estimates of averages of radiation and illumination. *Met. Mag. Lond.* **90,** 289–94.

TEMPLETON, W. C.; MOTT, G. O.; BULA, R. J. 1961. Some effects of temperature and light on growth and flowering of tall fescue, *Festuca arundinacea* Schreb. 2. Floral development. *Crop Sci.* **1,** 283–6.

THOMAS, R. G. 1961. The influence of environment on seed production capacity in white clover (*Trifolium repens* L.). 1. Controlled environment studies. *Aust. J. agric. Res.* **12,** 212–26.

WILSON, D.; COOPER, J. P. 1967. Assimilation in *Lolium* in relation to leaf mesophyll. *Nature, Lond.* **214,** 989–92.

WIT, C. T. DE 1959. Potential photosynthesis of crop surfaces. *Neth. J. agric. Sci.* **7,** 141–9.

CHAPTER 5

Definition of Nutritive Value in Grasses and Legumes

THOUGH grasses and legumes differ markedly in their anatomy, morphology and chemical composition, the general principles that govern their nutritive value as ruminant feeds are essentially similar and so are here discussed jointly. Nutritive value is a complex concept. Herbage from a given pasture species will not have the same value for all classes of livestock, the nutritive requirements of which vary with type of animal, whether they are growing, lactating, etc; also, herbage will differ in value according to whether it is grazed or conserved. Furthermore, its value as a feed may depend on other components in the ration; for example, deficiencies in minerals or other constituents can often be made good by supplementary feeding.

The nutritive value of herbage is thus most usefully considered under four headings:

(1) The detailed characterization of the herbage in significant nutritional terms, such as contents of energy, protein, minerals; the digestibility of these components; changes with maturity and season; effects of fertilizers or water regime.

(2) The voluntary intake of the herbage by ruminant animals.

(3) Possible changes in composition and intake characteristics resulting from method of conservation.

(4) Interactions between the herbage and other feed components.

In the following chapters the nutritive value of herbage is discussed mainly in terms of (1) and (2), which constitute its basic nutritional characters, though they may be modified to some extent by the processes involved in conservation (Chapter 42). But in interpreting this information, it must be borne in mind that, in practice, any deficiencies in a herbage will frequently be made good by combining it with other feeds, be it with other herbage species (e.g. grass with legume), with cereals, or with minerals or other essential factors. The information in the following chapters can thus be used both to indicate the nature of the feeds that should be fed in conjunction *with* a given herbage, as well as the basic nutritive value of that herbage.

In general, our starting point must be those characteristics of a herbage which determine the level of nutrient intake by ruminant animals consuming it. With this information and with knowledge of the nutritional requirements of the animals to be fed, it is then possible to determine whether supplementary feeding is needed to achieve the required level of nutrient intake. Thus nutrient intake is the key nutritional parameter, and can be most usefully regarded as a function of dry matter (DM) intake, digestibility and utilization (Raymond, 1964).

Intake

Intake is the weight (usually of DM) of feed voluntarily eaten by animals offered that feed *ad libitum*: it is often expressed as g DM/kg $LW^{0 \cdot 73}$ per day*, where LW is liveweight.

Digestibility

The term digestibility refers to the extent to which any feed or feed component is digested by an animal. The quantity digested, expressed as a percentage of the quantity consumed, is termed the digestibility coefficient. Coefficients of digestibility are determined by feeding animals a measured quantity of food and measuring the output of faeces. The difference between the input of food and the output of faeces is taken as the quantity digested, and from this a digestibility coefficient is determined. This coefficient is more correctly termed the apparent coefficient of digestibility, since besides undigested feed residues, faeces also contain endogenous and bacterial residues from the digestion process; the *true* digestibility of the feed is thus somewhat greater than its *apparent* digestibility. It is, however, customary to refer to the apparent digestibility simply as the digestibility of the feed.

The major value of forage crops is as a source of energy for ruminants; hence it is the digestibility of the energy, or the concentration of digestible energy in the forage, which is of principal interest. Determination of the gross energy of samples of feed and faeces is costly and much more information is available regarding the digestibility of the organic matter (percent OMD) or the digestibility of the dry matter (percent DMD). The OMD value of a forage is very closely related to the digestibility of the energy (Swift, 1957); even DMD is often sufficiently closely related to energy digestibility for it to be used as a basis for comparing forages, as the following

* Some workers have expressed intake in terms of LW to the power 0·75, but most of the available data reported in this and other chapters have been expressed using the power 0·73.

relationship for S23 ryegrass, derived by Armstrong, Blaxter and Waite (1964), shows:

$$\% \text{ Digestible energy} = 1{\cdot}09\% \text{ DMD} - 8{\cdot}9 \text{ (R.S.D.} < 2\%)$$

A more precise estimate of energy value, the metabolizable energy content (ME %) of the feed, can be derived from the digestible energy by means of the relationship: Metabolizable energy % = Digestible energy % × 0·81 (Armstrong, 1964).

Information on forage digestibility has been much advanced by the *in vitro* technique of measuring digestibility developed by Tilley and Terry (1963). This requires only 1 g of dried forage; it can thus be used to compare large numbers of samples and has provided much more comprehensive information on the changes that occur in different forage species as they mature than would have been possible with animal experiments alone. The results using this method are closely correlated with those of *in vivo* digestibility determinations made on the same forages fed to sheep, as is shown in the following relationship:

$$\% \text{ DMD } (\textit{in vivo}) = 0{\cdot}99\% \text{ DMD } (\textit{in vitro}) - 1{\cdot}01 \text{ (R.S.D.} \pm 2.31)$$ (Tilley & Terry, 1963)

Results obtained with the *in vitro* method can be expressed in terms either of DMD or of OMD. A further expression of digestibility, the percentage of digestible organic matter in the dry matter (D-value), is also used, particularly in relation to the practical rationing of livestock. In this, the content of digestible organic matter is used. This is independent of the insoluble ash content of forages, which may vary widely as a result of contamination by soil, and is expressed on the basis of DM, which is also the term in which intake of forage by the animal is normally expressed. The D-value provides a useful expression of the digestible energy or metabolizable energy in unit dry matter of a feed and is of particular value in advisory work. The high correlation between DMD, OMD and D-value as expressions of digestibility can be seen in the equations below, which were derived from the results of *in vivo* determinations made on 108 forage samples (Grassland Research Institute, Unpublished data):

$$\% \text{ OMD} = 1{\cdot}016\% \text{ DMD} + 0{\cdot}330\% \text{ ash} - 1{\cdot}669 \text{ (R.S.D.} \pm 0{\cdot}933)$$
$$\text{D-value} = 0{\cdot}926\% \text{ DMD} - 0{\cdot}403\% \text{ ash} + 4{\cdot}637 \text{ (R.S.D.} \pm 0{\cdot}875)$$

Digestibility is also a most important factor in determining *total* nutrient intake. As noted below, the digestibility of a feed largely determines how efficiently the intake of digestible energy is utilized by the animal. Digestibility also has a very marked effect upon voluntary food consumption (intake). There is much evidence that as the digestibility of a forage declines, so also does the quantity of it which a ruminant will eat (Balch & Campling, 1962). The following relationship between forage intake and digestibility was determined by Minson *et al.* (1964) with fresh grasses cut during primary growth and regrowth and fed to mature sheep after preservation by freezing:

$$I = 1{\cdot}36\%\,OMD - 25{\cdot}62, \text{ where } I = \text{intake in g DM/kg } LW^{0{\cdot}73} \text{ per day.}$$

Blaxter (1964) has related voluntary intake to the metabolizability of the gross energy in the ration (Q) in the following equation:

$$I = 9{\cdot}1\ Q - 0{\cdot}074\ Q^2 - 194\ (\text{R.S.D.} \pm 10)$$

Earlier work suggested that there was a single relationship between intake and digestibility for all forages. However, there is now much evidence that relationships can differ considerably between forage species and varieties. Thus Reid *et al.* (1962) have shown that the intake of lucerne is greater than that of grasses, and Minson *et al.* (1964) suggested that the intake of timothy was less than that of other grasses they examined (see Fig. 5.1). Differences

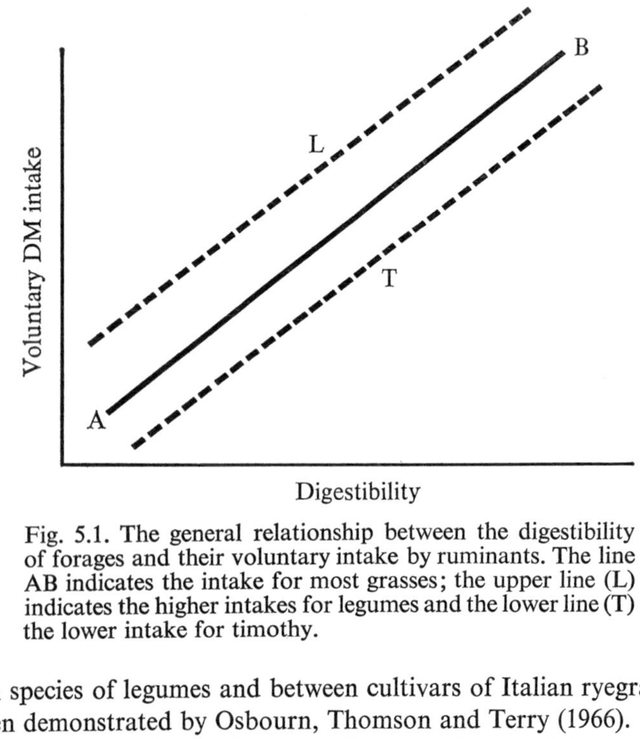

Fig. 5.1. The general relationship between the digestibility of forages and their voluntary intake by ruminants. The line AB indicates the intake for most grasses; the upper line (L) indicates the higher intakes for legumes and the lower line (T) the lower intake for timothy.

between species of legumes and between cultivars of Italian ryegrasses have also been demonstrated by Osbourn, Thomson and Terry (1966).

It must be accepted that the figure for the voluntary intake of a forage species is not absolute and repeatable, but varies according to the physiological condition and class of the animal and with the stage of maturity (digestibility) of the forage. Intrinsic differences in intake characteristics between forages can thus only be determined either by comparisons made at equal levels of digestibility, or by comparison of the regressions of intake upon digestibility, using a single class of animal. Few such comparisons have been made, and where voluntary intake of forage species is discussed in subsequent chapters, it is only possible to give comparative information for a limited number of forage species.

Utilization

Utilization refers to the usefulness of the digested (absorbed) nutrients in supplying the metabolic requirements of the animal. In general, the lower the digestibility of a feed, the less efficiently is the digestible organic matter (energy) intake utilized for either maintenance or production purposes by the animal. Low protein content of a forage may limit its utilization. Protein content can be considered in terms of either percentage content of crude protein (CP), or of digestible crude protein (DCP), calculated (for forages) from the following relationship:

$$\%\,DCP = 0{\cdot}929\,\%\,CP - 3{\cdot}48\ (R.S.D. \pm 0{\cdot}46)\ \text{(Holter \& Reid, 1959)} \qquad (1)$$

The concept of "utilization" also applies to mineral intake, but little is known about the availability of minerals in herbage and it is only possible to give total mineral contents (see below); it also takes account of vitamins and animal growth factors, and covers the presence or absence of toxic substances, all of which modify the nutritional usefulness of a feed.

To interpret the data on forages in terms of potential for animal feeding, it is also necessary to have information on the nutrient requirements of different classes of livestock. In the case of ruminants, these have been summarized in the Agricultural Research Council publication *Nutrient Requirements of Farm Livestock. No. 2. Ruminants* (1965). This gives information on the energy, protein and mineral requirements of dairy cows, beef cattle and sheep for maintenance and production. In that publication energy requirements are given (Tables 6.1–6.5) in terms of metabolizable energy which, for grasses, can be calculated from the percentage OMD by the equation:

ME (kcal/g DM) $= 0{\cdot}037$ OMD %(*in vitro*) $+ 0{\cdot}122$ (R.S.D. $\pm 0{\cdot}083$) (Armstrong, Alexander & McGowan, 1964).

The protein requirements of different classes of stock are given in the above publication as "available protein" (Tables 5.2–5.8). This does not include provision for the replacement of metabolic faecal loss of N; this loss must be allowed for and amounts to approximately 13·4 g digestible CP for cattle and 16·8 g for sheep for each kg of feed DM eaten. The *total* requirement of digestible CP can then be related to the intake of digestible CP (from equation (1)) to determine how adequate the feed is as a source of protein.

Mineral requirements are given in terms of total dietary intake (Tables 2.1–2.8). It must again be emphasized that the contribution made by each particular feed may depend on the availability of the minerals in that feed, about which little is known.

Thus, from the above-mentioned publication and from the data in the following chapters, an estimate can be made of how far each herbage species, depending on its physiological growth stage, can satisfy the nutrient requirements of different classes of farm livestock; at the same time, an indication can also be gained as to the levels of supplementary feeding needed to remedy any deficiencies. However, in applying these concepts to the practical feeding situation a number of qualifications must be borne in mind. In practice, the concept of *ad libitum* feeding on which voluntary intake is based is not always achieved or indeed aimed at; thus, herbage intake may be restricted by limitation of feed supply, and this may be accentuated by unpalatability resulting from fouling by excreta, from fungal infections on conserved herbage, or from soil contamination, which are not intrinsic features of the herbage. Similarly, under grazing conditions, animals may select their diet both within and between species, so that the feeding value of the herbage eaten may differ appreciably from that based on cuts of the whole plant. Furthermore, supplementary feeding may modify some of the parameters reported; for example, cereal supplements may reduce both the voluntary intake and digestibility of forages, while protein supplements may increase both the intake and digestibility of low-protein forages. Processing can markedly alter the nutritional characteristics of forages—ensilage generally reduces voluntary intake, while dehydration and milling can markedly increase intake but at the same time reduce digestibility. There is still much to learn about the interactions between supplementary feeds and forages conserved in different ways.

The following chapters can therefore indicate only the probable nutritive value of the different herbage species described; almost certainly, the differences *within* a given herbage species at different stages of physiological maturity are of as much practical significance as any differences *between* species at a similar stage of maturity.

References

ARMSTRONG, D. G. 1964. The evaluation of artificially dried grass as a source of energy for sheep. 2. The energy value of cocksfoot, timothy and two strains of ryegrass at varying stages of maturity. *J. agric. Sci., Camb.* **62,** 399–416.

ARMSTRONG, D. G.; ALEXANDER, R. H.; MCGOWAN, M. 1964. The use of *in vitro* digestibilities of dried grasses for the prediction of their energy value for ruminants. *Proc. 162nd Meet. Nutr. Soc. Abstr. Commun. XXVI.*

ARMSTRONG, D. G.; BLAXTER, K. L.; WAITE, R. 1964. The evaluation of artificially dried grass as a source of energy for sheep. 3. The prediction of nutritive value from chemical and biological measurements. *J. agric. Sci., Camb.* **62,** 417–24.

BALCH, C. C.; CAMPLING, R. C. 1962. Regulation of voluntary food intake in ruminants. *Nutr. Abstr. Rev.* **32,** 669–86.

BLAXTER, K. L. 1964. Utilization of the metabolizable energy of grass. *Proc. Nutr. Soc.* **23,** 62–70.

HOLTER, J. A.; REID, J. T. 1959. Relationship between the concentrations of crude protein and apparently digestible crude protein in forages. *J. Anim. Sci.* **18,** 1339–49.

MINSON, D. J.; HARRIS, C. E.; RAYMOND, W. F.; MILFORD, R. 1964. The digestibility and voluntary intake of S22 and H1 ryegrass, S170 tall fescue, S48 timothy, S215 meadow fescue and Germinal cocksfoot. *J. Br. Grassld Soc.* **19,** 298–305.

OSBOURN, D. F.; THOMSON, D. J.; TERRY, R. A. 1966. The relationship between voluntary intake and digestibility of forage crops using sheep. *Proc. 10th int. Grassld Congr., Helsinki, 1966* pp. 363–6.

RAYMOND, W. F. 1969. The nutritive value of forage crops. *Adv. Agron.* **21,** 1–108.

REID, R. L.; BURR, C.; WELCH, J. A.; DOSZA, L.; JUNG, G. A. 1962. Investigations of plant species and maturity stages on forage nutritive value as determined by *in vitro* digestion techniques. *Final Report USDA Contract No. 12-14-100-4524 (24), W. Va. Agric. Exp. Stn.*

SWIFT, R. N. 1957. The calorific value of TDN. *J. Anim. Sci.* **16,** 753–6.

TILLEY, J. M. A.; TERRY, R. A. 1963. A two stage technique for the *in vitro* digestion of forage crops. *J. Br. Grassld Soc.* **18,** 104–11.

PART II

THE GRASSES

CHAPTER 6

Origins of Cultivated Herbage Grasses

THE grasses, a large and ancient group of more than 10 000 species, are the most important and widely distributed members of the plant kingdom. However, in contrast to arable crops, their precise centres of origin are difficult to define. The characteristic features of our contemporary European species were moulded by climatic change and plant migration associated with successive Ice Ages. Under these conditions a large number of species, often polyploid, evolved. Concurrently, there was tremendous differentiation of ecotypes and adaptation to specific environmental conditions.

In Britain the climate favours a climax vegetation of forest cover, and natural grasslands were probably restricted to relatively few situations, such as the chalk uplands. The discovery of iron smelting, which ushered in the Iron Age and demanded large quantities of charcoal, gave great impetus to the destruction of the natural forests. This process continued in Britain during the Roman settlement and, with varying intensity, into the present century. A new pattern of indigenous grasslands, the nature of which depended on a complex of climatic, edaphic and biotic factors, replaced the forests. Some grasses of value in present-day agriculture were found in these communities, e.g. timothy and the fescues, whilst cocksfoot was often associated with woodland habitats. Ryegrass is perhaps of more recent origin, and is associated with arable cultivations and the spread of settled agriculture. The deliberate introduction of grass seeds began in the middle of the 18th century.

Perennial ryegrass (*Lolium perenne*)

In a comprehensive review of the development of ryegrass in British agriculture, Beddows (1953) provides evidence that perennial ryegrass has been sown as a forage crop for at least 300 years. In commerce, a high seed yield was important, but attempts were made by individual farmers to multiply material which gave good herbage production. "Varieties" associated with the names of these progressive farmers appeared in seedsmen's catalogues in the latter part of the 18th century, and these were followed by local races, such as Devon Eaver, which had attractive agronomic characteristics.

A major step forward in the improvement of ryegrass, as indeed of other grasses, came with the appreciation by Stapledon and Jenkin of the enormous genetic diversity, associated with defined ecological conditions, which existed in particular grass populations. They further recognized the adaptive influences of management and farming systems. A ryegrass which was regularly cut for hay and/or seed production developed into an early, low-tillering, stemmy and short-lived type with a high seed potential. At the other extreme, pastures which were heavily grazed and fertile, such as those situated near the homestead, gave rise to late-flowering, high-tillering, leafy and persistent plants with a lower potential for seed production. The synthetic cultivar S23 was produced from a combination of basic plants which had the inherent capacity to withstand heavy treading by stock, and could respond to increased soil fertility. As a breeding objective, it was also desirable to incorporate leafiness, high tillering capacity and more persistency in early-flowering varieties, and S24 was developed as a successful cultivar of this type.

More recent developments in the breeding of perennial ryegrass continue to recognize these basically different types, with greater attention being given to winter-hardiness and nutritive characteristics.

Italian ryegrass (*Lolium multiflorum*)

Introduced into Britain in the early 19th century, this species created considerable interest as a valuable grass for short leys. Breeding resulted in more leafy and slightly longer-lived varieties. More recently, cultivars incorporating greater winter-hardiness have been developed on the continent. Tetraploid cultivars have also been produced, but more significant advances can be expected from hybrids between perennial and Italian ryegrass at both diploid and tetraploid level.

Meadow fescue (*Festuca pratensis*)

This species is closely allied to the ryegrasses.

Tall fescue (*Festuca arundinacea*)

The biological potential of this species is very high, but existing cultivars demand very close attention to management if this potential is to be realized in terms of animal production. This is largely because palatability is low when the plant is mature.

The ryegrass/fescue group of grasses forms a wealth of material for the plant breeder and, with modern techniques, new combinations which should have a significant impact on British agriculture can be expected.

Cocksfoot (*Dactylis glomerata*)

Cocksfoot differs from ryegrass in that it is naturally adapted to a wider range of ecological habitats, including the fringes of woodlands where it forms large erect tussocks, well able to compete with other members of the plant community.

The tetraploid form has the widest distribution, and breeding has brought about a series of types analogous to those in ryegrass. Recent developments have involved combining the out-of-season growth of diploid cocksfoot from Portugal and northern Spain with the winter-hardiness of established tetraploid cultivars. Attempts are also being made to incorporate the high digestibility and succulence of leaf of a tetraploid species from the Portuguese coast into high-yielding indigenous varieties.

Timothy (*Phleum pratense*)

Timothy was introduced into Maryland by Timothy Hansen in 1720, where it quickly proved its worth as a hay producer. Seeds were sent to England in 1760 under the name Timothy. This grass is well adapted to the cooler, wetter parts of Britain, on peaty soils and low-lying heavy land, and commercial seed production has been centred in Scotland. It is a late-flowering grass giving high yields of leafy hay. The bred cultivars include good examples of the use of plants selected from old pastures to produce characteristic new strains. Among the late-flowering, persistent forms of timothy, the cultivar S50 belongs to a different species, *P. bertolonii.* It forms a dense, persistent sward and is valuable for special purposes such as the reseeding of hill land.

Reference

BEDDOWS, A. R. 1953. The ryegrasses in British agriculture: a survey. *Bull. H17 Welsh Pl. Breed. Stn* pp. 81.

CHAPTER 7

Morphology

THE morphology of a flowering plant is an expression of its genetic make-up and its response to environment. The Gramineae possess several distinctive morphological features. Some, such as the capacity for basal growth, are agriculturally important but others are mainly of taxonomic interest only. Within a given species there is a degree of morphological plasticity in its response to environmental changes, particularly management factors. It is convenient to examine the morphology of grasses at different stages of the life cycle.

The majority of the temperate forage grasses which are the subject of this work are upright tillering species and this growth habit is of primary interest because of the ease with which such species can be harvested.

The inflorescence

The gramineous inflorescence is complex and specialized. It consists of anemophilous flowers grouped in spikelets which may be borne on stalks on branches of the inflorescence (panicle) or are sessile and arranged directly on the main axis (spike).

The generalized structure of a grass spikelet is shown in Figs 7.1 and 7.2. Each spikelet has a short central axis (rachilla), which carries one or two barren bracts called glumes. Above the glumes, florets are arranged alternately, each consisting of two bracteoles, the lower a one-keeled lemma and the upper a two-keeled palea. In some species the lemma carries a terminal or dorsi-ventral bristle-like extension, called an awn.

The perianth is represented by two small scales, called lodicules, which lie at the base of the carpel and whose function is to open the lemma and palea at flowering. The gynaeceum consists of a superior ovary (a single carpel) bearing two feathery stigmas and enclosing a single ovule. There are three stamens in most genera, each with filaments and anthers.

Spikelets contain one or several florets. The lower florets are usually complete, but the upper florets are often imperfect or sterile. In most genera

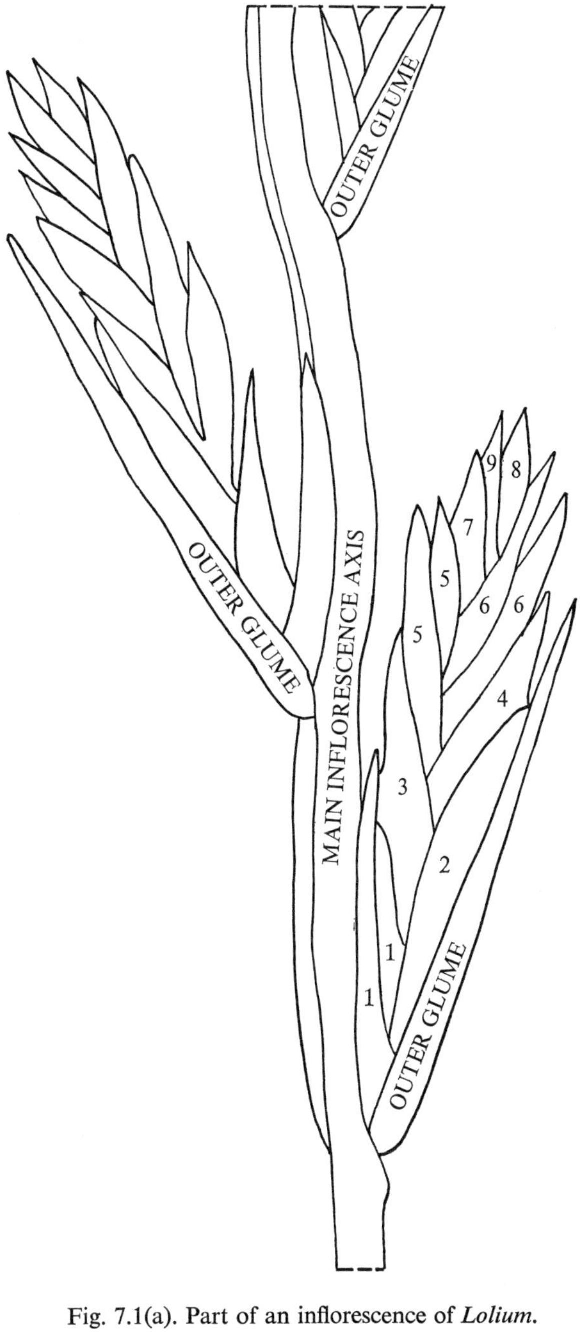

Fig. 7.1(a). Part of an inflorescence of *Lolium*.

the flowers are hermaphrodite, though in some they may be unisexual, e.g. in maize the male florets are carried in a terminal panicle and the female florets are borne in a lateral spike.

Aberrations occasionally occur during inflorescence development in some grasses (Langer & Ryle, 1958). Some of the floral organs (e.g. the lemma) may be modified to form leaves, and new vegetative shoots can develop by an extension of the rachilla. This vegetative proliferation is distinct from vivipary

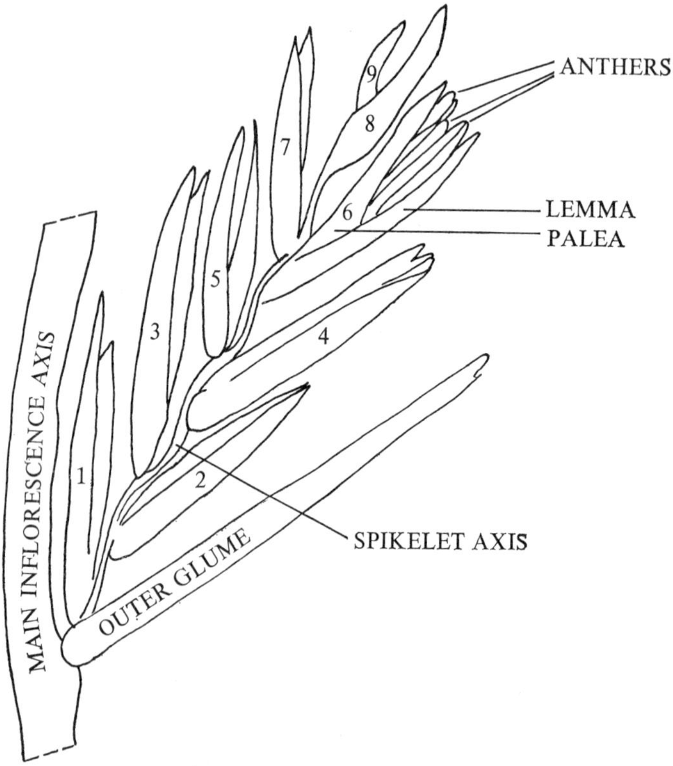

Fig. 7.1(b). Part of an inflorescence of *Lolium.* (Lower right spikelet of Fig. 7.1(a))

when the seeds germinate *in situ* on the inflorescence without a period of dormancy. Vegetative proliferation has been recorded in many native British grasses; in some of these its occurrence is determined genetically, while in others, including many of the important agricultural grasses, proliferations may develop when the photoperiodic requirements of the plants are incompletely satisfied. Other environmental factors, e.g. high humidity and temperature, may be conducive to continued vegetative growth.

The seed and seedling

In the Gramineae the characteristic fruit is a caryopsis—an achene in which the pericarp is united with the testa. Commercial grass 'seed' is variable and the caryopsis is often enclosed by persistent floral parts; in Yorkshire fog

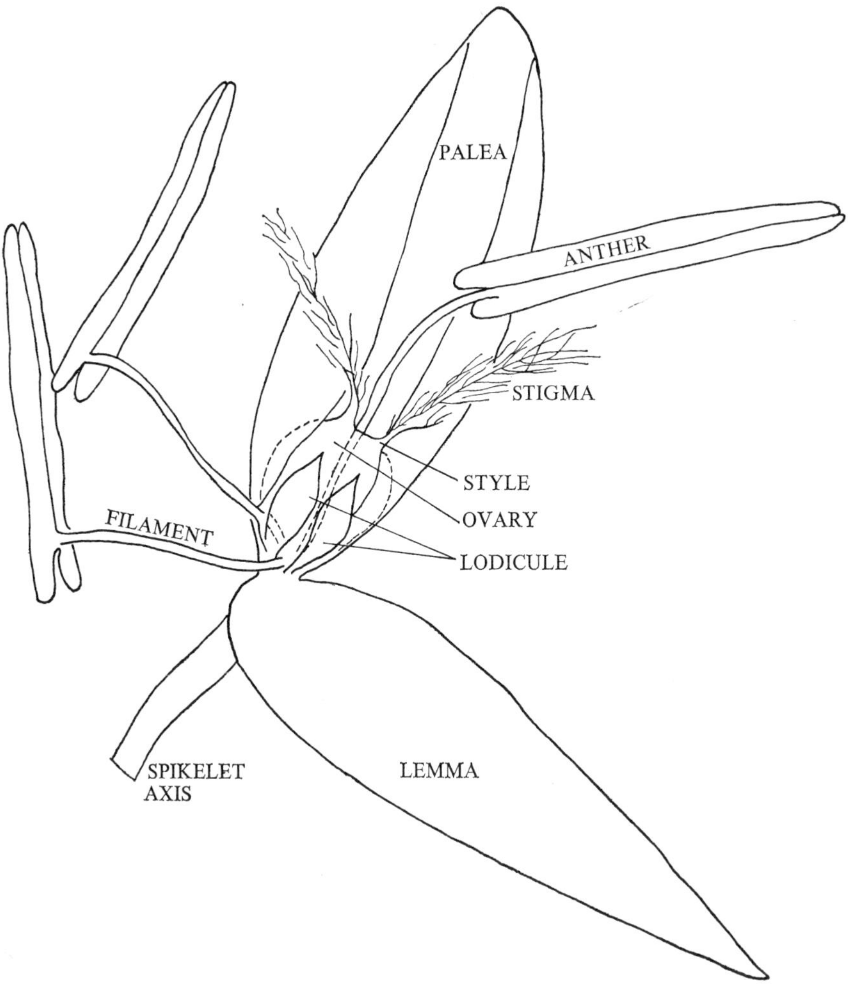

Fig. 7.2. Floret structure in *Lolium*.

the seed is an entire spikelet enclosed by glumes; in perennial ryegrass the caryopsis is enclosed by the lemma and palea; and in timothy and wheat the 'seed' is a naked caryopsis.

The generalized structure of a grass seed is shown in Fig. 7.3.

The bulk of the caryopsis consists of starchy endosperm, which is in greatest evidence in the major cereals. The embryo lies to one side of the endosperm and is partly covered by the scutellum which forms a vascular connection with the endosperm. The plumular bud, enclosed by the coleoptile, is attached to the scutellum by the mesocotyl. Arber (1934) considers that the single cotyledon is represented by the coleoptile, scutellum and part of the mesocotyl. The radicle, enclosed by the coleorhiza, lies at the lower end of the mesocotyl.

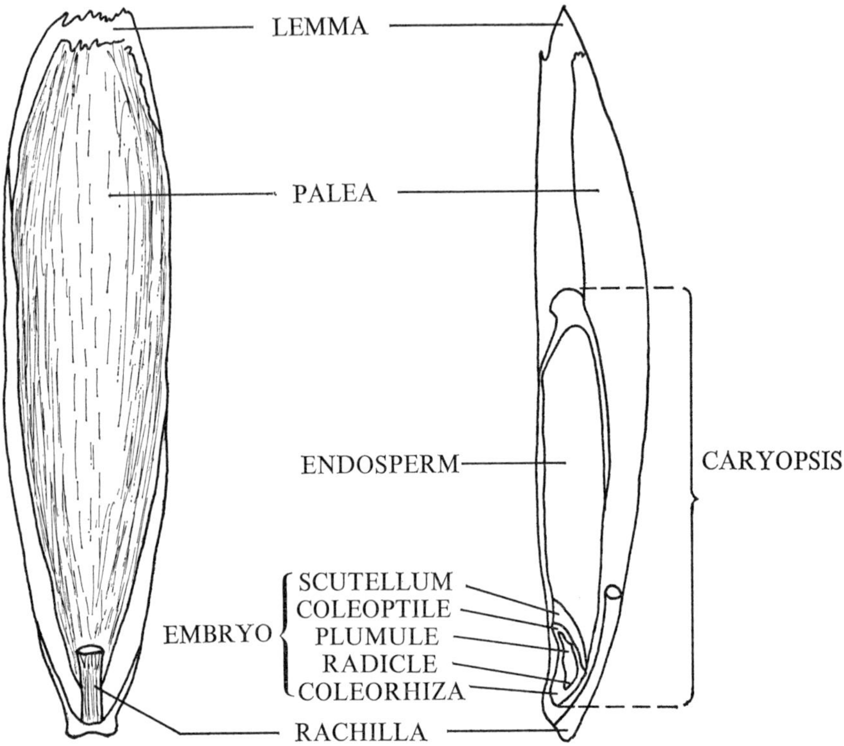

Fig. 7.3. Structure of a grass seed.

After germination the scutellum remains within the testa (Fig. 7.4) and acts as a secretory and absorbing organ supplying the developing embryo with food material from the endosperm. As the coleoptile elongates, the plumule grows, finally splitting the coleoptile. The mesocotyl elongates simultaneously and carries the plumular node to near the soil surface. The extent of mesocotyl elongation varies and is a function of species, seed size and depth of sowing (Jones & Arnott, 1967). The first tiller bud and adventitious roots arise at either the plumular or second node. The mesocotyl

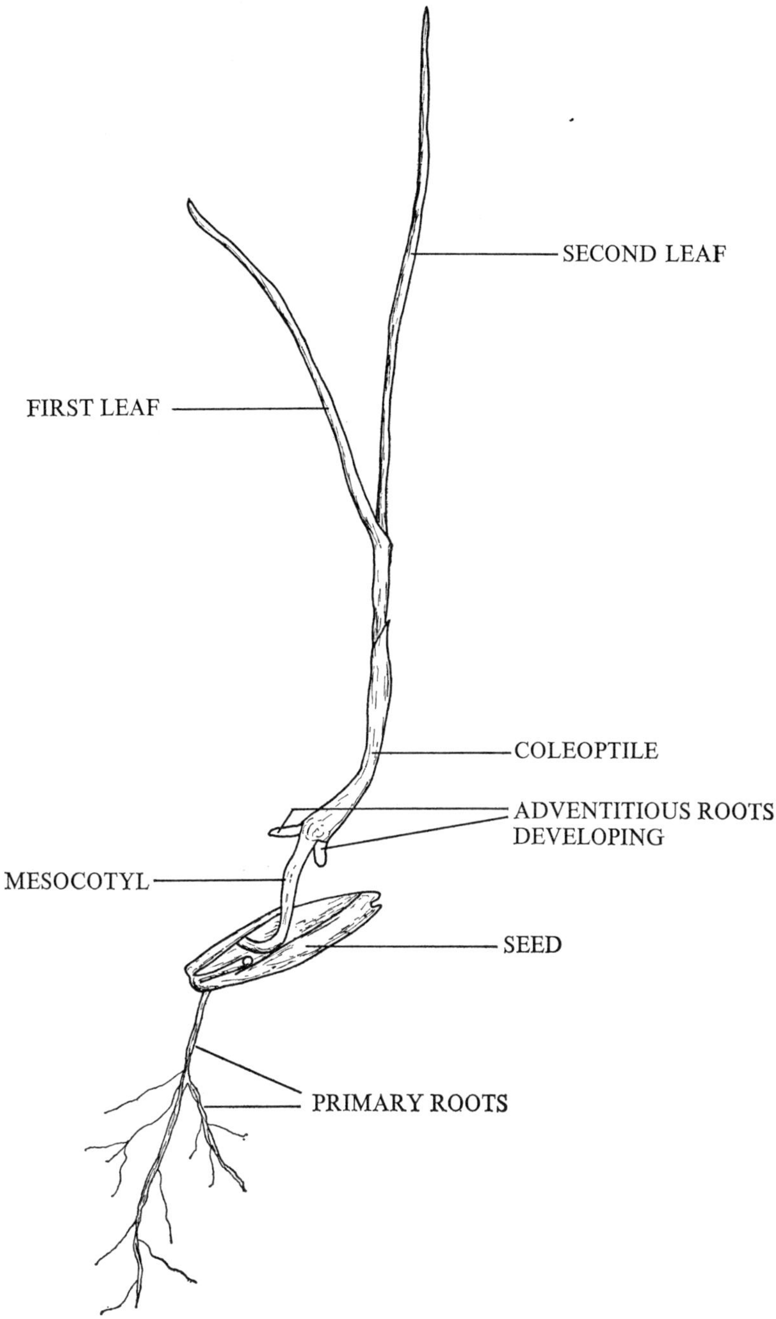

Fig. 7.4. A germinating seedling.

persists until adventitious roots are established at the second and third nodes, by which time the seminal root has disappeared.

The radicle grows through the coleorhiza. One or more seminal roots, the number being characteristic of the species, arise at the base of the radicle. The life of the radicle and seminal roots is limited and corresponds to that of the mesocotyl.

The vegetative and reproductive plant

The basic morphological unit of the grass plant is the tiller—a branch which originates from an axillary bud at the stem base. In grasses the tiller may be vegetative and barren, or may develop into a reproductive tiller bearing an inflorescence. The ratio of reproductive to vegetative tillers varies between species and varieties and within varieties and is also influenced by environment. In general, annual species have a higher proportion of reproductive tillers than have perennial species. A generalized tiller is shown in Fig. 7.5.

There is a basal stem with short internodes. From the basal nodes adventitious roots arise; these are branched, fibrous roots and although most exploit the upper 12 in. (30 cm) of the soil, some penetrate to 3–4 ft (91–122 cm) and have been recorded at depths of 10 ft (305 cm) or more. The vegetative stem (pseudostem) is a hollow cylinder formed by concentric leaf sheaths. Flowering stems, or culms, arise from apical primordia and have long, hollow internodes and swollen, solid nodes. Tiller buds form in the axils of the leaf sheaths. When fully developed, the tillers of most forage grasses have 4–6 expanded leaves.

Individual tillers are short-lived, generally persisting for a year or less. In upright perennial species tillers are replaced by new tillers which develop from the basal buds. This succession of tillers, each forming adventitious roots, gives rise to the tussock growth habit in spaced plants and the mat of tillers in swards. Buds at the upper nodes of flowering stems do not usually develop, although Kydd (1966) has recorded such tiller formation in perennial ryegrass when flowering stems were severed before flowering, under intensive grazing.

The stolons and rhizomes of creeping grasses are homologous with tillers, the elongation of basal internodes giving rise to the stoloniferous habit. Perennial ryegrass, normally an upright species, may assume a stoloniferous habit under intensive grazing (Kydd, 1966). In the rhizomatous grasses, of which couch grass (*Agropyron repens*) is typical, the rhizome has several nodes; the proximal internodes are long, while the 3 or 4 distal internodes are short and have a terminal apical bud. Tillers do not develop from the proximal nodes, which have scale leaves, unless the distal part of the rhizome is severed (Lambert, 1967). While adventitious roots develop at all nodes, it is only the distal nodes that produce tillers. The basal buds on any tiller may produce a new tiller, but generally give rise to new rhizomes.

There are two distinct parts to the grass leaf; the lower part is a cylindrical sheath which may have free margins or be entire at first and then split above, while the upper part is an expanded linear lamina. Venation is linear on the sheath and lamina, and both are photosynthetic. In some species the young

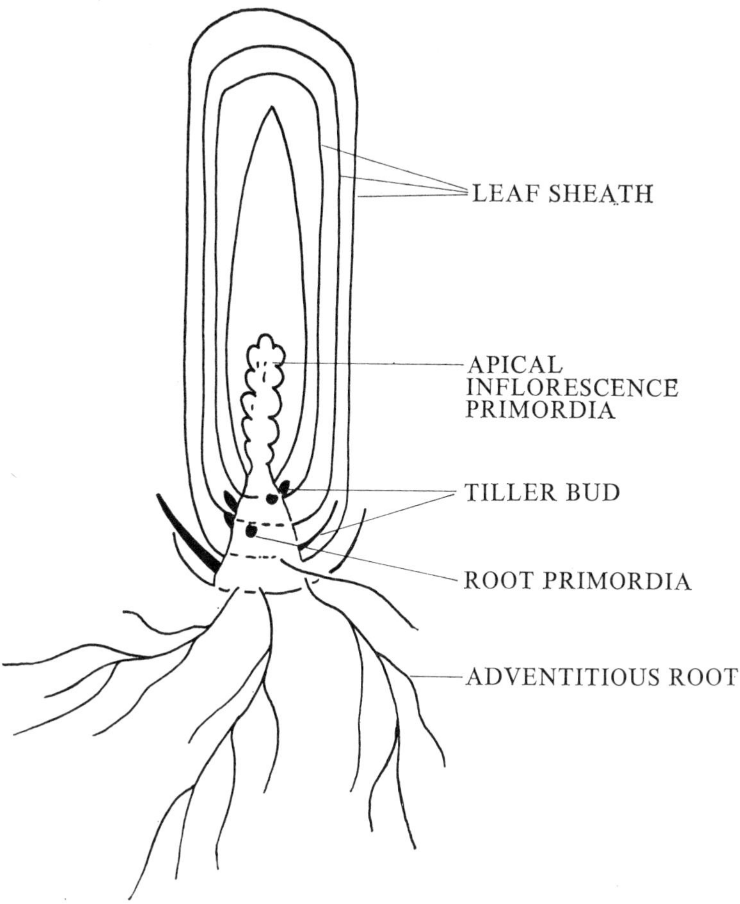

Fig. 7.5. Diagrammatic representation of a grass tiller.

leaves are rolled and in others flattened, giving an oval cross-section. There is often a prominent midrib on expanded leaves. In many species the upper leaf surface has distinct longitudinal ridges, the subepidermal layer of which is sclerenchyma, and the stomata are restricted to the hollows between these ridges. In some xerophytic grasses, e.g. red fescue, the leaves are rolled. The leaves of other species, e.g. Yorkshire fog, are covered with epidermal hairs. While these are prominent diagnostic features, they are also of agricultural significance as they may reduce the acceptability of the forage to livestock. Where the lamina joins the sheath two structures may be present; the auricles, which are twin outgrowths from the sheath and which clasp the stem, and a vertical membranous ligule which also encircles the stem. The ligule may be reduced to hairs or, occasionally, is absent (Hubbard, 1968).

Meristematic tissues occur in the buds and at the base of the leaf sheath. Leaf growth originates at a basal sheath meristem and there is some evidence of an intercalary meristem at the base of the lamina.

The morphological features of grasses which are of particular agricultural importance are basal growth originating from basal meristems and basal buds and, in many species, the ability to form a dense ground cover from a succession of rooted tillers. These characteristics make the grass plant particularly well adapted to withstand frequent defoliation by cutting or by grazing.

References

ARBER, A. 1934. The Gramineae. A study of cereal, bamboo and grass. London: Cambridge University Press, pp. 497.

HUBBARD, C. E. 1968. Grasses. Harmondsworth, Middlesex: Penguin Books Ltd, 2nd edn, pp. 483.

JONES, L.; ARNOTT, R. A. 1967. The Grassland Research Institute, Hurley. Private communication.

KYDD, D. D. 1966. The effect of intensive sheep stocking over a five-year period on the development and production of a sward. 1. Sward structure and botanical composition. *J. Br. Grassld Soc.* **21**, 284–8.

LAMBERT, D. A. 1967. The Grassland Research Institute, Hurley. Private Communication.

LANGER, R. H. M.; RYLE, G. J. A. 1958. Vegetative proliferations in herbage grasses. *J. Br. Grassld Soc.* **13**, 29–33.

CHAPTER 8

Principles of Establishment of Grasses

PLANT establishment from seed involves the processes of (a) germination, (b) underground elongation of the seedling shoot to allow it to emerge from the soil, and (c) the commencement of photosynthesis, resulting in growth and development. These processes are governed by the interaction of seed characteristics and environmental factors.

Germination

The germination of a viable seed is dependent on a favourable temperature and an adequate supply of water and oxygen. At temperatures below about 5°C, grass species differ in the extent to which the rate and percentage of germination are reduced (Chippindale, 1949). At higher temperatures, species differ only in their rate of germination. Under British conditions, variation in temperature rarely causes a complete failure of germination (Stapledon & Wheeler, 1948).

Since the volumes of air and water in a given volume of soil are complementary, in cultivated soils oxygen is likely to become a limiting factor to germination only where drainage is poor.

The top few millimetres of a soil profile are subjected to large diurnal fluctuations of temperature. High temperatures can dry out the surface centimetre of soil in a few days, but once this dry layer has formed, the rate of evaporation from the soil is appreciably reduced (Russell, 1961).

To maintain its moisture supply, the seed should be sown at a depth where it can benefit from the self-mulching action of the soil. The soil should be compressed to a degree that will give good contact between the seed and the available moisture and encourage upward capillary movement without restricting aeration and drainage.

Emergence

After germination, the endosperm and scutellum remain in the soil at the depth at which the caryopsis was placed, while the plumular bud and its ensheathing coleoptile are carried towards the soil surface by the elongation of the mesocotyl (Arber, 1934; Brown, 1960). Elongation of the mesocotyl is a response to darkness and ceases when the tip of the coleoptile emerges into the light. This process places the plumular bud just below the soil surface, irrespective of the depth at which the caryopsis is placed, within certain limits.

The potential mesocotyl elongation is generally least in smaller-seeded species, but within a given species the potential extension is similar for seeds of both low and high seed weight, the former producing a thinner structure. Elongation from a given endosperm reserve can, however, differ greatly between caryopses (Arnott, 1971). With seed sown deeply, elongation of the second internode can supplement that of the mesocotyl in carrying the plumular bud to a position near the soil surface.

While the unemerged seedling can absorb inorganic salts through its seminal root system, it is completely dependent on stored reserves for the supply of organic substances required for development.

In a favourable environment, emergence from the soil takes place in 6–8 days from a sowing depth of about 1 cm (Jones, 1971). Respiration losses during this period can account for over one-third of the caryopsis weight (Anslow, 1962). By delaying emergence, a less favourable environment or deeper sowing would increase the proportion of reserves utilized in respiration.

The depth from which emergence can take place is thus primarily determined by the amount of endosperm reserve, but can be modified by the extent of mesocotyl elongation and by the penetrability of the soil above the seed. Depth of sowing needs therefore to be related to seed size. The optimum sowing depth for the grasses commonly sown is within the range 0·5–3·0 cm (Herriott, 1958).

Growth and development

At coleoptile emergence, a shift from mesocotyl to plumular development occurs. The first leaf of the main shoot soon emerges from the coleoptile and chlorophyll formation within this leaf is rapid.

For a period after emergence, the efficiency of photosynthesis in the green tissue is low and only capable of balancing the loss due to respiration, the continued development of the shoot still being dependent on endosperm reserves. In perennial ryegrass developing at 20°C, an increase in total seedling weight first occurs five days after coleoptile emergence (Anslow, 1962).

The early post-emergence period is therefore critical for seedling development, as a rapid rate of photosynthesis must be established before endosperm reserves are exhausted. Adverse environmental conditions at this stage can seriously retard subsequent growth.

In normal development, when the first leaf is fully expanded the rate of photosynthesis is capable of supporting subsequent growth, and nodal roots soon develop from the lower nodes. In contrast to the seminal roots, these roots initially develop in a diageotropic manner. In some range grasses of the USA, lack of moisture in the surface soil can delay the initiation of nodal roots and this in turn retards further shoot development (Glendening, 1941). No similar studies have been reported on British grasses.

Subsequent development of the aerial shoot is dependent on the rate of leaf production, the size of leaves produced and the rate of tiller development.

The rate of leaf appearance for a genotype varies with the environment but remains constant for any given set of conditions, successive leaf blades becoming progressively longer and wider (Borrill, 1961). Depending on genotype, a certain minimum number of leaves is required before a tiller bud develops into a shoot (Jones, 1971). When this number of leaves has been produced, the axillary position from which the earliest tiller develops is governed by the environment (Patel & Cooper, 1961). When tiller buds are first prevented from developing by an unfavourable environment and then stimulated to develop by improved conditions, they show reduced vigour, depending on the length of time they have been inhibited (Mitchell, 1953).

Under favourable conditions of water, light and nutrient supply, the growth of young plants spaced so that no competition occurs between them, tends to progress geometrically by an increase in the number of tillers. However, a community in which plants compete with each other will give the highest dry matter production in the shortest possible time after sowing.

At a plant density capable of rapidly exploiting an environment in which water and nutrients are in adequate supply, competition for light can occur at an early stage in development. Since leaf initiation and development are the last to be affected by a shortage of assimilates (Begg & Wright, 1964), this shading of one plant by another—or competition for any other factor which reduces assimilation—will be reflected in reduced tillering and leaf size.

In common with other plants, grasses show extreme plasticity, and exhibit a remarkably varied response in size and form to environmental conditions (Donald, 1963). Thus, while initial yields are related to sowing density, yields eventually converge to a similar level despite wide initial variation in sowing densities (Holliday, 1953).

By a high rate of tillering at low plant densities and by relatively reduced tillering coupled with high plant mortality at high densities, a developing sward will adjust towards an equilibrium in terms of tiller number per unit area. The less favourable the environment, the smaller the number of tillers required to exploit it fully. A close relationship exists between seed weight and seedling vigour. This has been shown in comparisons between species (Herriott, 1958), and between cultivars of a species and within cultivars (Thomas, 1966). This effect is most marked during the early stages of development and is therefore an important factor in determining survival at high seedling density.

Successful establishment of the crop thus depends on providing optimum conditions for its rapid development before plant competition begins to limit growth, when unsown species, through better competitive ability, may gain the advantage.

References

Anslow, R. C. 1962. A quantitative analysis of germination and early seedling growth in perennial ryegrass. *J. Br. Grassld Soc.* **17,** 260–3.

Arber, A. 1934. The Gramineae. A study of cereal, bamboo and grass. London: Cambridge University Press, p. 233.

Arnott, R. A. 1971. The development and morphology of seedling grasses. Pt 1. *A. Rep. Grassld Res. Inst. Hurley, 1970* pp. 147–51.

Begg, J. E.; Wright, M. J. 1964. Relative effectiveness of top and basal leaves for the growth of vegetative shoots of reed canarygrass. *Crop Sci.* **4,** 607–9.

Borrill, M. 1961. The developmental anatomy of leaves in *Lolium temulentum. Ann. Bot.* **25,** 1–11.

Brown, W. V. 1960. The morphology of the grass embryo. *Phytomorphology* **10,** 215–23.

Chippindale, H. G. 1949. Environment and germination in grass seeds. *J. Br. Grassld Soc.* **4,** 57–61.

Donald, C. M. 1963. Competition among crop and pasture plants. *Adv. Agron.* **15,** 1–118.

Glendening, G. E. 1941. Development of seedlings of *Heteropogon contortus* as related to soil moisture and competition. *Bot. Gaz.* **102,** 684–98.

Herriott, J. B. D. 1958. The establishment of herbage species in Great Britain. *Herb. Abstr.* **28,** 73–82.

Holliday, R. 1953. Agronomic research in grassland husbandry problems. Pt 2. *Agric. Prog.* **28,** 113–23.

Jones, L. 1971. The development and morphology of seedling grasses. Pt 2. *A. Rep. Grassld Res. Inst. Hurley, 1970* pp. 151–7.

Mitchell, K. J. 1953. Influence of light and temperature on the growth of ryegrass. 2. The control of lateral bud development. *Physiologia Pl.* **6,** 425–43.

Patel, A. S.; Cooper, J. P. 1961. The influence of seasonal changes in light energy on leaf and tiller development in ryegrass, timothy and meadow fescue. *J. Br. Grassld Soc.* **16,** 299–308.

Russell, E. W. 1961. Soil conditions and plant growth. London: Longmans, Green & Co., 9th edn, pp. 383.

Stapledon, R. G.; Wheeler, D. E. 1948. An experiment to test the influence of date and depth of sowing on four of the chief grasses and white clover. *J. Br. Grassld Soc.* **3,** 263–72.

Thomas, R. L. 1966. The influence of seed weight on seedling vigour in *Lolium perenne. Ann. Bot.* **30,** 111–21.

CHAPTER 9

Vegetative Growth

WHEN established, a grass plant has already become multi-tillered and has both a seminal and an adventitious root system (see Chapter 8). Continued growth derives from the activity of meristematic centres at the stem apex and at the root tips. Subsequent growth is best understood by following events in a single stem apex and its derivatives. A diagrammatic half-section of one tiller showing the position of the stem apex is shown in Fig. 9.1. In all annual and many perennial temperate grasses, the stem internodes do not elongate during vegetative growth; the stem apex remains close to the ground and below the normal height for cutting or grazing, from where it may produce further leaves and new tillers.

The terminal part of the stem apex, known as the apical dome, is an area of rapidly dividing tissue which gives rise to an acropetal succession of alternating leaf primordia at a constant rate for a given environment. The leaf primordia first appear as crescentic protuberances, but further cell division converts the crescent into a collar and subsequently into a miniature cowl, which then grows upwards, enclosing the younger primordia and the apical dome (Sharman, 1945; Jewiss, 1966). The developing leaf is similarly ensheathed by older leaves. Soon after the leaf primordium extends around the stem apex, further cell division occurs in the deeper layers of cells on the opposite side of the apex and foreshadows the appearance of a bud in the axil of the next lower leaf primordium. Subsequent development of these buds forms the tillers.

Already within the stem apex, the structure of the grass plant is evident and may be illustrated as a succession of segments, or phytomers, having a leaf initial at the apical end and a bud, from which a tiller may later arise, in the opposite basal position (Fig. 9.2). Segments are joined at the nodes and adventitious roots may subsequently arise from cell division of deep-seated tissue just above each node.

Leaf growth

During the early development of the leaf primordium, meristematic activity becomes restricted to a zone of cells at its base, the intercalary meristem. The leaf emerges from within the encircling leaf sheaths by extension of cells cut off from this meristem. The region of rapid extension is confined to a few millimetres above the stem apex, so that extension growth is already complete in that part of the leaf emerging from the sheath of the next older leaf; the leaf matures from the tip downwards, the leaf sheath being the

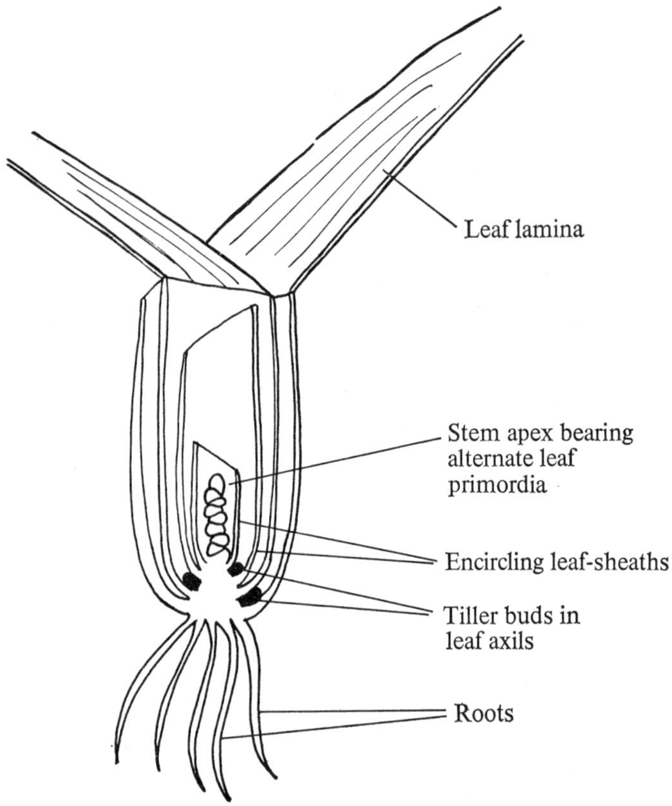

Fig. 9.1. Half-section of vegetative tiller showing position of stem apex and tiller buds.

last part to do so (Davidson & Milthorpe, 1966; Soper & Mitchell, 1956). Removal of part of the leaf lamina after the leaf is fully expanded (i.e. usually after the ligule has emerged) will not stimulate the recommencement of extension growth in the leaf lamina (Langer, 1954). Similarly, the leaf sheath appears incapable of further extension. As with the rate of production of primordia, the rate of leaf appearance may vary with environment but remains more or less constant under constant conditions. There are, however, large effects of environment on the size and dry weight of individual leaves (Anslow, 1966a). When the stem apex of a tiller changes to the reproductive condition, further development of the existing and new leaf primordia is inhibited (see Chapter 10, p. 65).

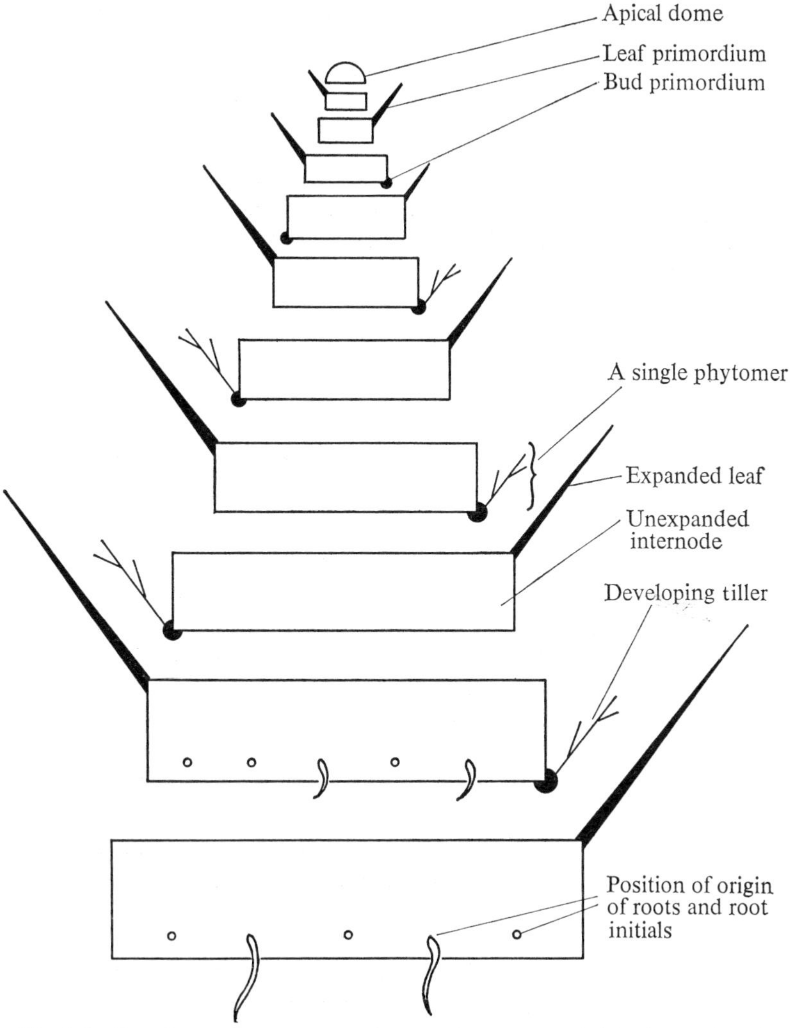

Fig. 9.2. The phytomer structure of vegetative grasses. (This shows the plant as an assemblage of individual segments each joined at the nodes. Note that there is no internode extension)

For plants in the vegetative phase, the area of individual leaves and the rate of leaf production generally increase to a maximum with rise in temperature to between 20 and 30°C, depending on species and climatic origin of the race or variety; specific leaf area (leaf area per unit dry weight) also increases with rise in temperature (Mitchell, 1956; Cooper, 1964; Robson, 1969). An increase in light intensity also increases the rate of leaf production but, in contrast to the response to temperature, leaf area and specific leaf area decrease as light intensity increases (Mitchell, 1953a; Robson, 1969). In direct contrast, an increase in daylength reduces the rate of leaf appearance and increases the area of individual leaves, although specific leaf area is decreased, the magnitude and nature of the effect varying with species and probably with variety also (Ryle, 1966 a, b). Of the major nutrient elements, N has the largest effect, increasing rates of application augmenting leaf area and dry weight (Langer, 1959). In comparison with N, the effects of P and K are small, but there is a marked positive interaction between N and P (Langer, 1959). Increased levels of N can also lead to an increase in rate of leaf production (Ryle, 1964).

Tillering

Since tillers arise in the axils of leaves, the potential rate of tillering is dependent on the rate of leaf production. The initiation of tillers has already been described, but there is little information on the subsequent development of tillers until they emerge from the sheath of the subtending leaves. The subsequent development of tiller buds is highly dependent on the environment and is affected by it to a much greater extent than is leaf production. For example, when environmental conditions result in carbohydrate being in short supply, leaf growth is favoured at the expense of the further development of tiller buds. This appears to continue until a new internal equilibrium is achieved within the plant. The nature of this check on tillering may be to increase the interval between the expansion of a leaf and the appearance of the tiller it subtends (Ryle, 1964).

In general, an increase in temperature increases the rate of tillering (Robson, 1969), although the acceleration is less than for rate of leaf production and the optimum temperature (about 15°C) is lower. Tillering is, however, decreased by a decrease in light intensity and an increase in photoperiod (Mitchell, 1953b; Ryle, 1966 a, b). N, P and K all increase tillering, N showing a positive interaction with the other two elements (Langer, 1959).

Root growth

Established plants possess both a seminal and an adventitious root system. Further root growth results from either extension and branching of the existing roots or from the production of new adventitious roots. One or more adventitious roots may arise from each internode. The sites of production of roots are at or near the soil surface, a zone which is particularly liable to drought. Both drought and high temperature are known to inhibit root production, and the removal of the crop by cutting leads to a high radiation receipt at the soil surface and a consequent increase in temperature.

Root growth and development are very closely related to the supply of carbohydrates from the shoot system, as has been pointed out by Brouwer (1966). The shoot/root ratio increases as the plant gets older, but for a given species, set of conditions and age of plant, this ratio will be constant. An increase in temperature increases the shoot/root ratio, as also does a fall in light intensity (Brouwer, 1966). Similarly, an increase in N supply also increases the shoot/root ratio, a change which conveys the false impression that root growth is greater under a low N regime. Soil moisture stress reduces shoot growth, but has a much smaller effect on root growth (Brouwer, 1966). Removal of the top of the plant dramatically reduces the shoot/root ratio and the source of carbohydrates; it also reduces root extension and activity (Davidson & Milthorpe, 1966). Subsequently the shoots grow faster than the roots until the shoot/root ratio existing before cutting is re-established (Brouwer, 1963; Brouwer & de Wit, 1969).

Growth in the sward

VEGETATIVE PHASE

For the purposes of practical agriculture, herbage grasses are usually sown in drills a few inches apart; with continued tillering of individual plants, the drill rows tend to coalesce to form a complete ground cover. Growth under sward conditions involves competition between plants for light, mineral nutrients and water, and in consequence the growth and morphology of the plant may be modified, e.g. the rate of tillering and the number of live leaves borne on a tiller are reduced. From a physiological viewpoint, growth may be regarded simply as the increase in dry weight of the whole plant resulting from carbon assimilation (photosynthesis) and nutrient uptake. From an agricultural point of view, growth is more usually thought of in terms of harvestable yield. Where herbage is concerned, this is less well defined than yield of grain or of a root crop and depends not only on the height of cutting or grazing but, in swards harvested by cutting, on the evenness of ground under the mowing machine. Most estimates of growth of herbage have dealt with harvestable yield.

The fact that the grass plant remains vegetative for most of the year and in consequence the stem apices, and therefore the leaf-producing, meristematic regions of the plant, remain close to the ground and below cutting or grazing height, makes the temperate herbage grasses particularly suited for these systems of utilization. In vegetative plants, regrowth of the shoot system following defoliation occurs immediately from the existing centres of meristematic activity.

EFFECTS OF REPRODUCTIVE GROWTH

In the indigenous temperate grasses, except in those which are stoloniferous, stem elongation usually occurs only in association with the onset of reproduction. Only at this stage of growth are the stem apices carried above ground level and thus become liable to be removed by cutting or grazing. Half the total annual yield of dry matter (DM) and digestible nutrients may be harvested in a single cut taken in the early part of the growing season when inflorescences have emerged (Table 9.1). The total accumulated yield obtained at the time of this harvest may be jeopardized if an earlier utilization has removed the stem apices before stem extension or inflorescence emergence are complete (Jones, 1959). Because of the effects of stem formation and inflorescence production on the yield of DM in the period from April to June, there is no close parallel between seasonal variation in temperature, light energy and photoperiod on the one hand and DM production on the other

Table 9.1. Yield of S24 perennial ryegrass swards when first cut at 70% digestible DM (at 10 days after inflorescence emergence), and subsequently at 6-week intervals

Year	1966		1967	
N applied (kg/ha)	457		522	
	Irrigated	Not irrigated	Irrigated	Not irrigated
First cut (25 May), '00 kg/ha	82·9	79·7	87·1	87·8
Total annual yield, '00 kg/ha	146·6	132·6	167·6	157·6
Yield at first cut, percentage of total	57	60	52	56

(From Green and Corrall, Grassland Research Institute, Unpublished data)

(Anslow & Green, 1967) (Fig. 9.3), as might be expected from a consideration of the effects of environment on the growth of single plants in the vegetative phase.

Regrowth following any utilization in which stem apices have been removed requires that further regrowth must come from new vegetative tillers produced at the base of the flowering stem, or from vegetative tillers originally present but which failed to become reproductive. The number and proportions of the various types of tiller present at a particular stage of growth appear to depend both on species and on previous management (Jewiss *et al.*, Unpublished). Thus in S48 timothy, cut first for hay and then twice in the aftermath, virtually no vegetative tillers present in early spring survived beyond the inflorescence emergence stage, whereas in S215 meadow fescue cut monthly, up to 60% of the population of tillers present when inflorescence emergence would normally occur and which survived beyond this stage of growth, were vegetative and were present in the following spring (see Fig. 11.4, p. 81). The replacement of reproductive by vegetative tillers after inflorescence production results in a poorer correlation between yield and environment than would otherwise be expected (Anslow, 1965) (Fig. 9.3).

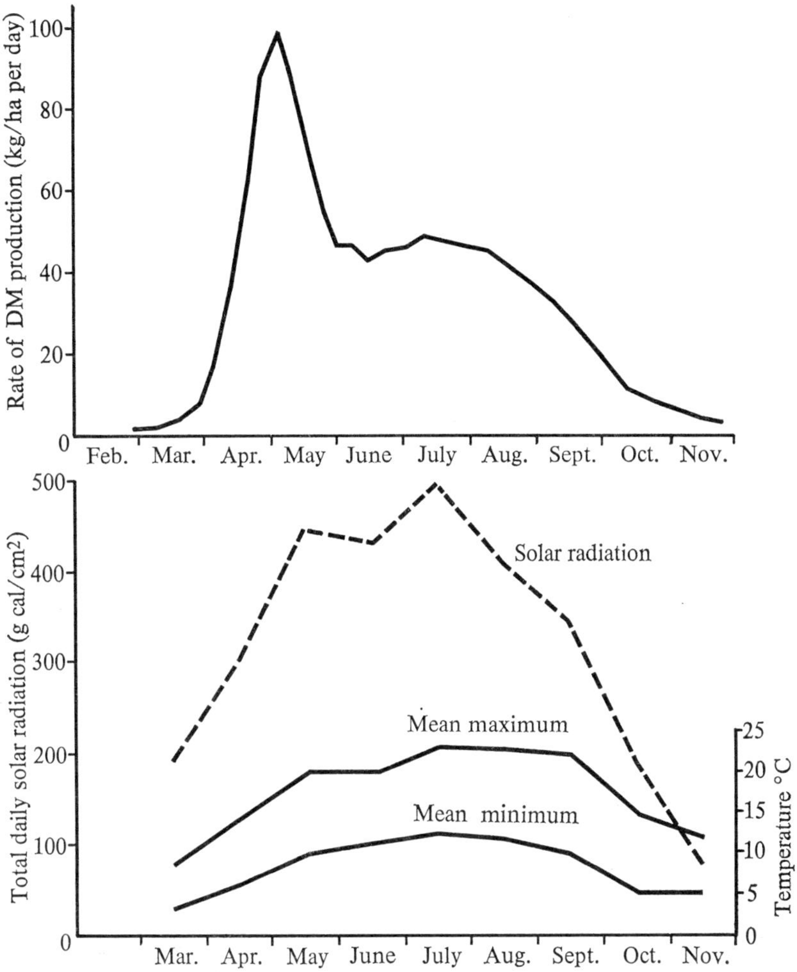

Fig. 9.3. Rate of production of S24 perennial ryegrass cut every 3 or 4 weeks during 1964; and changes in solar radiation and mean maximum and minimum temperature over the same period. (From Anslow and Green, 1967)

Clearly, the speed of regrowth after cutting at about the time of inflorescence emergence will depend on the structure of the sward and, in particular, on the number and size of any replacement tillers or old vegetative tillers below cutting height (Jewiss & Powell, 1966). For pure grass swards in a vegetative condition, in contrast to grass/legume mixtures, the severity of a single defoliation seems to have little effect on the rate of further growth (Anslow, 1966a, cf. Donald, 1961). Thus when either all or no green leaves were removed, the subsequent accumulation of green leaf (expressed as dry weight) was linear in both instances, though when losses due to senescence were taken into account, the actual rate of DM production was greater where no leaf was removed.

YIELD, EFFICIENCY OF PRODUCTION AND HARVESTING

In addition to the control of growth, and thus yield, imposed by environment and the special effects of inflorescence production on the annual pattern

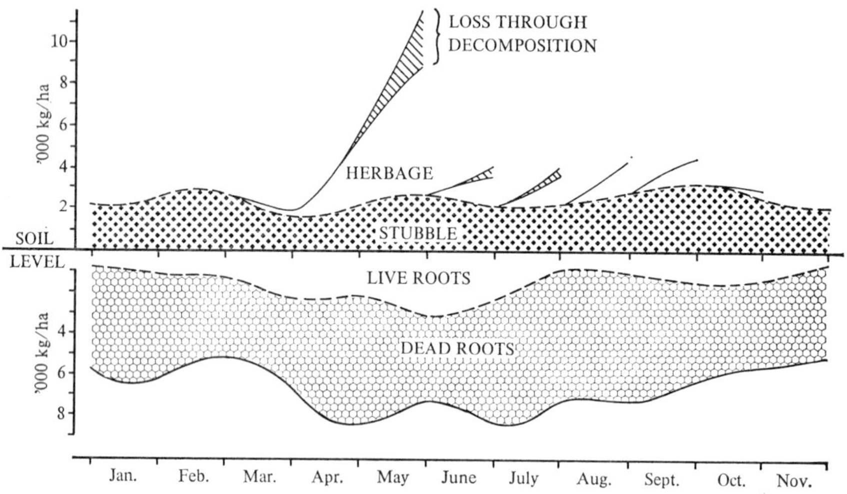

Fig. 9.4. Distribution of DM throughout a perennial ryegrass sward, uncut until late May, then cut at 5 cm above soil level once a month. (From Garwood, Grassland Research Institute, Unpublished data)

of growth, yield may also be restricted in swards, as in single plants, by lack of N and moisture supplies.

It is interesting to note that for the growth of the whole plant, the efficiency of utilization of total solar energy is less than 2%. For grass, however, only about one-third to one-half of the plant is harvested, representing a conversion of some 0·7 to 1% of the actual radiation receipt into harvestable DM. Garwood (1967) has indicated the large amounts of unharvested DM present in the stubble and roots throughout the year (Fig. 9.4). It would be erroneous to conclude that there is a large static pool of DM remaining untouched, however; swards exist in a dynamic state, with a continual turnover in the number of living tillers and their roots. There is appreciable evidence of death of some existing roots after the tops of the plants have been cut; also, on any one tiller, leaves have only a finite life. Rate of leaf production in summer varies between one leaf every 5 to one every 11 days according to species and conditions. Rate of leaf death is about equal to rate of leaf appearance (Bean, 1964), so that under sward conditions an established tiller will support a relatively constant number of living leaves, varying between 3 and 6 according to genotype and environment.

To avoid losses due to senescence of leaves and therefore an uneconomic use of photosynthetically-fixed solar radiation, harvest intervals for plants in the vegetative stage ought not to be longer than the length of life of a leaf produced since the last utilization. Clearly, however, though losses due to leaf senescence may be minimized by frequent cutting, such a system will entail the frequent exposure of bare ground and a consequent loss of ability to fix solar radiation.

Also, during the formation of stems and inflorescences, any premature harvest will remove rapidly expanding sinks for assimilates and will lead to a lower harvestable yield. Thus, when swards of single grass species are repeatedly harvested, cutting or grazing frequencies are likely to be determined by considerations of maximum yield and more particularly by the parameters of interest to the animal nutritionist, such as herbage intake and digestibility.

References

ANSLOW, R. C. 1965. Grass growth in midsummer. *J. Br. Grassld Soc.* **20,** 19–26.

ANSLOW, R. C. 1966a. Light interception and growth rate of a perennial ryegrass sward. *Proc. 9th int. Grassld Congr., São Paulo, 1965* pp. 403–5.

ANSLOW, R. C. 1966b. The rate of appearance of leaves on tillers of the Gramineae. *Herb. Abstr.* **36,** 149–55.

ANSLOW, R. C.; GREEN, J. O. 1967. The seasonal growth of pasture plants. *J. agric. Sci., Camb.* **68,** 109–22.

BEAN, E. W. 1964. The influence of light intensity upon the growth of S37 cocksfoot (*Dactylis glomerata*). *Ann. Bot.* **28,** 427–43.

BROUWER, R. 1963. Some aspects of the equilibrium between overground and underground plant parts. *Jaarb. Inst. biol. scheik. Onderz. LandbGewass.* pp. 31–9.

BROUWER, R. 1966. Root growth of grasses and cereals. In: Milthorpe, F. L.; Ivins, J. D. [Eds] The growth of cereals and grasses. Proceedings of the twelfth Easter School in Agricultural Science, University of Nottingham, 1965. London: Butterworth and Co., pp. 153–66.

BROUWER, R.; DE WIT, C. T. 1969. A simulation model of plant growth with special attention to root growth and its consequences. In: Whittington, W. J. [Ed.] Root growth. Proceedings of the fifteenth Easter School in Agricultural Science, University of Nottingham, 1968. London: Butterworth and Co., pp. 224–44.

COOPER, J. P. 1964. Climatic variation in forage grasses. 1. Leaf development in climatic races of *Lolium* and *Dactylis*. *J. appl. Ecol.* **1,** 45–61.

DAVIDSON, J. L.; MILTHORPE, F. L. 1966. The effect of defoliation on the carbon balance in *Dactylis glomerata*. *Ann. Bot.* **30,** 185–98.

DONALD, C. M. 1961. Competition for light in crops and pastures. In: Milthorpe, F. L. [Ed.] Mechanisms in biological competition. Symposia of the Society for Experimental Biology. No. 15. Cambridge: Univ. Press, pp. 282–313.

Garwood, E. A. 1967. Studies on the roots of grasses. *A. Rep. Grassld Res. Inst. Hurley, 1966* pp. 72–9.

Jewiss, O. R. 1966. Morphological and physiological aspects of growth of grasses during the vegetative phase. In: Milthorpe, F. L.; Ivins, J. D. [Eds] The growth of cereals and grasses. Proceedings of the twelfth Easter School in Agricultural Science, University of Nottingham, 1965. London: Butterworth and Co., pp. 39–54.

Jewiss, O. R.; Powell, C. E. 1966. The growth of S48 timothy swards after cutting in relation to carbohydrate reserves and leaf area. *A. Rep. Grassld Res. Inst. Hurley, 1965* pp. 67–72.

Jewiss, O. R.; Ryle, S. M.; Langer, R. H. M. The changing plant and tiller populations of timothy and meadow fescue swards. 2. Tiller longevity. (Unpublished).

Jones, L. I. 1959. Varietal characteristics of herbage plants in relation to their agronomic assessment. In: Ivins, J. D. [Ed.] The measurement of grassland productivity. Proceedings of the sixth Easter School in Agricultural Science, University of Nottingham, 1959. London: Butterworth and Co., pp. 34–50.

Langer, R. H. M. 1954. A study of leaf growth in timothy (*Phleum pratense*). *J. Br. Grassld Soc.* **9,** 275–84.

Langer, R. H. M. 1959. Growth and nutrition of timothy (*Phleum pratense* L.). 4. The effect of nitrogen, phosphorus and potassium supply on growth during the first year. *Ann. appl. Biol.* **47,** 211–21.

Mitchell, K. J. 1953a. Influence of light and temperature on the growth of ryegrass (*Lolium* spp.). 1. Pattern of vegetative development. *Physiol. Pl.* **6,** 21–46.

Mitchell, K. J. 1953b. Influence of light and temperature on the growth of ryegrass (*Lolium* spp.). 2. The control of lateral bud development. *Physiol. Pl.* **6,** 425–43.

Mitchell, K. J. 1956. Growth of pasture species under controlled environment. 1. Growth at various levels of constant temperature. *N.Z. Jl Sci. Technol. Sect. A* **38,** 203–16.

Robson, M. J. 1969. Light, temperature and the growth of grasses. *A. Rep. Grassld Res. Inst. Hurley, 1968* pp. 111–23.

Ryle, G. J. A. 1964. A comparison of leaf and tiller growth in seven grasses as influenced by nitrogen and temperature. *J. Br. Grassld Soc.* **19,** 281–90.

Ryle, G. J. A. 1966a. Effects of photoperiod in the glasshouse on the growth of leaves and tillers in three perennial grasses. *Ann. appl. Biol.* **57,** 257–68.

Ryle, G. J. A. 1966b. Effects of photoperiod in growth cabinets on the growth of leaves and tillers in three perennial grasses. *Ann. appl. Biol.* **57,** 269–79.

Sharman, B. C. 1945. Leaf age and bud initiation in the Gramineae. *Bot. Gaz.* **106,** 269–89.

Soper, K.; Mitchell, K. J. 1956. The developmental anatomy of perennial ryegrass (*Lolium perenne* L.). *N.Z. Jl Sci. Technol. Sect. A* **37,** 484–505.

CHAPTER 10

Reproduction and Seed Formation

THE early growth of all grass shoots, whether derived from a tiller bud or a seed, is vegetative, and is thus restricted to the production of leaves, tillers and roots. Reproduction, i.e. the formation of an inflorescence bearing florets and the subsequent development, if any, of seed within the florets, occurs by an abrupt alteration of the form of growth within the vegetative shoot.

The first morphological sign of reproduction is the precocious expansion of the buds in the axils of leaf primordia along the flanks of the shoot apex of older shoots (Fig. 10.1 and Plate 1). Usually this condition is preceded by an increased rate of apical growth. For a short time, both leaf primordia and their axillary structures, the inflorescence branch primordia, develop simultaneously (double-ridges stage); very soon, however, further development of leaf primordia is inhibited and the whole of the shoot apex is given over to the production of inflorescence branches (Plates 2 and 3). This activity at the shoot apex generally precedes internode expansion in the sub-apical zone which carries up the incipient inflorescence within the enveloping leaf sheaths (Fig. 10.1B). A wave of elongation moving upwards through successively younger internodes causes stem elongation; generally, the ear-bearing stem consists of 3–5 elongated internodes. Once the inflorescence has emerged it stops elongating, but the stem continues to elongate, albeit slowly.

The early stages of inflorescence development are similar in all grasses (see Fig. 10.1). The very varied form which the mature inflorescence takes in different grasses is determined by variations in the number of florets which are grouped together into a spikelet, by whether there is simple or complex branching and by variations in the length of the branches. These differences result from variations in the amount of growth or branching occurring in those parts of the incipient inflorescence common to all grasses. Essentially, therefore, the grass inflorescence consists of an assemblage of florets arranged in one form or another. The floret, with minor structural variations, is the unit common to all grasses.

Some weeks after inflorescence emergence, the florets open and pollen is released from the anthers (anthesis). Because of differences in morphology, there is considerable variability among species in the pattern of anthesis along inflorescences. Where the florets are borne on stalked branches, some spreading apart of branches generally precedes pollen release. The grasses considered here are predominantly cross-fertilized in nature, but in some species

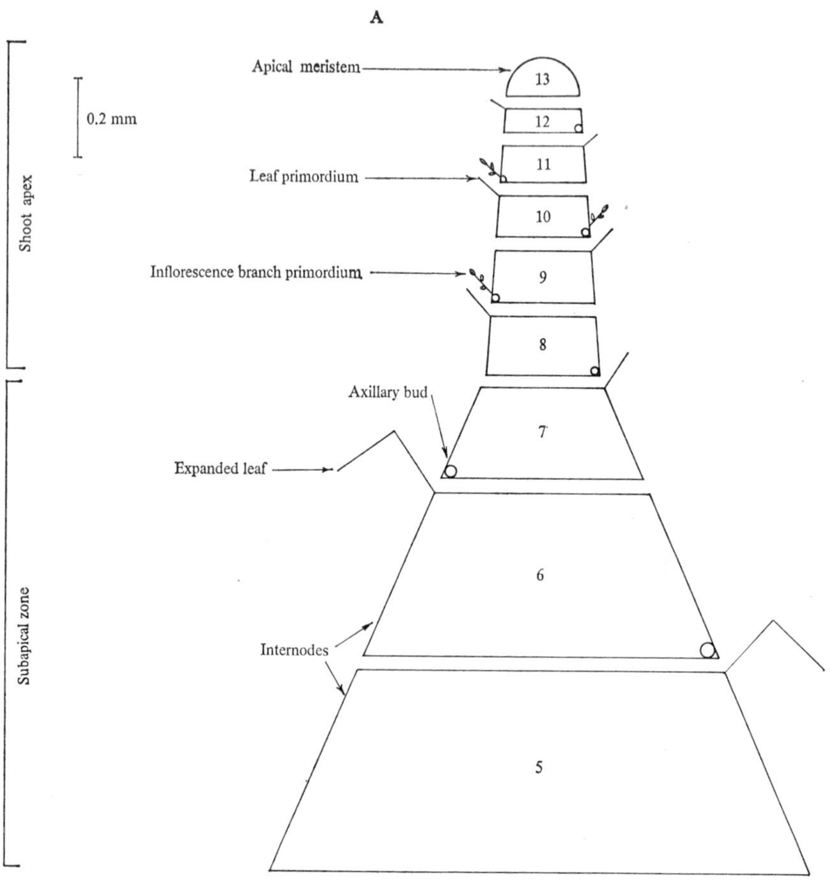

Fig. 10.1A. Diagrammatic representation of the shoot apex and subapical zone of a ryegrass shoot at transition to the reproductive stage.

self-fertilization can and presumably does occur (Beddows, 1931). Anthesis may continue intermittently over a period of about two weeks, depending on the weather (Anslow, 1963), and in most species fertilization occurs concurrently because stigmata are exserted at the same time. Not all florets set seed; even in the grasses which produce relatively large amounts of seed, such

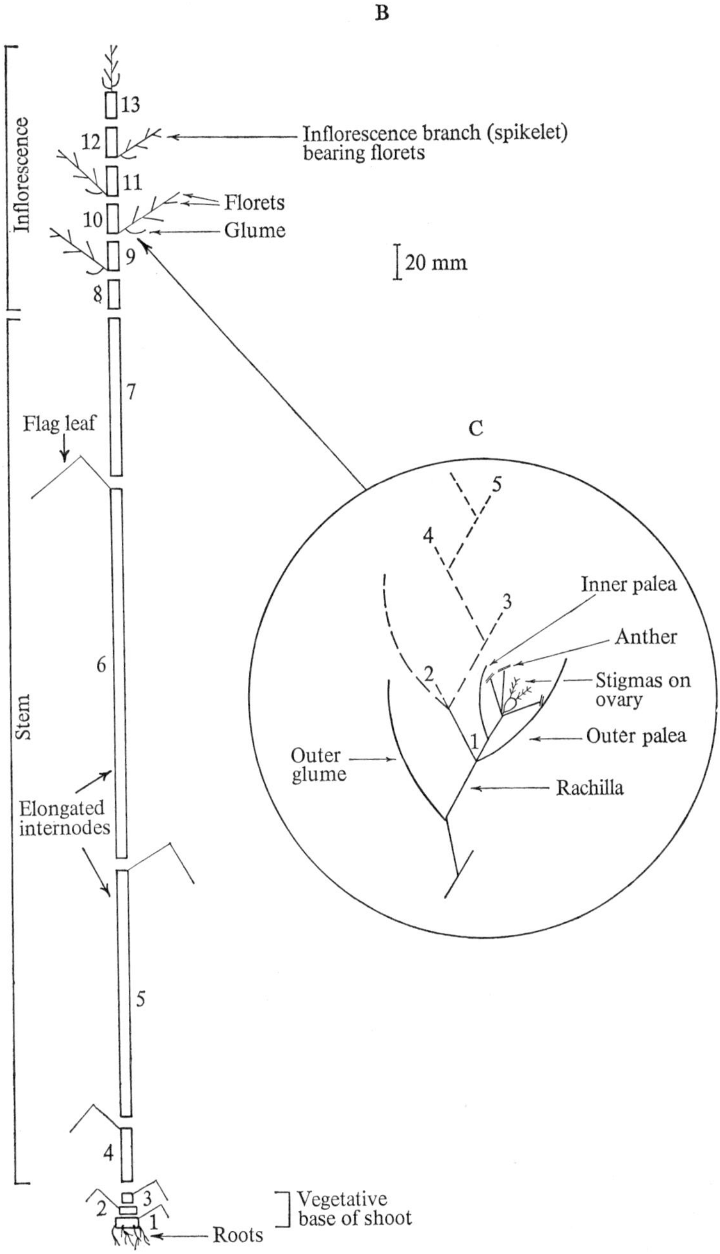

Fig. 10.1B, C. Diagrammatic representation of the shoot apex and subapical zone of a ryegrass shoot.

B. After stem elongation and inflorescence development.

C. Representation of the first floret on the inflorescence branch (spikelet) on internode 10.

In diagrams 10.1A and 10.1B, numbers denote corresponding internodes.

D

as ryegrass, the proportion seldom exceeds 75% and is often much lower. Some florets may be morphologically sterile and thus incapable of developing a seed (Johnston, 1960); however, environmental factors and management are also reported to affect seed-set. Some embryos may die soon after fertilization (Johnston, 1960), and even when embryo development continues, there is great variation between individual seeds in the final weight they attain.

Environmental control of reproduction

The development of the inflorescence in grasses is usually determined by responses to local conditions of light (both duration and intensity) and temperature at each stage of development. These requirements vary greatly with species and cultivar and are related to the climatic or agronomic origin of the material (Cooper, 1960). Before the shoot apex can change over from leaf production to inflorescence production, most perennial and biennial grasses of temperate regions must be exposed to winter conditions (low temperatures and/or short daylengths) for a period characteristic for the species or cultivar. In common with annual grasses such as Westerwolds ryegrass, timothy differs from other perennial species in having no such requirements. Exposure to winter conditions causes no morphological change in the shoot, and it is only by the ability of the shoot to advance to the next stage of development, the production of inflorescence branch primordia, that a requirement for winter induction can be recognized.

Once the shoot is induced, the morphological change to reproductive growth generally depends on exposure to a suitable daylength, although light intensity and temperature may modify the response (Ryle, 1966). Most temperate grasses are long-day plants and will not initiate an inflorescence until a critical daylength is reached in the spring (Fig. 10.2). This critical daylength differs between species and even between cultivars; in ryegrass cultivars it may vary from 8 to 13 h, while in timothy a daylength of up to 15 or 16 h may be required.

After an inflorescence is initiated, photoperiods of appropriate length are still necessary for normal development to proceed, but spring temperatures are also important in determining the rate of development of the inflorescence and the time of its emergence from the leaf sheaths. Exposure to photoperiods shorter than the critical daylength or to excessively high temperatures at this stage may cause a reversion to vegetative growth, leading to the development of abnormal florets and proliferation of leaf-like bracts or vegetative shoots. The latter are capable of rooting and leading an independent existence if detached from the parent shoot (Langer & Ryle, 1958).

There is little firm experimental evidence to indicate which factors are important for fertilization of the floret and seed formation in the field. While it seems likely that daylength may control the basic rhythm of floret activity at seed-setting, local conditions such as low temperatures, wet and dull

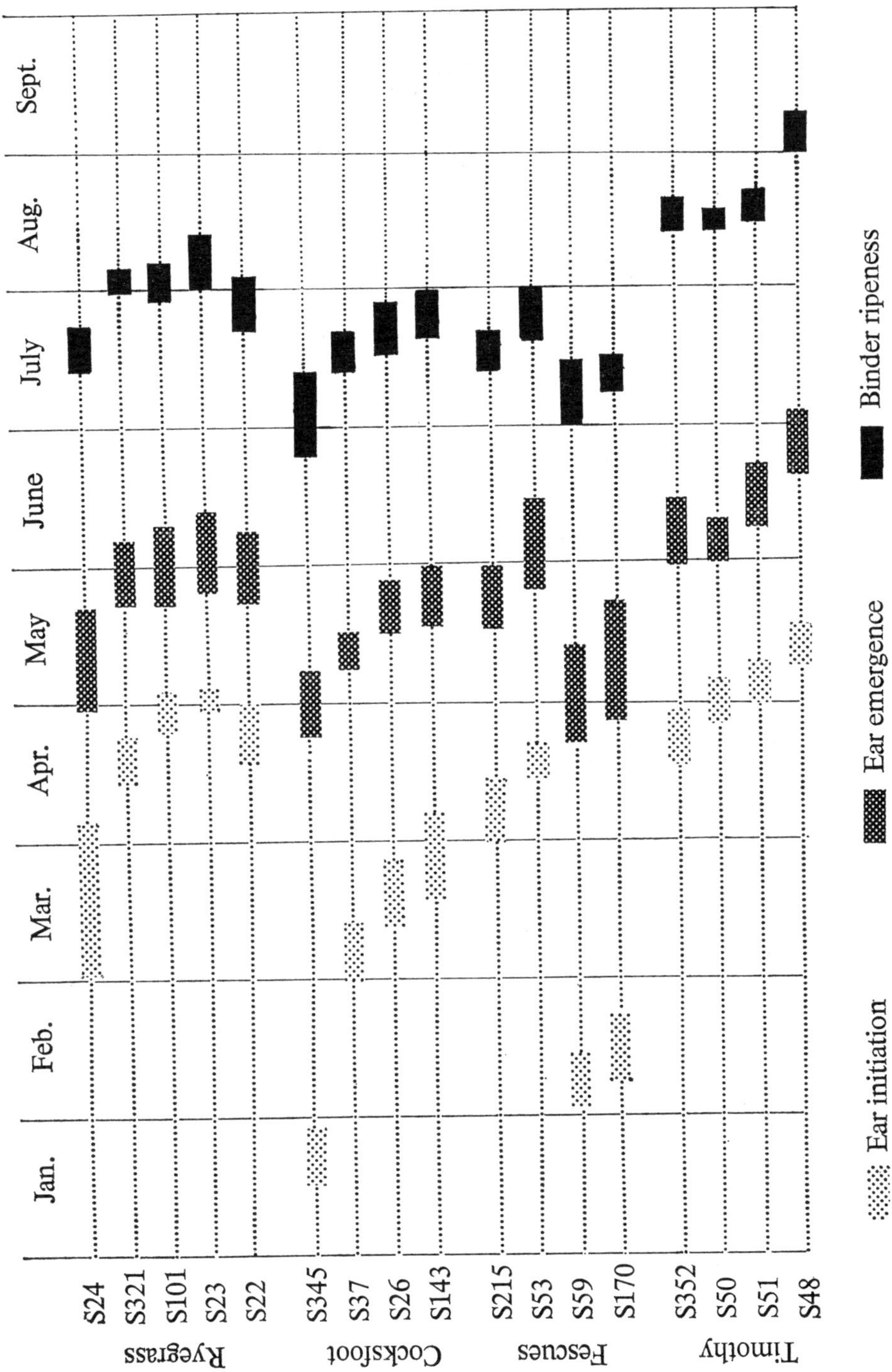

Fig. 10.2. Sequence and approximate times of ear initiation, emergence and seed ripeness.

weather and strong winds delay anthesis and the distribution of pollen and thus affect fertilization; in addition, the effects of these factors on anthesis vary in different species (Emecz, 1962; Evans, 1964). It is not known whether conditions preventing the separation of the paleae indirectly promote self-fertilization in normally cross-pollinated grasses. It has been suggested (Anslow, 1963) that the simultaneous shedding of large numbers of viable pollen grains of optimum size into a crop is a necessary prerequisite if a high proportion of florets is to be fertilized. However, little is known as to how this may best be achieved.

It has already been pointed out that after fertilization has occurred, some embryos may die; the reasons for this are still largely unknown, although competition for metabolites between embryos in the same inflorescence may be involved. At a still later stage, the variation in individual seed weights can be related to the location of the floret in the inflorescence, and to the time of origin and the position of the parent shoot within the plant (Anslow, 1964). Once again, internal regulation by growth factors is presumably the cause.

Reproduction and dry matter production

Growth in the vegetative phase is restricted to increases in leaf and root weight and to the expansion of new tillers. Two further potentially very active centres of growth, the stem and the inflorescence, are added to these at reproduction. While the development of an inflorescence terminates the ability of a shoot to initiate further leaves, some overlapping of leaf and inflorescence growth is ensured by the expansion of already initiated but as yet unexpanded leaves. Further, these last-formed leaves on the reproductive shoot are carried up by internode extension which results in increased vertical separation of their assimilating surfaces.

The relationships between growth of the stem and ear and that of the other parts of a shoot during inflorescence development in darnel (*Lolium temulentum*) are shown in Fig. 10.3; although the relationships may be accentuated in this extreme annual grass, the comparison serves to indicate the high rates of dry matter (DM) production in stem and ear. In swards, the close association between reproduction and the very rapid accumulation of DM in spring and early summer has prompted several investigators to suggest that reproductive shoots are capable of faster growth than shoots growing vegetatively (Hunt, 1956; Langer, 1958; Knight, 1965; Silsbury, 1965). The vertical extension of the leaf canopy following stem extension, together with the added assimilating capacity of stems and ears, may increase the efficiency of utilization of incident radiation.

Alternatively, the efficiency of the assimilating system of the shoot itself may be influenced by the internal physiological changes consequent upon reproduction. It is also possible that reproduction may be the prelude to a marked shift in the internal organization of the shoot, possibly through an effect on the internal partitioning of energy resources between root and lateral and terminal meristems. However, such hypotheses require substantiation.

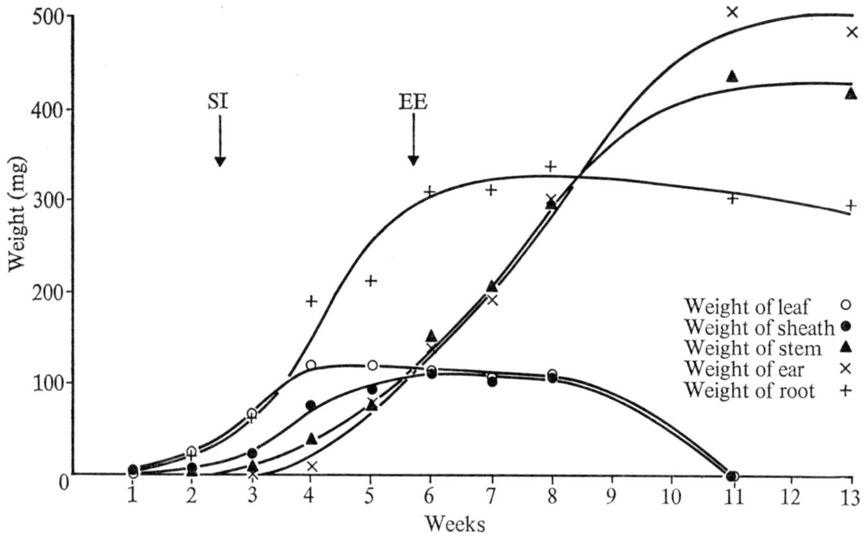

Fig. 10.3. Weight of leaf, sheath, stem and ear on the main shoot during inflorescence development in *Lolium temulentum*. The root weight of the whole plant is also shown. Light intensity 1200 f.c. Daylength 16 h. Temperature 20°C day/15°C night. Arrows denote onset of inflorescence initiation (SI) and ear emergence (EE)

References

ANSLOW, R. C. 1963. Seed formation in perennial ryegrass. 1. Anther exsertion and seed set. *J. Br. Grassld Soc.* **18,** 90–6.

ANSLOW, R. C. 1964. Seed formation in perennial ryegrass. 2. Maturation of seed. *J. Br. Grassld Soc.* **19,** 349–57.

BEDDOWS, A. R. 1931. Seed setting and flowering in various grasses. *Bull. H12 Welsh Pl. Breed. Stn* pp. 5–99.

COOPER, J. P. 1960. The use of controlled life-cycles in the forage grasses and legumes. *Herb. Abstr.* **30,** 71–9.

EMECZ, T. I. 1962. The effect of meteorological conditions on anthesis in agricultural grasses. *Ann. Bot.* **26,** 159–72.

EVANS, L. T. 1964. Reproduction. In: Barnard, C. [Ed.] Grasses and grasslands. London: Macmillan, pp. 126–53.

HUNT, I. V. 1956. Comparison of production from six strains of perennial ryegrass. *Res. Bull. 13 W. Scotl. agric. Coll.*

JOHNSTON, M. E. H. 1960. Investigations into seed setting in cocksfoot seed crops in New Zealand. *N.Z. Jl agric. Res.* **3,** 345–57.

KNIGHT, R. 1965. The relation between yield and the reproductive phase in cocksfoot (*Dactylis glomerata L.*) in a winter rainfall environment. *Aust. J. agric. Res.* **16,** 505–15.

LANGER, R. H. M. 1958. A study of growth in swards of timothy and meadow fescue. 1. Uninterrupted growth. *J. agric. Sci., Camb.* **51,** 347–52.

LANGER, R. H. M.; RYLE, G. J. A. 1958. Vegetative proliferations in herbage grasses. *J. Br. Grassld Soc.* **13,** 29–33.

RYLE, G. J. A. 1966. Physiological aspects of seed yield in grasses. In: Milthorpe, F. L.; Ivins, J. D. [Eds] The growth of cereals and grasses. Proceedings of the twelfth Easter School in Agricultural Science, University of Nottingham, 1965. London: Butterworth and Co., pp. 106–20.

SILSBURY, J. H. 1965. Interrelations in the growth and development of *Lolium.* 1. Some effects of vernalization on growth and development. *Aust. J. agric. Res.* **16,** 903–13.

CHAPTER 11

Perenniality and Senescence

General considerations

BY DEFINITION a perennial plant is one which continues its growth from year to year, whereas an annual completes its life cycle, from seed germination to seed production, followed by death, in a single season; similarly, a biennial takes two years to complete its life-cycle. In many instances in the plant kingdom, perennation involves a period of enforced dormancy to enable the plant to survive the rigours of an adverse season such as winter; often, special organs are developed for perennation. In other species, including the cultivated herbage grasses, perennation is no more than the continued growth of the plant from year to year, without either the need for, or the intervention of, any specialized organ of perennation. Grass plants comprise a population of relatively short-lived tillers each of which dies after flowering or, if vegetative, must eventually senesce for some other reason. Evans (1964) has stated that "perenniality of grasses depends on the rate at which new tillers are initiated, grow and become self-sufficient in relation to the rate at which they die following flowering or senescence."

Ryegrasses (*Lolium* spp.) range from annual to extreme perennial in habit, and Cooper (1960, 1961) has shown that in a population originating from a single cultivar it is possible to select for these extremes. Whether a plant is annual or perennial seems to be genetically controlled and depends on (a) how rapidly the main stem derived from the embryo can produce an inflorescence, and thus what opportunity exists for subsidiary tillering, and (b) the degree of control exerted by the developing inflorescence on metabolic activity and survival of subsidiary tillers. At the time of writing there is a paucity of information on the degree of control the developing inflorescence has on tiller production in typical annual and perennial grasses, although it is known that the growth of tiller initials is inhibited by the presence of a developing inflorescence (Skripčinskiĭ, 1958—cited by Evans, 1964); certainly, vegetative tillers do not survive after seed harvest in cereals and the developing barley inflorescence is alleged to control the ability of the root system to take up nutrients (Gregory, 1937). Research in progress at the Grassland Research

Institute, Hurley, and elsewhere indicates fundamental differences between perennial and annual species of *Lolium* in the allocation of assimilates to the major growth centres (Ryle, 1969) (Fig. 11.1).

Clearly the longevity of the plant is determined by whether the species and cultivar in question is an annual, a biennial, or a short- or long-lived perennial. Whether a grass is annual or biennial rather than perennial seems to be determined simply by the genetic control of flowering and by the proportion of reproductive to vegetative tillers. In perennials, other factors in addition to ease of flowering and proportion and ease of production of vegetative tillers around the time of inflorescence production, play an important role in determining their length of life.

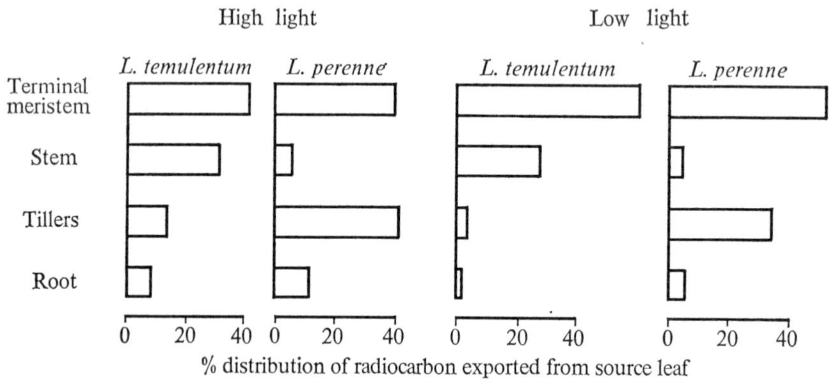

Fig. 11.1. Partitioning of assimilates between meristems of a single shoot in an annual (*Lolium temulentum*) and a perennial (*L. perenne*) ryegrass.

Labelled CO_2 fed to the most recently fully expanded leaf on the main stems of vegetative single plants grown in controlled environment, distribution of labelled products of photosynthesis determined 24 h later. Note the difference between species, enhanced in low light, in the allocation of these products to stems, tillers and roots. (From Ryle, 1969)

In the agricultural situations with which this book is primarily concerned, grasses are grown in swards in competition with genotypes of the same and/or different species. The ability of a single genotype to survive depends on its success in competition with other genotypes in relation to the management applied and the environment experienced. The susceptibility of the genotype to microbiological attack also depends on the nature of the genotype and any interaction with management and environment. The agriculturalist is concerned that sufficient genotypes should survive from a sown population to maintain yield above a certain arbitrary level set by him under his particular management. Harberd (1961) has claimed that individual genotypes of red fescue (*Festuca rubra*) have survived by vegetative propagation for several centuries on the hills of Scotland, but the longevity of individual genotypes of sown species under grassland management does not appear to have been studied, in spite of the numerous references to the decline in productivity of sown swards with time.

There is no known physiological reason why an individual plant of a perennial grass should not survive for an unlimited period, given the right cultural conditions. There is no conclusive evidence of physiological ageing in a grass plant. However, the morphological construction of most of our indigenous cultivated grasses, which show a tufted habit of growth resulting from the repeated production of tillers on the main stem and on successive orders of tillers, makes them susceptible to degradation. In a relatively short period the soil volume available to the roots is completely exploited, and the nutrients become exhausted unless replaced at regular intervals. The repeated turnover in leaf, tiller and root material as a result of senescence makes the area surrounding the plant a substrate for many saprophytes, some of which may become pathogenic; the area occupied by the plant may also become attractive to pathogens. Some forms of pathological attack, such as viruses, may not themselves kill the plant, but by stunting further growth may cause the plant to succumb to competition from its more healthy neighbours (Catherall & Griffiths, 1966).

Senescence of leaves

There appears to be little detailed information on the senescence of grass leaves. Rates of leaf production have, however, been studied in some detail and observations have been made on the number of live leaves carried on a tiller. From these data reasonable inferences can be made about leaf senescence.

EFFECT OF ENVIRONMENT

The leaves of grasses are only transient organs. Leaf production is continuous and is controlled both by the genetic make-up of the plant and by the environment in which that plant grows. Since the number of leaves borne on a tiller appears to be constant under constant conditions, leaf senescence must occur at the same rate as leaf production. For any one species, more leaves are generally present on a tiller of a single isolated plant than may be found on a comparable tiller from the sward situation (Table 11.1), although leaf number per tiller in multi-tillered single plants may be reduced to a level similar to that occurring in the sward. During the life cycle of any one tiller the number of live leaves it supports increases as inflorescence development occurs (Hunt & Brougham, 1966). In both the single plant and the elongating inflorescence-bearing tiller, each individual tiller experiences a higher level of illumination than that experienced by vegetative tillers growing in competition in the sward. It is known that plants grown under high light intensities have a higher number of live leaves per tiller than plants grown at low levels of illumination, and that this is due to an increased rate of leaf production and to a longer leaf life in the higher light intensity (Robson & Ryle, Personal communication). Also, when individual leaves are shaded on plants maintained in conditions of high light intensity, the life-span of the shaded leaf is reduced (Woledge, Personal communication). Although similar effects have been looked for in the sward situation (Bean, 1964), they have not yet been demonstrated. Temperature, photoperiod and nitrogen supply are also known to influence the number of live leaves on a tiller (Mitchell, 1953; Ryle, 1964, 1966a).

Table 11.1. Number of living leaves per tiller for 4 grasses, either grown as single plants or in swards

	Cocksfoot	Meadow fescue	Perennial ryegrass	Timothy	Reference
Single plants	5·8	4·7	4·9	5·2	Ryle, 1964
	...	...	...	5·0–2·7	Langer, 1957
	6·8	5·0	4·7	...	Ryle, 1966b
Swards	3·7	...	...	...	Ryle, 1966b
	3·4–2·7	...	...	...	Bean, 1964
	...	3·0–1·4	...	4·0–1·5	Langer, 1958
	...	4·0–2·0	...	...	Lambert, 1962
	...	...	3·0	...	Anslow, 1968

EFFECT OF INTERNAL FACTORS

The effect of environment on leaf duration is small in comparison with the internal control of leaf life exerted by the plant itself. Observations on the distribution of the products of assimilation in the plant (Williams, 1964) show that until a leaf is fully emerged it imports assimilates and also retains those made by the leaf itself. As soon as the leaf is fully expanded, assimilates are exported to meristematic and actively expanding areas of the plant, such as the stem apex, emerging leaves, young dependent tillers, and to the root tips. As the leaf ages, much less assimilate is translocated to expanding aerial parts of the plant and at maturity most assimilate flows into the root system (Plates 4, 5, 6, 7). Also, over this period the photosynthetic capacity of the leaf is known to decline (Fig. 11.2) (Woledge & Jewiss, 1969). It would appear that the growth of the plant is sustained by the photosynthetic activity of the most recently produced leaves. There appears to be a finely balanced control system within the plant, mediated by growth substances derived from the stem apex and from the leaves themselves. It is known for cereals that as leaves age their auxin content increases (Wheeler, 1967; Sheldrake & Northcote, 1968); over the same period nitrogen and phosphorus compounds are gradually withdrawn from the leaf and utilized elsewhere in the plant (Williams, 1957, 1968), and other minerals may be leached out. There is, however, no indication that carbon compounds are re-utilized.

Actual figures for the duration of leaf life have little meaning unless accompanied by a description of both the species and the environment. However, as a guide, leaves of S24 perennial ryegrass live a total of approximately 35 days from first appearance to senescence in summer and approximately 74 days in winter. (Data derived from Ryle, 1964, 1966a.)

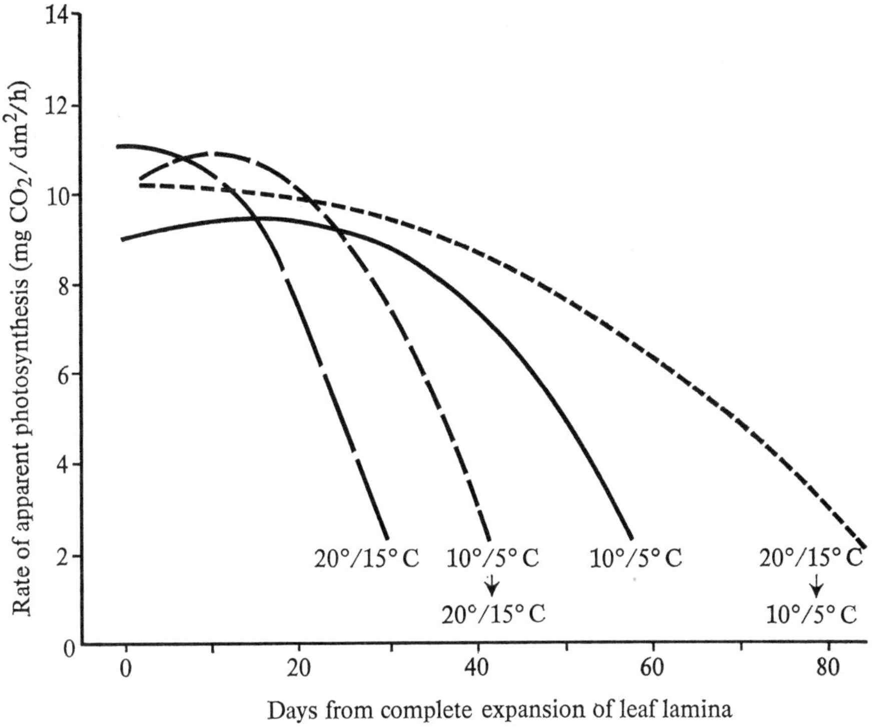

Fig. 11.2. Fall in rate of apparent photosynthesis of the 4th fully expanded leaf lamina on the main stem of tall fescue plants grown at high or low temperatures, and also after transfer between temperature regimes at full leaf expansion.

Lines are regressions through a series of measurements made on new samples of plants on successive days. All observations at 20°C. (From Woledge and Jewiss, 1969)

Tiller senescence

For the purposes of this discussion the term 'tiller' is regarded as meaning an aerial shoot and may be used to refer to the main stem, or to any shoot derived from the main stem or from any other shoot. Tiller senescence follows damage or removal of its terminal meristem. So long as the terminal meristem remains undifferentiated and undamaged, it retains the propensity for continued growth. Tiller death may result from (a) attack on the plant by pests

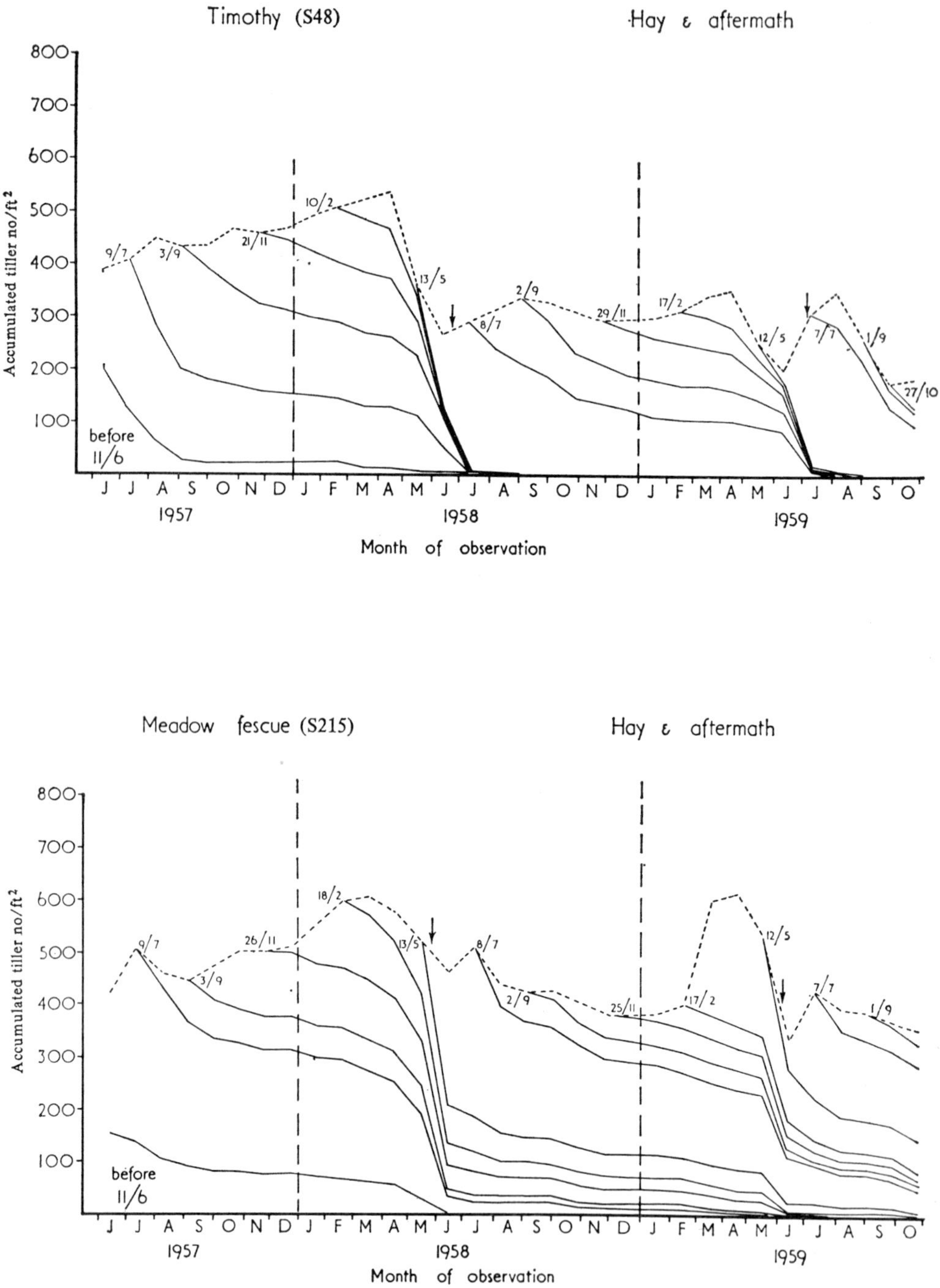

Fig. 11.3. Life history of tillers in timothy and meadow fescue swards under hay/aftermath and Dotted line shows total number of tillers in successive months. Dates shown are first observations on the vertical distances between lines. Arrows indicate dates of first cut in each year. (From Jewiss, 1966)

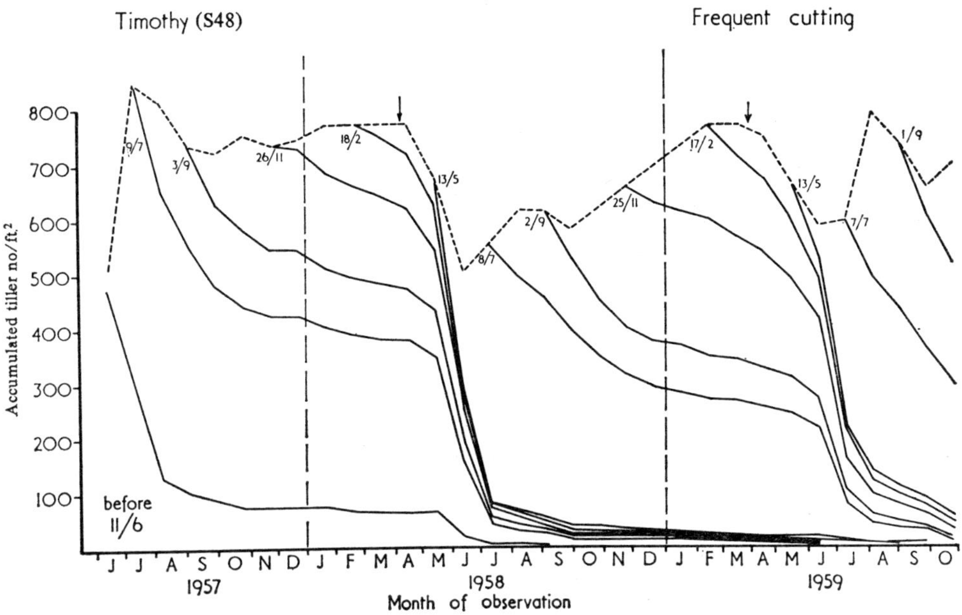

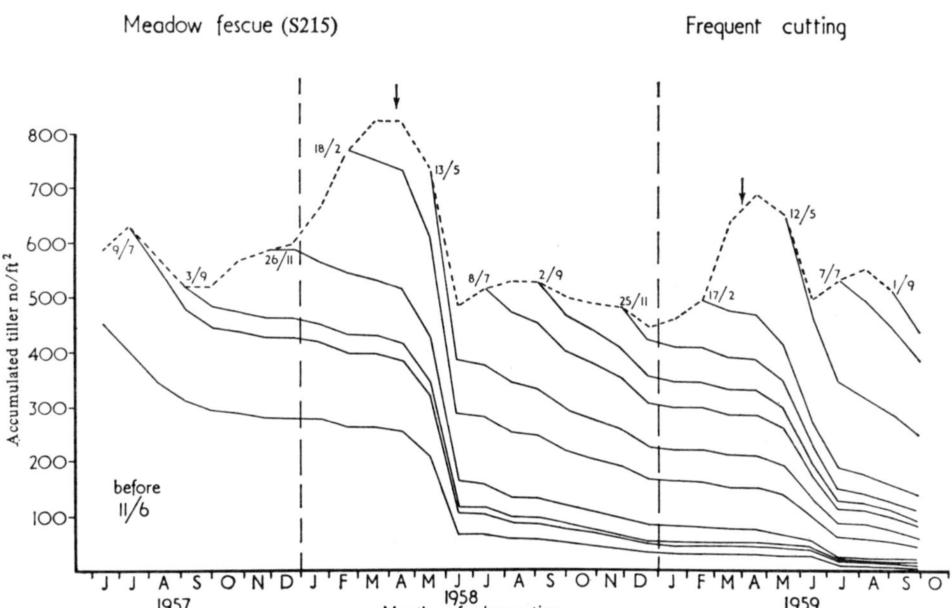

frequent-cutting systems of management.
tillers arising in the previous two months; numbers present in each successive month are represented by

or diseases, (b) inflorescence production, and (c) premature senescence consequent on other factors of the environment or management which impose an internal or external competitive stress on the plant. Chapter 12 deals with pests and diseases and no further reference to this topic will be made here.

The effects of inflorescence production are straightforward. For the important pasture grasses, winter conditions and/or long days result in inflorescence initiation and stem extension (see Chapter 10). Without the intervention of cutting or grazing, the tiller bearing the inflorescence does not senesce until the seed which has been set in the ear begins to be shed; however, because of the limited life of the leaves, their total number and dry weight decline from ear emergence onwards (Langer, 1957; Williams, 1957). In agricultural practice, all grasses apart from those grown for herbage seed production, are cut or grazed at least once by, or soon after, the ear emergence stage.

Once the inflorescence-bearing stems begin to elongate, both the yield (see Chapter 9) and longevity of these tillers depend on whether the reproductive stem apex is removed during harvesting or defoliation. Premature removal of this apex stops any further expansion of leaves on the tiller. Once the existing leaves and their sheaths have senesced, the remainder of the tiller will die. The speed with which death occurs is dependent on the activity of the leaves that remain on the stump of the flowering stem at the time of cutting and on the mobilization of the reserves in this stem to form new tillers in the axils of the senescing leaves. From work on herbage seed production it is known that the greater proportion of ears present at harvest originate from tillers formed in the autumn or early winter (Wilson, 1959); in cocksfoot up to 90% of the fertile tillers derive from the previous September (Lambert, 1963). A similar situation obtains for pure grass swards of timothy and meadow fescue (Fig. 11.3) (Lambert & Jewiss, 1970; Jewiss *et al.*, Unpublished).

Tiller production, like the production of leaves in whose axils the tillers occur, is a continuous process but, in contrast to leaf production, is more susceptible to changes in the environment (see Chapter 9). Leaf production is greater from April to October than in the winter months and, in general, so also is tillering. However, in the highly competitive situation of the sward or a large single plant, not all the tillers that are produced are able to survive until such time as they produce inflorescences. Fig. 11.3 shows the number of tillers produced by S48 timothy and S215 meadow fescue in each two-month period of the year and the number which survived in successive months under two systems of management—cut at the hay stage of growth and twice in the aftermath, or cut at monthly intervals in the summer (Jewiss *et al.*, Unpublished). The data for tiller death in timothy in successive months for 4 representative groups of tillers from each management regime are presented as histograms in Fig. 11.4. Similar data exist for swards of meadow fescue (Jewiss *et al.*, Unpublished) and large plants of tall fescue (Robson,

1968). The intensity of competition is less in smaller single plants and in consequence tiller death is less marked than in a sward (Langer, 1956).

In timothy the tillers recorded as dead in July under infrequent cutting, and in June and July under frequent cutting, probably represent removal by cutting of tillers bearing inflorescences; corresponding periods for meadow fescue are June, and May and June, for the respective cutting treatments. Apart from these tillers, the main periods of premature tiller death are from August to October and immediately before inflorescence production. It seems likely that many of the tillers produced in summer never become properly established as independent units, as they would if they developed their own

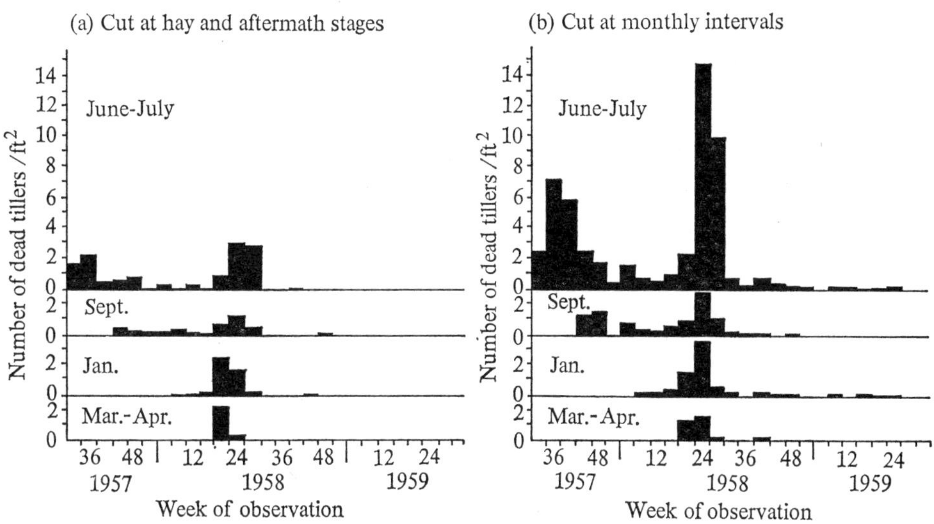

Fig. 11.4. Pattern of tiller death in successive months in swards of S48 timothy cut at either (a) hay and aftermath stages, or (b) monthly intervals during summer, for tillers arising in the 30 days up to 9 July, 1 Oct., 21 Jan. and 15 Apr. (From Jewiss *et al.*, Unpublished)

root systems. It is known that daughter tillers are initially dependent on their parents for both carbohydrates and minerals; indeed several leaves may have emerged before adventitious roots are formed (Langer, 1957; Williams, 1964). Intensive competition for light or for minerals may cause the loss of many tillers in summer at this stage of growth. The loss of tillers which precedes inflorescence production occurs before there is any recognizable physiological change in the plant heralding flowering. Again, tiller death appears to derive from the inability of certain small tillers to survive the intense competition from established tillers, which increases in severity as a result of their faster rates of growth, particularly leaf production, and large leaf size resulting from the improved environmental conditions in spring (Jewiss *et al.*, Unpublished). Because the life of these vegetative tillers may vary between one and

12 months (and exceptionally up to 36 months), data for mean tiller life are rather meaningless. The life-span of such tillers is extended by more frequent cutting and depends on species.

Root death

Studies on the growth of grass roots have long been bedevilled by the difficulty of distinguishing between living and dead roots. It is not clear how closely root death below ground is associated with tiller senescence above ground. Williams (1957) has shown that the nutrient content of inflorescence-bearing tillers does not go on increasing after ear emergence and it seems likely that the roots of the inflorescence-bearing tiller perish with the tiller. However, Williams (1962) has also shown that minerals taken up by a single root may be distributed around the whole of the shoot system of the plant and similarly (Williams, 1964) that carbohydrates from part of the shoot system may be distributed around the whole of the root system. Young tillers derived from the base of the inflorescence-producing stems rely on the root system of the parent tiller for water and minerals, until their own root systems are fully developed.

Brouwer (1966) has shown that the growth of the root and shoot system greatly depends on the availability of carbohydrates and minerals, respectively. Cutting has been shown to result in root death, presumably through a reduction in the availability of carbohydrate (Crider, 1955; Oswalt *et al.*, 1959). Davidson and Milthorpe (1966) have related the effects of severity of cutting to the degree of carbohydrate starvation and the consequent severity of the effect on root extension and nutrient uptake. In general then, root death may be regarded as paralleling tiller death and it may also occur following cutting or grazing, depending on their severity. It is also affected by environment; at the onset of drought root growth follows the drying of the soil profile, but not all roots are able to keep pace with the rate of drying and some root death may result (Garwood & Williams, 1967).

Root growth, like leaf growth, is also a dynamic state. As the root grows and branches, so the region of absorption is extended, and it has until recently been thought that the outer cortex of the root soon loses its ability to absorb minerals and water and is sloughed off, leaving the endodermis and the stele connecting the shoot system with the absorbing area of the root. However, recent evidence of Clarkson *et al.* (1968) shows that the ability of the older parts of the root to absorb and translocate minerals is only lost when the root becomes desiccated. In this respect, depending on environment, senescence of part of the root may follow closely behind the active growth of that root and, in so doing, bring into sharp contrast the dual functions of the root, namely as an organ for absorbing and conducting water and minerals. Alternatively, under favourable conditions the whole of the root system may continue to be able to absorb and translocate minerals and, presumably, water.

In considering these aspects of senescence the inter-relationships between the various parts of the plant should be borne in mind. Not only does a fine balance exist in the shoot system controlling the amount of photosynthetic surface carried by the plant, but similarly the proportion of root to tiller is also controlled via the carbohydrate status of the plant, which itself is controlled by the supply of mineral nutrients and (probably) amino acids to the shoot, the whole, no doubt, being mediated by auxin control mechanisms.

References

ANSLOW, R. C. 1968. The production of dry matter by swards of perennial ryegrass differing in average age of foliage. *J. Br. Grassld Soc.* **23,** 195–201.

BEAN, E. W. 1964. The influence of light intensity upon the growth of S37 cocksfoot (*Dactylis glomerata*). *Ann. Bot.* **28,** 427–43.

BROUWER, R. 1966. Root growth of grasses and cereals. In: Milthorpe, F. L.; Ivins, J. D. [Eds] The growth of cereals and grasses. Proceedings of the twelfth Easter School in Agricultural Science, University of Nottingham, 1965. London: Butterworth and Co., pp. 153–66.

CATHERALL, P. L.; GRIFFITHS, E. 1966. Influence of cocksfoot streak virus on the growth of cocksfoot swards. *Ann. appl. Biol.* **57,** 149–54.

CLARKSON, D. T.; SANDERSON, J.; RUSSELL, R. S. 1968. Ion uptake and root age. *Nature, Lond.* **220,** 805–6.

COOPER, J. P. 1960. Short-day and low temperature induction in *Lolium. Ann. Bot.* **24,** 232–46.

COOPER, J. P. 1961. Selection and population structure in *Lolium*. 5. Continued response and associated changes in fertility and vigour. *Heredity* **16,** 435–53.

CRIDER, F. J. 1955. Root growth stoppage resulting from defoliation of grass. *Tech. Bull. 1102 U.S. Dep. Agric.* pp. 1–23.

DAVIDSON, J. L.; MILTHORPE, F. L. 1966. The effect of defoliation on the carbon balance in *Dactylis glomerata. Ann. Bot.* **30,** 185–98.

EVANS, L. T. 1964. Reproduction. In: Barnard, C. [Ed.] Grasses and grasslands. London: Macmillan and Co., pp. 126–53.

GARWOOD, E. A.; WILLIAMS, T. E. 1967. Soil water use and growth of a grass sward. *J. agric. Sci., Camb.* **68,** 281–92.

GREGORY, F. G. 1937. Mineral nutrition of plants. *A. Rev. Biochem.* **6,** 557–78.

HARBERD, D. J. 1961. Observations on population structure and longevity of *Festuca rubra* L. *New Phytol.* **60,** 184–206.

HUNT, L. A.; BROUGHAM, R. W. 1966. Some aspects of growth in an undefoliated stand of Italian ryegrass. *Jnl appl. Ecol.* **3,** 21–8.

JEWISS, O. R. 1966. Morphological and physiological aspects of growth of grasses during the vegetative phase. In: Milthorpe, F. L.; Ivins, J. D. [Eds] The growth of cereals and grasses. Proceedings of the twelfth Easter School in Agricultural Science, University of Nottingham, 1965. London: Butterworth and Co., pp. 39–54.

JEWISS, O. R.; RYLE, S. M.; LANGER, R. H. M. The changing plant and tiller populations of timothy and meadow fescue swards. 2. Tiller longevity. (Unpublished)

LAMBERT, D. A. 1962. A study of growth in swards of timothy and meadow fescue. 3. The effect of two levels of nitrogen under two cutting regimes. *J. agric. Sci., Camb.* **59,** 25–32.

LAMBERT, D. A. 1963. The influence of density and nitrogen in seed production stands of S37 cocksfoot (*Dactylis glomerata* L.). *J. agric. Sci., Camb.* **61,** 361–73.

LAMBERT, D. A.; JEWISS, O. R. 1970. The position in the plant and the date of origin of tillers which produce inflorescences. *J. Br. Grassld Soc.* **25,** 107–12.

LANGER, R. H. M. 1956. Growth and nutrition of timothy (*Phleum pratense*). 1. The life history of individual tillers. *Ann. appl. Biol.* **44,** 166–87.

LANGER, R. H. M. 1957. Growth and nutrition of timothy (*Phleum pratense*). 2. Growth of the plant in relation to tiller development. *Ann. appl. Biol.* **45**, 528–41.

LANGER, R. H. M. 1958. A study of growth in swards of timothy and meadow fescue. 1. Uninterrupted growth. *J. agric. Sci., Camb.* **51**, 347–52.

MITCHELL, K. J. 1953. Influence of light and temperature on the growth of ryegrass (*Lolium* spp.). 1. Pattern of vegetative development. *Physiologia Pl.* **6**, 21–46.

OSWALT, D. L.; BERTRAND, A. R.; TEEL, M. R. 1959. Influence of nitrogen fertilization and clipping on grass roots. *Proc. Soil Sci. Soc. Am.* **23**, 228–30.

ROBSON, M. J. 1968. The changing tiller population of spaced plants of S170 tall fescue (*Festuca arundinacea*). *Jnl appl. Ecol.* **5**, 575–89.

RYLE, G. J. A. 1964. A comparison of leaf and tiller growth in seven grasses as influenced by nitrogen and temperature. *J. Br. Grassld Soc.* **19**, 281–90.

RYLE, G. J. A. 1966a. Effects of photoperiod in the glasshouse on the growth of leaves and tillers in three perennial grasses. *Ann. appl. Biol.* **57**, 257–68.

RYLE, G. J. A. 1966b. Effects of photoperiod in growth cabinets on the growth of leaves and tillers in three perennial grasses. *Ann. appl. Biol.* **57**, 269–79.

RYLE, G. J. A. 1969. Growth and its limitation and control by internal factors. *A. Rep. Grassld Res. Inst. Hurley, 1968* pp. 125–34.

SHELDRAKE, A. R.; NORTHCOTE, D. H. 1968. Production of auxin by detached leaves. *Nature, Lond.* **217**, 195.

SKRIPČINSKIĬ, V. V. 1958. The relation between the work of organs of vegetative propagation in perennial and annual cereals and the process of seed formation. *Dokl. Akad. Nauk uzbek. SSR* **118**, 1046–8.

WHEELER, A. W. 1967. Bound auxins in leaves. *Rep. Rothamsted exp. Stn, 1966* pp. 91–2.

WILLIAMS, R. D. 1957. Growth and nutrition of timothy (*Phleum pratense*). 3. Absorption and distribution of nitrogen, phosphorus and potassium during the first year of growth. *Ann. appl. Biol.* **45**, 664–73.

WILLIAMS, R. D. 1962. On the physiological significance of seminal roots in perennial grasses. *Ann. Bot.* **26**, 129–36.

WILLIAMS, R. D. 1964. Assimilation and translocation in perennial grasses. *Ann. Bot.* **28**, 419–26.

WILLIAMS, R. D. 1968. Translocation in grasses. *A. Rep. Grassld Res. Inst. Hurley, 1967* pp. 86–91.

WILSON, J. R. 1959. The influence of time of tiller origin and nitrogen level on the floral initiation and ear emergence of four pasture grasses. *N.Z. Jl agric. Res.* **2**, 915–32.

WOLEDGE, J.; JEWISS, O. R. 1969. The effect of temperature during growth on the subsequent rate of photosynthesis in leaves of tall fescue (*Festuca arundinacea* Schreb.). *Ann. Bot.* **33**, 897–913.

CHAPTER 12

Pests and Diseases

Introduction

DAMAGE caused by fungi, viruses or invertebrate pests can markedly decrease the yield and quality of herbage and may result in disorders in grazing animals. Such damage is occasionally the reason for a sward being ploughed up earlier than originally intended. When there is a fall in the yield of grass or of animal products, attack by pests or diseases may at times be suspected, though not diagnosed with certainty. The recognition and measurement of losses is complicated by the fact that weed grasses may invade a sward damaged in this way. Such weed grasses make their own contribution to the yield of the sward, even though their presence may in fact conceal an overall reduction in both yield and quality.

Yields of grass are, however, less affected by pests and diseases than are those of many other crops, and such yield losses can often be completely countered by applying N fertilizer. This is because grass has many more growing points per unit area of land than most other crops and if the soil is fertile, the surviving healthy plants or shoots can often quickly restore the density of the sward after an attack. There is, however, little information on the optimum amounts of fertilizer to apply to grass which is being, or has been, attacked. Where high levels of production are sought from grass, the practice of monoculture and heavy manuring may predispose the plants to certain diseases.

Tables 12.1 and 12.2 list the most important pests and diseases of grasses in Britain in their estimated order of importance. Also shown are their distribution and prevalence, their effect on yield, measures for their control and references to literature on these subjects.

Further information on the importance of individual organisms in Britain and on chemical control measures is given by Gair and Roberts (1969). Bulletin 154 of the Ministry of Agriculture, Fisheries and Food (1966) contains a chapter on herbage pests and diseases and their control, and Moore (1959) has given an excellent account of the distribution and prevalence

Table 12.1. Pests of grasses in Britain

Pest	Crop	Prevalence	Damage	Notes and control measures*	References
Aphids: *Metopolophium festucae* (Theo.), *Rhopalosiphum insertum* (Wlk.), *R. padi* (L.).	Most grasses.	Common, especially in warm summers following mild winters and in meadows and seed crops.	Leaves destroyed and yield of seed reduced.	*R. insertum* and *R. padi* transmit barley yellow dwarf virus to grasses. Other aphids transmit cocksfoot streak virus. Control: Many aphids removed by taking a silage cut; after harvesting seed, burn the stubble. For seed crops, apply demeton-S-methyl or dimethoate.	A.L. †106 (*R. insertum*) Gair (1953) Janson (1959
Leatherjackets: *Tipula* spp., especially *T. paludosa* Mg.	All grasses. Timothy damaged more than ryegrass or cocksfoot.	Common, especially in cool, wet regions and under wet soil conditions. Damage greatest in resown grassland. Attacks occur in spring, in wet summers and (in Scotland) in autumn. Numbers fluctuate widely from year to year and site to site.	Bare patches occur. Newly resown swards destroyed Severe attacks can reduce growth made by June by 40%.	Control: Cultivate thoroughly before sowing, harrow and roll after seedling emergence, to kill the larvae; re-sow in summer rather than spring or autumn. When established swards are attacked, yields can be increased by the application of extra N fertilizer in early spring. Apply γ-BHC (or fenitrothion is sometimes effective) or, except in rainy weather, bran bait with Paris Green, which also controls slugs.	A.L. 179 Cameron (1940) French (1969) La Croix & Newbold (1968) White (1967a) White & French (1968)
Grass seed midges (larvae): *Contarinia* spp., *Dasyneura* spp., *Mayetiola dactylidis* Kieffer, *Stenodiplosis* spp.	Cocksfoot, perennial ryegrass, timothy, fescues, meadow grasses.	Common. Populations increase with age of sward.	Yield and viability of seed reduced.	Seed-borne. Control: For seed production use a good crop rotation; choose windswept areas; isolate from other stands; cut before flowering crops not required for seed; plough up old stands.	Jones (1945) Stokes (1957)
Chafers (larvae): *Melolontha melolontha* (L.), *Phyllopertha horticola* (L.).	All grasses. Cocksfoot and perennial ryegrass partly resistant.	Fairly common, especially on light land, in wet hilly areas. Attacks occur in late summer and are favoured by rank growth.	Swards destroyed.	Control: Disc and roll the seed-bed to kill larvae; graze new sowings to keep them short and to consolidate the soil. Apply BHC to control adults.	A.L. 235 Cameron (1941) Milne (1956) Raw (1951)
Slugs: *Agriolimax reticulatus* (Müll.), *Arion hortensis* (Fer.).	All grasses.	Common on heavy soils. Troublesome in new sowings after leys, peas, cereals or brassicas. In mixed swards *A. reticulatus* is associated with cocksfoot because the grass affords protection.	Young plants destroyed. Damage underground or at soil surface, mainly occurs at night, in spring and autumn and during mild periods in winter.	Control: Remove trash, especially before ploughing for new sowing so soil surface can dry. Cultivate thoroughly before sowing. Use bran bait with metaldehyde, with or without addition of Paris green; or methiocarb. Best applied in warm, humid, but not rainy weather.	A.L. 115 Cameron (1938) South (1965) Stephenson (1968)

(*Cont.*)

Table 12.1. (cont.)

Pest	Crop	Prevalence	Damage	Notes and control measures*	References
Antler moth (larvae), *Cerapteryx graminis* (L.).	All grasses.	Occasional. Important in uplands.	Large areas of pastures destroyed.	Control: Apply DDT.	Cameron (1938)
Dipterous stemborers (larvae): *Chlorops pumilionis* (Bjerk.), *Geomyza* spp., *Meromyza* spp., *Opomyza* spp., *Oscinella* spp.	All grasses. Cocksfoot partly resistant.	Common. Overwintering populations of 100–200 larvae/m² common in S. England. Most numerous in fine-leaved grasses and where grass is kept short in summer and autumn.	Summer sowings occasionally destroyed.	*Oscinella frit* L. is an important pest of oats. Grasses are the host in which it overwinters. Where grass is followed in the autumn by winter wheat, larvae migrate from the grass stems into the wheat. At Hurley, the average loss of grain yield from this cause is 350 kg/ha.	A.L. 110 A.L. 174 Heard & Hopper (1963) Jepson & Heard (1959) Nye (1959) Webley (1958)
Common leaf-weevil (larvae & adults), *Phyllobius pyri* (L.).	Perennial ryegrass, fescues.	Occasional on light soils.	Larvae feed on roots	Control: Use heavy roller when weevils first seen. Apply DDT.	
Wireworms, *Agriotes* spp.	All grasses.	Common, especially in S. and E. England. Most numerous in grass resown directly, or after only 1 or 2 years' arable cropping. Numbers and size of larvae increase with age of sward.	Young plants destroyed older plants weakened.	Control: When sowing, increase seed-rate; cultivate thoroughly and manure heavily. Apply γ-BHC as seed-dressing and later.	A.L. 199 *Bull. 128 Min. Agric., Fish. Fd.*
Swift moths (larvae): *Hepialus lupulinus* (L.), *H. humuli* (L.).	All grasses.	Common. Numerous in resown grass.	Meadows and establishing swards seriously damaged. Larvae feed on roots.	Control: Some larvae can be killed by cultivations. Apply DDT.	A.L. 160 Cameron (1941) Edwards (1964)
Cocksfoot moth (larvae), *Glyphipterix cramerella* (Fab.).	Cocksfoot.	Fairly common in S. England. Population increases with age of sward.	Yield of seed reduced.	Seed-borne. Control: For seed crops, isolate from other cocksfoot swards and when harvesting cut low or burn stubble immediately after.	Empson (1956)
Timothy fly (larvae), *Amaurosoma* spp.	Timothy.	Common in areas producing much timothy seed.	In 20 fields in S.E. England surveyed in 1966, there was a 25% reduction in seed yield in two.	Early-flowering strains of timothy least susceptible. Recent experiments suggested that dimethoate is effective if applied at egg-laying stage. This stage is difficult to recognize.	Coghill & Gair (1954) Collingwood (1967) Newbold (1966)

* A warning concerning the use of pesticides, especially DDT, is given on p. 91

† A.L. = Advisory Leaflet of the Ministry of Agriculture, Fisheries and Food

Table 12.2. Diseases of grasses in Britain

Disease	Crop	Prevalence	Damage	Notes and control measures	References
FUNGI					
Rusts: *Puccinia* spp., *Uromyces dactylidis* Otth.	All herbage grasses, especially the ryegrasses.	Common, especially in S. half of Britain. Most active in warm summers and autumns.	Affects establishment and reduces yields of herbage and seed. Late silage crops particularly affected.	Severely infected herbage is unpalatable to livestock and may cause digestive disorders. In cocksfoot, plants with high concentrations of water-soluble carbohydrate are especially susceptible to infection by *P. striiformis*. Also rust may depress the ultimate level of this carbohydrate in the plant. Control: Rusts develop more if plants are checked. High N (e.g. 300 kg N/ha per year) plus frequent defoliation gives some control. There are perennial ryegrass cultivars with some resistance to crown rust, timothy cultivars resistant to timothy rust and cocksfoot cultivars resistant to stripe rust.	Carr (1966)
Choke, *Epichloe typhina* (Pers. ex Fr.) Tul.	Cocksfoot, timothy, fescues, meadow grasses.	Common, except in Scotland.	Reduces yield of seed. In a survey of England and Wales, choke reduced yield of cocksfoot seed by 1% in the 2nd crop year, 7% in the 3rd and 21% in the 5th–7th years.	Control: Choke is seed-borne in red fescue, therefore sow seed from healthy crop.	Large (1952) Large (1954) Western & Cavett (1959)
Powdery mildew, *Erysiphe graminis* DC. ex Mérat. Many physiologic races	All herbage grasses except timothy.	Common. Most active following dry period in late spring.	Rarely lethal, but herbage yields are reduced. Seed yields of S345 cocksfoot sometimes severely reduced.		
Leaf fleck, *Mastigosporium rubricosum* (Dearn. & Barth.) Sprague.	Cocksfoot, timothy.	Fairly common, mainly in S.W. England, Wales and Ireland. Most active in wet autumns.	Rarely lethal, but causes partial defoliation. Seed yields of cocksfoot S345 sometimes severely reduced. Severe attack can halve the concentration of water-soluble carbohydrate in cocksfoot herbage.	In cocksfoot, early cultivars (e.g. S345) may escape spring attack, but are very susceptible in autumn.	Carr (1962)

(*Cont.*)

Table 12.2. (cont.)

Disease	Crop	Prevalence	Damage	Notes and control measures	References
Blind-seed of ryegrass, *Gloeotinia temulenta* (Prill. & Delacr.) Wilson, Noble & Gray.	Perennial and Italian ryegrasses.	Common in cool areas with high rainfall, notably Ireland and S.W. and N.E. Scotland.	In N. Ireland and Scotland this disease sometimes reduces yield of viable seed by 50%.	Seed-borne. Principal means of spread is by infected seed. Control: Where the disease is endemic only crops in their 1st harvest year should be taken for seed. Sow healthy seed (affected seed can be disinfected by hot-water treatment); rotate with other crops; improve soil nutrient status, notably by applying Nitro-chalk, which reduces both the level of infection and the effect of the disease; at harvest, burn seed-screenings and stubbles. There are resistant varieties.	Calvert & Muskett (1945) Chestnutt (1958) Hardison (1963) Wilson, Noble & Gray (1945)
Damping-off: *Cylindrocarpon radicicola* Wollenw., *Drechslera* spp., *Fusarium* spp., *Pythium* spp., *Rhizoctonia* spp.	All herbage grasses.	Occasional. Mainly in N. uplands.	Can affect establishment.	Seed-borne and soil-borne. Control: Seed dressing with Difolatan or Captan.	Carr (1967) Gray (1964) Michail & Carr (1966) Tribe & Herriott (1969)
Ergot, *Claviceps purpurea* (Fr.) Tul.	Most herbage grasses.	Fairly common in S.W. England and in Scotland. Most common in old pastures. In seed crops, occurs most when conditions for pollination are poor.	Reduces yield of seed.	Seed-borne. In hay, and in late heads in pasture, can cause poisoning of stock. Control: If field infected, plough deeply before resowing; prevent grasses flowering in pastures and meadows.	A.L.* 548 Jenkinson (1958) Woods, Bradley, Jones & Mantle (1966)
(i) Leaf spot and (ii) Net blotch of ryegrass; (i) *Drechslera siccans* (Drechs.) Shoemaker, (ii) *D. dictyoides* (Drechs.) Shoemaker.	Perennial and Italian ryegrasses, meadow fescue.	Fairly common. Common in ryegrass seed stocks in Scotland.	Causes leaf browning. Can affect establishment in cold humid areas.	Seed-borne.	
Take-all: *Ophiobolus graminis* (Sacc.) Sacc., *O. graminis* var. *avenae* E.M. Turner.	Perennial and Italian ryegrasses, cocksfoot, timothy, meadow and red fescues.	Common.	Can affect establishment on heavily infected areas. In Ireland reported as doing much damage to established ryegrass.	Soil-borne. Control: Keep down rhizomatous grass weeds which carry the disease; avoid over-liming.	Brooks (1965) Doling & Hepple (1959) O'Rourke (1968)

* A.L. = Advisory Leaflet of the Ministry of Agriculture, Fisheries and Food

(*Cont.*)

Table 12.2. (cont.)

Disease	Crop	Prevalence	Damage	Notes and control measures	References
Halo spot, *Selenophoma donacis* (Pass.) Sprague & C. G. Johnson.	Cocksfoot, timothy, rough-stalked meadow grass.	Fairly common in wet areas.	Causes partial defoliation. Reduces yield and quality of cocksfoot seed.		
VIRUSES					
Cocksfoot mottle (CFMV).	Cocksfoot.	Common. Incidence increases rapidly with age of sward. At Gleadthorpe there was more CFMV in swards cut 3 times than in those grazed 8 times a year.	Lethal. Causes serious reductions in yields of herbage and seed. At Gleadthorpe, almost total destruction of cocksfoot by CFMV in 3rd year.	Vector: the cereal leaf beetle, *Lema melanopa* (L.) is possibly important in bringing the virus into the crop. Some other and very efficient means of spread is involved after this—probably involving livestock or machinery. From grasses infection is spread to cereals.	Serjeant (1964) Upstone (1969)
Cocksfoot streak (CFSV).	Cocksfoot, perennial and Italian rye-grasses, many other grasses.	Common.	Seedlings destroyed and there is increased susceptibility to winter-kill. At Aberystwyth and Hurley about half the number of plants infected died within 2 years. However, CFMV may also have been present and the main agent. The virus modifies plant habit, causing an abnormally high ratio of fertile to vegetative tillers.	Vectors: a number of aphids.	Catherall & Griffiths (1966 a, b)
Barley yellow dwarf (BYDV).	Perennial and Italian rye-grasses, cocks-foot, timothy, annual and smooth-stalked meadow grasses.	Common. From a survey of 112 fields of perennial ryegrass in England and Wales, 93% carried the virus.	In one experiment, with half the perennial ryegrass plants in-fected, yield was reduced by 9%. This loss was doubled when all plants were infected.	In perennial ryegrass there may be a year's delay before symptoms appear. Vectors: a number of aphids. From the grasses infection is spread to cereals.	Catherall (1966) Doodson (1967)
Ryegrass mosaic (RMV).	Perennial and Italian rye-grasses, cocks-foot, meadow fescue.	Common in S. England and perhaps elsewhere. In 1968–9 in S. England 49 crops of Italian ryegrass for seed were surveyed. In 13 of these more than 20% of plants were infected.	In perennial ryegrass at Hurley, RMV has reduced yield in the first cut on 31 May from 8500 to 5200 kg DM/ha. It also causes reductions in yields of seed.	In a survey, in the first harvest year there was more infection in crops sown under cereals in spring, than in those sown direct in autumn. Vector: a mite, *Abacarus hystrix* (Nalepa) which is probably wind-borne. From the grasses infection is spread to oats. The virus is probably also spread by machinery.	Mulligan (1960

of diseases. Life histories and other details of pests have been given by Jones and Jones (1964) and Edwards and Heath (1964), and of diseases by Sampson and Western (1954). White (1967b) discusses those pests that are also of importance in sports turf. Notes on seed-borne diseases are given by Noble and Richardson (1968). Slykhuis (1958, 1967) has reviewed the literature on the virus diseases which attack grass. The principal pests and diseases of grass seed crops and measures for their control are described in bulletins issued by the Welsh Plant Breeding Station (1967) and by the Ministry of Agriculture, Fisheries and Food (1968). Many of the pests and diseases of grasses found in Britain also occur in France. Accounts of these have been given by Courtillot (1961) and the Institut Technique des Céréales et des Fourrages (1964). Excellent coloured illustrations of pests and diseases are to be found in the advisory leaflets of the Institut National de la Recherche Agronomique/I.T.C.F./F.N.A.M.S. (1961), and in a book by Gram, Bovien and Stapel (1969).

WARNING ON THE USE OF PESTICIDES

An adequate period of time, depending upon the pesticide used (e.g. a minimum period of 2 weeks for DDT) should elapse between application and allowing animals to graze treated crops. A UK Government committee has recently advised that because of its long persistence and the consequent risk to animal life, DDT be withdrawn from use on grassland (Department of Education and Science, 1969). This general recommendation has been accepted by the UK Government and the regulations in force should therefore be consulted.

Pests and diseases of new sowings

In south-east England at least, failures or partial failures of new sowings are more often a result of poor drainage or else of smothering either by lodging of the cereal companion crop or by straw or chaff deposited by combine harvesters, than of pests or disease. However, new sowings may be particularly vulnerable to pests and diseases, especially when the seedlings grow slowly, for instance in cold wet soils or on poorly prepared seedbeds, and also when low seed-rates are used, e.g. for seed crops.

Failures caused by pests are especially likely when new sowings are made on ploughed-up old grassland. Leatherjackets, wireworms, stem-borers, slugs, chafer-grubs and the caterpillars of the swift (*Hepialus* spp.), common rustic (*Mesapamea secalis*) and crambid (*Crambus hortuellus*) moths may be present in large numbers in old grassland and can seriously damage new sowings. Ploughing and cultivating kills some of these pests, and still more will die if the interval between ploughing and sowing can be extended to 4 weeks or more. Slugs are especially troublesome in new sowings after crops which leave large amounts of crop residues and litter, and after so-called 'chemical ploughing'. For example, a number of fields distributed over several farms in south-west Scotland were reseeded to grass in 1968 using the herbicide paraquat to destroy the old grass. A few days after the herbicide treatment, seed was sown in shallow slits cut in the dead turf. The establishment of a sward on those fields sown after 1 August failed as a result of the late sowing and of attack by slugs and young leatherjackets moving along the slits.

When old grassland is being resown, any grasses remaining from the old sward are a potential source of infection, particularly of virus diseases, and it is therefore good practice to ensure that all such plants are killed.

Fungi such as *Drechslera*, *Fusarium*, *Pythium* and *Rhizoctonia* spp. are often recorded in samples of seed and may sometimes be responsible for poor establishment. In cool, wet areas there has therefore been some interest in extending the use of fungicidal seed-dressings to include herbage crops (Grainger, 1965). Experiments under a wide range of conditions are required to determine whether such seed-dressings confer any practical benefit; from experiments at the Welsh Plant Breeding Station (Michail & Carr, 1966; Carr, 1967) and in Scotland (Tribe & Herriott 1969) it appears that they do not.

Establishment of grasses is sometimes poor on peaty soils in East Anglia. Extensive failures in 1967 and 1968 were caused by the migratory plant parasitic nematode *Longidorus elongatus* (Winfield, Personal communication). This problem has been alleviated by sowing in the autumn.

Pests and diseases of established swards

Even when swards are well established, large areas can occasionally be destroyed. The larvae of the antler moth for instance can cause considerable damage to upland swards. At the Grassland Research Institute, Hurley, attacks by cocksfoot streak virus (possibly accompanied by cocksfoot mottle virus) have frequently almost destroyed swards by the end of the second year. Another example is the damage done to perennial and Italian ryegrasses by the snow moulds *Fusarium nivale* and *Typhula incarnata* in northern Scotland (Gray, 1964) and in England during severe winters such as that of 1962–3.

However, in established swards it is unusual for large areas to be destroyed in a short time by pests or diseases. In swards that have been only thinned the ability of the surviving healthy plants to restore the leaf cover quickly depends on the maintenance of fertilizer and defoliation management favourable to the healthy plants. Certain virus diseases affect the growth habit of the plant, and for this reason the degree of compensation made by healthy plants may depend on the frequency with which the sward is defoliated. For example, plants of the early-flowering cocksfoot cultivar S345 infected with cocksfoot streak virus acquired an abnormally early-flowering habit and an abnormally high ratio of fertile to vegetative tillers (Catherall & Griffiths, 1966a,b). Infected plants were weakened by frequent defoliation and healthy plants became dominant in the sward, at least for two years, during which period they largely compensated for the poor yield from the infected plants.

In swards defoliated less frequently, the fertile tillers of infected plants grew tall and shaded the healthy plants, and yields were severely reduced.

Two of the most serious pests, particularly in cool, high-rainfall areas, are leatherjackets and slugs. In north-east England, epidemics of leatherjackets occur about once in every 5–8 years. In epidemic years about 10% of grass fields are seriously damaged. In most of the past few years there have been serious outbreaks in south-west Scotland. In England damage usually occurs in the early months of the year, up to June, but in Scotland the young larvae also cause damage in the autumn.

Apart from the pests noted in the Tables, migratory plant parasitic nematodes are worth mentioning although their overall significance is not yet known. Populations of *Paratylenchus*, *Helicotylenchus* and *Rotylenchus* spp. increase greatly under grass (Bridge, 1971) and occasionally appear to reduce herbage yields (Lewis & Webley, 1966).

Large numbers of Bibionid larvae are sometimes found feeding on dead and dying plants. Their role requires clarifying; occasionally they appear to be responsible for the death of some plants.

There is little information as to the way in which viruses such as ryegrass mosaic and cocksfoot mottle are spread to new grass fields and within the field. Within the field, sap-inoculation as a result of crushing of leaves by machinery or the grazing animal may be of greater importance than the activities of insect vectors (Upstone, 1969).

Measures to control pests and diseases of grassland may be of benefit to other crops on the farm. Some organisms present in grassland can affect neighbouring arable crops or cereal companion crops. Some weevils and stem-borers pass the winter in grassland and migrate to field crops in spring. Virus diseases can be transferred by insect vectors between grasses and cereals. When grassland is ploughed for an arable crop, many pests of the grass (e.g. leatherjackets) are a danger to the arable crop (Osborne, 1960).

Pests and diseases of seed crops

When grasses are grown for seed, considerable losses in yield can be caused by aphids and by larvae of seed midges. Cocksfoot moth and timothy fly are only occasionally important. The most important diseases are the rusts, choke, ergot, the viruses and, in Ireland and Scotland, blind-seed of ryegrass (Welsh Plant Breeding Station, 1967; Ministry of Agriculture, Fisheries & Food, 1968).

Chemical and other control measures are more frequently employed in grass crops grown for seed than in those grown for herbage because (a) the seed crop has a high cash value, (b) it often provides a more favourable habitat for pests and diseases than does grassland grown for forage only, and (c) fertile culms or spikelets, if damaged, are less easily replaced than vegetative tillers.

However, with the use of chemical control measures on seed crops there is a risk of perpetuating susceptible genes in the plants and so other methods may be preferred. It is important that old swards be ploughed up and a good rotation chosen, because, apart from the other pests of old grassland, larvae of seed midges and cocksfoot moth and the diseases choke, ergot, blind-seed of ryegrass and the viruses can carry-over from the old sward to the new. For the same reasons deep ploughing can also be beneficial where direct resowing is practised. Healthy seed should be sown and the crop sufficiently isolated from other similar crops. For preference a wind-swept site should be chosen to avoid attacks by seed midges and aphids. The grasses in hedgerows should be kept under control and should be cut to prevent flowering, thus preventing the build-up and spread of pests and diseases. Immediately after harvest, if the crop was not cut close to the ground, the stubble should be trimmed, or if there is damage by cocksfoot moth, aphids, or blind-seed disease of ryegrass, it should be burnt.

Effects on seasonal production and quality of grass

Aphids and dipterous stem-borers selectively destroy young growth, highly nutritious to stock. The leaf diseases leaf fleck, stripe rust and powdery mildew and the viruses can all severely decrease sward photosynthetic area and affect the concentration of water-soluble carbohydrates in the leaves. Thus in cocksfoot, stripe rust and leaf fleck reduce the concentration of water-soluble carbohydrate (Carr, 1962, 1966); in contrast, virus infection often slows down the translocation of certain sugars from the leaf and leads to their accumulation (Goodman, Watson & Hill, 1965).

Disorders of grasses can affect animal health. Rusts sometimes render herbage uneatable and can cause digestive and other disturbances. Ergot can poison stock (Woods *et al.*, 1966). Slugs are the intermediate hosts of the larval stages of certain parasites of sheep and poultry.

In crops grown for seed the most important organisms affecting the viability of the seed are the larvae of seed midges and blind-seed disease.

References

BRIDGE, J. 1971. The ecology and biology of nematodes associated with grasses. Ph.D. Thesis, Univ. London.

BROOKS, D. H. 1965. Wild and cultivated grasses as carriers of the take-all fungus (*Ophiobolus graminis*). *Ann. appl. Biol.* **55,** 307–16.

CALVERT, E. L.; MUSKETT, A. E. 1945. Blind-seed disease of ryegrass (*Phialea temulenta* Prill. & Delacr.) *Ann. appl. Biol.* **32,** 329–43.

CAMERON, A. E. 1938. Insects and pests of 1937. *Trans. R. Highld agric. Soc. Scotl.* **50,** 94–127.

CAMERON, A. E. 1940. Insects and pests of 1939. *Trans. R. Highld agric. Soc. Scotl.* **52,** 91–115.

CAMERON, A. E. 1941. Insects and pests of 1940. *Trans. R. Highld agric. Soc. Scotl.* **53,** 77–97.

CARR, A. J. H. 1962. Plant pathology. *Rep. Welsh Pl. Breed. Stn 1961* pp. 100–3.
CARR, A. J. H. 1966. Plant pathology. *Rep. Welsh Pl. Breed. Stn 1965* p. 95.
CARR, A. J. H. 1967. Plant pathology. *Rep. Welsh Pl. Breed. Stn 1966* pp. 117–19.
CATHERALL, P. L. 1966. Effects of barley yellow dwarf virus on the growth and yield of single plants and simulated swards of perennial ryegrass. *Ann. appl. Biol.* **57**, 155–62.
CATHERALL, P. L.; GRIFFITHS, E. 1966a. Influence of cocksfoot streak virus on the growth of cocksfoot swards. *Ann. appl. Biol.* **57**, 149–54.
CATHERALL, P. L.; GRIFFITHS, E. 1966b. Influence of cocksfoot streak virus on the growth of single cocksfoot plants. *Ann. appl. Biol.* **57**, 141–8.
CHESTNUTT, D. M. B. 1958. Susceptibility of perennial ryegrass to blind seed disease (*Gloeotinia temulenta*). *J. Br. Grassld Soc.* **13**, 275–8.
COGHILL, K. J.; GAIR, R. 1954. The estimation in the field of the damage caused by timothy flies (*Amaurosoma* spp.). *J. Br. Grassld Soc.* **9**, 329–34.
COLLINGWOOD, C. A. 1967. *Rep. Advisory Entomologist N.A.A.S., S.E. Region, 1966.*
COURTILLOT, M. 1961. Les maladies des graminées fourragères. Paris: Institut Technique des Céréales et des Fourrages.
DEPARTMENT OF EDUCATION AND SCIENCE. 1969. Further review of certain persistent organo-chlorine pesticides used in Great Britain. Report of the Advisory Committee on pesticides and other toxic chemicals. London: H.M.S.O., p. 34 and p. 65.
DOLING, D. A.; HEPPLE, S. 1959. Occurrences of take-all in cocksfoot and other grasses. *Pl. Path.* **8**, 73–5.
DOODSON, J. K. 1967. A survey of barley yellow dwarf virus in S.24 perennial ryegrass in England and Wales, 1966. *Pl. Path.* **16**, 42–5.
EDWARDS, C. A. 1964. The bionomics of swift moths. 1. The ghost swift moth, *Hepialus humuli* (L.). *Bull. ent. Res.* **55**, 147–60.
EDWARDS, C. A.; HEATH, G. W. 1964. The principles of agricultural entomology. London: Chapman and Hall Ltd.
EMPSON, D. W. 1956. Cocksfoot moth investigations. *Pl. Path.* **5**, 12–18.
FRENCH, N. 1969. Assessment of leatherjacket damage to grassland and economic aspects of control. *Proc. 5th Br. Insectic. Fungic. Conf.* pp. 511–21.
GAIR, R. 1953. Observations on grass aphids in Derbyshire, 1950–52. *Pl. Path.* **2**, 117–21.
GAIR, R.; ROBERTS, E. T. 1969. Pest and disease control in grass and fodder crops. In: Martin, H. [Ed.] Insecticide and fungicide handbook. Oxford: Blackwell Scientific Publications, 3rd edn.
GOODMAN, P. J.; WATSON, M. A.; HILL, A. R. C. 1965. Sugar and fructosan accumulation in virus-infected plants; rapid testing by circular-paper chromatography. *Ann. appl. Biol.* **56**, 65–72.
GRAINGER, J. 1965. Disease and grass. *Fmg Rev.* No. 29, 23–8.
GRAM, E.; BOVIEN, P.; STAPEL, C. 1969. Recognition of diseases and pests of farm crops. London: Blandford Press, 2nd edn.
GRAY, E. G. 1964. Snow moulds in grassland in the North of Scotland. *Scott. Agric.* **44**, 52–3.
HARDISON, J. R. 1963. Control of *Gloeotinia temulenta* in seed fields of *Lolium perenne* by cultural methods. *Phytopathology* **53**, 460–4.
HEARD, A. J.; HOPPER, M. J. 1963. Populations of stem-boring Diptera in leys and in subsequent winter wheat and the effects of aldrin and phorate. *Ann. appl. Biol.* **51**, 301–11.
INSTITUT NATIONAL DE LA RECHERCHE AGRONOMIQUE. 1961. Maladies des graminées fourragères. Paris: Association de Coordination Technique Agricole.
INSTITUT TECHNIQUE DES CÉRÉALES ET DES FOURRAGES. 1964. Étude sur la nature, la fréquence et la gravité des differents parasites des plants fourragères. Paris: Inst. Tech. Céréales, Fourrages.
JANSON, H. W. 1959. Aphids on cereals and grasses in 1957. *Pl. Path.* **8**, 38.
JENKINSON, J. G. 1958. Ergot infection of grasses in the south-west of England. *Pl. Path.* **7**, 81–5.
JEPSON, W. F.; HEARD, A. J. 1959. The frit fly and allied stem-boring Diptera in winter wheat and host grasses. *Ann. appl. Biol.* **47**, 114–30.

JONES, D. P. 1945. Gall midges and grass seed production. *Agriculture, Lond.* **52,** 248–51.

JONES, F. G. W.; JONES, M. G. 1964. Pests of field crops. London: Edward Arnold Ltd.

LACROIX, E. A. S.; NEWBOLD, J. W. 1968. Autumn treatments against leatherjackets. *Pl. Path.* **17,** 78–82.

LARGE, E. C. 1952. Surveys for choke (*Epichloe typhina*) in cocksfoot seed crops, 1951. *Pl. Path.* **1,** 23–8.

LARGE, E. C. 1954. Surveys for choke (*Epichloe typhina*) in cocksfoot seed crops, 1951–53. *Pl. Path.* **3,** 6–11.

LEWIS, S.; WEBLEY, D. 1966. Observations on two nematodes infesting grasses. *Pl. Path.* **15,** 184–6.

MICHAIL, S. H.; CARR, A. J. H. 1966. Effect of seed treatment on establishment of grass seedlings. *Pl. Path.* **15,** 60–4.

MILNE, A. 1956. Biology and ecology of the garden chafer, *Phyllopertha horticola* (L.). II The cycle from egg to adult in the field. *Bull. ent. Res.* **47,** 23–45.

MINISTRY OF AGRICULTURE, FISHERIES AND FOOD. 1966. Grass and grassland. *Bull. 154 Min. Agric. Fish. Fd* 4th edn, pp. 91–7.

MINISTRY OF AGRICULTURE, FISHERIES AND FOOD. 1968. Grass and clover crops for seed. *Bull. 204 Min. Agric. Fish. Fd* pp. 35–42.

MOORE, W. C. 1959. British parasitic fungi. Cambridge: Univ. Press, pp. 430.

MULLIGAN, T. E. 1960. The transmission by mites, host-range and properties of ryegrass mosaic virus. *Ann. appl. Biol.* **48,** 575–9.

NEWBOLD, J. W. 1966. The timothy fly problem. *Proc. Conf. Timothy Seed Prod. W. Scot. Agric. Coll.* pp. 19–21.

NOBLE, M.; RICHARDSON, M. J. 1968. An annotated list of seed-borne diseases. *Phytopathological Papers, No. 8* Kew, England: Commonwealth Mycological Institute, 2nd edn.

NYE, I. W. B. 1959. The distribution of shoot-fly larvae (Diptera, Acalypterae) within pasture grasses and cereals in England. *Bull. ent. Res.* **50,** 53–62.

O'ROURKE, C. 1968. Survey of diseases of grasses and clovers. *Res. Rep. An Foras Talúntais, Pl. Sci. Crop Husb., 1967* pp. 98–101.

OSBOURNE, P. 1960. Pests associated with ploughed-out grassland. *Scott. Agric.* **39,** 150–3.

RAW, F. 1951. The ecology of the garden chafer *Phyllopertha horticola* (L.) with preliminary observations on control measures. *Bull. ent. Res.* **42,** 605–46.

SAMPSON, K.; WESTERN, J. H. 1954. Diseases of British grasses and herbage legumes. Cambridge: Univ. Press, pp. 118.

SERJEANT, E. P. 1964. Cocksfoot mottle virus. *Pl. Path.* **13,** 23–4.

SLYKHUIS, J. T. 1958. A survey of virus diseases of grasses in northern Europe. *F.A.O. Pl. Prot. Bull.* **6,** 129–34.

SLYKHUIS, J. T. 1967. Virus diseases of cereals. *Rev. appl. Mycol.* **46,** 401–29.

SOUTH, A. 1965. Biology and ecology of *Agriolimax reticulatus* (Müll.) and other slugs: spatial distribution. *J. Anim. Ecol.* **34,** 403–17.

STEPHENSON, J. W. 1968. A review of the biology and ecology of slugs of agricultural importance. *Proc. malac. Soc. Lond.* **38,** 169–76.

STOKES, B. M. 1957. *Mayetiola dactylidis* Kieffer, in cocksfoot grass. *Pl. Path.* **6,** 127–30.

TRIBE, A. J.; HERRIOTT, J. B. D. 1969. Fungicidal treatment for herbage seed. *Exp. Wk Edinb. Sch. Agric. 1968* p. 61.

UPSTONE, M. E. 1969. Epidemiology of cocksfoot mottle virus. *Ann. appl. Biol.* **64,** 49–55.

WEBLEY, D. 1958. Effect of frit fly on the development of a ley. *Pl. Path.* **7,** 73–4.

WELSH PLANT BREEDING STATION. 1967. Principles of herbage seed production. *Tech. Bull. 1 Welsh Pl. Breed. Stn* pp. 110–16.

WESTERN, J. H.; CAVETT, J. J. 1959. The choke disease of cocksfoot (*Dactylis glomerata*) caused by *Epichloe typhina* (Fr.) Tul. *Trans. Br. mycol. Soc.* **42,** 298–307.

WHITE, J. H. 1967a. The use of organochlorine insecticides and possible alternatives against leatherjackets. *Pl. Path.* **16,** 83–8.

WHITE, J. H. 1967b. Pests of turf and their control. *Proc. 1st Symp. Control Weeds, Pests, Diseases cultivated Turf.* British Weed Control Council and British Insecticide and Fungicide Council.

White, J. H.; French, N. 1968. Leatherjacket damage to grassland. *J. Br. Grassld Soc.* **23**, 326–9.

Wilson, M.; Noble, M.; Gray, E. G. 1945. The blind-seed disease of ryegrass and its causal fungus. *Trans. R. Soc. Edinb.* **61**, 327–40.

Woods, A. J.; Bradley Jones, J.; Mantle, P. G. 1966. An outbreak of gangrenous ergotism in cattle. *Vet. Rec.* **78**, 742–9.

CHAPTER 13

Chemical Composition

Introduction

IN THIS chapter information available on the main chemical constituents of grass herbage will be discussed under the following headings:

Structural carbohydrates (cellulose, hemicellulose, pectin)
Non-structural carbohydrates
Lignin
Nitrogen compounds (proteins, non-protein organic compounds, inorganic nitrogen)
Lipids
Organic acids
Vitamins, oestrogens and goitrogenic compounds
Mineral elements

The mineral elements considered in this chapter are those known to be essential for either plants or ruminant animals. In addition, lead and fluorine are included because of their possible toxicity to animals, and also silicon, which is accumulated by grasses and may, like lignin, reduce the digestibility of their structural carbohydrates. All numerical values quoted for the contents of various constituents refer to a dry matter (DM) basis except where otherwise stated.

The contents of all the constituents listed can vary widely, depending on such factors as stage of maturity of the herbage. The changes occurring with advancing maturity can be attributed, at least in part, to changes in plant morphology and to the fact that leaf material differs in composition from that of head and stem. Thus leaves contain more protein, lipids and ash constituents but less soluble carbohydrate and fibre than stems. As grass plants mature they become more fibrous and the percentage contents of other constituents are therefore necessarily reduced. These changes which occur with advancing maturity introduce a difficulty inherent in any comparison between species, namely whether to make the comparison between samples cut on the same date, or between samples cut at the same physiological stage of growth.

With constituents such as cellulose and phosphorus which change markedly with advancing maturity, the choice between these alternatives can influence considerably the result of the comparison. Environmental factors, such as nutrient supplies and light intensity, can also have major effects on chemical composition.

The results obtained can also be influenced by the sampling and analytical procedures adopted, and these vary considerably in the investigations reported in the literature. First of all, the height above ground level at which the herbage is cut governs the proportions of leaf and stem, and these differ in the contents of some constituents. Secondly, the method of drying the sample can influence the contents of certain constituents. For example, as long as conditions allow respiration to continue, there is inevitably some loss of non-structural carbohydrate material.

Complications in inorganic element analysis at these early stages are most liable to arise through contamination of the sample by soil or by dust during the drying and grinding operations. Precautions which should be adopted for trace-element analysis have been described by Mitchell (1960). Variations in results as a consequence of analytical procedures are generally greater in organic analysis owing to the more imprecise nature of the components being analysed. The fraction known as 'crude fibre' is a prime example of a chemically imprecise constituent; nevertheless it has been reported on in many investigations. As pointed out by Ferguson (1963), this fraction contains most, though not all, of the cellulose, together with part of the hemicellulose and lignin.

Structural carbohydrates

CELLULOSE

Cellulose is a major constituent of plant cell walls. It consists of long chains of glucose units, forming micelles or micro-fibrils which are encrusted with lignin, hemicelluloses and silica.

Grass herbage cut at a vegetative stage of growth normally contains 15 to 25% cellulose. The absolute level of the value reported depends on whether or not the method of determination included xylan associated with the cellulose.

Cellulose content increases with advancing maturity (Armstrong *et al.*, 1950; Phillips *et al.*, 1954; MacKenzie & Wylam, 1957; Jarrige & Minson, 1964; Jones, 1964; Waite *et al.*, 1964). When herbage is cut at regular intervals throughout the growing season, there is little change in cellulose content. For example, Jarrige and Minson (1964) reported values around 20% for monthly and two-monthly regrowths of perennial ryegrass and cocksfoot. Differences between grass species and cultivars at a young vegetative stage of growth appear to be relatively small, but increase somewhat as the plants mature (Jarrige & Minson, 1964; Jones, 1964) (see Fig. 13.1). Data from

Table 13.1. Cellulose contents of grass species—mean and range values (% of DM)

Description of samples	Perennial ryegrass	Italian ryegrass	Hybrid ryegrass	Timothy	Cocksfoot	Meadow fescue	Tall fescue	References
5 cuts on same dates	34·1–46·2 38·6	...	...	...	35·2–43·4 39·4	...	33·5–39·9 37·0	Armstrong *et al.* (1950)
Cuts at 5 growth stages	...	...	...	29·6	31·1	...	29·4	Phillips *et al.* (1954)
5 cuts on same dates	11·2–18·8 14·5	10·9–14·9 12·2	...	8·3–12·8 9·7	10·3–23·7 15·9	...	...	Jones (1964)
6 cuts at monthly intervals	13·1	10·3	10·6	...	...	...	...	Bailey (1964)
Late leafy stage	20·3	...	...	26·2	18·0	...	...	Waite *et al.* (1964)
First 4 cuts taken at grazing height	33·0	28·0	...	36·0	39·0	38·0	...	Waite (1965)

investigations in which various species were compared are given in Table 13.1.

The application of fertilizer N has a negligible effect on the content of cellulose in herbage (Blaser, 1964; Bryant & Ulyatt, 1965; Reid *et al.*, 1966).

Low light intensities, and increasing temperatures over the range 10–25°C, were found by Deinum (1966) to increase the crude fibre content of perennial ryegrass, presumably due largely to changes in cellulose content.

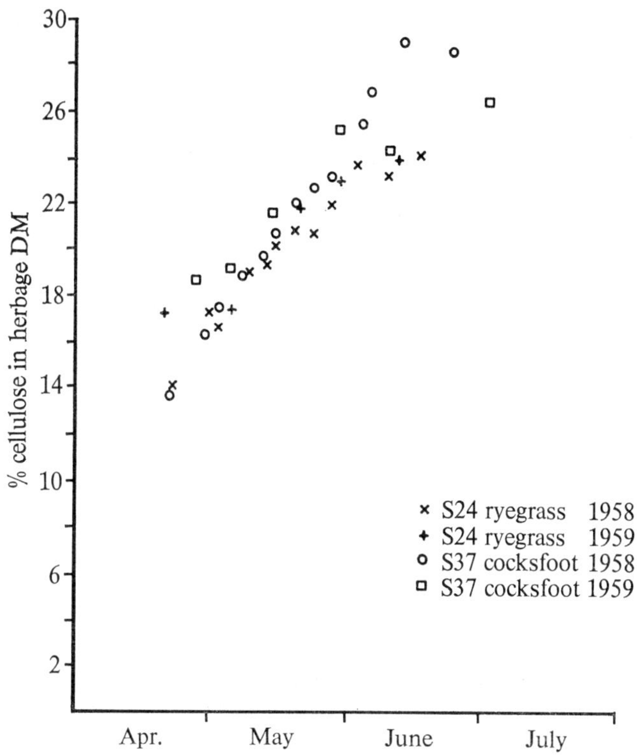

Fig. 13.1. Changes in cellulose content of 2 grasses with advancing maturity. (From Jarrige and Minson, 1964)

HEMICELLULOSE

The hemicellulose fraction of herbage consists of polysaccharides made up mainly from pentose sugars, but with a small proportion of hexose and uronic acid units incorporated in the structure.

Hemicellulose increases with advancing maturity (Deriaz, 1961; Waite, 1963; Jarrige & Minson, 1964). The values obtained for S24 perennial ryegrass and S37 cocksfoot by Jarrige and Minson are shown in Fig. 13.2.

Deriaz (1961), who determined only pentosans, reported an increase for S24 ryegrass from 8·3% on 22 April to 20·6% on 16 June.

When herbage is cut at regular intervals there is a tendency for hemicellulose contents to be lower in the August–November period than earlier in the year (Jarrige & Minson, 1964).

An increase in the rate of N fertilizer was found by Bryant and Ulyatt (1965) to cause a reduction in the hemicellulose content of hybrid ryegrass.

There is very little information on differences in hemicellulose content

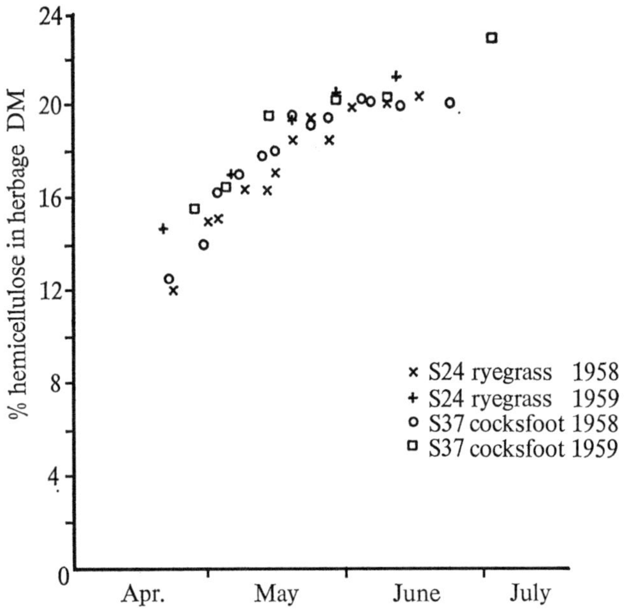

Fig. 13.2. Changes in hemicellulose content of 2 grasses with advancing maturity. (From Jarrige and Minson, 1964)

among species. Fig. 13.2 shows no significant difference between S24 ryegrass and S37 cocksfoot, but Bailey (1964) found perennial ryegrass to have a somewhat higher content (9·8%) than Italian (8·0%) or hybrid ryegrass (8·5%).

PECTIN

Pectin consists mainly of polyuronide material. It is a minor constituent of grass herbage and its content changes little with maturity. Waite (1963) reported values between 2·1 and 2·4% in S23 ryegrass at 4 stages of maturity and Johns (1966) in New Zealand found mean values of 1·1–1·7% for 4 types of ryegrass.

Non-structural carbohydrates

The non-structural carbohydrates (hexoses, oligosaccharides and fructosan) form the water-soluble fraction of grass herbage. Fructosan generally constitutes a greater proportion of this fraction than the hexoses (glucose and fructose) and the oligosaccharides (mainly sucrose) (Ferguson, 1963; McIlroy, 1967). Reported values for water-soluble carbohydrate range from 1·6 to 38·3% of the herbage DM.

The content of water-soluble carbohydrate is higher in stems than in leaves, the difference sometimes being as much as 3- to 4-fold.

Differences among species and variations within species are greater for soluble carbohydrate than for the structural carbohydrates cellulose and hemicellulose (see Table 13.2). The highest concentrations generally occur just before the time of flowering (Waite & Boyd 1953a; Jones, 1962; ap Griffith, 1963), although Jones (1962) found that the pattern of change of soluble carbohydrate content during the period 1 May–13 June was similar for both flowering and non-flowering cultivars of perennial ryegrass. With regrowth herbage cut at monthly intervals, Jarrige and Minson (1964) found that contents were lowest in August or September (Table 13.3).

Environmental factors have a large influence on the content of non-structural carbohydrate and, in some circumstances, may mask the effects of other factors such as stage of growth. Light intensity and temperature are particularly important (Jones, 1962; Deinum, 1966; Nowakowski & Cunningham, 1966). Table 13.4 shows the values obtained by Deinum for water-soluble carbohydrate in ryegrass grown in pots in glasshouse conditions.

Diurnal changes in carbohydrate content occur, due mainly to the alternation of light and darkness. Holt and Hilst (1969) reported that the water-soluble carbohydrate content of three grass species increased continuously from an average of 5·3% at 0600 h to 8·0% at 1800 h.

The application of fertilizer N results in a considerable decrease in the content of water-soluble carbohydrate (Jones *et al.*, 1961; Jones, 1962; Nowakowski, 1962; Gordon *et al.*, 1962; Blaser, 1964; Jones *et al.*, 1965; Bryant & Ulyatt, 1965; Reid, 1966). Rhykerd *et al.* (1966) reported that N fertilizer at rates up to 1200 lb N/ac (1345 kg/ha) per year resulted in a highly significant decrease in percentage soluble carbohydrate in cocksfoot, but did not significantly affect the content in brome grass or timothy. Nowakowski (1962) found that the reduction in Italian ryegrass was due to a reduction in fructosan content and not in that of glucose, fructose and sucrose. This was also found by Jones (1962).

Grass herbage contains little or no starch, although Bailey (1958) found a content of 0·7% in ryegrass leaves harvested in bright sunlight.

In a comparison of 8 cultivars of 6 species, ap Griffith (1963) found a two-fold difference in soluble carbohydrate percentage between the highest,

Table 13.2. Water-soluble carbohydrate contents of grass species—mean and range values (% of DM)

Description of samples	Perennial ryegrass	Italian ryegrass	Hybrid ryegrass	Timothy	Cocksfoot	Meadow fescue	Tall fescue	References
6 or 7 cuts on different dates	4·6–21·6 13·4	...	...	5·5–12·7 10·4	2·0–9·1 5·5	3·5–14.0 9·7	...	Waite & Boyd (1953b)
Monthly cuts on same dates	7·1–31·3	...	...	4·3–22·9	4·0–20·6	5·7–21·2	...	Dent & Aldrich (1963)
8 successive growth stages	17·2–34·5	19·2–38·3	...	15·9–25·7	10·5–22·1	15·1–26·3	14·9–26·1	ap Griffith (1963)
4 cuts on same dates (primary and 3 re-growth cuts)	14·0–23·0 19·5	...	...	...	7·7–14·0 11·4	9·3–19·3 15·0	8·3–21·0 15·8	Miles *et al.* (1964)
6 cuts at monthly intervals	13·7	17·7	16·8	...	...	...	...	Bailey (1964)
First 4 cuts taken at grazing height	19·0	27·0	...	12·0	7·0	15·0	...	Waite (1965)
Monthly cuts over 4 years at 2 sites	16·4	19·4	...	11·4	9·4	13·1	...	Dent (1967) (Personal communication)

Table 13.3. Water-soluble carbohydrate contents in monthly regrowths (% of DM)

		June	July	Aug.	Sept.	Oct.
S24 ryegrass	1958	10·7	11·3	4·8	5·3	7·1
	1959	15·9	14·1	9·5	11·6	13·5
S37 cocksfoot	1958	5·4	6·9	3·1	1·6	3·7
	1959	11·7	9·0	6·0	4·8	8·0

From Jarrige & Minson (1964)

Table 13.4. Water-soluble carbohydrate contents of ryegrass (% of DM)

Light intensity	Day and night temperature		
cal/cm² per day	25–20°C	20–15°C	15–10°C
490	21·2	26·7	33·2
350	18·8	21·2	28·4
90	8·2	7·9	9·0

From Deinum (1966)

S22 Italian ryegrass, and the lowest, Bc 4651 cocksfoot. The changes in water-soluble carbohydrate that occurred with advancing maturity in these 8 cultivars are shown in Fig. 13.3. The earlier work of Waite and Boyd (1953b) had shown that perennial ryegrass generally had a higher content of soluble carbohydrate than timothy and meadow fescue, and cocksfoot had a particularly low content (Table 13.3). In a study of a number of cultivars of perennial ryegrass, cocksfoot, timothy and meadow fescue, Dent and Aldrich (1963) found that, on average, perennial ryegrass had a considerably higher content of water-soluble carbohydrate than the other species. Tetraploid ryegrass cultivars had higher than average contents of soluble carbohydrate.

Lignin

Lignin is a phenolic complex of high molecular weight, the structure of which has not yet been fully elucidated. In the plant it occurs in secondary thickening of the cell walls. Absolute values for the lignin content of herbage are impossible to obtain owing to the imprecision of the analytical methods in use, but a certain amount of comparative data is available.

Lignin content increases with advancing maturity, as illustrated in Fig. 13.4, which is based on data drawn from the results of Jarrige and Minson (1964). Waite *et al.* (1964) obtained the values shown in Table 13.5 (p. 107) for S23 ryegrass.

A large increase in the lignin content of perennial ryegrass at the seed setting stage was also noted by Armstrong *et al.* (1950). Their values were generally higher than those of Waite *et al.* (1964), probably due to the method of analysis. Higher values are found in stem than in leaf tissue (Johnston

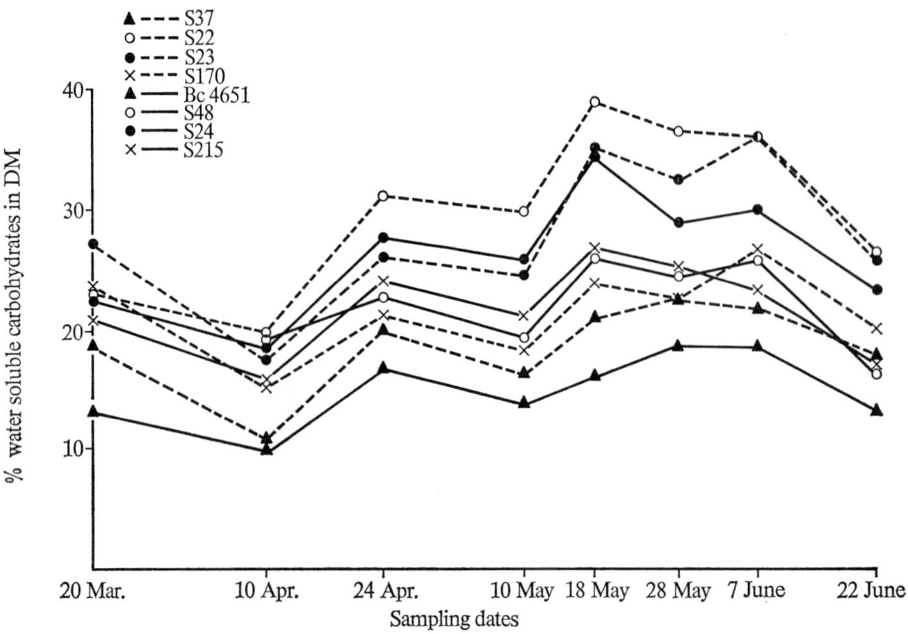

Fig. 13.3. Water-soluble carbohydrate content of 8 grasses. First cuts from early spring to hay stage. (From ap Griffith, 1963)

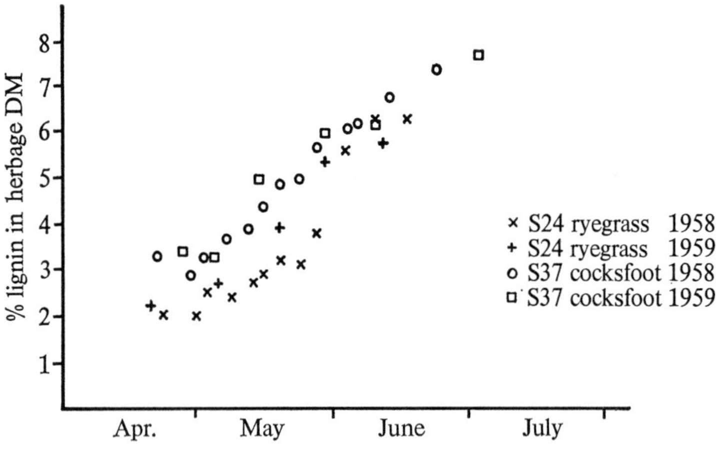

Fig. 13.4. Changes in lignin content of 2 grasses with increasing maturity. (From Jarrige and Minson, 1964)

& Waite, 1965). In herbage cut at regular intervals throughout the season, lignin contents of perennial ryegrass were rather lower in late summer and autumn than earlier in the year, but this trend was not apparent with cocksfoot (Jarrige & Minson, 1964).

Table 13.5. Lignin contents of S23 perennial ryegrass

Maturity stage	Date	% of DM
Young leafy	11 May '59	2·7
Late leafy	25 May '59	3·6
Ear emergence	1 June '59	4·3
Seed setting	29 June '59	7·3

From Waite *et al.* (1964)

The limited amount of information available on species differences is summarized in Table 13.6. Cocksfoot and timothy appear to have higher contents than the other grasses.

Nitrogen compounds

Differences in contents of total N (often expressed as crude protein by multiplying the total N content by 6·25) between species grown under similar conditions and sampled at the same stage of growth are relatively small (see Table 13.7). Differences are least at young leafy stages of growth and tend to increase as flowering stems develop ('t Hart, 1967). Generalizations, however, are difficult to make, since sampling at the same growth stage, but on different dates, results in differences in the time interval since the date of fertilizer application. Moreover, there are differences among cultivars within species, as reported for Italian ryegrass by Hunt and Alexander (1961).

The content of total N in uncut grass herbage declines steadily with advancing maturity from about 4·0–5·0% in the DM at a young leafy stage of growth in early April, to about 1·0–1·5% at the flowering stage, assuming that fertilizer N has been applied in late February or March. With frequent defoliation, seasonal changes in N content are less marked, but relatively high values may occur in late summer and autumn (Jarrige & Minson, 1964).

N contents are generally increased by the application of fertilizer N, the extent of this depending on the time interval between application and sampling as illustrated in Fig. 13.5.

Increasing temperature tended to reduce total N content in experiments conducted by Deinum (1966), whereas Nowakowski *et al.* (1965) found that the effect of temperature varied with the supply of N. Increasing light intensity tends to decrease N content (Nowakowski & Cunningham, 1966; Deinum, 1966).

PROTEINS

Normally about 70–90% of the total herbage N is in the form of protein,

Table 13.6. Lignin contents of grass species—mean and range values (% of DM)

Description of samples	Perennial ryegrass	Italian ryegrass	Hybrid ryegrass	Timothy	Cocksfoot	Meadow fescue	Tall fescue	References
5 cuts on same dates	3·7–12·7 6·5	...	...	...	4·9–9·7 6·9	...	3·8–7·5 4·5	Armstrong *et al.* (1950)
6 cuts at monthly intervals	1·63	1·78	1·74	...	...	...	...	Bailey (1964)
Cuts at various stages of maturity over 2 years	2·0–6·3 3·6	...	...	...	2·9–7·7 4·9	...	...	Jarrige & Minson (1964)
4 cuts taken at grazing height	3·1–5·0 4·1	2·8–4·4 3·5	...	Scots 4·3–4·7 4·5 S48 5·1–6·1 5·4	4·7–5·4 5·2	3·3–4·5 4·0	...	Waite (1965)

Table 13.7. Total N contents of grass species—mean and range values (% of DM)

Description of samples	Perennial ryegrass	Italian ryegrass	Hybrid ryegrass	Timothy	Cocksfoot	Meadow fescue	Tall fescue	References
6 stages of growth on different dates	...	...	...	0·92	1·36	...	1·84	Phillips *et al.* (1954)
Cut on the same dates; primary growth	1·1–4·7 2·3	...	...	1·5–4·6 2·6	1·4–4·6 2·5	...	...	Fleming & Coulter (1963)
4 cuts on same dates (primary and 3 re-growth cuts)	2·24–3·79 3·09	...	...	...	2·56–4·32 3·31	2·51–4·27 3·34	2·35–3·73 3·01	Miles *et al.* (1964)
50% ear emergence, on different dates	1·4–1·7	1·5	...	1·6	2·0	...	S70 2·5 Syn. 1 1·7	Green & Corrall (1967)
First 4 cuts taken at grazing height	2·02–2·12	3·08	...	Scots 3·02 S48 2·52	3·24	3·04	...	Waite (1965)
Monthly cuts over 4 years at 2 sites	3·05	2·82	...	3·39	3·26	3·33	...	Dent (1967) (Personal communication)

the remainder occurring in soluble organic compounds or as nitrate. When K supplies are low, the percentage of total N in protein form is reduced (see Table 13.8).

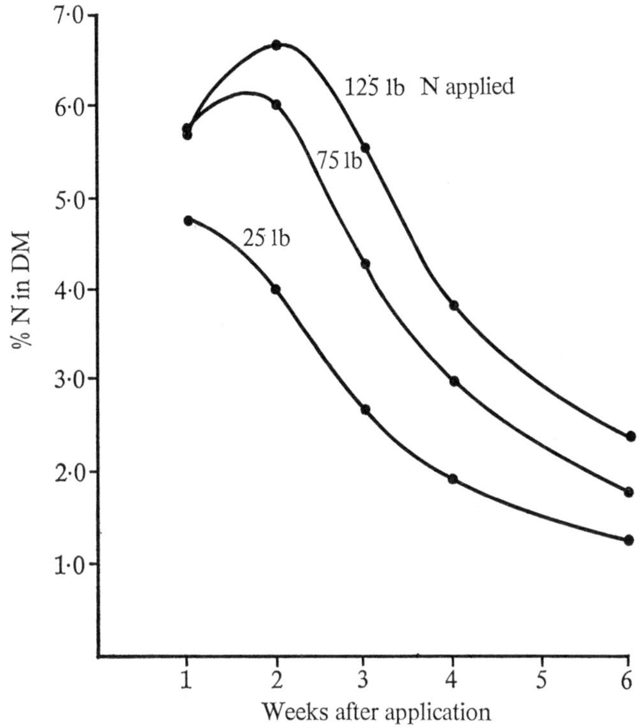

Fig. 13.5. The influence of increasing N supply on N content of Italian ryegrass sampled at weekly intervals for 6 weeks following N application. (From Wilman, 1965)

Table 13.8. Effect of K supply on N partition in Italian ryegrass given 112 lb N/ac as ammonium nitrate

K applied, lb/ac	0	95	190
(K applied, kg/ha)	(0	105	213)
K content of herbage (% of DM)	0·80	0·96	1·29
Total N content (% of DM)	2·30	2·31	2·29
Protein N (% of total N)	71·1	74.6	76.8
Soluble organic N (% of total N)	24·9	21·4	18·7
Inorganic N (% of total N)	4·0	4·0	4·5

From Nowakowski (1964)

The proteins extracted from the leaves of various plants, including certain grasses and legumes, have been found to be similar in amino-acid composition (Gerloff *et al.*, 1965; Chibnall *et al.*, 1963). Wilson and Tilley (1965) found that protein preparations from perennial ryegrass, Italian ryegrass, meadow fescue, timothy and cocksfoot all contained a similar pattern of amino acids. Details of amino-acid composition will not be given in this account, since the dietary requirement of ruminants for individual amino acids is uncertain (McDonald, 1968).

SOLUBLE ORGANIC NITROGEN COMPOUNDS

The soluble organic N compounds comprise mainly peptides, amino acids, amides and purines.

The work of Ferguson and Terry (1954) indicated that very approximately half the soluble organic N in ryegrass herbage was present in peptide form, a larger proportion than was found by Synge (1951) using a different fractionation procedure. Free amino acids accounted for about 15–20% of the soluble organic N in the herbage samples examined by Ferguson and Terry; the proportion of amide-N varied, with a maximum of about 10%, and purine-, betaine- and choline-N amounted to about 12% of the soluble organic N. However, the fraction consisting of non-amino-acid bases such as purines and choline obtained from ryegrass by Synge (1951) accounted for 15–25% of the soluble non-protein N.

Traces of various other organic N compounds such as alkaloids and amines also occur in herbage and some of these may have physiological significance for livestock; there is some evidence that alkaloids, particularly perloline, may be involved in 'ryegrass staggers' in sheep. Aasen *et al.* (1969) have reported from Australia that under normal pasture conditions, the alkaloid content of ryegrass is in the range 0·02–0·05%, but that seedlings and young shoots may contain 0·1–0·25%. It has also been suggested that relatively high levels of histamine, a compound which can occur in concentrations up to 20 ppm. in grass herbage, may be associated with tetany in animals (O'Sullivan, 1968).

The application of N and K in fertilizers influences the herbage content of soluble organic N. Pot experiments with Italian ryegrass reported by Nowakowski *et al.* (1965) and Nowakowski and Cunningham (1966) showed that the proportion of soluble organic N in the total herbage N was much greater when no fertilizer N was applied, or when fertilizer N was applied as ammonium, than when nitrate was applied. Increasing the supply of ammonium up to 400 ppm. had little effect on the proportion of organic non-protein N in the first-mentioned experiment (Nowakowski *et al.*, 1965), but approximately doubled this fraction in the second (Nowakowski & Cunningham, 1966). Carey *et al.* (1952) found that applying 300 lb N as ammonium nitrate to brome grass produced a peak amide-N content of 0·07% of the DM (equivalent to about 2% of the total herbage N) compared with about 0·01% of the DM in grass given no fertilizer. As shown in Table 13.8, soluble organic N is also increased by K deficiency, with amides such as asparagine tending to accumulate (Griffith *et al.*, 1964).

Most of the soluble free amino-acid fraction in grass herbage appears to be composed of relatively few amino acids, mainly dicarboxylic acids, alanine and serine, whereas the peptide fraction has an amino-acid composition intermediate between that of the free amino-acid fraction and the herbage protein (Bathurst, 1953).

INORGANIC NITROGEN

Of the inorganic forms of N, nitrate can occur in quantities amounting, exceptionally, to about 50% of the total herbage N, whereas ammonium occurs in trace amounts only. Nowakowski *et al.* (1965) found that less than 2% of the total N in Italian ryegrass was in ammonium form even when large amounts of ammonium together with a nitrification inhibitor were added to the soil.

The nitrate content of grass herbage is governed mainly by the supply of nitrate from fertilizer and soil, and by the time interval between the application of fertilizer and sampling. This is illustrated by Fig. 13.6, prepared from

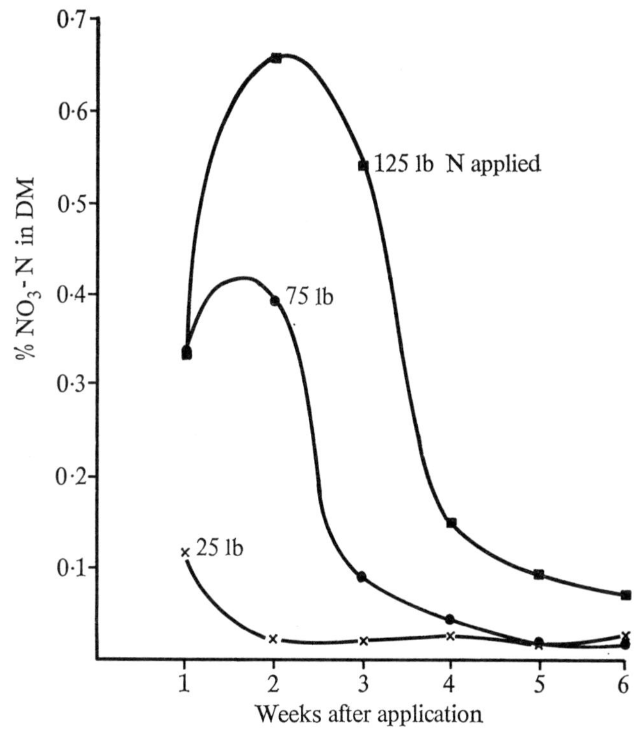

Fig. 13.6. The influence of 3 levels of N, applied as Nitro-chalk, on the content of nitrate-N in Italian ryegrass sampled at weekly intervals for 6 weeks following N application. (From Wilman, 1965)

data obtained by Wilman (1965). Large and Spedding (1966) reported that a perennial ryegrass sward given Nitro-chalk at the rate of 100 lb N/ac (112 kg/ha) per month had nitrate-N contents ranging from 0·024 to 0·66% in samples taken 4 weeks after the application, whereas with a similar sward given only 20 lb N/ac (22 kg/ha) per month, nitrate-N contents did not exceed 0·038% of the DM. With a perennial ryegrass/timothy sward treated with various levels of Nitro-chalk, Reid (1966) found that at rates between 400 lb and 800 lb N/ac per year, most of the additional N in the herbage was in the form of nitrate.

Nitrate contents are considerably increased by low light intensities (Deinum, 1966; Nowakowski & Cunningham, 1966). Other factors, such as S deficiency (Adams & Sheard, 1966), which restrict protein synthesis without restricting nitrate uptake, also tend to increase nitrate contents in herbage.

Differences in the extent of nitrate accumulation between species and cultivars grown under sward conditions appear to be small. Thus, ap Griffith and Johnston (1960) found that with spaced plants, S23 perennial ryegrass and S143 cocksfoot had consistently higher nitrate contents than S24 perennial ryegrass, S48 and S50 timothy, S170 tall fescue, S59 red fescue, and Br 616 bent grass, although with swards there were no consistent differences between cultivars. Murphy and Smith (1967) found that of several grasses sampled at various growth stages, cocksfoot and tall fescue had higher nitrate contents than timothy and *Poa* spp. However, since samples were taken at the same growth stages rather than on the same dates, the time intervals between the application of fertilizer N and sampling would have been different for the different species.

Lipids

The lipids comprise a fairly complex mixture of fats, waxes, phospholipids, glycolipids and sterols, all of which, with the exception of sterols, contain fatty acids in their chemical structure.

Grass herbage normally contains 4–8% of lipid material, as assessed by extraction with ether or a similar solvent. The total lipid content tends to decline with advancing maturity: in samples of S24 perennial ryegrass analysed by Jarrige and Minson (1964), lipid content dropped from 9·0% on 22 April to 2·9% on 16 June. A similar effect has been reported by Lefebvre *et al.* (1967).

Fats (esters of fatty acids with glycerol) are present in only small amounts, but galactolipids constitute a major proportion (about 60%) of the total lipid material (Shorland, 1961).

The fatty-acid composition of the lipid material from mixed grass herbage was studied by Garton (1960), who found that linolenic acid amounted to 61%, palmitic acid 16% and linoleic acid 13% of the total. The proportion of unsaturated acids (linolenic and linoleic) is therefore high. Several other acids were identified in quantities of less than 4%. Deijs and Harberts (1966) have confirmed that grass lipid contains a high proportion of unsaturated fatty acids. A change in fatty-acid composition with advancing maturity was shown by the data obtained by Oksanen and Thafvelin (1965). In timothy cut at a relatively young hay stage, 50% of the total fatty acid was linolenic acid, but only 22% at a later stage of growth.

There appear to be very few comparative data on species for total lipid content or for fatty-acid composition. The values reported for total lipid contents by Jarrige and Minson (1964) show a tendency for S37 cocksfoot to have slightly higher contents than S24 perennial ryegrass.

Fertilizer N at 240 kg/ha, but not at lower rates, increased the lipid content of cocksfoot, but did not affect fatty-acid composition (Lefebvre *et al.*, 1967). Changes in the fatty-acid composition of dehydrated ryegrass during storage have been reported by Czerkawski (1967).

Evidence from brome grass indicates that the content of total sterols increases with advancing maturity, that the proportion in combined form increases from about 15% to about 45% of the total from late April to late June, and that the application of fertilizer N increases the content of both free and combined sterols (Blair *et al.*, 1951).

Organic acids

The main organic acids present in grass herbage are malic, citric, quinic and succinic acids. It can be calculated from the data given by van Tuil (1965) that the total content of these acids in young perennial ryegrass is normally about 6% of the dry weight. This figure is obtained by expressing as malic acid the value of 900 m.eq./kg DM given as the normal content for the total of organic acids salts. Similarly calculated values ranging from 5·6 to 9·6% for young perennial ryegrass cut at various times during the summer were quoted by Ferguson (1963).

There is some evidence of differences among species in total acid content and in the proportions of individual acids, but very little specifically comparative data. Jones and Barnes (1967) estimated the quantities of 6 organic acids in samples of 6 species cut on each of two dates. S215 meadow fescue had the highest mean total organic-acid content of 2·90%, followed by S37 cocksfoot with 2·54%, S48 timothy 2·39%, S24 perennial ryegrass 2·20%, S143 cocksfoot 2·05%, red fescue 1·92%, S23 perennial ryegrass 1·85%, and *Agrostis* 1·31%. Meadow fescue was found by Fauconneau (1962) to have a higher content of total organic acids than cocksfoot.

Davies and Hughes (1954) reported a decrease in total acid content with increasing maturity in perennial ryegrass and timothy, but a slight increase in cocksfoot; unfortunately, they gave no actual values.

The following organic-acid contents in S23 perennial ryegrass at advancing stages of maturity were reported by Waite *et al.* (1964): young leafy 4·2%, late leafy 4·9%, ear emergence 4·6%, seed setting 2·9%. A decrease over the period 1–29 May was also reported by Jones and Barnes (1967). Values for 5 individual acids at 3 stages of maturity in Hl ryegrass are given by Melvin (1965).

Environment can influence total acid contents and individual acids. In a pot experiment with perennial ryegrass, high levels of chloride application were found to decrease the content of organic acids, while by depressing chloride uptake, nitrate application increased the organic-acid content (van Tuil, 1965).

Evidence of a diurnal variation in the ratio of malic to citric acid was obtained by Davies and Hughes (1954). Wilting and slow drying result in a considerable loss of acids (Hirst & Ramstad, 1957).

Some grass species can accumulate up to 3% or more of *trans*-aconitic acid on a DM basis (Stout *et al.*, 1967). Among British grasses, Yorkshire fog tends to accumulate *trans*-aconitic acid (Molloy, 1969) but perennial ryegrass, Italian ryegrass, Westerwolds ryegrass and cocksfoot do not (Molloy, 1969; Stout *et al.*, 1967).

Vitamins, oestrogens and goitrogenic compounds

The vitamin content of herbage is generally of little significance for grazing ruminant animals, but is of considerable importance when conserved herbage is fed to animals indoors, particularly to pigs and poultry, but also to cattle and sheep. The requirements of the different classes of stock for vitamins have been reviewed by the UK Agricultural Research Council (1965).

VITAMIN A AND CAROTENE

Herbage contains no vitamin A, but its precursor, carotene, occurs in considerable quantities, values of 20–40 mg/100 g in the DM being normal for grass at the grazing stage (Ferguson, 1963). The content is usually highest in spring and autumn and lowest in midsummer (Moon, 1939a; Ferguson, 1963), and decreases as the herbage matures (Smith & Wang, 1941). Application of fertilizer N may increase carotene content by up to 40% (Moon, 1939b; Smith & Wang, 1941). The very limited amount of data available shows no appreciable difference between species (Cowlishaw & Eyles, 1957).

VITAMIN B COMPLEX

The small amount of information on the B vitamins has been summarized by Cowlishaw and Eyles (1957).

VITAMIN C

Few data on contents of vitamin C in herbage have been obtained, presumably because dietary supplies of vitamin C appear to be unimportant for ruminants (Agricultural Research Council, 1965). Ferguson (1963) quotes a report of 140 mg/100 g DM in fresh summer grass and 40 mg in winter grass.

VITAMIN D

Assessments based on biological assay have indicated that vitamin D does not generally occur in fresh herbage, but the results may have been influenced by the presence, demonstrated in other work, of rachitogenic substances (Ferguson, 1963; Agricultural Research Council, 1965). Vitamin D is present in sun-cured hays as a result of ultraviolet irradiation, and also in small dead areas of leaf tissue (Ferguson, 1963).

VITAMIN E

The vitamin E activity of 4 grass species, assessed by chemical determination of α-tocopherol, was reported by Brown (1953). There were no significant differences in content between perennial ryegrass, cocksfoot, meadow fescue and timothy, but in each species the content fell markedly with increasing maturity, from 20–35 mg/100 g in mid-April to 1·5–5·0 mg/100 g in mid-August. Booth (1964) reported the leaves to contain about 4 times as much α-tocopherol as the stems and to contain most when growth rate was slow.

OESTROGENS

The important references to work on oestrogenic activity in herbage are given in Chapter 31.

Oestrogenic compounds, which occur quite widely in leguminous plants, are only rarely present in appreciable quantities in grasses, although slight activity was detected in 6 of the 14 British species examined by Pope *et al.* (1959) and in cocksfoot by Kitts *et al.* (1959), and considerable activity in grass of unspecified type by Bartlett *et al.* (1948).

GOITROGENIC COMPOUNDS

Although cyanogenetic glucosides have been reported to occur in grasses (Moran, 1954) and such compounds are potentially goitrogenic (Russell & Duncan, 1956), there appears to be no evidence of appreciable goitrogenic activity in grass herbage of the kind reported for legumes and brassicas (Calderbank, 1963; Agricultural Research Council, 1965).

Mineral elements

This section comprises brief accounts of the mineral elements significant for the nutrition of ruminant animals. These elements include many which are essential for both plants and animals (P, K, Ca, Mg, S, Na, Cl, Fe, Mn, Zn, Cu and Co), others known to be essential for animals only (I and Se) and some which, while not known to be essential, may have undesirable effects on animals if present in excess (Mo, Pb, F and Si).

Where appropriate in the sections dealing with plant nutrient elements, percentage contents are quoted as 'critical values'. These refer to the minimum content of the element in the herbage DM at which the highest DM yield is produced. It is assumed that growth is not limited by a lack of any other nutrient element. Although the critical values given apply only to the conditions under which they were determined and are subject to some variation due to varietal differences, stage of growth and environmental factors, they are nevertheless useful in the diagnosis of nutrient deficiencies (Andrew, 1968; Melsted *et al.*, 1969).

PHOSPHORUS

The P content of herbage is significant in the nutrition of both grass and livestock and appears to be adequate for grass growth when the content in young vegetative tissue is greater than about 0·25% (de Wit *et al.*, 1963; Hylton *et al.*, 1965). Cultivated grass species appear to be similar in P content, as indicated in Table 13.9.

Table 13.9. P contents of grass species—mean and range values (% of DM)

Perennial ryegrass	Hybrid ryegrass	Timothy	Cocksfoot	Meadow fescue	Tall fescue	References
0·18	...	0·20	0·31	...	0·24	Beeson *et al.* (1947)
0·18–0·31 0·26	... ...	0·19–0·27 0·23	0·15–0·31 0·26	0·19–0·33 0·25	0·21–0·30 0·24	Thomas *et al.* (1952)
0·22–0·50 0·33	... ...	0·27–0·53 0·38	0·22–0·50 0·35	... ...	... ...	Fleming & Coulter (1963)
0·36–0·50 0·42	0·34–0·45 0·40	0·30–0·46 0·37	0·43–0·59 0·48	0·42–0·63 0·51	0·27–0·43 0·37	Coppenet (1964)
0·23–0·30 0·26	... ...	0·18–0·24 0·22	0·22–0·34 0·29	0·23–0·26 0·25	... ...	Hurley data (see Whitehead, 1966b)
0·23–0·34 0·29	... ...	0·27–0·36 0·32	0·26–0·38 0·30	0·25–0·28 0·26	0·24–0·27 0·25	Welsh Pl. Breed. Stn data (see Whitehead, 1966b)

The main factors affecting P content are (a) stage of maturity of the herbage and (b) supply of P from the soil. P contents decline steadily with advancing maturity, a typical pattern of change being shown in Fig. 13.7. When herbage undergoes frequent defoliation there is no consistent pattern of change with season, although considerable fluctuations may occur.

On P-deficient soils, contents of P in mature herbage may be as low as 0·10%. Alderman (1963) has reported that over the period 1958–62, an average of 5% of hays from South Wales contained less than 0·15% P and that up to 40% of the samples analysed contained less than 0·26%. With a plentiful supply of P, the content in herbage can be as high as 0·66%P. This value was reported for meadow fescue by Coppenet (1964).

Other factors which affect herbage P contents include moisture supply and the application of N and K fertilizers and lime (Whitehead, 1966a). Several workers have reported that phosphate contents in grass herbage were higher when supplies of moisture were high. The effect of fertilizer N on herbage P content varies with the supply of P; N fertilizers tend to reduce P content in herbage when P supplies are low and to increase it when supplies are high. With moderate supplies there may be no effect.

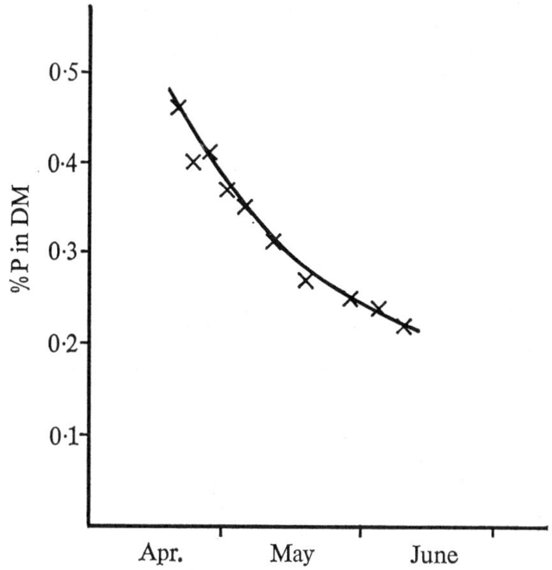

Fig. 13.7. Influence of stage of maturity on P content of S24 perennial ryegrass. (From Whitehead, 1966b)

POTASSIUM

K deficiency is common in grass, but rarely if ever occurs in livestock. Cation imbalances in livestock due to excess K are, however, fairly common.

K contents reported for grass herbage in Britain range from 0·4% for a mixed sward of perennial ryegrass, cocksfoot, *Agrostis* and red fescue (Hemingway, 1963), to 4·8% in Italian ryegrass (Widdowson *et al.*, 1965). Reports of critical levels for grass vary considerably. Working with perennial ryegrass and cocksfoot in culture solution and pot experiments, de Wit *et al.* (1963) obtained a value equivalent to 1·0% K. However, field experiments have indicated that critical K contents for grasses cut at 4- to 6-week intervals are at least 2% K (MacLeod, 1965; Clement & Hopper, 1968).

In an investigation by ap Griffith *et al.* (1965), differences between the mean K contents of perennial ryegrass, cocksfoot, meadow fescue and tall fescue were small, but in all except meadow fescue there were approximately two-fold differences between cultivars within species. Data from various investigations in which species were compared are given in Table 13.10.

Important factors affecting the K content of herbage are stage of maturity and K supply. Although the pattern of change with maturity is less consistent than with phosphate, K contents at the flowering stage are always lower than at the young vegetative stage (Fleming & Coulter, 1963; ap Griffith & Walters, 1966). With repeated defoliation, seasonal differences in herbage K content depend on the availability of soil K. When supplies are low, K content tends to decline during the season but can be maintained by applying fertilizer K (Hemingway, 1963).

Increases in herbage K content brought about by the application of K fertilizers have been reported in many investigations (see Whitehead, 1966a). The effect of fertilizer N on herbage K content depends to a considerable extent on the K supply; N tends to increase herbage K content when K supplies are plentiful and to depress it when K supplies are limited.

Table 13.10. K content of grass species—mean and range values (% of DM)

Perennial ryegrass	Hybrid ryegrass	Timothy	Cocksfoot	Meadow fescue	Tall fescue	References
1·45–2·32 1·98	... ...	1·89–2·33 2·17	1·83–2·81 2·32	1·51–2·63 2·06	1·38–2·51 1·99	Thomas *et al.* (1952)
1·7–3·1 2·5	... ...	2·7–3·4 3·1	2·6–4·0 3·3	... ...	... ...	Fleming & Coulter (1963)
1·56–2·70 2·35	1·90–2·58 2·21	2·07–2·74 2·30	2·35–3·43 2·64	3·02–4·41 4·01	2·57–3·43 2·98	Coppenet (1964)
1·77–2·32 2·01	... ...	1·55–1·83 1·72	1·72–2·49 2·27	1·84–2·18 1·99	... ...	Hurley data (see Whitehead, 1966b)
2·02–3·08 2·35	... ...	2·43–3·06 2·74	2·26–3·06 2·59	2·70–3·01 2·89	2·98–3·08 3·00	Welsh Pl. Breed. Stn data (see Whitehead, 1966b)

CALCIUM

The Ca content of grass herbage is usually in the range 0·4–1·0%. According to the data of de Wit *et al.* (1963) the critical level for grass is about 0·1%.

There appear to be no consistent differences in Ca content among the commonly cultivated grass species (Fleming & Coulter, 1963; Coppenet, 1964; Whitehead, 1966b). The Ca content of primary growth falls from early May to July (Fleming & Coulter, 1963; Whitehead, 1966b); with frequent defoliation, however, Ca content tends to rise to a peak in late summer.

The application of fertilizers can affect herbage Ca content, although not all reported effects are consistent. Nitro-chalk at up to 300–400 lb N/ac per year (336–448 kg/ha) has often caused some decrease in Ca content but resulted in an increase at higher rates of application (Whitehead, 1966 a, b). N fertilizers containing no Ca generally decrease herbage Ca contents and K applications also generally decrease Ca contents (Whitehead, 1966a).

MAGNESIUM

Mg deficiency in grass is very rare. The critical level for grass, as indicated by the results of de Wit *et al.* (1963), is equivalent to 0·06%, and no values below this appear to have been reported in Britain. However, Mg deficiency in livestock is relatively common. Although there is no direct relationship between herbage Mg content and the incidence of hypomagnesaemia, the disorder is rare when herbage contents are above 0·2% Mg, but is a definite possibility with contents below 0·2%. There appears to be a tendency for timothy to have rather lower Mg contents than the other cultivated grasses (see Table 13.11).

Table 13.11. Mg content of grass species—mean and range values (% of DM)

Perennial ryegrass	Hybrid ryegrass	Timothy	Cocksfoot	Meadow fescue	Tall fescue	References
0·15–0·31	...	0·22–0·31	0·19–0·25	0·22–0·31	0·28–0·39	Thomas *et al.* (1952)
0·21	...	0·26	0·22	0·27	0·33	
0·14–0·27	...	0·11–0·21	0·16–0·27	...	...	Fleming & Coulter
0·18	...	0·14	0·18	...	...	(1963)
0·11–0·37	0·17–0·34	0·15–0·26	0·18–0·34	0·16–0·28	0·26–0·36	Coppenet (1964)
0·25	0·26	0·18	0·26	0·23	0·30	
0·07–0·11	...	0·06–0·08	0·09–0·11	...	...	Hurley data (see
0·09	...	0·07	0·10	...	...	Whitehead 1966b)
0·12–0·18	...	0·10–0·15	0·15–0·18	0·14–0·15	0·15–0·15	Welsh Pl. Breed. Stn
0.15	...	0·13	0·17	0·15	0·15	data (see Whitehead, 1966b)

The Mg content of primary growth tends to decrease with increasing maturity, but when swards are defoliated at intervals, Mg contents are higher in August, September and October than earlier in the season (Whitehead, 1966b).

Application of N and K fertilizers can appreciably affect herbage Mg levels. N fertilizers often increase the total Mg content of grass herbage, but may reduce Mg availability to livestock (see Whitehead, 1966a). K fertilizers often depress Mg contents, although Reith (1963, 1967) has stated that applications of K in amounts sufficient to maintain yields have practically no effect on Mg contents.

SULPHUR

The S content of grass herbage is almost always within the range 0·13–0·75% reported for Italian ryegrass from various parts of Britain by Cunningham (1964). The usual range is 0·20–0·45%.

Dijkshoorn *et al.* (1960) considered that S deficiency is indicated when the S content is less than 0·0617 times the organic N content (i.e. an organic N: S ratio wider than 16·2 : 1). For young grass herbage with 3·0% N content, the critical S content would therefore be 0·18%. A critical level of 0·1% S has been reported for perennial ryegrass by Saalbach and Judel (1966).

There appear to be no comparative data for the various grass species, but large differences are unlikely. The N : S ratio of the protein in herbage tends to be fairly constant at about 17 : 1 owing to the constancy of the proportion of the S-containing amino acids in grass proteins. However, it is possible for variations in total N : S ratio to occur owing to the presence of non-protein forms of N or S (Dijkshoorn *et al.*, 1960).

S deficiency in grass grown in the field has not been reported in Britain, but with the increasing intensification of herbage production and the diminishing use of fertilizers containing S, it may well occur in the future.

Moir (1966) has obtained evidence that ruminants require a level of S in the diet such that the N : S ratio is no wider than about 10 : 1, although results obtained by other workers indicated that a ratio of 15 : 1 was satisfactory for dairy cows (Davis *et al.*, 1954).

Fertilizer applications can influence S contents in herbage. Ammonium sulphate causes an appreciable increase, while calcium ammonium nitrate has no effect (Conroy, 1961). Application of Cl reduces the S content of herbage (Dijkshoorn, 1959).

SODIUM

There are important differences in Na content between species. Some, such as the ryegrasses and cocksfoot, show wide variations in content, depending on environmental conditions, whereas others, such as timothy, are consistently low in Na content (ap Griffith & Walters, 1966).

The extent of the variation in a species potentially high in Na is illustrated by the range from a trace, up to a maximum of 2·12%, reported for Italian ryegrass sampled at various times from a number of sites in Britain (Cunningham, 1964). The value of 2·12% appears to be the highest reported for grass in Britain, although investigations in which various species have been compared have generally shown perennial ryegrass and cocksfoot to have higher contents than Italian ryegrass (see Table 13.12). In addition to differences between species, ap Griffith and Walters (1966) found large differences between cultivars within species potentially high in Na. In cocksfoot, the cultivar richest in Na contained 9 times more of that element than did the cultivar containing least Na.

Factors influencing the Na content of grass have been reviewed by Henkens (1965). An increased content results from an increase in Na supply and a decreased content from an increase in K supply. Application of N fertilizer generally causes an increase in Na content, particularly in the high-potential species, but to a much smaller extent in the low-potential species (Whitehead, 1966a, b). When both N and K fertilizers are applied there may be little effect on Na content.

Table 13.12. Na contents of grass species—mean and range values (% of DM)

Perennial ryegrass	Italian ryegrass	Hybrid ryegrass	Timothy	Cocksfoot	Meadow fescue	Tall fescue	References
0·10–0·19 0·14	...	...	0·13–0·29 0·22	0·10–0·15 0·13	0·07–0·21 0·15	0·07–0·16 0·10	Thomas *et al.* (1952)
0·27–0·42 0·33	...	...	0·06–0·10 0·07	0·20–0·36 0·30	...	...	Fleming & Coulter (1963)
0·40–0·76 0·57	...	0·16–0·64 0·33	0·04–0·16 0·11	0·25–0·77 0·45	0·03–0·14 0·07	0·19–0·28 0·24	Coppenet (1964)
0·04–0·22 0·10	...	...	0·01–0·02 0·02	0·03–0·19 0·09	0·03–0·05 0·04	...	Hurley data (see Whitehead, 1966b)
...	0·06–0·16	0·04–0·08	...	...	...	0·10–0·16	Hurley data (unpublished)
0·20–0·66 0·48	...	...	0·06–0·13 0·08	0·14–0·63 0·40	0·07–0·10 0·09	0·15–0·17 0·16	Welsh Pl. Breed. Stn data (see Whitehead, 1966b)
0·13–0·36 0·23	0·06–0·14 0·10	...	0·02–0·06 0·03	0·21–0·30 0·25	0·03–0·08 0·04	0·06–0·15 0·10	ap Griffith *et al.* (1966)

Grasses show no consistent trend in Na content with increasing maturity or with season in swards that are regularly cut or grazed (Whitehead, 1966a, b).

CHLORINE

The Cl content of herbage varies widely. Orr (1929) reported that in 200 samples analysed at the Rowett Research Institute it ranged from 0·015 to 1·71%, and Cunningham (1964) reported a range of 0·41 to 1·99% for Italian ryegrass from various sites.

There is very little information on species differences, but the following average contents were found in an experiment at Hurley: S23 perennial ryegrass 0·91%, S24 perennial ryegrass 0·94%, S48 timothy 0·81%, S37 cocksfoot 0·99%, and S215 meadow fescue 0·85% (Whitehead, 1966b). Thomas *et al.* (1952), however, reported cocksfoot to have an appreciably lower average content (0·31%) than perennial ryegrass (0·51%), timothy (0·56%), meadow fescue (0·66%) and tall fescue (0·47%). Application of K fertilizer, especially the commonly used form of KCl, is an important factor influencing herbage Cl content. Ryegrass herbage from cut swards given 0, 52, 104 and 208 lb Cl/ac (0, 58, 117 and 233 kg/ha) as KCl had mean Cl contents of 0·39, 0·75, 1·04 and 1·30%, respectively (Whitehead, 1966b). There is some evidence that N fertilizers tend to reduce herbage Cl contents (Whitehead, 1966a).

Cl contents are not appreciably influenced by stage of maturity, but with repeated defoliation Cl contents tend to be lower in late summer and autumn than earlier in the year, an effect possibly due to the greater time interval since the application of KCl fertilizer.

IRON

It is exceptional for herbage grasses to suffer from Fe deficiency, although instances have been reported on blanket peat soils in Ireland (O'Toole, 1966).

Fe contents of herbage are very susceptible to overestimation due to traces of soil in the samples. When precautions are taken to avoid contamination, the contents in grass are usually in the range 50–200 ppm. Differences between species are not consistent (Thomas *et al.*, 1952; Fleming, 1965; Whitehead, 1966b). Fe contents are higher in young vegetative herbage than in material at the flowering stage (Thomas *et al.*, 1952; Loper & Smith, 1961).

Soil pH has a considerable effect on the availability of Fe in the soil and consequently on its content in herbage. Availability is increased by acid conditions (Thomas & Thompson, 1948).

MANGANESE

Reported values for Mn contents of herbage vary widely, the usual range being 30–300 ppm. Data from various investigations in which several species have been compared are summarized in Table 13.13. In most instances cocksfoot had a higher content than the other grasses. Thomas *et al.* (1952) found

Mn contents to be highest in young herbage, but the reduction in content with advancing maturity was greater in perennial ryegrass and cocksfoot than in timothy, meadow fescue and tall fescue.

Mn contents are very much influenced by soil pH. An increase in soil pH by one unit in the range pH 5·0–7·5 reduces the Mn content of herbage by about 50% (Dijkshoorn, 1962; Stewart & McConaghy, 1963; Reith & Mitchell, 1964).

ZINC

The Zn content of grass herbage is usually within the range 15–60 ppm. However, there appear to be no data on the basis of which the Zn contents of complete herbage of the commonly cultivated grass species can be compared. The mean values for leaf, stem and head fractions for 4 species sampled at early maturity by Fleming (1963) were: perennial ryegrass 23 ppm., timothy 24 ppm., cocksfoot 24 ppm., and meadow fescue 22 ppm. A slight decline in Zn content with advancing maturity in brome grass was reported by Loper and Smith (1961).

COPPER

Cu deficiency sufficiently severe to restrict grass growth has not been reported in the UK, but it has been diagnosed on blanket peat in Ireland (Grennan, 1966).

Cocksfoot generally has a rather higher Cu content than other grass species (see Table 13.14). The Cu content of grass herbage declines fairly steadily with advancing maturity. Fleming (1963) reported decreases from 12·9 to 4·0 ppm. for cocksfoot and 8·3 to 3·4 ppm. for perennial ryegrass over the period 4 April to 28 June. With a mixed grass sward cut on 4 successive occasions at the silage stage, Hemingway (1962) noted an increase in Cu content during the growing season.

Conflicting results have been obtained on the influence of fertilizer N on herbage Cu contents (see Whitehead, 1966a, b).

COBALT

Co contents of grass herbage free from soil contamination are usually in the range 0·06–0·20 ppm., though Thomas *et al.* (1952) and Andrews (1966) reported somewhat inconsistent results relating to differences between species (see Table 13.15, p. 126).

Soil pH is an important factor influencing the Co content of herbage. Values are increased by increasing acidity. Fertilizer applications appear to have little if any effect unless they change soil pH (Reith & Mitchell, 1964; Whitehead, 1966a).

IODINE

There is relatively little recent information on the I content of grass

Table 13.13. Mn contents of grass species—mean and range values (ppm.)

Perennial ryegrass	Hybrid ryegrass	Timothy	Cocksfoot	Meadow fescue	Tall fescue	References
...	...	192	564	...	288	Beeson *et al.* (1947)
16–31 22	...	25–35 30	33–61 46	22–43 29	17–40 23	Thomas *et al.* (1952)
96–270 188	111–174 138	86–110 102	109–208 149	102–173 141	174–274 231	Coppenet (1964)
200	...	130	314	...	...	Fleming (1965)

Table 13.14. Cu contents of grass species—mean and range values (ppm.)

Perennial ryegrass	Hybrid ryegrass	Timothy	Cocksfoot	Meadow fescue	Tall fescue	References
...	...	5·5	9·9	...	8·4	Beeson *et al.* (1947)
4·8–15·2 8·5	...	6·8–8·2 7·5	7·0–15·0 10·0	5·5–23·0 9·5	2·2–5·5 4·1	Thomas *et al.* (1952)
5·8–10·1 7·3	5·8–8·6 7·0	6·2–9·7 7·9	7·4–8·3 7·9	6·3–10·0 8·2	4·8–10·9 7·5	Coppenet (1964)
3·4–8·3 5·4	...	4·2–9·7 6·3	4·0–12·9 7·7	...	...	Fleming (1963)
6·9	...	6·6	8·6	6·5	6·0	ap Griffith (1966)

herbage. Alderman (1963) reported an average value of 0·2 ppm. I for pastures in South Wales, with contents as low as 0·08 ppm. on some farms. The two investigations summarized in Table 13.16 in which species were compared produced very different results. The data obtained by Alderman and Jones (1967) showed that higher contents occurred in early April and in November than during the remainder of the growing season and that the application of fertilizer N depressed I contents.

MOLYBDENUM

Although the essentiality of Mo for ruminants has not been clearly demonstrated, it is well known that when present at relatively high levels, it can have a harmful effect by interfering with Cu metabolism. According to Cunningham (1960), when the Cu content of pasture is only 3 ppm., Cu deficiency can be induced by 3 ppm. Mo and when the Cu content is 5 ppm. by 7 ppm. Mo. The disorder of cattle known as teart is due to Mo toxicity and occurs on pastures containing 20–100 ppm. Such high levels are restricted to certain alkaline soils with high Mo contents. More usual values range from 0·1 to 4·0 ppm., but there are virtually no published investigations in which grass species are compared. Soil pH is an important factor in influencing herbage Mo content, and increases up to several-fold can be induced by liming (Reith & Mitchell, 1964).

SELENIUM

There is little information on the Se content of grass herbage in Britain, but values normally appear to be in the range 0·1–1·0 ppm. (Whitehead, 1966a, b). No instances of Se deficiency in animals appear to have been reported in Britain, and toxic levels seem to occur only in Ireland, where on areas of seleniferous soils, grasses may contain more than 50 ppm. Se. Lane and Fleming (1966) found 11–21 ppm. in perennial ryegrass growing on a seleniferous soil, with a steady increase in content as the grass matured over the period early June to late July.

LEAD

The Pb content of grass herbage during active growth appears to lie normally within the range 0·3–3·5 ppm. (Mitchell and Reith, 1966; Whitehead, 1966b). However, the content in senescent herbage is several times greater than that in actively growing material and values for winter herbage may reach 30–40 ppm. (Mitchell & Reith, 1966).

FLUORINE

Allcroft *et al.* (1965) reported the F content of 19 pasture samples from a rural area to range between 2 and 16 ppm., with a mean value of 5·3 ppm. Samples of herbage from a farm in an industrial area at Stoke-on-Trent contained up to 292 ppm.

Table 13.15. Co contents of grass species—mean and range values (ppm.)

Perennial ryegrass	Hybrid ryegrass	Timothy	Cocksfoot	Meadow fescue	Tall fescue	References
...	...	0·08	0·08	...	...	Beeson *et al.* (1947)
0·14–0·17 0·15	...	0·12–0·18 0·15	0·07–0·20 0·14	0·12–0·20 0·16	0·08–0·18 0·11	Thomas *et al.* (1952)
0·12–0·20 0·16	0·10–0·17 0·13	0·07–0·11 0·09	0·08–0·14 0·11	0·09–0·14 0·12	...	Andrews (1966)

Table 13.16. I contents of grass species—mean and range values (ppm.)

Perennial ryegrass	Italian ryegrass	Hybrid ryegrass	Timothy	Cocksfoot	Meadow fescue	Tall fescue	References
1·30–1·86 1·45	0·60–1·22 0·85	0·16–0·52 0·31	...	0·08–0·27 0·17	...	...	Johnson & Butler (1957)
0·08–0·40 0·22	0·10–0·48 0·25	...	0·10–0·69 0·31	0·16–0·59 0·32	0·12–0·45 0·22	0·12–0·41 0·20	Alderman & Jones (1967)

SILICA

Values reported for the silica content of herbage generally include both the silica present in the plant tissue and the silica in any soil contamination. In a study reported by Johnston *et al.* (1967) in which herbage was cut by hand 2·5 cm above the soil surface, plant silica amounted to about half the total present. It is clear that where no distinction is made between sand and plant silica, variations in total content may merely reflect differences in the extent of contamination.

There are species differences in the extent to which plants take up silica, the Gramineae often containing 10 to 20 times as much as legumes and other dicotyledons (Jones & Handreck, 1967). Information on differences in silica content between the common British grasses, however, appears to be lacking. As examples of values found, Rook and Wood (1960) reported 1·78–3·86% in mixed herbage sampled by hand to minimize soil contamination, and Waite *et al.* (1964) found 0·6–1·2% in samples of ryegrass, cocksfoot and timothy dried in a commercial grass drier.

Van Soest and Jones (1968) have recently found that in forage grasses, silica is an important factor in reducing the digestibility of cell-wall carbohydrates. Their results show that its importance may rival that of lignin in reed canarygrass (containing 0·6–5·4% SiO_2) and possibly Coastal Bermuda grass (5·2% SiO_2) and tall fescue (4·4% SiO_2).

There is evidence indicating that the silica content of grasses increases with advancing maturity in parallel with the increase in lignin contents (Jones & Handreck, 1967). Silica content is also influenced by soil and by nutrient supply, being decreased by liming and by applications of fertilizer N and P.

References

AASEN, A. J.; CULVENOR, C. C. J.; FINNIE, E. P.; KELLOCK, A. W.; SMITH, L. W. 1969. Alkaloids as a possible cause of ryegrass staggers in grazing livestock. *Aust. J. agric. Res.* **20,** 71–86.

ADAMS, C. A.; SHEARD, R. W. 1966. Alterations in the nitrogen metabolism of *Medicago sativa* and *Dactylis glomerata* as influenced by potassium and sulfur nutrition. *Can. J. Pl. Sci.* **46,** 67–80.

AGRICULTURAL RESEARCH COUNCIL. 1965. The nutrient requirements of farm livestock. No. 2. Ruminants. London: H.M.S.O., pp. 264.

ALDERMAN, G. 1963. Mineral nutrition and reproduction in cattle. *Vet. Rec.* **75,** 1015–8.

ALDERMAN, G.; JONES, D. I. H. 1967. The iodine content of pastures. *J. Sci. Fd Agric.* **18,** 197–9.

ALLCROFT, R.; BURNS, K. N.; HEBERT, C. N. 1965. Fluorosis in cattle. 2. Development and alleviation: experimental studies. Min. Agric. Fish. Fd. Animal Disease Surveys. Report No. 2 Part II. London: H.M.S.O., pp. 58.

ANDREW, C. S. 1968. Problems in the use of chemical analyses for diagnosis of plant nutrient deficiencies. *J. Aust. Inst. agric. Sci.* **34,** 154–62.

ANDREWS, E. D. 1966. Cobalt concentrations in some New Zealand fodder plants grown on cobalt-sufficient and cobalt-deficient soils. *N.Z. Jl agric. Res.* **9,** 829–38.

ARMSTRONG, D. G.; COOK, H.; THOMAS, B. 1950. The lignin and cellulose contents of certain grassland species at different stages of growth. *J. agric. Sci., Camb.* **40,** 93–9.

BAILEY, R. W. 1958. Carbohydrates in pasture species. 1. The starch content of clovers and ryegrass. *J. Sci. Fd Agric.* **9,** 743–7.

BAILEY, R. W. 1964. Pasture quality and ruminant nutrition. 1. Carbohydrate composition of ryegrass varieties grown as sheep pastures. *N.Z. Jl agric. Res.* **7,** 496–507.

BARTLETT, S.; FOLLEY, S. J.; ROWLAND, S. J.; CURNOW, D. H.; SIMPSON, S. A. 1948. Oestrogens in grass and their possible effects on milk secretion. *Nature, Lond.* **162,** 845.

BATHURST, N. O. 1953. The free amino acids and peptides of plant tissues. *J. Sci. Fd Agric.* **4,** 221–6.

BEESON, K.C.; GRAY, L.; ADAMS, M. B. 1947. The absorption of mineral elements by forage plants. 1. The phosphorus, cobalt, manganese and copper content of some common grasses. *J. Am. Soc. Agron.* **39,** 356–62.

BLASER, R. E. 1964. Symposium on forage utilization: effects of fertility levels and stage of maturity on forage nutritive value. *J. Anim. Sci.* **23,** 246–53.

BOOTH, V. H. 1964. The α-tocopherol content of forage crops. *J. Sci. Fd Agric.* **15,** 342–4.

BROWN, F. 1953. The tocopherol content of farm feeding stuffs. *J. Sci. Fd Agric.* **4,** 161–5.

BRYANT, A. M.; ULYATT, M. J. 1965. Effects of nitrogenous fertilizer on the chemical composition of short-rotation ryegrass and its subsequent digestion by sheep. *N.Z. Jl agric. Res.* **8,** 109–17.

CALDERBANK, G. 1963. Iodine. In: Worden, A. N.; Sellers, K. C.; Tribe, D. E. [Eds] Animal health, production and pasture. London: Longmans, Green and Co. Ltd, pp. 681–736.

CAREY, V.; MITCHELL, H. L.; ANDERSON, K. 1952. Effect of nitrogen fertilization on the chemical composition of bromegrass. *Agron. J.* **44,** 467–9.

CHIBNALL, A. C.; REES, M. W.; LUGG, J. W. H. 1963. The amino acid composition of leaf proteins. *J. Sci. Fd Agric.* **14,** 234–9.

CLEMENT, C. R.; HOPPER, M. J. 1968. The supply of potassium to high-yielding cut grass. *N.A.A.S. q. Rev.* No. 79, 101–9.

CONROY, E. 1961. Effects of heavy applications of nitrogen on the composition of herbage. *Ir. J. agric. Res.* **1,** 67–71.

COPPENET, M. 1964. [Changes in the mineral composition of pasture grasses under a grazing regime.] *C.r. hebd. Séanc. Acad. Agric. Fr.* **50,** 330–42.

COWLISHAW, S. J.; EYLES, D. E. 1957. The nutritive value of herbage for poultry. *Nutr. Abstr. Rev.* **27,** 983–96.

CUNNINGHAM, I. J. 1960. Molybdate topdressing and animal health. *N.Z. Jl Agric.* **100,** 419–28.

CUNNINGHAM, R. K. 1964. Cation-anion relationships in crop nutrition. 1. Factors affecting cations in Italian ryegrass. *J. agric. Sci., Camb.* **63,** 97–101.

CZERKAWSKI, J. W. 1967. Effect of storage on the fatty acids of dried ryegrass. *Br. J. Nutr.* **21,** 599–608.

DAVIES, C. W.; HUGHES, R. B. 1954. The organic acids of grass extracts. *J. Sci. Fd Agric.* **5,** 200–5.

DAVIS, R. F.; WILLIAMS, C.; LOOSLI, J. K. 1954. Studies on sulfur to nitrogen ratios in feeds for dairy cows. *J. Dairy Sci.* **37,** 813–8.

DEIJS, W. B.; HARBERTS, C. L. 1966. Lipid fractions in feed and faeces and their apparent digestibility in cattle. *Jaarb. Inst. biol. scheik. Onderz. LandbGewass. 1966* pp. 79–89.

DEINUM, B. 1966. Influence of some climatological factors on the chemical composition and feeding value of herbage. *Proc. 10th int. Grassld Congr., Helsinki, 1966* pp. 415–8.

DENT, J. W. 1967. National Institute of Agricultural Botany. Personal communication.

DENT, J. W.; ALDRICH, D. T. A. 1963. The inter-relationship between heading date, yield, chemical composition and digestibility in varieties of perennial ryegrass, timothy, cocksfoot and meadow fescue. *J. natn. Inst. agric. Bot.* **9,** 261–81.

DERIAZ, R. E. 1961. Routine analysis of carbohydrates and lignin in herbage. *J. Sci. Fd Agric.* **12,** 152–60.

DIJKSHOORN, W. 1959. The rate of uptake of chloride, phosphate and suphate in perennial ryegrass. *Neth. J. agric. Sci.* **7,** 194–201.

DIJKSHOORN, W. 1962. The effect of soil pH on manganese absorption by *Lolium perenne* L. *Jaarb. Inst. biol. scheik. Onderz. Landbgewass. 1962* pp. 131–3.

DIJKSHOORN, W.; LAMPE, J. E. N.; VAN BURG, P. F. J. 1960. A method of diagnosing the sulphur nutrition status of herbage. *Pl. Soil* **13,** 227–41.

FAUCONNEAU, G. 1960. [The nitrogenous fractions and organic acids in grasses and legumes.] *Proc. 8th int. Grassld Congr., Reading, 1960* pp. 617–21.

FERGUSON, W. S. 1963. Pasture chemistry. In: Worden, A. N.; Sellers, K. C.; Tribe, D. E. [Eds] Animal health, production and pasture. London: Longmans Green and Co. Ltd, pp. 92–127.

FERGUSON, W.S.; TERRY, R.A. 1954. The fractionation of the non-protein nitrogen of grassland herbage. *J. Sci. Fd Agric.* **5**, 515–24.

FLEMING, G. A. 1963. Distribution of major and trace elements in some common pasture species. *J. Sci. Fd Agric.* **14**, 203–8.

FLEMING, G. A. 1965. Trace elements in plants with particular reference to pasture species. *Outl. Agric.* **4**, 270–85.

FLEMING, G. A.; COULTER, B. S. 1963. Mineral elements in pasture plants. *Proc. 1st Reg. Conf. int. Potash Inst., Wexford (Ireland) 1963* pp. 63–70.

GARTON, G. A. 1960. Fatty acid composition of the lipids of pasture grasses. *Nature, Lond.* **187**, 511–2.

GERLOFF, E. D., LIMA, I. H.; STAHMANN, M. A. 1965. Amino acid composition of leaf protein concentrates. *J. agric. Fd Chem.* **13**, 139–43.

GORDON, C. H.; DECKER, A. M.; WISEMAN, H. G. 1962. Some effects of nitrogen fertilizer, maturity and light on the composition of orchardgrass. *Agron. J.* **54**, 376–8.

GREEN, J. O.; CORRALL, A. J. 1967. The Grassland Research Institute, Hurley. Unpublished data.

GRENNAN, E. J. 1966. Copper deficiency in pasture on blanket peat. *Ir. J. agric. Res.* **5**, 147–53.

GRIFFITH, G. AP. 1963. Varietal differences in chemical composition during primary growth. *Rep. Welsh Pl. Breed. Stn 1962* pp. 87–8.

GRIFFITH, G. AP. 1966. Copper content of grass species and varieties. *Rep. Welsh Pl. Breed. Stn 1965* pp. 99–100.

GRIFFITH, G. AP; JOHNSTON, T. D. 1960. The nitrate nitrogen content of herbage. 1. Observations on some herbage species. *J. Sci. Fd Agric.* **11**, 622–6.

GRIFFITH, G. AP; JONES, D. I. H.; WALTERS, R. J. K. 1965. Specific and varietal differences in sodium and potassium in grasses. *J. Sci. Fd Agric.* **16**, 94–8.

GRIFFITH, G. AP; WALTERS, R. J. K. 1966. The sodium and potassium content of some grass genera, species and varieties. *J. agric. Sci., Camb.* **67**, 81–9.

GRIFFITH, W. K.; TEEL, M. R.; PARKER, H. E. 1964. Influence of nitrogen and potassium on the yield and chemical composition of orchardgrass. *Agron. J.* **56**, 473–5.

HART, M. L. 'T. 1967. [The effect of climate, manuring, age and genetic origin on the chemical composition of grass.] *Z. Acker- u. Pflbau* **125**, 47–56.

HEMINGWAY, R. G. 1962. Copper, molybdenum, manganese and iron contents of herbage as influenced by fertilizer treatments over a three-year period. *J. Br. Grassld Soc.* **17**, 182–7.

HEMINGWAY, R. G. 1963. Soil and herbage potassium levels in relation to yield. *J. Sci. Fd Agric.* **14**, 188–95.

HENKENS, C. H. 1965. Factors influencing the sodium content of meadow grass. *Neth. J. agric. Sci.* **13**, 21–47.

HIRST, E. L.; RAMSTAD, S. 1957. Changes in organic acid content of perennial ryegrass during conservation. *J. Sci. Fd Agric.* **8**, 727–32.

HOLT, D. A.; HILST, A. R. 1969. Daily variation in carbohydrate content of selected forage crops. *Agron. J.* **61**, 239–42.

HUNT, I. V.; ALEXANDER, R. H. 1961. The crude protein content of Italian ryegrass. *J. Br. Grassld Soc.* **16**, 226–9.

HYLTON, L. O.; ULRICH, A; CORNELIUS, D. R.; OKHI, K. 1965. Phosphorus nutrition of Italian ryegrass relative to growth, moisture content and mineral constituents. *Agron. J.* **57**, 505–8.

JARRIGE, R.; MINSON, D. J. 1964. [Digestibility of the constituents of S24 perennial ryegrass and S37 cocksfoot, with special reference to the carbohydrates.] *Annls Zootech.* **13**, 117–50.

JOHNS, A. T. 1966. Some aspects of the ruminal digestion of herbage. *Proc. 9th int. Grassld Congr., São Paulo, 1965* pp. 843–8.

JOHNSON, J. M.; BUTLER, G. W. 1957. Iodine content of pasture plants. *Physiologia Pl.* **10,** 100–11.

JOHNSTON, A.; BEZEAU, L. M.; SMOLIAK, S. 1967. Variation in silica content of range grasses. *Can. J. Pl. Sci.* **47,** 65–71.

JOHNSTON, M. J.; WAITE, R. 1965. Studies in the lignification of grasses. 1. Perennial ryegrass (S24) and cocksfoot (S37). *J. agric. Sci., Camb.* **64,** 211–19.

JONES, D. I. H. 1962. Studies on the soluble carbohydrate content of grasses. *Rep. Welsh Pl. Breed. Stn 1961* pp. 157–64.

JONES, D. I. H. 1964. Structural carbohydrates in grasses. *Rep. Welsh Pl. Breed. Stn 1963* pp. 88–9.

JONES, D. I. H.; GRIFFITH, G. AP; WALTERS, R. J. K. 1961. The effect of nitrogen fertilizer on the water-soluble carbohydrate content of perennial ryegrass and cocksfoot. *J. Br. Grassld Soc.* **16,** 272–5.

JONES, D. I. H.; GRIFFITH, G. AP; WALTERS, R. J. K. 1965. The effect of nitrogen fertilizers on the water-soluble carbohydrate content of grasses. *J. agric. Sci., Camb.* **64,** 323–8

JONES, E. C.; BARNES, R. J. 1967. Non-volatile organic acids of grasses. *J. Sci. Fd Agric.* **18,** 321–4.

JONES, L. H. P.; HANDRECK, K. A. 1967. Silica in soils, plants and animals. *Adv. Agron.* **19,** 107–49.

KITTS, W. D.; SWIERSTRA, E.; BRINK, V. C.; WOOD, A. J. 1959. The estrogen-like substances in certain legumes and grasses. 2. The effect of stage of maturity and frequency of cutting on the estrogenic activity of some forages. *Can. J. Anim. Sci.* **39,** 158–63.

LANE, J. C.; FLEMING, G. A. 1966. Selenium, molybdenum and copper levels in perennial ryegrass. *Res. Rep. An Foras Talúntais Soils Division 1965* pp. 72–3.

LARGE, R. V.; SPEDDING, C. R. W. 1966. The health and performance of the grazing animal in relation to fertilizer nitrogen usage. 2. Weaned lambs. *J. agric. Sci., Camb.* **67,** 41–52.

LEFEBVRE, J.-M.; JAILLET, C.; DI COSTANZO, G.; MERIAUX, S. 1967. [The lipids of *Dactylis glomerata* leaves. Influence of N fertilization.] *C.r. hebd. Séanc. Acad. Sci., Paris (D)* **265,** 459–61. [Cited in *Herb. Abstr.* **38**: 394].

LOPER, G. M.; SMITH, D. 1961. Changes in micronutrient composition of the herbage of alfalfa, medium red clover, Ladino clover and bromegrass with advance in maturity. *Res. Rep. 8 Wisconsin agric. Exp. Stn* pp. 19.

MCDONALD, I. W. 1968. The nutrition of grazing ruminants. *Nutr. Abstr. Rev.* **38,** 381–400.

MCILROY, R. J. 1967. Carbohydrates of grassland herbage. *Herb. Abstr.* **37,** 79–87.

MACKENZIE, D. J.; WYLAM, C. B. 1957. Analytical studies on the carbohydrates of grasses and clovers. 7. Changes in carbohydrate composition during the growth of perennial ryegrass. *J. Sci. Fd Agric.* **8,** 38–45.

MACLEOD, L. B. 1965. Effect of nitrogen and potassium on the yield and chemical composition of alfalfa, bromegrass, orchardgrass and timothy grown as pure species. *Agron. J.* **57,** 261–6.

MELSTED, S. W.; MOTTO, H. L; PECK, T. R. 1969. Critical plant nutrient composition values useful in interpreting plant analysis data. *Agron. J.* **61,** 17–20.

MELVIN, J. F. 1965. Changes in the non-volatile organic acids of ryegrass during air-drying. *J. Sci. Fd Agric.* **16,** 612–4.

MILES, D. G.; GRIFFITH, G. AP; WALTERS, R. J. K. 1964. Variation in the chemical composition of four grasses. *Rep. Welsh Pl. Breed. Stn 1963* pp. 110–14.

MITCHELL, R. L. 1960. Contamination problems in soil and plant analysis. *J. Sci. Fd Agric.* **11,** 553–60.

MITCHELL, R. L.; REITH, J. W. S. 1966. The lead content of pasture herbage. *J. Sci. Fd Agric.* **17,** 437–40.

MOIR, R. J. 1966. Personal communication. [Cited by Allaway, W. H.; Thompson, J. F. 1966. Sulfur in the nutrition of plants and animals. *Soil Sci.* **101,** 240–7.]

MOLLOY, L. F. 1969. Determination of *trans*-aconitic acid in forage grasses. *J. Sci. Fd Agric.* **20,** 238–41.

MOON, F. E. 1939a. The carotene content of some grass and clover species with a note on pasture weeds. *Emp. J. exp. Agric.* **7,** 235–43.

MOON, F. E. 1939b. The influence of manurial treatment on the carotene content of poor pasture grass and on the relationship of this constituent to the ash and organic fractions. *J. agric. Sci., Camb.* **29,** 524–43.

MORAN, E. A. 1954. Cyanogenetic compounds in plants and their significance in animal industry. *Am. J. Vet. Res.* **15,** 171–6.

MURPHY, L. S.; SMITH, G. E. 1967. Nitrate accumulation in forage crops. *Agron. J.* **59,** 171–4.

NOWAKOWSKI, T. Z. 1962. Effects of nitrogen fertilizers on total nitrogen, soluble nitrogen and soluble carbohydrate contents of grass. *J. agric. Sci., Camb.* **59,** 387–92.

NOWAKOWSKI, T. Z. 1964. Mineral fertilization and organic composition of herbage. *Proc. 2nd Reg. Conf. int. Potash Inst. Morat, Switzerland, 1964* pp. 63–73.

NOWAKOWSKI, T. Z.; CUNNINGHAM, R. K. 1966. Nitrogen fractions and soluble carbohydrates in Italian ryegrass. 2. Effects of light intensity, form and level of nitrogen. *J. Sci. Fd Agric.* **17,** 145–50.

NOWAKOWSKI, T. Z.; CUNNINGHAM, R. K.; NIELSEN, K. F. 1965. Nitrogen fractions and soluble carbohydrates in Italian ryegrass. 1. Effects of soil temperature, form and level of nitrogen. *J. Sci. Fd Agric.* **16,** 124–34.

OKSANEN, H. E.; THAFVELIN, B. 1965. Fatty acid composition of hay. *J. Dairy Sci.* **48,** 1305–9.

ORR, J. B. 1929. Minerals in pastures and their relation to animal nutrition. London: H. F. Lewis and Co., pp. 165.

O'SULLIVAN, M. 1968. Grass tetany—effects of environment and fertilisers on pasture histamine levels. *J. Sci. Fd Agric.* **19,** 12–15.

O'TOOLE, M. A. 1966. Differences in susceptibility to iron chlorosis of grass species grown on blanket peat. *Nature, Lond.* **212,** 213.

PHILLIPS, T. G.; SULLIVAN, J. T.; LOUGHLIN, M. E.; SPRAGUE, V. G. 1954. Chemical composition of some forage grasses. 1. Changes with plant maturity. *Agron. J.* **46,** 361–9.

POPE, G. S.; MCNAUGHTON, M. J.; JONES, H. E. H. 1959. Oestrogens in British pasture plants. *J. Dairy Res.* **26,** 196–202.

REID, D. 1966. The response of herbage yields and quality to a wide range of nitrogen application rates. *Proc. 10th int. Grassld Congr., Helsinki, 1966* pp. 209–13.

REID, R. L.; JUNG, G. A.; MURRAY, S. J. 1966. Nitrogen fertilization in relation to the palatability and nutritive value of orchardgrass. *J. Anim. Sci.* **25,** 636–45.

REITH, J. W. S. 1963. The magnesium contents of soils and crops. *J. Sci. Fd Agric.* **14,** 417–26.

REITH, J. W. S. 1967. Effects of soil magnesium levels and of magnesium dressings on crop yield and composition. *Tech. Bull. 14 Min. Agric. Fish. Fd* London: H.M.S.O., pp. 97–109.

REITH, J. W. S.; MITCHELL, R. L. 1964. The effect of soil treatment on trace element uptake by plants. In: Bould, C.; Prevot, P.; Magness, J. R. [Eds] Plant analysis and fertilizer problems. Vol. 4. E. Lansing, Michigan: Amer. Soc. Hort. Sci., pp. 241–54.

RHYKERD, C. L.; DILLON, J. E.; NOLLER, C. H.; BURNS, J. C. 1966. The influence of nitrogen fertilization and drying method on yield and chemical composition of *Dactylis glomerata, Bromus inermis* and *Phleum pratense*. *Proc. 10th int. Grassld Congr., Helsinki, 1966* pp. 214–18.

ROOK, J. A. F.; WOOD, M. 1960. Mineral composition of herbage in relation to the development of hypomagnesaemia in grazing cattle. *J. Sci. Fd Agric.* **11,** 137–43.

RUSSELL, F. C.; DUNCAN, D. L. 1956. Minerals in pasture; deficiencies and excesses in relation to animal health. *Tech. Commun. 15, Commonw. Bur. Anim. Nutr.* 2nd edn, pp. 170.

SAALBACH, E.; JUDEL, G. K. 1966. [Investigations on the effect of sulphur fertilization on the yield of fodder plants.] *Agrochimica* **10,** 114–25. [Cited in *Herb. Abstr.* **38:** 405].

SHORLAND, F. B. 1961. Acetone-soluble lipids of grasses and other forage plants. 2. General observations on the properties of the lipids with special reference to the yield of fatty acids. *J. Sci. Fd Agric.* **12,** 39–43.

SMITH, A. M.; WANG, T. 1941. The carotene content of certain species of grassland herbage. *J. agric. Sci., Camb.* **31,** 370–8.

SOEST, P. J. VAN; JONES, L. H. P. 1968. The effect of silica in forages upon digestibility. *J. Dairy Sci.* **51,** 1644.

STEWART, J. W. B.; MCCONAGHY, S. 1963. Some effects of reaction (pH) changes in a basaltic soil on the mineral composition of growing crops. *J. Sci. Fd Agric.* **14,** 613–21.

STOUT, P. R.; BROWNELL, J.; BURAU, R. G., 1967. Occurrence of *trans*-aconitate in range forage species. *Agron. J.* **59,** 21–4.

SYNGE, R. L. M. 1951. Non-protein nitrogenous constituents of ryegrass: ionophoretic fractionation and isolation of a "bound amino acid" fraction. *Biochem. J.* **49,** 642.

THOMAS, B.; THOMPSON, A. 1948. The ash constituents of some grasses and herbs on the Palace Lea plots at Cockle Park. *Emp. J. exp. Agric.* **16,** 221–30.

THOMAS, B.; THOMPSON, A.; OYENUGA, V. A.; ARMSTRONG, R. H. 1952. The ash constituents of some herbage plants at different stages of maturity. *Emp. J. exp. Agric.* **20,** 10–22.

TUIL, H. D. W. VAN. 1965. Organic salts in plants in relation to nutrition and growth. *Agric. Res. Reps* [*Versl. landbouwk. Onderz*] No. 657, pp. 83.

WAITE, R. 1963. Botanical and chemical changes in maturing grass and their effects on its digestibility. *Agric. Prog.* **38,** 50–6.

WAITE, R. 1965. The chemical composition of grasses in relation to agronomical practice. *Proc. Nutr. Soc.* **24,** 38–46.

WAITE, R.; BOYD, J. 1953a. The water-soluble carbohydrates of grasses. 1. Changes occurring during the normal life cycle. *J. Sci. Fd Agric.* **4,** 197–204.

WAITE, R.; BOYD, J. 1953b. The water-soluble carbohydrates of grasses. 2. Grasses cut at grazing height several times during the growing season. *J. Sci. Fd Agric.* **4,** 257–61.

WAITE, R.; JOHNSTON, M. J.; ARMSTRONG, D. G. 1964. The evaluation of artificially dried grass as a source of energy for sheep. 1. The effect of stage of maturity on the apparent digestibility of ryegrass, cocksfoot, and timothy. *J. agric. Sci., Camb.* **62,** 391–8.

WHITEHEAD, D. C. 1966a. Nutrient minerals in grassland herbage. *Review Series 1/1966 Commonw. Bur. Past. Fld Crops* Farnham Royal: Commonw. Agric. Bur., pp. 86.

WHITEHEAD, D. C. 1966b. Data on the mineral composition of grassland herbage from the Grassland Research Institute, Hurley, and the Welsh Plant Breeding Station, Aberystwyth. *Tech. Rep. 4 Grassld Res. Inst.* pp. 55.

WIDDOWSON, F. V.; PENNY, A.; WILLIAMS, R. J. B. 1965. An experiment measuring effects of N, P and K fertilizers on yield and N, P and K contents of grass. *J. agric. Sci., Camb.* **64,** 93–9.

WILMAN, D. 1965. The effect of nitrogenous fertilizer on the rate of growth of Italian ryegrass. *J. Br. Grassld Soc.* **20,** 248–54.

WILSON, R. F.; TILLEY, J. M. A. 1965. Amino acid composition of lucerne and of lucerne and grass protein preparations. *J. Sci. Fd Agric.* **16,** 173–8.

WIT, C. T. DE; DIJKSHOORN, W.; NOGGLE, J. C. 1963. Ionic balance and growth of plants. *Versl. landbouwk. Onderz.* 69. 15, pp. 69.

CHAPTER 14

Response to Management and Environment

Response to fertilizer

THE main factors which are liable to restrict crop growth are soil and air temperature, incident solar radiation, depth and structure of soil, soil moisture, nutrients and plant diseases. In practice, the supply of nutrients from the soil may often be inadequate for growth to proceed within the limits set by the other factors mentioned and it has become common practice to add nutrients in the form of fertilizers. The major nutrients commonly applied are Ca (as lime, etc.), N, P, K and, to a lesser extent, Mg. In particular circumstances, minor nutrients known to be in short supply in the soil and thus limiting to growth have been added with spectacular results, as for example Mo on pastures in South Australia (Anderson, 1942).

One of the earliest hypotheses suggested for the relationship between growth and a factor affecting growth was the Law of the Minimum (Liebig, 1855). This implied that the amount of growth or yield is controlled by the factor present in minimum quantity. As the supply of this factor is increased, so also is yield. Eventually, a point is reached where no further increase in growth or yield occurs; there may be a toxic accumulation of the factor and growth may decrease again. Although this hypothesis has been of considerable value, the law is not valid in all circumstances because of interaction between factors and because the absolute limitation of yield by a single factor is not always clearly demonstrable.

Numerous data have been collected from field experiments with a range of crops in which all growth factors except for one nutrient element have been adequately provided, and in which growth (or crop yield) has been measured and related to an increasing supply of the missing nutrient. The resultant response curve has usually been sigmoid or, where the treatments have involved extremely low levels of the nutrient under study, exponential-sigmoid.

Mitscherlich (1905) expressed the form of the curves mathematically. He assumed that decrement from a maximum yield where all factors were adequately supplied was directly related to the inadequacy of supply of a particular nutrient. He also considered that the increase in yield produced by a measured increase in supply of the nutrient concerned is proportional to the decrement from the maximum. The relationship can be expressed as:

$$y = A(1-e^{-cx})$$

where y is yield obtained when x is the amount of factor present (and $y = 0$ when $x = 0$), A is the maximum yield obtained when the nutrient concerned is present in excess and c is a constant for that nutrient and crop. This equation is useful but not perfect. For instance, the curve obtained has been sigmoid in some instances and in others has sometimes shown a decline where excessive quantities of the nutrient being studied have been given.

Crowther and Yates (1941) used an equation of the exponential form to examine results from a great many experiments on a variety of crops in the UK. They developed recommendations for optimum fertilizer usage from the response curves obtained by using an equation of the form:

$$y = y_0 + d(1-10^{-kx})$$

where y is the yield and x the amount of nutrient given as fertilizer, y_0 is the yield without fertilizer, d is the limiting response and k a constant for the particular nutrient.

There have been problems in the general application of response curves even with annual crops. For example, it has been difficult to predict the yield which might be expected from a control crop given no fertilizer from measurement of a given soil parameter. Also, it has been difficult to predict the degree of inadequacy of the supply of a nutrient by examining the concentration of that nutrient in the plant tissues.

The difficulties of determining the level of nutrients required to maintain a given level of production are generally greater with grasses than with arable crops. A succession of harvests is involved, each taken at a different time of the year, each differing in morphological structure, and each likely to influence subsequent growth. The combinations of possible managements of the sward in terms of the time and frequency at which crops are harvested are great and there is the added complication that utilization by grazing does not involve the complete removal of plant nutrients in the harvested parts of the crop because some nutrients are returned to the soil in excreta.

Much of the experimental evidence concerning the response of grassland to the application of fertilizers has been collected from swards which consisted of a mixture of various grass species, often with leguminous forage plants and various herbs and weed species. The response of herbage dry matter (DM) yield to the application of a nutrient may be dependent on the type of plant involved, as for instance the response of grass/clover swards to N (see Chapter 32).

There are relatively few data showing the response of single-species swards to a particular fertilizer nutrient. Also, because there are differences among grass species and cultivars in their normal growth cycle, exact comparisons may be difficult; thus it has been demonstrated that the growth, and hence the yield, of a given cultivar of grass is influenced by a number of variables, particularly the timing of harvests (Green & Eyles, 1960; Cowling, 1966). Where different species or cultivars of grass are harvested in the same way, the harvesting system interacts with their growth rhythm and with weather conditions and may obscure the response to fertilizer application.

RESPONSE TO LIME

The grass species used in agriculture tolerate a fairly wide range in soil pH, if adequate nutrients are also available. However, the application of lime to extremely acid soils to raise pH increases the availability of some nutrients, e.g. P and Mo, and decreases the availability of others, e.g. Mn, Fe and Co. Arrhenius (1926) found that timothy tolerated a range of pH from 4·7 to 8·5, though its yield showed a sharp optimum at 4·8. Small (1946) suggests that both cocksfoot and meadow fescue prefer the pH range 5·5–7·9. McNeur (1953) found little difference in yield of clones of perennial ryegrass, Italian ryegrass and cocksfoot grown in a soil modified to give a pH range of 3·6–7·2; the optimum pH was 5·4.

The Rothamsted Park Grass experiment (Warren & Johnston, 1964), in which a mixed permanent grassland sward received various combinations of nutrients with or without lime, showed a general increase in yield with lime mainly because of an increase in legume content. In plots receiving N, in which legumes were suppressed, there appeared to be no response to lime, although without it the soil was very acid, pH 3·7–3·8.

RESPONSE TO NITROGEN

Most grass species respond well to N, and although Bradshaw *et al.* (1964) have demonstrated differences among various cultivated and non-cultivated species, the evidence from a number of field trials indicates that the differences among commonly cultivated species are small and inconsistent.

Response to N is subject to considerable variation because of different harvesting frequencies and methods, the timing of N application and the form of N given, in addition to the many variations due to soil and weather differences. Generally, such factors as timing of cuts and timing of application of N have a greater effect on yield response to N than have the differences between grass species.

It has not been possible to relate the response of grasses to N to a measured soil-N parameter; the rate of mineralization of soil N is more important than the total soil N supply, but is influenced considerably by soil disturbance. On the other hand, van Burg (1962) has shown that the internal concentration of nitrate in the DM of grasses provides a useful measure of the adequacy of supply of N when this is applied as fertilizer in the nitrate form. There were no increases in yield of DM when the herbage contained between 700 and 1400 ppm. nitrate-N. Unpublished results from the Grassland Research Institute tend to support van Burg's findings and suggest that in many situations 90% of the maximum yield is obtained by the use of 500 kg N/ha per year.

RESPONSE TO PHOSPHORUS

Bradshaw *et al.* (1960), working with sand cultures, found that perennial ryegrass gave a greater response to P than did a range of non-cultivated grasses. In both Scotland and Northern Ireland, experiments have shown no yield response of grasses to P even where a legume has been present in the sward (Reith *et al.*, 1961; Chestnutt, 1965).

There is no evidence from experiments in the United Kingdom regarding the 'critical level'* of P in grasses. Knauer (1966), working in Germany, suggested limiting values (critical levels) for the content of P in herbage which varied with the content of N. Where grass contained 2·4–3·4% N, responses to P might be expected if the content of P was less than 0·35%. This value is confirmed by American work with grasses, including tall fescue (Lunt *et al.*, 1965).

RESPONSE TO POTASSIUM

Responses to K in terms of grass DM yield are generally smaller and less frequently found than are responses to N, although results from Scotland (Reith *et al.*, 1961) demonstrated large responses to K in mixed grass swards given high rates of N and harvested by cutting. The indications are that responses to K may be expected if the concentration of K in the DM is less than 1·8–2·0% (McNaught, 1958; Clement, 1967).

Response to irrigation

In general, grasses show a measurable response to irrigation in all parts of Britain in some years, and in some parts of Britain in all years. Whether the response is sufficient to outweigh the cost of providing irrigation depends on the frequency and severity of drought, the nature of the water supply (i.e. whether expensive source works such as deep boreholes or large reservoirs have to be provided), and also on such variables as cost of equipment and fuel, value of the crop and on managerial skill. In the face of so many variables, any statement as to where in Britain irrigation is likely to be useful must inevitably be based on a large element of conjecture.

Pearl *et al.* (1954) have prepared a map of the calculated frequency of

* For definition see Chapter 13, p. 115–16.

irrigation need, which shows a frequency exceeding 5 years in 10 for the area south-east of a line through Leeds and Exeter, and exceeding 9 years in 10 in the extreme south-east of England. However, to relate the calculated frequency of a need to the practical benefit of meeting that need requires knowledge of (a) the response of yield to irrigation, for which data satisfactory in quality, but only meagre in quantity, are available, and (b) the cost factors mentioned above, for which data are wholly inadequate.

The surveys of the cost of irrigation in the 1950s, summarized by the Natural Resources (Technical) Committee (1962), show an extraordinary range in each item of cost, e.g. depreciation and interest varied from 0·3 to 9·6 new pence per m^3. This wide variation is probably due to the fact that irrigation practice was in the pioneering stage at the time of the surveys, but whatever the cause, the surveys are inadequate as a guide to present-day costs. More recently, North (1964) quoted an example in which the total costs of a carefully designed system with no source works was 1·25p/m^3. It ought to be possible to take careful designs for granted in future installations, and in the face of rising prices, a cost range of 1·2–2p/m^3 is likely to remain valid for some years to come.

The introduction of the concept of soil moisture deficit (i.e. the quantity of water, expressed in mm, required to bring a soil profile to field capacity), together with methods for calculating it (Penman 1948, 1963), has rendered unnecessary the repetition of experiments in every part of the country in order to give information relating to each local range of weather conditions. The response of a crop to a particular pattern of moisture deficit will be nearly the same in any part of the country, and the range experienced in a few years at one site is comparable with that experienced across the country in a given year. The calculated soil moisture deficit is also of great value in planning and managing practical irrigation, enabling wastage of water to be minimized, while avoiding exposure of the crop to drought.

The various experiments that have been carried out on the response of grass growth to irrigation (e.g. Low & Armitage, 1959; Munro, 1958; Penman, 1962; Reid & Castle, 1965; Stiles & Williams, 1965) have shown qualitative agreement. In general, the largest yields have been produced by regimes that kept the soil moisture closest to field capacity. The few exceptions were the small decreases in yield produced by irrigation in wet seasons, an effect probably attributable to loss of nutrients from light soils by leaching when rain following irrigation produced loss of soil water in the form of drainage. There were no marked differences in response between grass species, though a tendency for white clover to be more drought-sensitive than grasses often led to greater responses to irrigation from grass/clover swards than from pure grass ones. Responses to irrigation and to fertilizer N were very nearly additive. Both positive and negative interactions occurred during short periods, but became small or negligible when averaged over a whole growing season.

Fig. 14.1 illustrates the range of response to irrigation obtained in experiments in which soil moisture deficit was maintained at less than 25 mm. The points, taken from a variety of sources, show a broad scatter which is the combined effect of several causes, among which may be mentioned botanical and nutritional differences, differences in the irrigation treatments imposed, stage of growth at cutting and errors in determining moisture deficit and quantity of irrigation water. An additional cause, introduced in the preparation of the figure, is the use of the maximum potential deficit as a measure of the dryness of the season, and hence as a basis for comparison between seasons. The maximum deficit is the best simple index of the dryness of the season, but is imperfect since the curves showing the change of deficit with time vary widely in shape as well as in height.

The efficiency with which irrigation water is used in the production of DM (Fig. 14.2) is subject to the same sources of variability as mentioned above; in addition, there is the random occurrence of rain in relation to times of irrigation. It is probably true, as Fig. 14.2 suggests, that in wet seasons (maximum deficit under 100 mm) the efficiency of irrigation is nearly always low. Since the quantity of irrigation used in wet seasons is small, this does not accurately express the mean efficiency obtained over several years in the drier areas, but it is a discouragement to the use of irrigation in the wetter parts of the country.

When larger, less frequent, irrigations are used, production is reduced, but in the normal unsettled weather of Britain the total irrigation requirement is also reduced, so that efficiency of water use is little changed. The practical maximum deficit is about 50 mm, beyond which, at times of severe drought, serious loss of production occurs. Penman's results (1962) indicate that the maximum acceptable deficit within the range 25–50 mm increases as the quantity of applied fertilizer N is increased. This does not imply that the quantity of available water in the soil profile is changed by the nutrient status, since Garwood and Williams (1967) have shown that the capacity of soils to hold water available for transpiration is much greater than the soil moisture deficit at which active growth ceases. There is no indication that these quantities of water are related.

When irrigation is managed so as to allow the soil profile to dry slowly by applying at each irrigation somewhat less water than has been transpired since the previous watering in order to economize on water supplies, production is less than that of swards watered to field capacity, but the efficiency of water use tends to be greater. More experiments are required to elucidate the full implications of this, but it is not to be expected that there is a universal optimum regime. The optimum at any time and place will depend on the relative costs of land, water and labour, and the value of production. When the first application of irrigation is delayed until the soil water is so depleted that growth is about to cease, and irrigation is then applied in small quantities so as to satisfy only the current demand for water, the total demand for

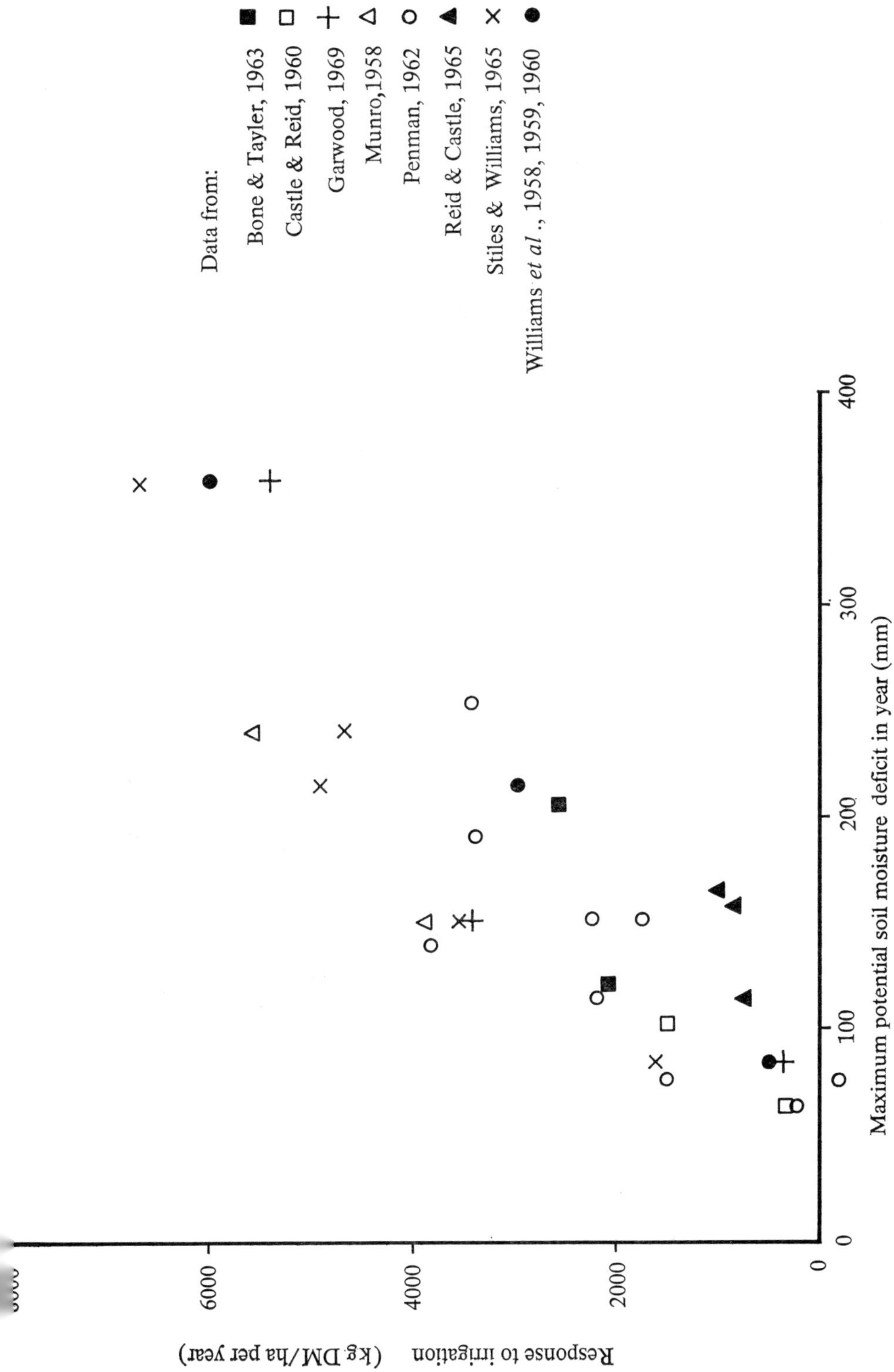

Fig. 14.1. Response of grass to irrigation according to dryness of summer.

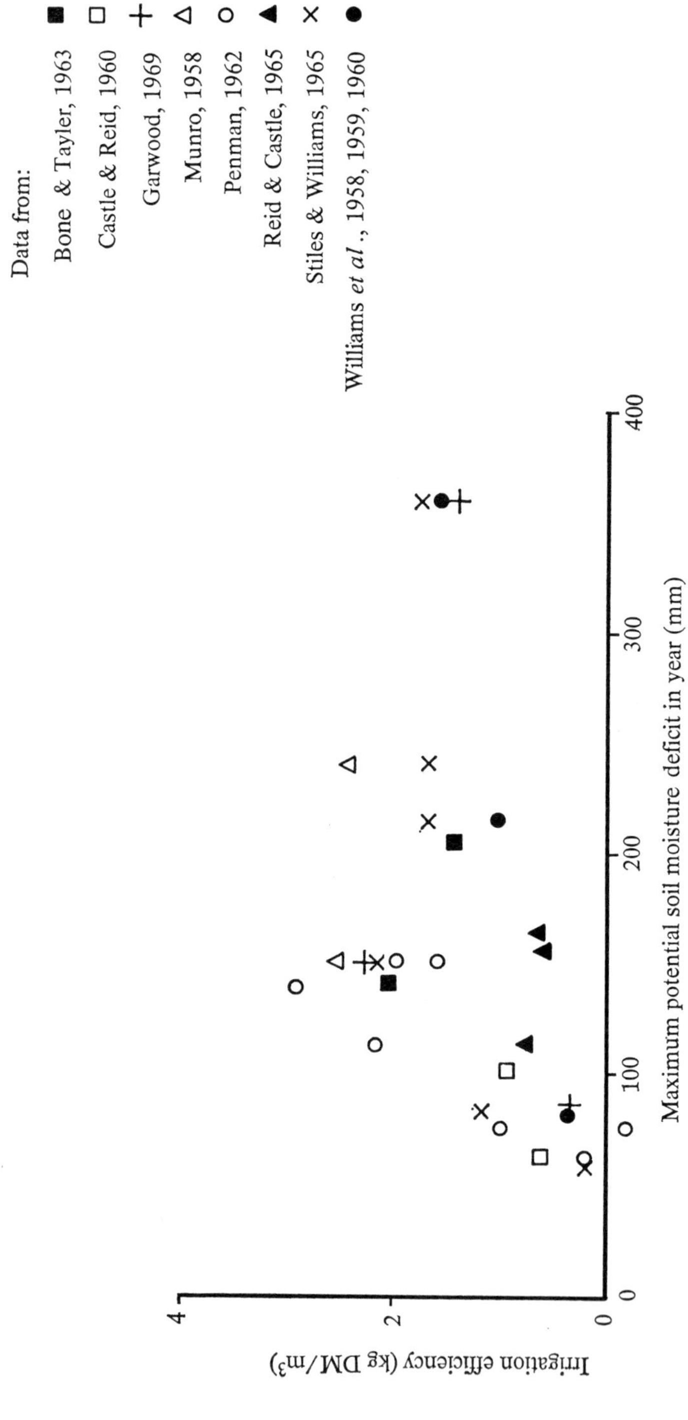

Fig. 14.2. Efficiency of use of irrigation water.

irrigation water is minimized. Nevertheless this practice is to be avoided, as growth is so reduced that both efficiency and total production are reduced compared with swards watered to field capacity.

Response to temperature

There is little direct experimental evidence of the effect of temperature on the yield of the grass crop. Circumstantial evidence is available from controlled-environment work on spaced plants, and from attempts to correlate fluctuations in the external environment with yield by the use of multiple regressions, although, as Morley (1961) points out, such correlations are often suspect. It is thus not possible to estimate accurately the DM yield that a given grass will produce at a particular temperature, and still less can this be said of digestible organic matter, protein or energy. However, it is possible to indicate the general response of grass to temperature, usually in terms of DM yield, and to cite optimum temperatures where these have been determined.

During winter, growth is restricted more by low temperature than by any other factor of the environment. Soil temperature is particularly important, since it is at or near the soil surface that much of the meristematic tissue of the vegetative plant is concentrated. Blackman (1936) suggested 5·5°C (42°F) as the limiting soil temperature for grass growth. Calder and Davies (1965) found that the growth of simulated swards of cocksfoot and ryegrass continued at soil temperatures below 5·5°C, but they did note a marked increase in the erectness of the sward (perhaps associated with the onset of flowering), and hence an increase in harvestable DM, when spring temperatures were in the region of 5·5°C. Although many north-temperate grasses can make some growth at lower temperatures, a soil temperature of 5·5°C may be a useful agronomic guide to the onset of spring growth.

During summer, temperature continues to influence growth, although some correlation analyses have suggested that above about 10°C, light is a more important determinant of yield. Controlled-environment studies show an optimum ambient temperature for growth usually between 20° and 30°C (Evans *et al.*, 1964); this optimum increases with increasing light intensity. Such studies have almost always been carried out on small, spaced plants, which tend to be more sensitive to temperature than the larger plants found in swards. Moreover, Mitchell (1956) points out that for several grasses, increased growth in response to temperature comes largely from increased tillering; the rate of growth of established tillers is less affected in the range of about 15° to 30°C. Extrapolating from controlled-environment studies to the sward, it might be expected that when plants are large and swards are dense, so that competition for light confines most of what growth there is to already established tillers, temperature may have only a small effect on yield, particularly if water and/or nutrients are in short supply, and shoot : root ratios are therefore low (Alberda, 1966). When plants are small and the sward is open, such as in the establishment phase or following a severe defoliation, tillering is not restricted by competition for light; in this situation temperature may be a more important determinant of yield than in established swards, since tiller bases will be directly exposed to radiation, and temperature fluctuations greatly in excess of those experienced at the base of a dense sward will result (Mitchell & Bieleski, 1964).

A final assessment of the importance of temperature in determining crop yield must await growth-room studies on simulated swards, or more precise control of environmental factors in the field.

Response to defoliation

The effect of defoliation on a grass sward may be considered as the response to three factors: (a) frequency of defoliation (number of defoliations during the season and the time intervals between successive defoliations), (b) severity of defoliation (height above ground level at which the crop is defoliated), and (c) the pattern of defoliation. Uniform cutting heights have been studied in numerous cutting experiments but the results will not necessarily apply to grazing situations. There is little information on the effect of heterogeneous cutting heights or of the pattern of defoliation within a sward. However, if a comparison between cutting and grazing on a mixed sward, with return of dung and urine in both instances, can be accepted as a guide, large changes in botanical composition may occur without much effect on DM yield (Miles, 1960); consequential changes in crude protein (CP) content and digestibility were not assessed in this experiment.

The largest total annual yields are obtained from infrequent cutting, and it is accepted that a hay and aftermath system gives the highest yields of DM (National Institute of Agricultural Botany, 1966). There should be a period of uninterrupted growth from April at the latest until about 10 days after inflorescence emergence before the first cut is taken, after which the aftermath can be cut at intervals of about 8 weeks. The rapid decrease in digestibility at about the time of inflorescence emergence (Minson *et al.*, 1964) may necessitate slightly earlier harvesting to obtain herbage of acceptable digestibility even though the absolute yield of digestible organic matter is still increasing (Green, 1966). DM yield is only slightly reduced by this earlier cutting.

As the frequency of defoliation is increased, the annual yield of DM generally decreases, so that the proportion of the potential annual yield which is produced is determined by the management imposed. Average figures for 6 species (excluding Westerwolds and hybrid ryegrass) showed that harvesting a grass crop on 9 as opposed to 4 occasions during the season caused a 50% reduction in yield (National Institute of Agricultural Botany, 1966). This effect of defoliation frequency is influenced by the level of applied N, the yield reduction apparently becoming less as the level of N increases (e.g. Cowling & Lockyer, 1963).

High growth rates are associated with stem elongation (Silsbury, 1965) and result in a major proportion of the total annual yield being produced during a short period of time (Anslow & Green, 1967). Should defoliation remove the elongating stem and its apex, thus preventing further growth, the new tillers produced will not exhibit the same high growth rate and serious loss in yield will result. This accounts for much of the loss in yield resulting from frequent defoliation, but there is apparently no published information to show whether frequency of defoliation has such a large effect on yield in the latter part of the growing season.

The height above ground level at which the grass crop is harvested has large effects, particularly on the yield of DM harvested. Close cutting (to 2·5 cm) increases the yield of DM when compared with a more lenient height of cutting (5-0, 7·5 or 10·0 cm), without reducing the subsequent regrowth (Blood, 1963). However, defoliation at ground level reduces yields compared with defoliation at 5·9 cm (Stapledon, 1924) or 15·0 cm (Stapledon & Milton, 1930), and higher yields have been obtained by defoliating at 2·5 cm than at 1·25 cm or at ground level (Roberts & Hunt, 1936). Lazenby and Rogers (1965) obtained greater yields when defoliating to a tiller length of 5·0 cm than to 2·5 cm, but a tiller length of only 2·5 cm represents a closer cut than 2·5 cm above ground level, since vegetative tillers do not grow in an upright manner.

A wide variety of defoliation frequencies has been used, from fortnightly defoliations (12 per season) to cutting for conservation (three cuts per season) (Reid, 1966); close cutting (2·5 cm) always gave higher yields than lenient cutting (6·25 cm). There was an interaction between frequency and severity of defoliation, and swards cut very frequently (at 14-day intervals) produced an even greater increase in yield when defoliated to 2·5 cm than swards defoliated less frequently (Reid, 1966). This in no way affects the overall loss in yield as the frequency of cutting increases.

Much of the work has been carried out on pure swards of perennial ryegrass (Chestnutt, 1960) or mixed swards of perennial ryegrass and white clover (Frame, 1965; Reid, 1966). In the latter instances it was mainly the grass component of the sward that increased as a result of close cutting. However, S48 timothy and S37 cocksfoot responded in a similar manner to close cutting, even if less dramatically than S23 perennial ryegrass (Green *et al.*, 1960). Evidence from outside Britain (e.g. Wilson & McGuire, 1961) confirms that increased yields are obtained from cocksfoot as a result of a close cutting regime. Anslow (1966) found that in S24 perennial ryegrass, leaves remaining in the stubble after defoliation were of no assistance to subsequent regrowth, possibly because the older leaves which remain are of low photosynthetic efficiency (Jewiss & Woledge, 1967). There is, therefore, sufficient evidence available to suggest that the optimum height of cutting, regardless of the frequency of defoliation, is about 2·5 cm above ground level.

The digestibility of herbage harvested at frequent intervals has been found to be higher than that of herbage harvested less frequently. Average figures for 6 species (excluding Westerwolds and hybrid ryegrass) showed that harvesting a grass crop on 9 as opposed to 4 occasions during the season increased the digestibility of the harvested crop by 7% (National Institute of Agricultural Botany, 1966). Digestibility falls with increasing age and is usually greater in leaves than stems (Terry & Tilley, 1964), so that the increase in digestibility associated with frequent cutting results from younger and more leafy material being harvested; Walters (1966) has shown that, within a species, the digestibility of a grass crop can be predicted from its leaf content.

The proteins in the herbage of various grasses have been determined (e.g. Waite *et al.*, 1953) but the likely, and probably important, effects of cutting (Davidson & Milthorpe, 1966) on the nature of the protein present have not been investigated. Although CP as calculated from the total N content includes non-protein nitrogenous fractions (ap Griffith & Johnston, 1960), it is usually a satisfactory estimate (Minson & Brown, 1959). The content of CP in leaf material is higher than in stem material (Tilley & Deriaz, 1959) and is higher in young herbage than in more mature herbage, although this is possibly a result of the proportion of leaf lamina in the material (Minson *et al.*, 1960). The overall result is that the more frequently grass is harvested, the higher is its content of CP (Woodman *et al.*, 1931), though this is not unrelated to the fact that frequent harvesting may lower yield. However, the above effects are limited by, and operate within, the large effects of N fertilizer (e.g. Holmes, 1949) and seasonal variation in CP content (Chestnutt, 1960).

The calorific or gross energy value of grass is relatively constant, whether expressed on the basis of the DM or digestible organic matter (DOM) content of the herbage. A value of 4·4 kcal/g DOM can be accepted for a hay crop (Morgan & Jones, 1965) but a higher value might be expected from less mature herbage harvested at more frequent intervals, since Hunt (1967) found that young leaves had a higher gross energy value than either mature leaves or leaf sheaths. However, Armstrong (1964) found little difference in energy value between young leafy herbage and very mature herbage, but his value was rather higher, 4·92 kcal/g DOM (Armstrong *et al.*, 1964), than reported elsewhere in the literature. There is insufficient information at present to resolve this apparent conflict between the findings of Armstrong (1964) and Hunt (1967). The amount of energy utilized by the animal consuming the herbage (the metabolizable and net energy values) depends upon factors such as the level of intake; defoliation only affects the latter in so far as it is related to digestibility (Armstrong *et al.*, 1964).

There are also some effects of defoliation at critical times on the survival of grass plants. This is particularly so in the autumn. If too much herbage is left at this time of year, plants of perennial ryegrass and Italian ryegrass in particular may be completely killed in subsequent periods of hard frost (Baker & David, 1963). The short-duration ryegrasses appear to be more sensitive to winter damage than perennial ryegrass, and the latter is more sensitive than cocksfoot (except for some Mediterranean forms) and timothy.

For information on the autumn management of grasses, see Grassland Research Institute (1960, 1964), Baker (1957) and Baker, Chard and Jenkins (1961).

References

ALBERDA, T. 1966. Responses of grasses to temperature and light. In: Milthorpe, F. L.; Ivins, J. D. [Eds] The growth of cereals and grasses. Proceedings of the twelfth Easter School in Agricultural Science, University of Nottingham, 1965. London: Butterworth and Co., pp. 200–12.

ANDERSON, A. J. 1942. Molybdenum deficiency on a South Australian Ironstone soil. *J. Aust. Inst. agric. Sci.* **8,** 73–5.

ANSLOW, R. C. 1966. Light interception and growth rate of a perennial ryegrass sward. *Proc. 9th int. Grassld Congr., São Paulo, 1965* pp. 403–5.

ANSLOW, R. C.; GREEN, J. O. 1967. The seasonal growth of pasture grasses. *J. agric. Sci., Camb.* **68,** 109–22.

ARMSTRONG, D. G. 1964. Evaluation of artificially dried grass as a source of energy for sheep. 2. The energy value of cocksfoot, timothy and two strains of ryegrass at varying stages of maturity. *J. agric. Sci., Camb.* **62,** 399–416.

ARMSTRONG, D. G.; BLAXTER, K. L.; WAITE, R. 1964. Evaluation of artificially dried grass as a source of energy for sheep. 3. The prediction of nutritive value from chemical and biological measurements. *J. agric. Sci., Camb.* **62,** 417–24.

ARRHENIUS, O. 1926. Kalkfrage, Bodenreaktion und Pflanzenwachstum. Leipzig.

BAKER, H. K. 1957. Studies on the root development of herbage plants. 3. The influence of cutting treatments on the root, stubble and herbage production of a perennial ryegrass sward. *J. Br. Grassld Soc.* **12,** 197–208.

BAKER, H. K.; CHARD, J. R. A.; JENKINS, D. G. 1961. The production of early spring grass. *J. Br. Grassld Soc.* **16,** 146–53.

BAKER, H. K.; DAVID, G. L. 1963. Winter damage to grass. *Agriculture, Lond.* **70,** 380.

BLACKMAN, G. E. 1936. The influence of temperature and available nitrogen supply on the growth of pasture in the spring. *J. agric. Sci., Camb.* **26,** 620–47.

BLOOD, T. F. 1963. Effect of height of cutting on the subsequent regrowth of a sward. *N.A.A.S. q. Rev.* **14,** 139–43.

BONE, J. S.; TAYLER, R. S. 1963. The effect of irrigation and stocking rate on the output from a sward. 1. Methods and herbage results. *J. Br. Grassld Soc.* **18,** 190–6.

BRADSHAW, A. D.; CHADWICK, M. H.; JOWETT, D.; LODGE, R. W.; SNAYDON, R. W. 1960. Experimental investigation into the mineral nutrition of several grass species. 3. Phosphate level. *J. Ecol.* **48,** 631–7.

BRADSHAW, A. D.; CHADWICK, M. H.; JOWETT, D.; SNAYDON, R. W. 1964. Experimental investigations into the mineral nutrition of several grass species. 4. Nitrogen level. *J. Ecol.* **52,** 665–76.

BURG, P. F. J. VAN. 1962. Internal nitrogen balance, production of dry matter and ageing of herbage and grass. *Versl. landbouwk. Onderz.* No. 68. 12, pp. 131.

CALDER, D. M.; DAVIES, A. 1965. Spring growth of perennial ryegrass and cocksfoot. *J. Br. Grassld Soc.* **20,** 194–6.

CASTLE, M. E.; REID, D. 1960. Irrigation of grassland in south-west Scotland and its influence on the utilisation of fertilizer nitrogen. *Proc. 8th int. Grassld Congr., Reading, 1960* pp. 146–50.

CHESTNUTT, D. M. B. 1960. The effect of cutting treatment on the yield of dry matter and crude protein of eight perennial ryegrass varieties. *Res. exp. Rec. Minist. Agric. N. Ire.* **10,** 35–52.

CHESTNUTT, D. M. B. 1965. The effects of nitrogen, potash and phosphate on the yield and composition of a grass/clover sward. *Rec. agric. Res.* (*Nth. Ireld*) **14,** 71–81.
CLEMENT, C. R. 1967. Potassium for the grass crop, its supply from soil and fertilizers. *A. Rep. Grassld Res. Inst. Hurley, 1966* pp. 80–5.
COWLING, D. W. 1966. The effect of the early application of nitrogenous fertilizer and of the time of cutting in spring on the yield of rye-grass/white clover swards. *J. agric. Sci., Camb.* **66,** 413–31.
COWLING, D. W.; LOCKYER, D. R. 1963. The contribution to yield made by the white clover and fertilizer nitrogen at various cutting frequencies. *Exps Prog. 15 Grassld Res. Inst., 1961–2* pp. 17–18.
CROWTHER, E. M.; YATES, F. 1941. Fertilizer policy in war-time: the fertilizer requirements of arable crops. *Emp. J. exp. Agric.* **9,** 77–97.
DAVIDSON, J. L.; MILTHORPE, F. L. 1966. The effect of defoliation on the carbon balance in *Dactylis glomerata. Ann. Bot.* **30,** 185–98.
EVANS, L. T.; WARDLAW, I. F.; WILLIAMS, C. N. 1964. Environmental control of growth. In: Barnard, C. [Ed.] Grasses and grasslands. London: Macmillan & Co. Ltd, pp. 102–25.
FRAME, J. 1965. The effects of cutting and grazing techniques on herbage production. *Exp. Rec. 10 W. Scot. agric. Coll.* pp. 43.
GARWOOD, E. A. 1969. Seasonal tiller populations of grass and grass/clover swards with and without irrigation. *J. Br. Grassld Soc.* **24,** 333–44.
GARWOOD, E. A.; WILLIAMS, T. E. 1967. Soil water use and growth of a grass sward. *J. agric. Sci., Camb.* **68,** 281–92.
GRASSLAND RESEARCH INSTITUTE. 1960. *Exps Prog. 12 Grassld Res. Inst. 1958–9* p. 13.
GRASSLAND RESEARCH INSTITUTE. 1964. *Exps Prog. 16 Grassld Res. Inst. 1962–3* p. 15.
GREEN, J. O. 1966. Plant introduction and evaluation. *A. Rep. Grassld Res. Inst. Hurley, 1965* pp. 15–18.
GREEN, J. O.; DAVID, G. L.; EYLES, J. C.; ASTLE, W. R. 1960. Level and frequency of cutting: pure grass swards at two levels of nitrogen. *Exps Prog. 12 Grassld Res. Inst. 1958–9* pp. 19–21.
GREEN, J. O.; EYLES, J. C. 1960. A study in methods of grass variety testing. *J. Br. Grassld Soc.* **15,** 124–32.
GRIFFITH, G. AP; JOHNSTON, T. D. 1960. The nitrate nitrogen content of herbage. 1. Observations on some herbage species. *J. Sci. Fd Agric.* **11,** 622–9.
HOLMES, W. 1949. The intensive production of herbage for crop drying. 2. A study of the effect of massive dressings of nitrogenous fertilizer and of their time of application on the yield, chemical and botanical composition of two grass leys. *J. agric. Sci., Camb.* **39,** 128–41.
HUNT, L. A. 1967. Ash and energy content of seven forage grasses. *Crop Sci.* **6,** 507–9.
JEWISS, O. R.; WOLEDGE, J. 1967. The effect of age on the rate of apparent photosynthesis of leaves of tall fescue (*Festuca arundinacea* Schreb.) *Ann. Bot.* **31,** 661–71.
KNAUER, N. 1966. The use of plant analysis for determining the phosphorus and potassium needs of grassland. *Proc. 10th int. Grassld Congr., Helsinki, 1966* pp. 171–4.
LAZENBY, A.; ROGERS, H. H. 1965. Selection criteria in grass breeding. 6. Effects of defoliation on plants growing in small plots in field and controlled environment conditions. *J. agric. Sci., Camb.* **65,** 397–404.
LIEBIG, J. 1855. Die Grundsätze der Agricultur-chemie mit Rüchksicht auf die in England angestellten Untersuchungen. Braunschweig: Vieweg.
LOW, A. J.; ARMITAGE, E. R. 1959. Irrigation of grassland. *J. agric. Sci., Camb.* **52,** 256–62.
LUNT, O. R.; BRANSON, R. L.; CLARK, S. B. 1965. Response of five grass species to phosphorus on six soils. *Proc. 9th int. Grassld Congr., São Paulo, 1965* pp. 1697–701.
MCNAUGHT, K. J. 1958. Potassium deficiency in pastures. 1. Potassium content of legumes and grasses. *N.Z. Jl agric. Res.* **1,** 148–81.
MCNEUR, A. J. 1953. Effect of soil acidity on growth of pasture plants. *N.Z. Jl Sci. Technol. Sect. A* **35,** 311–26.
MILES, D. G. 1960. Herbage assessment techniques. *Rep. Welsh Pl. Breed. Stn 1959* pp. 43–5.

MINSON, D. J., BROWN, S. A. 1959. Herbage nitrogen as an index of herbage organic matter digestibility. *Exps Prog. 11 Grassld Res. Inst. 1957–8* pp. 99–103.

MINSON, D. J.; HARRIS, C. E.; RAYMOND, W. F.; MILFORD, R. 1964. The digestibility and voluntary intake of S22 and H1 ryegrass, S170 tall fescue, S48 timothy, S215 meadow fescue and Germinal cocksfoot. *J. Br. Grassld Soc.* **19**, 298–305.

MINSON, D. J.; RAYMOND, W. F.; HARRIS, C. E. 1960. Studies in the digestibility of herbage. 8. The digestibility of S37 cocksfoot, S23 ryegrass and S24 ryegrass. *J. Br. Grassld Soc.* **15**, 174–80.

MITCHELL, K. J. 1956. Growth of pasture species under controlled environment. 1. Growth at various levels of constant temperature. *N.Z. Jl Sci. Technol. Sect. A* **38**, 203–15.

MITCHELL, K. J.; BIELESKI, R. L. 1964. Note on the temperatures of leaf and meristematic tissue of plants of short-rotation ryegrass in summer conditions. *N.Z. Jl agric. Res.* **7**, 761–5.

MITSCHERLICH, E. A. 1905. Bodenkunde für Landwirte, Forstwirte und Gartner. Berlin: Paul Parey.

MORGAN, D. E.; JONES, M. G. S. 1965. The energy value of barn-dried hay. *J. Br. Grassld Soc.* **20**, 174–6.

MORLEY, F. H. W. 1961. Subterranean clover. *Adv. Agron.* **13**, 57–123.

MUNRO, L. A. 1958. Irrigation of grassland. *J. Br. Grassld Soc.* **13**, 213–21.

NATIONAL INSTITUTE OF AGRICULTURAL BOTANY. 1966. Varieties of grasses. *Fmrs' Leafl. 16, 1966.*

NATURAL RESOURCES (TECHNICAL) COMMITTEE. 1962. Irrigation in Great Britain. London: H.M.S.O., pp. 90.

NORTH, J. J. 1964. Irrigation of crops and grassland. *Agric. Prog.* **39**, 89–96.

PEARL, R. T. 1954. The calculation of irrigation need. *Tech. Bull. 4 Minist. Agric. Fish. Fd* London: H.M.S.O.

PENMAN, H. L. 1948. Natural evaporation from open water, bare soil, and grass. *Proc. R. Soc. (A)* **193**, 120–45.

PENMAN, H. L. 1962. Woburn irrigation 1951-59. 2. Results for grass. *J. agric. Sci., Camb.* **58**, 349–64.

PENMAN, H. L. 1963. Vegetation and hydrology. *Tech. Comm. 53 Commonw. Bur. Soils* Farnham Royal, Bucks: Commonw. Agric. Bur., pp. 124.

REID, D. 1966. Studies on cutting management of grass-clover swards. 4. The effects of close and lax cutting on the yield of herbage from swards cut at different frequencies. *J. agric. Sci., Camb.* **66**, 101–6.

REID, D.; CASTLE, M. E. 1965. The response of grass-clover and pure-grass leys to irrigation and fertilizer nitrogen treatment. 1. Irrigation effects. *J. agric. Sci., Camb.* **64**, 185–94.

REITH, J. W. S.; INKSON, R. H. E.; STEWART, A. B.; HOLMES, W.; MACLUSKY, D. S. *et al.* 1961. The effects of fertilizers on herbage production. 1. The effects of nitrogen, phosphate and potash on yield. *J. agric. Sci., Camb.* **56**, 17–30.

ROBERTS, R. A.; HUNT, I. V. 1936. The effect of shoot cutting on the growth of root and shoot of perennial ryegrass (*Lolium perenne* L.) and of timothy (*Phleum pratense* L.). *Welsh J. Agric.* **12**, 158–74.

SILSBURY, J. H., 1965. Interrelations in the growth and development of *Lolium*. 1. Some effects of vernalisation on growth and development. *Aust. J. agric. Res.* **16**, 903–13.

SMALL, J. 1964. pH and plants. London: Baillière, Tindall and Cox.

STAPLEDON, R. G. 1924. The seasonal productivity of herbage grasses, with particular reference to the influence of different systems of cutting on indigenous and non-indigenous strains, respectively. *Bull. H3 Welsh Pl. Breed. Stn* pp. 5–130.

STAPLEDON, R. G.; MILTON, W. E. J. 1930. The effect of different cutting and manurial treatments on the tiller and root development of cocksfoot. *Welsh J. Agric.* **6**, 166–74.

STILES, W.; WILLIAMS, T. E. 1965. The response of a ryegrass/white clover sward to various irrigation regimes. *J. agric. Sci., Camb.* **65**, 351–64.

TERRY, R. A.; TILLEY, J. M. A. 1964. The digestibility of the leaves and stems of perennial ryegrass, cocksfoot, timothy, tall fescue, lucerne and sainfoin as measured by an *in vitro* procedure. *J. Br. Grassld Soc.* **19**, 363–72.

TILLEY, J. M. A.; DERIAZ, R. E. 1959. Soluble constituents in herbage. *Exps Prog. 11 Grassld Res. Inst. 1957–8* p. 65.

WAITE, R.; FENSON, A.; LOVETT, S. 1953. The amino acid composition of extracted grass protein. 1. The basic acids. *J. Sci. Fd Agric.* **4,** 28–33.

WALTERS, R. J. K. 1966. Prediction of digestibility. *Rep. Welsh Pl. Breed. Stn 1965* pp. 70–2.

WARREN, R. G.; JOHNSTON, A. E. 1964. The Park Grass Experiment. *Rep. Rothamsted exp. Stn 1960* pp. 240–62.

WILLIAMS, T. E.; STILES, W.; TURNER, P. J. 1958. *Exps Prog. 10 Grassld Res. Inst. 1956–7* p. 54.

WILLIAMS, T. E.; STILES, W.; TURNER, P. J. 1959. *Exps Prog. 11 Grassld Res. Inst. 1957–8* p. 58.

WILLIAMS, T. E.; STILES, W.; TURNER, P. J. 1960. *Exps Prog. 12 Grassld Res. Inst. 1958–9* p. 57.

WILSON, D. B.; MCGUIRE, W. S. 1961. Effects of clipping and nitrogen on competition between three pasture species. *Can. J. Pl. Sci.* **41,** 631–42.

WOODMAN, H. E.; NORMAN, D. B.; FRENCH, M. H. 1931. Nutritive value of pasture. 7. The influence of the intensity of grazing on the yield, composition and nutritive value of pasture herbage. (Part III). *J. agric. Sci., Camb.* **21,** 267–321.

CHAPTER 15

Perennial ryegrass *(Lolium perenne)*

Morphology

PERENNIAL ryegrass (Plate 8) is a loosely- to densely-tufted perennial. Root system highly branched, with adventitious roots arising from the basal nodes of tillers. Vegetative stems more or less erect, slender, branching freely, intravaginally. Leaf sheaths somewhat flattened, glabrous, red at base, becoming fibrous, persistent. Ligule short, auricles small. Leaf blade glabrous, generally folded in the sheath abaxially, shiny green with prominent keel, prominently ribbed adaxially. Culms slender, 2 to 4 elongated internodes. Inflorescence a spike of alternate, sessile, awnless, 3- to 10-flowered spikelets, edgeways on to the axis; 2 glumes present in terminal spikelet; in others, the inner glume is missing. Caryopsis enclosed in hardened lemma and palea.

Garwood (1967c) studied the root distribution in the soil profile of the perennial ryegrass cultivars S23 and S24 (Table 15.1).

Table 15.1. Distribution of the root mass of S24 and S23 perennial ryegrass in the soil profile

Horizon in.	Horizon cm	S23 % of total root weight	S24 % of total root weight
0–2	0–5	66·5	69·5
2–4	5–10	11·7	13·1
4–6	10–15	6·9	7·1
6–12	15–30	7·3	7·2
12–18	30–45	5·7	1·8
18–24	45–60	1·5	0·9
24–30	60–75	0·4	0·4

After Garwood (1967c)

Vegetative stems are almost erect at first but become decumbent as the primary axes grow. Each stem has an apical region about 1 cm long producing leaves at successive nodes and tiller buds in each leaf axil. Internodes just below this apical region elongate, separating the bases of individual tillers by up to 2 cm. Vegetative stems can reach a length of 20 cm in exceptional circumstances, and under intensive grazing the rooting of these vegetative tillers produces a stoloniferous growth habit (Kydd, 1966). Ryle (1964), studying the influence of N and temperature, recorded that perennial ryegrass leaves were relatively narrow (4 mm) and up to about 180 mm long, increasing steadily in size from the first to the 7th leaf of a tiller. Each tiller, however, had no more than three leaves at any one time.

Anslow (Personal communication) recorded tiller weights of 2·0–2·9 g/1 000 tillers for S24 and 1·0–3·0 g/1 000 tillers for S23 at population densities of 77–103 and 77–223 tillers/dm^2, respectively, under conditions of irrigation and high N input over the whole year.

There are many ecotypes and cultivars of perennial ryegrass, including some artificially induced tetraploids (2n = 28). Morphological differences between cultivars are very difficult to quantify as there is great phenotypic variation within any genotype in response to environmental factors, e.g. fertilizer input and frequency and intensity of defoliation.

The date of flowering can be used to classify cultivars (National Institute of Agricultural Botany, 1965). Early-flowering types tend to be taller, more erect and less densely tillered than late-flowering types. Tetraploid cultivars tend to have larger leaves and larger seeds and to be less densely tillered than diploid types.

Changes in the proportion of leaf as the first crop develops have been studied by Terry and Tilley (1964), and are shown in Table 15.2.

Table 15.2. Proportion of leaf blade, sheath, stem and inflorescence in the DM of S24 perennial ryegrass during the growth of the first crop

	Percentage of dry weight		
Date	Leaf blade	Sheath	Stem + inflorescence + dead material
27 Apr.	70	25	5
11 May	40	30	30
19 May	31	16	53
29 May	18	12	70
11 June	11	9	80

Date of 50% ear emergence: 19 May
From Terry and Tilley (1964)

Green and Corrall (Unpublished) showed that, at the same stage of growth, S24 had a lower proportion of leaf than S23, falling to minima of 20 and 30%, respectively.

Distribution

Lolium perenne is a native species occurring in all vice-counties of Britain (Clapham *et al.*, 1962; Beddows, 1967) though rather less common in the north and west of Scotland (Perring & Walters, 1962). It occurs on roadsides and in waste land (Hubbard, 1954), where sown for fodder (Clapham *et al.*, 1962), in old pastures and meadows (Hubbard, 1954), and tolerates exposure to salt spray (Beddows, 1967).

It occurs naturally at nearly all altitudes in Britain, though less frequently above 750 m (Perring & Walters, 1962). If used agriculturally above 500 m it requires soil amelioration (including liming) and control of grazing (Beddows, 1967).

The distribution of perennial ryegrass bears no apparent relation to the summer isotherms (Perring & Walters, 1962), but the optimum temperature for leaf growth is 20°C (Beddows, 1967). Similarly, in England its distribution shows no relation to the January isotherms, but in Scotland it tends to be less frequent where the February mean minimum is less than 2°C (Perring & Walters, 1962). Winter cold is necessary before there is a photoperiodic induction of floral initials under long days (13 h or more). Mediterranean varieties do not need cold, and flowers are initiated under a 9- to 10-h day-length (Beddows, 1967). It is vulnerable to prolonged snow cover (Beddows, 1967) though this is not normally a problem in Britain; distribution can, however, be limited by adverse conditions, even if they occur only infrequently. In Britain, light or temperature are rarely limiting factors but winter mortality may be influenced by management; for example, excessive autumn growth causes the stem bases to become elevated, thus reducing the ability of the plants to withstand severe cold (Beddows, 1967).

The species occurs on most soil types, especially the rich, heavy soils of the lowlands (Hubbard, 1954) and distribution is not apparently related to the presence of chalk or limestone (Perring & Walters, 1962). It is sparser on upland soils where these are wet and of low pH (Perring & Walters, 1962). It is found on soils with pH values of 5·0 to 8·0, but its optimum pH range is 6·0–7·0 (Spurway, 1941).

Distribution in Britain is unrelated to isohyets or relative humidity (Perring & Walters, 1962), although Beddows (1967) considered a minimum annual rainfall of 900 mm to be desirable. Roots show vigorous growth in soils of high moisture content (Baumann & Klauss, 1955), but surface drainage must be good and it will not tolerate stagnant water (Hughes *et al.*, 1962). It can withstand temporary inundation by sea water and can be established in soil containing up to 0·5% of salt (Beddows, 1967).

Early-flowering types can produce flowers under a daylength of 6 h, whereas late-flowering types need more than 12 h (Tincker, 1925); thus day-length is not a limiting factor in Britain. Perennial ryegrass can thus be grown in most agricultural areas of Britain.

Compatibility

Perennial ryegrass is aggressive, develops very quickly from seed and, once established, is the most productive of the perennial grasses. Because of its rapid growth in the establishment phase, it will suppress any slower-growing grass or legume sown with it (Davies, 1928).

COMPATIBILITY WITH OTHER GRASSES

Several workers have reported suppression of other grasses by perennial ryegrass, e.g. Lampeter (1960) for timothy, meadow fescue and cocksfoot; Crocker and Martin (1964) for meadow fescue; Harkess (1966) for tall fescue. Anslow (1955) and Arens (1962) both noted that perennial ryegrass was more aggressive to cocksfoot in the seedling condition than when the grasses were fully established. The degree of competition offered to other grasses may be modified by the following factors:

Management

Frequent defoliation favours the increase of perennial ryegrass relative to other grasses such as cocksfoot. England (1965) compared the performance of cocksfoot and perennial ryegrass in pure swards with that in 50 : 50 mixtures. When the swards were kept vegetative, ryegrass was the more competitive species, but when they were allowed to "pass out of the vegetative phase" before the first cut, cocksfoot was competitively superior.

Cultivar

The more leafy and persistent cultivars are more competitive against other species and for a longer period than the less leafy and more short-lived types. Jones (1958) found that S23 was much more competitive than Northern Irish perennial ryegrass towards Danish cocksfoot and Italian ryegrass.

Manuring

Jones (1958) reported that when no fertilizer was added to a sown mixture of perennial ryegrass and *Agrostis tenuis* (New Zealand browntop) in experiments at Aberystwyth, the two grasses remained present in about equal proportions for three years. However, where lime and slag had been applied, the proportion of *Agrostis* was reduced and where a compound fertilizer was also applied, it was reduced to a point at which it comprised less than 5% of the herbage. In another trial, the same author reported that raising the level of N fertilizer application decreased the proportion of perennial ryegrass in a mixture with Italian ryegrass.

COMPATIBILITY WITH WHITE CLOVER

In confirmation of the results of earlier work at the Grassland Research Institute, Hurley (1952), Cowling and Lockyer (1965) found that the amounts of white clover which grew with perennial ryegrass and meadow fescue swards were similar and greater than the amounts which grew with cocksfoot swards. There were indications of an initial period of greater competition associated with the greater speed of establishment of perennial ryegrass than of meadow fescue. In this work the grasses were each represented by a bred cultivar from the Welsh Plant Breeding Station. Green and Corrall (1965) reported that a larger quantity of white clover had grown with Northern Irish commercial than with either S23 or S24 perennial ryegrass. In their work, which was under grazing conditions, parity of clover content had been obtained with a seed rate of 22 kg/ha for the Northern Irish material and only 5 or 6 kg/ha for S23 or S24. When the seed rate of S23 and S24 was reduced from 22 kg to 5 kg/ha, there was only a small effect on yield of ryegrass, but this was offset by the increase in white clover.

Seed

CHARACTERS

The seed of perennial ryegrass is ovate-oblong with a shallow groove along its ventral surface; seed length is 5–8 mm, and the average width at the mid-point is 1·0–1·5 mm (Armstrong, 1948; Hubbard, 1954; Peto, 1961). The lemma is blunt or slightly pointed at the apex, and is convex, awnless and smooth. The palea is similar in length to the lemma. The rachilla is compressed close to the palea, and is elliptical or rhomboidal in cross-section (Armstrong, 1948; Hubbard, 1954).

The 1 000-seed weight varies considerably between and within cultivars. In 11 varieties Hunt (1954) found a range of 1·328g (for Cornish Eaver 'A') to 2·299 g (for S24); generally, the 1 000-seed weight was high in hay and low in pasture varieties. Within varieties, Evans (1959) found a range of 1·56 to 1·78 g/1 000 seeds in S23 ryegrass, and Green (1965) reported a variation of 1·80 to 2·18 g in S24. Individual seeds with average weights of 1·49 mg (1·49 g/1 000) and 3·14 mg have been selected for experimental work in S24 (Woodford, 1966); in this cultivar weight per seed can vary from 1·0 to 4·0 mg (Arnott, 1969). The tetraploid cultivars recently introduced have larger seeds than the diploids, 1 000-seed weights quoted for tetraploid seed being 2·85 g (Gerdes, 1952) and 3·49 g (Twyford Seeds, 1964).

PRODUCTION

Sowing in spring is advised, but comparable yields may be obtained from sowings made up to the end of August (National Institute of Agricultural Botany, 1959; Sonneveld & Evers, 1954); cover crops reduce yields (Evans, 1937; Roberts, 1958). Broadcasting, or sowing in drill rows up to 35 cm apart, have been recommended, in each case with a seed rate of 10–15 kg/ha, or a lower rate if in wider spaced drills (National Institute of Agricultural Botany, 1959; Evans, 1939, 1959; Green & Evans, 1957; Sonneveld & Evers, 1954).

An application of 100–155 kg N/ha is advised in late February or early

March; and after harvest, application of 30 kg P, 60 kg K and 75 kg N is recommended following close trimming of the stubble. Crops may be lightly grazed in winter, from October to December for early cultivars or until February for late cultivars (National Institute of Agricultural Botany, 1959). Grazing can increase the yield of crops sown in drills, but may reduce them in broadcast stands (Green & Evans, 1957). To ensure adequate isolation, crops should be separated by at least 180 m during anthesis (National Institute of Agricultural Botany, 1959). Principles and procedures of harvesting are well explained elsewhere (Ministry of Agriculture, Fisheries and Food, 1968).

The annual yield of seed per hectare can vary greatly with mean yields ranging from 400 to 1 090 kg, depending on the variety and the year (Lincs Seed Growers' Association, 1965).

Yield

DRY MATTER

When grown with high inputs of fertilizer and with adequate water, perennial ryegrass has given yields up to 15 000 kg dry matter (DM)/ha (e.g. Williams, Stiles & Turner, 1959). However, yields of only about half this value have been recorded, also under circumstances of considerable fertilizer application and with only three cuts per year (Henderson, Edwards & Hammerton, 1962). On upland peat soil the average annual yield of three cultivars of perennial ryegrass over a 5-year period was only 2 500 kg DM/ha and, of this, more than half was contributed by unsown species (Hunt, 1964). Most reported annual yields for perennial ryegrass on fertile lowland soil under a sequence of relatively infrequent cutting have been around 10 000 kg DM/ha (e.g. Lazenby & Rogers, 1964). Dent and Aldrich (1963b) measured the production of 26 perennial ryegrass varieties under high fertilizer application. At Cambridge, yields of 7 700 kg/ha were obtained from plots cut every month and 13 900 kg/ha from plots cut at ear emergence and again 8 weeks later. At Trawscoed, Wales, yields were 7 600 and 10 800 kg/ha, respectively.

Perennial ryegrass has generally been regarded as early in growth in spring, although in certain circumstances its ability to yield well in summer has been in doubt (e.g. Scott-Watson, 1938). The seasonal distribution of the yield of perennial ryegrass in Scotland has been described by Hunt (1957), who showed that when cut 5 times per year, early cultivars attained their maximum rate of production by early May and late ones by the middle to the end of May. Production was low in June but increased again from early July to mid-August. From mid-September onwards, growth was negligible. Anslow and Green (1967) drew curves showing the changes in rate of production of cultivars of perennial ryegrass throughout three years. Under a regime of approximately monthly cuts, maximum rates of production occurred in May;

midsummer values were lower, but depended to a great extent on the rate of N fertilizer used and on the efficiency of the irrigation routine. Production declined continuously after mid-July. Differences between cultivars were not great; a late-heading cultivar was up to 10 days later than an early-heading one, but production in June and July was slightly more rapid from the former. Williams, Stiles and Turner (1960) showed that a satisfactory pattern of growth for perennial ryegrass depended on the maintenance of adequate soil moisture.

Hunt (1964) has given the percentage distribution of yield throughout the season under a monthly cutting regime. Approximately 40% of the total annual yield was harvested in the first two cuts. With less frequent defoliation, the proportion of the yield contributed by the spring cuts was generally greater than this (Green & Corrall, 1967).

The continued high yield of a perennial ryegrass sward from year to year depends on the use of an inherently persistent variety. Lack of recognition of this was stated to be one of the chief reasons for the former unpopularity of ryegrass in long-term swards (Scott-Watson, 1951), and data which demonstrate the point have been provided by Green and Corrall (1965).

ENERGY AND PROTEIN

With frequent cutting (i.e. monthly), DM digestibility remains consistently high (75%–78%), and under fertile conditions DM yields of 7 700–10 000 kg/ha will give 20–27 Mcal/ha of metabolizable energy (Anslow & Green, 1967; Dent & Aldrich (1963a).

With less frequent cutting, higher yields of DM are attained at DM digestibility values within the range 64–73%, the actual digestibility of the regrowths being partly dependent on the time of cutting of the first growth (Green & Corrall, 1967, Unpublished data). The total yield of metabolizable energy in this instance can be as high as 33 Mcal/ha (Table 15.3).

Compared with other species, perennial ryegrasses have been shown to be outstanding in terms of yield of DM and metabolizable energy.

Table 15.3. Potential annual yields of energy and protein from perennial ryegrass (DM basis)

	Frequent cutting	Infrequent cutting
DM digestibility (%)	75–78	70 or less
Yield of digestible OM (kg/ha)	5 850–7 200	9 000
Yield of metabolizable energy (Mcal/ha)	20·0–27·0	33·4
Crude protein (%)	20·0	10·0
Yield of crude protein (kg/ha)	1 540	1 390
Yield of digestible crude protein (kg/ha)	1 160	820

Data compiled by D. J. Thomson (Grassland Research Institute, Hurley) from literature cited in the text. For conversion formulae see Chapter 5.

Response to management and environment

RESPONSE TO FERTILIZER

Nitrogen

The response to N in terms of DM yield has been measured in experiments at a number of research centres in Britain. Holmes and MacLusky (1955) found that the cultivars S101 and Ayrshire gave near-linear response in total annual yield when N rates totalling 190 or 380 kg/ha were applied in divided dressings for each of 3 to 6 cuts per year. The yield of DM in the absence of N (2 240 kg/ha) and the response to the first increment of N (17·1 kg DM/kg N) were similar for the two cultivars. Responses to the second N increment were 19 and 16 kg DM/kg N for S101 and Ayrshire, respectively. Holliday and Wilman (1962) found the response to N in the variety New Zealand was equivalent to 30·1 kg DM/kg N when the sward was cut 4–6 times annually and was given 59 kg N/ha before each cut; the yield without applied N was 1 610 kg DM/ha. An average yield response to N of 26·2 kg DM/kg N was measured by Herriott and Wells (1960) over a three-year period from a sward of S23 cut 4 times and given 77 kg N/ha each year; the yield without applied N was 2 260 kg DM/ha. Davies *et al.* (1966) examined the early-flowering varieties Irish, S24 and tetraploid Reveille in a two-year experiment. The grasses were cut 10 days after ear emergence and three times subsequently. Annual application of N was varied, 78, 156 or 234 kg N/ha being divided between the cuts. The cultivars differed only in the second year, when Reveille gave the highest yield; the response to N did not differ greatly between cultivars. The yield of DM with 78 kg N/ha averaged 8 500 kg/ha and the response to extra N averaged 38·8 and 30·2 kg DM/kg N for the intermediate and high N levels, respectively.

Widdowson *et al.* (1963) examined the effect of 0, 38 or 76 kg/ha, applied before each cut, on the yield of DM of S24. The grass was cut 7 times over a 2·5-year period. The aggregate yield without N was 7 270 kg DM/ha and the response was 36·4 and 30·8 kg DM/kg N for the low and high level, respectively. The response of ryegrass was similar to that of meadow fescue, but both were inferior in this respect to timothy and cocksfoot.

Response of S24 perennial ryegrass to N has been measured in a number of studies at the Grassland Research Institute, Hurley. Where the first growth was harvested about 14 days after ear emergence, response has been approximately linear up to 140 kg N/ha. Yield has increased by 30 to 60 kg DM/kg N, depending on seasonal effects and differences in time of application of the fertilizer. In experiments with the same cultivar, Cowling *et al.* (1962) found that response to N decreased with increasing frequency of defoliation (see Table 15.4). In the same experiment the influence of frequency of defoliation in the period up to flowering was also examined; results are given in Table 15.5.

Table 15.4. Yield of DM and response to N of S24 ryegrass at different frequencies of defoliation

Frequency of defoliation cuts/year	Yield without N '00 kg DM/ha	Response to levels of N* 126 kg N/ha	252 kg N/ha	504 kg N/ha
		kg DM/kg N		
3	30·5	27	25	19
6–8	12·4	13	16	15
53–59	12·2	12	11	10

* Response measured above no N

From Cowling *et al.* (1962)

Table 15.5. Response of S24 ryegrass to N in the period to end of May under different frequencies of defoliation

Number of defoliations	Response to levels of N* 56 kg N/ha	112 kg N/ha	224 kg N/ha
	kg DM/kg N		
1	32	30	21
2	20	19	18
16	12	10	9

* Response measured above no N

From Cowling *et al.* (1962)

Table 15.6. Response to N of annual DM yield between different N levels (weighted means)

	Response to N 0–182 kg N/ha	182–364 kg N/ha
	kg DM/kg N	
S24	26·2	20·6
S23	24·0	23·0
Irish	22·4	14·2

From Cowling and Lockyer (1965)

The response of S23, S24 and Irish perennial ryegrass to N was measured over a three-year period by Cowling and Lockyer (1965). The grasses were given 0, 91, 182 or 364 kg N/ha each year, split into 4 or 5 applications, one for each cut. S23 yielded slightly less than S24, but the response to N was similar in both and almost linear. Yields of Irish ryegrass and S24 were similar at lower rates of N, but the response of the former to N fell off markedly at the highest N rate. The weighted mean response measured in annual DM yields for the three cultivars is given in Table 15.6.

In a similar experiment in 1963 and 1964 (Cowling & Lockyer, 1964), the yield and N response of S24 perennial ryegrass were compared with those of

S22 Italian ryegrass. The two grasses gave almost identical results. Response to N rates of up to 200 kg/ha was almost linear, DM yield increasing by 23·6 kg/kg N; the response to rates between 200 and 400 kg N/ha declined to 15·3 kg DM/kg N.

The response to N of the tetraploid cultivar Petra was compared with that of two diploids of a similar maturity group, S23 and Melle, over a two-year period (Green & Corrall, 1967). The swards were cut 6 times each year and were given 0, 105, 210 or 420 kg N/ha per year, divided equally for the cuts. Yields were similar for each cultivar and the response to N was approximately linear at 19·0 kg DM/kg N.

In recent studies at the Grassland Research Institute, Hurley, the form of the response of an irrigated sward of S23 perennial ryegrass to N at various periods of the year has been examined. Non-linear responses were recorded in each of 4 periods in the year when up to 350 kg N/ha was given in each period. These data are well represented by the asymptotic curves shown in Fig. 15.1. The response to N in DM yield was greatest in the first period when flowering stems were developing; responses in the subsequent 3 periods were similar and somewhat lower than in the first.

Phosphorus and potassium

There are very few studies in which the response of perennial ryegrass to P and K has been examined.

One of the few reports on a response to phosphate measured on an all-grass sward is given by Wheeler (1960). A sward, predominantly ryegrass, which received dung, urine and applications of 306 kg N/ha each year, was cut 8 or 9 times in each of two years. The response to a rate of 50 kg P/ha was equivalent to 51 kg DM/kg P.

The response of S23 to K has been examined at the Grassland Research Institute, Hurley (Clement, 1967). The sward, almost devoid of white clover because of continuous high rates of N application, was given 0, 62·5, 125 or 250 kg K/ha per year and was harvested by cutting over a 4-year period. The K response in terms of annual DM yield declined as the rate of applied K increased. The increase in yield with the first increment of K was equivalent to 8 kg DM/kg K; the additional response to the next increment was 4 kg DM/kg K, and to the final increment 2 kg DM/kg K.

RESPONSE TO IRRIGATION

The typical response of perennial ryegrass to irrigation is described in Chapter 14 (Williams *et al.*, 1959, 1960; Reid & Castle, 1965).

RESPONSE TO TEMPERATURE

The response of perennial ryegrass to temperature follows the general pattern described in Chapter 14. It is clear that low temperatures are the

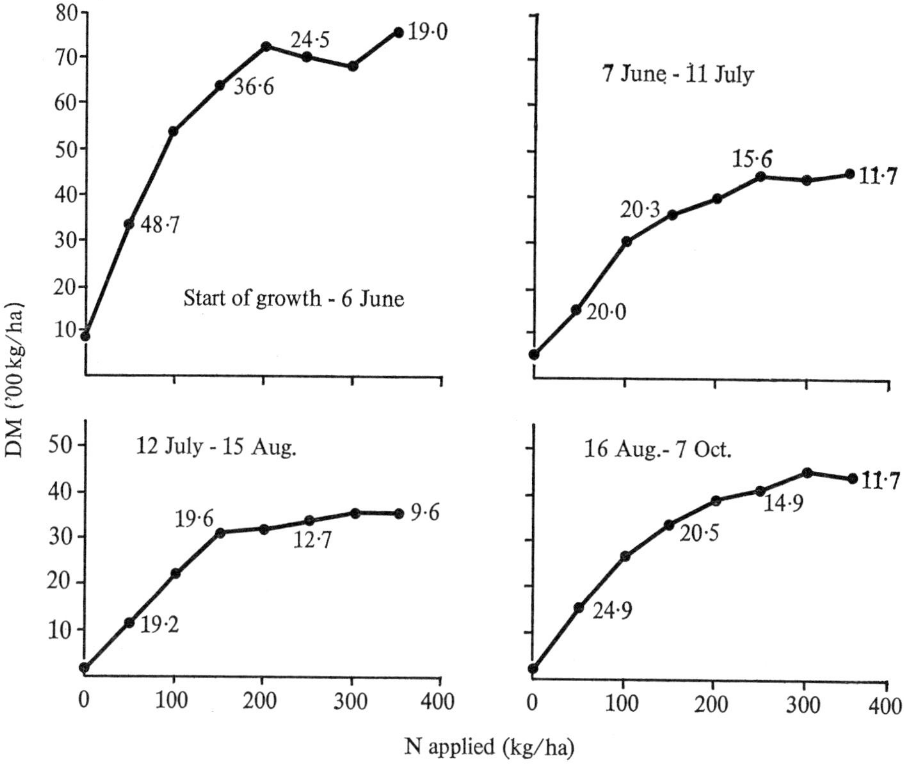

Fig. 15.1. Response to N of an irrigated sward of S23 perennial ryegrass in various periods of the year. Values on curves represent response above N_0 (no applied N) in kg DM/kg N.

major cause of low yields in winter, whereas in summer a more rapid rate of growth is possible, depending largely on the greater amount of light energy received. Precise data on the effect of temperature on the growth of perennial ryegrass can only be gained at present from controlled-environment studies, the results of some of which have been drawn on in this section.

In constant-environment conditions, the optimum temperature for the growth of perennial ryegrass is generally between 20° and 25°C, depending on plant age, nutrient status and other environmental conditions. Troughton (1961), for example, reported an optimum of 21°C for a New Zealand cultivar grown in a light intensity of 4 000 f.c. (40 000 lux); similarly, Mitchell (1956) reported an optimum of 20°C for another grown at 2 700 f.c. (27 000 lux). Mitchell and Lucanus (1962) found a slightly higher optimum temperature for growth in 16- than in 8-h days, while Alberda (1965) reported an optimum of 25°C for a tetraploid ryegrass given 5×10^4 erg cm^{-2} sec^{-1} of light energy, falling to 20°C when the light energy level was reduced by 60%.

Generally, where different day and night temperatures have been given, a maximum growth rate has been achieved with a night temperature equal to or slightly less than that of the day, e.g. the 21°/16°C regime cited by Sullivan and Sprague (1949). Night temperature appears to be less important as a determinant of growth than day temperature.

RESPONSE TO DEFOLIATION

Anslow and Green (1967) found little difference in the productivity of Aberystwyth cultivars cut approximately every month. However, Dent and Aldrich (1963b) found varietal differences when a much wider range of material (26 varieties in all) was tested. They found larger differences when the varieties were cut for hay than from the aggregate dry matter (DM) yields of monthly cutting. At one experimental site, monthly cutting on average reduced the annual yield by 44% compared with taking one cut at the hay stage and one aftermath cut 8 weeks later; at another site the difference in yield was 30%.

Green *et al.* (1960) found that yields of S23 cut at intervals of 10 days were about 45% lower than those of S23 cut every 30 days. Cowling and Lockyer (1963) compared the yield of S24 perennial ryegrass cut (a) twice-weekly, (b) 6 times during the season, and (c) cut at the hay stage and twice subsequently. The reduction in yield with more frequent cutting varied from 50% (plots cut twice-weekly; 280 kg N/ha applied) to a reduction of only 20% (cut 6 times in the season; 560 kg N/ha applied).

Baker (1957) found that severe autumn defoliation of S24 perennial ryegrass swards temporarily reduced growth early in the following spring. At all other times of the year the yields of perennial ryegrass were increased by severe defoliation (to 2·5 cm) compared with more lenient defoliation (e.g. Reid, 1966).

Quality

CHEMICAL COMPOSITION

Perennial ryegrass is relatively high in water-soluble carbohydrate compared with other grasses and has a potentially high Na content. Otherwise it is typical in chemical composition (see Chapter 13).

WATER CONTENT

Cultivar, polyploidy, stage of maturity and the amount of N fertilizer applied have all been found to influence the water content of perennial ryegrass. Johnston and Waite (1965) found that when S24 perennial ryegrass was allowed to grow without interruption, its water content decreased from 85·4% on 23 May to 60% on 16 Aug. De Vries (1957) harvested leafy late-pasture and less-leafy early perennial ryegrass grown in a glasshouse at progressively later growth stages. A period of constant water content was followed by a period of decreasing water content. It was found that the earlier the ears emerged, the shorter was the period of constant water content.

Ivins (1952) found that the percentage of water in Irish and S23 perennial ryegrass when given 224 lb N/acre (246 kg/ha) was higher than when it was given no N.

NUTRITIVE VALUE (DRY MATTER DIGESTIBILITY)

It has long been known that as grasses mature, their nutritive value falls, and that this fall is accompanied by a decline in digestibility. Any assessment of the nutritive value of a species must therefore take account of the changes in quality and yield with increasing maturity at different times of the year.

Young leafy grass frequently cut

When swards of perennial ryegrass are defoliated throughout the year at intervals of 3 to 4 weeks, DM digestibility normally ranges from 74% to 82%, with the highest values in spring and lower values in July and August. Differences between cultivars under this frequency of cutting are small (Dent & Aldrich, 1966).

Uninterrupted primary growth

Several studies have been made of changes in digestibility with growth up to and beyond ear emergence.

In an early-heading cultivar such as S24, DM digestibility at 50% ear emergence is usually about 75%, after which stage the digestibility falls by approximately 0·5 percentage units per day, reaching 70% DM digestibility 12 to 14 days later (Fig. 15.2, top).

In a late-heading cultivar such as S23, DM digestibility has usually already fallen to 70% by the time the 50% ear-emergence stage is reached. These contrasting patterns seem to be typical of most early- and late-heading perennial ryegrass cultivars, with mid-season ones showing an intermediate pattern. However, some cultivars give digestibility values at variance with those expected from a knowledge of their heading date. For example, the tetraploid Reveille heads within a few days of S24, but shows a consistently higher digestibility in its primary growth. This is frequently associated with a higher soluble carbohydrate content (Dent & Aldrich, 1963a) and a lower fibre content.

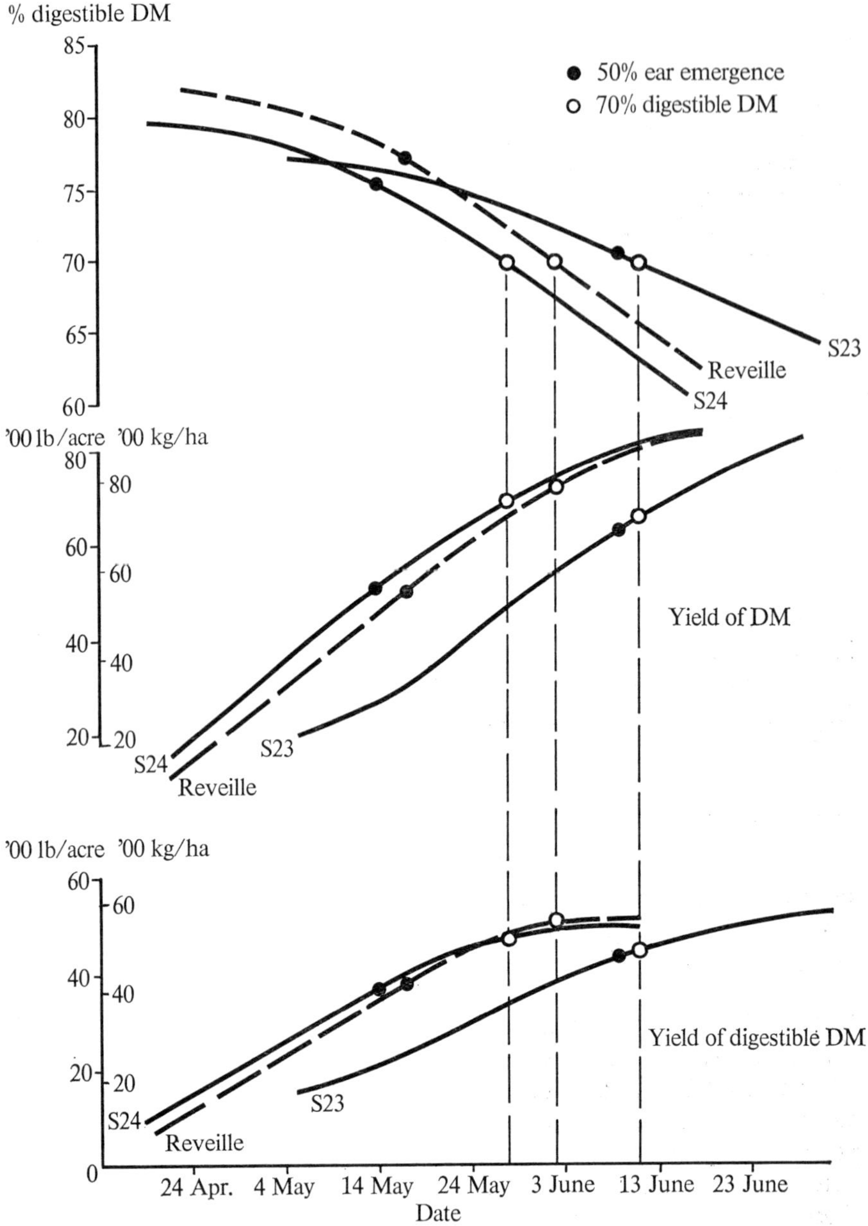

Fig. 15.2. Perennial ryegrass S23, S24 and Reveille. Seasonal changes in DM yield and digestibility. (After Aldrich and Dent, 1967)

Although S23 heads about 26 days later than S24 (Fig. 15.2, top), S23 falls to 70% DM digestibility only two weeks later than S24.

Fig. 15.2, middle, shows that the yield of DM typically increases most rapidly when digestibility has begun to fall steeply. The highest yield of digestible DM is thus often obtained when digestibility has fallen to a low level; hence, to achieve maximum production of nutrients some compromise is necessary between quality and yield. For conservation purposes it has been suggested that a DM digestibility of 70% gives a feed of reasonable quality without too great a sacrifice in yield, but in perennial ryegrass this digestibility level in the primary growth is associated with low protein content (Dent & Aldrich, 1963b). This in turn is associated with a high soluble carbohydrate percentage, which is a characteristic of the species. Crude protein content at 70% digestibility may range from 7% in the latest-maturing cultivars, to 12% in the earliest. There are varietal differences in soluble carbohydrate content within the species; for example, the Canadian cultivar Norlea has been shown to have a consistently higher soluble carbohydrate content than others in the same maturity group.

Regrowth after cutting

In the regrowth following a primary cut in May or June, most perennial ryegrass cultivars will produce a second crop of flowering stems and after a two-month regrowth period still frequently have a DM digestibility of about 70%.

In this first regrowth, early-heading cultivars are usually less digestible than the late pasture types, and in some seasons stemmy regrowth in these early cultivars may result in DM digestibility values of well below 70%. In two-month regrowths later in the year, the fall in digestibility is less marked and values between 70% and 75% DM digestibility are common.

ANIMAL INTAKE CHARACTERISTICS

Indirect determinations (Greenhalgh, 1966) have shown that the intake of perennial ryegrass by ruminants is greater than that of cocksfoot, the difference being largely attributable to differences in digestibility. Evidence from New Zealand workers using sheep (Rae *et al.*, 1963; Rae, Brougham & Barton, 1964) and dairy cattle (Wilson & McDowell, 1966; Wilson, 1966) suggests by inference that (a) the voluntary intake of perennial ryegrasses is less than that of Italian ryegrasses and (b) intake of the hybrid ryegrasses is variable, in some instances being similar to the *L. perenne* parent and in others to the *L. multiflorum* parent.

Longevity and persistency

Perennial ryegrass is a relatively long-lived plant made up of relatively short-lived components. Individual tillers are annual or winter-annual (Beddows, 1967) but detailed data on their longevity are lacking. There appear to be no published data on leaf longevity *per se*. Anslow (1968) found that in plants examined in June, very few of the third most recently expanded leaves were still green, and some senescence occurred in second leaves in September. Since leaves emerged every 11 days in June and every 20 days in September, it may be deduced that the leaves lived for 30 to 50 days. These observations confirm those of Davies (1963), who concluded that perennial ryegrass possesses a mechanism for adjusting the life-span of individual leaves to the rate of leaf appearance. Cooper, Harrington and Mills (1965) reported a maximum rate of leaf appearance of 1·75/week in controlled-environment experiments, while Hunt (1965) has recorded the appearance of new leaves every 6 to 8 days in summer in New Zealand, implying minimum leaf longevities of 12 and 18 days, respectively.

Average root longevity (Garwood, 1967a) has been found to vary from 2 months for roots formed during March–May to 6·2 months for those formed in October. Maximum recorded longevity was 8 months (Garwood, 1967b). Plant types range from erect individuals with relatively few tillers to prostrate compact cushions with numerous tillers (Beddows, 1967), the former types being usually early-heading (beginning of April) and short-lived, and the latter late-heading (July) and very persistent. Plants which develop heads in their first autumn tend to be less persistent. Charles (1961) noted that 25% of viable seeds sown in March produced plants which survived for two months, and only 11% produced plants which survived the first winter. The potential life span of the more perennial types is not known.

The population in a well-grazed old pasture is apparently ageless; yet plants from such a habitat, vegetatively propagated and grown on as spaced plants which were allowed to develop inflorescences and reach the hay stage, did not usually survive in a vigorous condition for much longer than two years (Beddows, 1967). In such old pastures it is rare for all the inflorescences to be grazed.

Survival of perennial ryegrass under sward conditions depends on adequate supplies of phosphate and lime (Thomas, 1936, cited by Beddows, 1967) and a pH of 5–8, with an optimum of 6–7 (Spurway, 1941). When growth of the species is poor due to a lack of available plant nutrients, high soil acidity, poor drainage, etc., a sown stand is soon invaded by other species more suited to the environment. Even under below-optimum conditions, however, small quantities of perennial ryegrass may persist in swards which are grazed. When the environment is conducive to a more vigorous growth of the species, for example in sown swards yielding 6 000 kg DM/ha per year and above, and when stocking rates are appropriately high, the effects of grazing are most marked. It is under these conditions that returns of urine affect the highest proportion of the sward at any time during the grazing season (Petersen *et al.*, 1956) and in doing so increase the supply of plant nutrients (mainly N and K). The effect of treading will also be at its greatest, and treading has been found to encourage ryegrass in relation to other species (Davies, 1938).

Perennial ryegrass persists at high stocking rates associated with close defoliation; under these conditions a horizontal as opposed to an erect tiller formation may develop and the sward may be invaded by other species (Kydd, 1966). Ryegrass is adversely affected when swards are allowed to remain tall in the late autumn, when they may carry appreciable top growth into the winter (Baker, 1956), a situation easily remedied by grazing the surplus.

Varietal differences in persistence have been demonstrated in perennial ryegrass; for example, over a period of ten years Aberystwyth 'S' cultivars were found to be more persistent than commercial varieties (Alder, 1963).

In general, perennial ryegrass is persistent under grazing, contributing most when fertility is high and grazing is adjusted to production. Under these conditions persistence is likely to suffer more from insufficient usage than from overstocking (Jones, 1939).

References

ALBERDA, T. 1965. The influence of temperature, light intensity and nitrate concentration on dry-matter production and chemical composition of *Lolium perenne* L. *Neth. J. agric. Sci.* **13**, 335–60.

ALDER, F. E. 1963. A comparison of the productivity of Aberystwyth-bred and commercial grass varieties in small farm units. *J. Br. Grassld Soc.* **18**, 146–53.

ALDRICH, D. T. A., DENT, J. W. 1967. The relationship between yield and digestibility in the primary growth of nine grass varieties. *J. natn. Inst. agric. Bot.* **11**, 104–13.

ANSLOW, R. C. 1955. The effects of competition of the shoot and root growth of three grass species. M.Sc. Thesis, Univ. of Reading.

ANSLOW, R. C. Grassland Research Institute. Personal communication.

ANSLOW, R. C. 1968. The production of dry matter by swards of perennial ryegrass differing in average age of foliage. *J. Br. Grassld Soc.* **23**, 195–201.

ANSLOW, R. C.; GREEN, J. O. 1967. The seasonal growth of pasture grasses. *J. agric. Sci., Camb.* **68**, 109–22.

ARENS, R. 1962. The effect of rate of seeding on competition between species and the productivity of the first year's pasture. *Z. Acker- u. PflBau* **115**, 357–74.

ARMSTRONG, S. F. 1948. British grasses and their employment in agriculture. Cambridge Univ. Press, 3rd edn.

ARNOTT, R. A. 1969. The effect of seed weight and depth of sowing on the emergence and early seedling growth of perennial ryegrass. *J. Br. Grassld Soc.* **24**, 104–10.

BAKER, H. K. 1956. Note on the influence of previous management on the death of perennial ryegrass during winter. *J. Br. Grassld Soc.* **11**, 235–7.

BAKER, H. K. 1957. Studies on the root development of herbage plants. 3. The influence of cutting treatments on the root, stubble and herbage production of a perennial ryegrass sward. *J. Br. Grassld Soc.* **12**, 197–208.

BAUMANN, H.; KLAUSS, M. L. 1955. [Root development in an environment with a high water-table.] *Z. Acker- u. Pflbau* **99**, 410–26.

BEDDOWS, A. R. 1967. Biological Flora of the British Isles, *Lolium perenne* L. *J. Ecol.* **55**, 567–87.

CHARLES, A. H. 1961. Differential survival of cultivars of *Lolium*, *Dactylis* and *Phleum*. *J. Br. Grassld Soc.* **16**, 69–75.

CLAPHAM, A. R.; TUTIN, T. G.; WARBURG, E. F. 1962. Flora of the British Isles. Cambridge: Univ. Press, 2nd edn.

CLEMENT, C. R. 1967. Potassium for the grass crop, its supply from soil and fertilisers. *A. Rep. Grassld Res. Inst. Hurley 1966* pp. 80–5.

COOPER, J. P.; HARRINGTON, G. N.; MILLS, P. A. 1965. Leaf development. *Rep. Welsh Pl. Breed. Stn 1964* pp. 19–21.

COWLING, D. W.; LEIGH, J. H.; LOCKYER, D. R. 1962. Nitrogenous fertilizers and the role of legumes. *Exps Prog. 14 Grassld Res. Inst. 1960–61* pp. 17–18.

COWLING, D. W.; LOCKYER, D. R. 1963. The contribution to yield made by the white clover and fertilizer nitrogen at various cutting frequencies. *Exps Prog. 15 Grassld Res. Inst. 1961–62* pp. 17–18.

COWLING, D. W.; LOCKYER, D. R. 1964. Nitrogenous fertilizers and the role of legumes. *Exps Prog. 16 Grassld Res. Inst. 1962–63* pp. 21–3.

COWLING, D. W.; LOCKYER, D. R. 1965. A comparison of the reaction of different grass species to fertilizer nitrogen and to growth in association with white clover. *J. Br. Grassld Soc.* **20,** 197–204.

CROCKER, R. L.; MARTIN, P. M. 1964. Competition between perennial ryegrass and meadow fescue under field-plot conditions. *J. Br. Grassld Soc.* **19,** 27–9.

DAVIES, H.; THOMAS, B. V.; ALDRICH, D. T. A. 1966. The effects of different levels of nitrogen fertilizer on the yields of varieties of perennial ryegrass and cocksfoot. *J. Br. Grassld Soc.* **21,** 245–9.

DAVIES, I. 1963. Life-span of individual leaves of perennial ryegrass. *Rep. Welsh Pl. Breed. Stn 1962* pp. 27–9.

DAVIES, W. 1928. The factor of competition between one species and another in seeds mixtures. *Bull. H8 Welsh Pl. Breed. Stn.*

DAVIES, W. 1938. The vegetation of grass verges and other excessively trodden habitats. *J. Ecol.* **26,** 38–49.

DENT, J. W.; ALDRICH, D. T. A. 1963a. Soluble carbohydrate content of varieties of tetraploid ryegrass. *Nature, Lond.* **198,** 905.

DENT, J. W.; ALDRICH, D. T. A. 1963b. The inter-relationships between heading date, yield, chemical composition and digestibility in varieties of perennial ryegrass, timothy, cocksfoot, and meadow fescue. *J. natn. Inst. agric. Bot.* **9,** 261–81.

DENT, J. W.; ALDRICH, D. T. A. 1966. The *in vitro* digestibility of herbage species and varieties and its relationship with cutting treatment, stage of growth and chemical composition. *Proc. 10th int. Grassld Congr., Helsinki, 1966* pp. 419–24.

ENGLAND, F. J. W. 1965. Interactions in mixtures of herbage grasses. *Rec. Scot. Pl. Breed. Stn* pp. 125–49.

EVANS, G. 1937. Seed production of a pasture type of ryegrass. *Welsh J. Agric.* **13,** 195–211.

EVANS, G. 1939. Perennial ryegrass; seed production of bred strains. *Agriculture, Lond.* **46,** 355–62.

EVANS, G. 1959. Seed rates of grasses for seed production. 1. Pasture varieties of ryegrass, cocksfoot, and timothy. *Emp. J. exp. Agric.* **27,** 291–9.

GARWOOD, E. A. 1967a. Studies on the roots of grasses. *A. Rep. Grassld Res. Inst. Hurley 1966* pp. 72–9.

GARWOOD, E. A. 1967b. Seasonal variation in appearance and growth of grass roots. *J. Br. Grassld Soc.* **22,** 121–30.

GARWOOD, E. A. 1967c. Some effects of soil water conditions and soil temperature on the roots of grasses. 1. The effect of irrigation on the weight of root material under various swards. *J. Br. Grassld Soc.* **22,** 176–81.

GERDES, G. 1952. [Studies on the absolute weight (1 000 seed weight) of the most important fodder grass species and their varieties in regard to the performance of grass breeding.] *Züchter* **22,** 353–66.

GRASSLAND RESEARCH INSTITUTE. 1952. *Exps Prog. 5 Grassld Res. Inst. 1952* pp. 23–31.

GREEN, J. 1965. Harvesting relationship of moisture, maturity and weight in herbage seeds. *Rep. Lincs. Seed Growers' Assn 1964* Agric. House, Woodhall Spa, pp. 9–13.

GREEN, J. O.; CORRALL, A. J. 1965. The testing of grass varieties in swards with clover: the effect of grass seed-rate on comparisons of gross yield. *J. Br. Grassld Soc.* **20,** 207–11.

GREEN, J. O.; CORRALL, A. J. 1967. Grassland Research Institute, Hurley. Personal communication.

GREEN, J. O.; CORRALL, A. J. 1967 Grassland Research Institute, Hurley. Unpublished.

GREEN, J. O.; CORRALL, A. J.; TAYLER, J. C. 1960. Level and frequency of cutting: pure grass swards at two levels of nitrogen. *Exps Prog. 12 Grassld Res. Inst. 1958–9* pp. 19–21.

GREEN, J. O.; EVANS, T. A. 1957. Grazing management for seed production in leafy strains of grasses. *J. Br. Grassld Soc.* **12,** 4–9.

GREENHALGH, J. F. D. 1966. Studies of herbage consumption and milk production in grazing dairy cows. *Proc. 10th int. Grassld Congr., Helsinki, 1966* pp. 351–5.

HARKESS, R. D. 1966. The assessment of inter and intra-specific competition by the wire grid technique. *Proc. 9th int. Grassld Congr., São Paulo, 1965* pp. 417–19.

HENDERSON, J. L.; EDWARDS, R. S.; HAMMERTON, J. L. 1962. The productivity of five grass species at six levels of compound fertilizer application. 1. Dry-matter production. *J. agric. Sci., Camb.* **59,** 5–11.

HERRIOTT, J. B. D.; WELLS, D. 1960. Clover nitrogen and sward productivity. *J. Br. Grassld Soc.* **15,** 63–9.

HOLLIDAY, R.; WILMAN, D. 1962. The effect of white clover, fertilizer nitrogen and simulated animal residues on yield of grassland herbage. *J. Br. Grassld Soc.* **17,** 206–13.

HOLMES, W.; MACLUSKY, D. S. 1955. The intensive production of herbage for crop-drying. *J. agric. Sci., Camb.* **46,** 267–86.

HUBBARD, C. E. 1954. Grasses. Harmondsworth, Middlesex: Penguin Books.

HUGHES, H. D.; HEATH, M. E.; METCALF, D. S. 1962. Forages. Ames, Iowa: Iowa State Univ. Press, 2nd edn.

HUNT, I. V. 1954. Seed establishment in the West of Scotland. *J. Br. Grassld Soc.* **9,** 85–98.

HUNT, I. V. 1957. Comparative productivity of 27 strains of grass in the West of Scotland. 4. Seasonal growth. *J. Br. Grassld Soc.* **12,** 127–37.

HUNT, I. V. 1964. Comparative productivity of herbage varieties on upland peat. *J. Br. Grassld Soc.* **19,** 55–61.

HUNT, L. A. 1965. Some implications of death and decay in pasture production. *J. Br. Grassld Soc.* **20,** 27–31.

IVINS, J. D. 1952. The relative palatibility of herbage plants. *J. Br. Grassld Soc.* **7,** 43–54.

JOHNSTON, M. J.; WAITE, R. 1965. Studies in the lignification of grasses. 1. Perennial ryegrass (S24) and cocksfoot (S37). *J. agric. Sci., Camb.* **64,** 211–19.

JONES, I. 1939. Pasture management and its effect on the sward. *Bull. H15 Welsh Pl. Breed. Stn* pp. 40–129.

JONES, L. I. 1958. Grassland agronomy. *Rep. Welsh Pl. Breed. Stn 1950–56* pp. 97–9.

KYDD, D. D. 1966. The effect of intensive sheep stocking over a five year period on the development and production of the sward. 3. Ecological discussion. *J. Br. Grassld Soc.* **21,** 298–302.

LAMPETER, W. 1960. [Mutual influences of higher plants regarding growth of shoots and roots, mineral content and water consumption studied in some economically important forage plants.] *Wiss. Z. Karl-Marx-Univ. Lpz. 1959-60* **9,** 611–722.

LAZENBY, A.; ROGERS, H. H. 1964. Selection criteria in grass breeding. 2. Effect, on *Lolium perenne*, of differences in population density, variety and available moisture. *J. agric. Sci., Camb.* **62,** 285–98.

LINCOLNSHIRE SEED GROWERS' ASSOCIATION. 1965. Report for 1964. Agric. House, Woodhall Spa, pp. 3–7.

MINISTRY OF AGRICULTURE, FISHERIES AND FOOD. 1968. Grass and clover crops for seed. *Bull. 204* London: H.M.S.O., pp. 42–60.

MITCHELL, K. J. 1956. Growth of pasture species under controlled environments. 1. Growth at various levels of constant temperature. *N.Z. Jl Sci. Technol. Sect. A* **38,** 203–15.

MITCHELL, K. J.; LUCANUS, R. 1962. Growth of pasture species under controlled environment. 3. Growth at various levels of constant temperature with 8 and 16 hours of uniform light per day. *N.Z. Jl agric. Res.* **5,** 135–44.

NATIONAL INSTITUTE OF AGRICULTURAL BOTANY. 1959. Growing grasses for seed. *Herbage Seed Growers' Leaflet No. 2.*

National Institute of Agricultural Botany. 1965. Varieties of ryegrass. *Farmers' Leaflet No. 16*.

Perring, F. H.; Walters, S. M. 1962. Atlas of the British flora. London: Thomas Nelson and Sons Ltd, pp. 456.

Petersen, R. G.; Lucas, H. L.; Woodhouse, W. W. 1956. The distribution of excreta by freely grazing cattle and its effect on pasture fertility. 1. *Agron. J.* **48,** 440–4.

Peto, D. H. 1961. Methods and problems of cleaning herbage seeds. *Rep. Lincs. Seed Growers' Assn. 1960* Agric. House, Woodhall Spa, pp. 19–21.

Rae, A. L.; Brougham, R. W.; Barton, R. A. 1964. A note on liveweight gains of sheep grazing different ryegrass pastures. *N.Z. Jl agric. Res.* **7,** 491–5.

Rae, A. L.; Brougham, R. W.; Glenday, A. C.; Butler, G. W. 1963. Pasture type in relation to live-weight gain, carcass composition, iodine nutrition and some rumen characteristics of sheep. 1. Live-weight growth of sheep. *J. agric. Sci., Camb.* **61,** 187–90.

Reid, D. 1966. Studies on cutting management of grass-clover swards. 4. The effects of close and lax cutting on the yield of herbage from swards cut at different frequencies. *J. agric. Sci., Camb.* **66,** 101–6.

Reid, D.; Castle, M. E. 1965. The response of grass-clover and pure-grass leys to irrigation and fertilizer nitrogen treatment. 1. Irrigation effects. *J. agric. Sci., Camb.* **64,** 185–94.

Roberts, H. M. 1958. The effect of defoliation on the seed-producing capacity of bred strains of grasses. 1. Timothy and perennial ryegrass. *J. Br. Grassld Soc.* **13,** 255–61.

Ryle, G. J. A. 1964. A comparison of leaf and tiller growth in seven perennial grasses as influenced by nitrogen and temperature. *J. Br. Grassld Soc.* **19,** 281–90.

Scott-Watson, J. A. 1938. Grassland improvement. *Trans. R. Highld agric. Soc. Scotl.* **50,** 160–83.

Scott-Watson, J. A. 1951. A century of farming progress. *Agriculture, Lond.* **58,** 58–63.

Sonneveld, A.; Evers, A. 1954. Some results of recent research on grass seed production in the Netherlands. *Proc. Europ. Grassld Conf. 1954* pp. 141–9.

Spurway, C. H. 1941. Soil reaction (pH) preference of plants. *Spec. Bull. 306 Mich. agric. Exp. Stn* pp. 36.

Sullivan, J. T.; Sprague, V. G. 1949. The effect of temperature on the growth and composition of the stubble and roots of perennial ryegrass. *Pl. Physiol., Lancaster* **24,** 706–19.

Terry, R. A.; Tilley, J. M. A. 1964. The digestibility of leaves and stems of perennial ryegrass, cocksfoot, timothy, tall fescue, lucerne and sainfoin, as measured by an *in vitro* procedure. *J. Br. Grassld Soc.* **19,** 363–72.

Thomas, M. T. 1936. Investigations on the improvement of hill grazing. 2. The introduction and maintenance of nutritious and palatable species and strains. *Bull. H14 Welsh Pl. Breed. Stn* pp. 4–57.

Tincker, M. A. H. 1925. The effect of length of day upon the growth and reproduction of some economic plants. *Ann. Bot.* **39,** 721–54.

Troughton, A. 1961. The effect of photoperiod and temperature on the relationship between the root and shoot systems of *Lolium perenne. J. Br. Grassld Soc.* **16,** 291–5.

Twyford Seeds. 1964. Tetraploid establishment. *Spring catalogue 1964* Twyford Seeds Ltd, Adderbury, Oxon., p. 9.

Vries, P. de. 1957. The relationship between dry weight, green weight and growth stages of three species and varieties of grasses. *Jaarb. Inst. biol. scheik. Onderz. LandbGewass. 1957* pp. 155–8.

Wheeler, J. L. 1960. A response to phosphate on an all-grass sward. *J. Br. Grassld Soc.* **15,** 300–1.

Widdowson, F. V.; Penny, A.; Williams, R. J. B. 1963. An experiment comparing responses to nitrogen fertilizer of four grass species. *Expl Husb.* No. 9, 28–36.

Williams, T. E.; Stiles, W.; Turner, P. J. 1959. *Exps Prog. 11 Grassld Res. Inst. 1957–8* p. 58.

Williams, T. E.; Stiles, W.; Turner, P. J. 1960. *Exps Prog. 12 Grassld Res. Inst. 1958–9* p. 57.

Wilson, G. F. 1966. Ryegrass varieties in relation to dairy cattle performance. 2. The influence of ryegrass varieties on intake, digestibility and some characteristics of rumen fermentation. *N.Z. Jl agric. Res.* **9,** 1053–63.

WILSON, G. F.; MCDOWALL, F. H. 1966. Ryegrass varieties in relation to dairy cattle performance. 1. The influence of ryegrass varieties on milk yield and composition. *N.Z. Jl agric. Res.* **9,** 1042–52.

WOODFORD, E. K. 1966. Crop physiology. *A. Rep. Grassld Res. Inst. Hurley 1965* p. 35.

CHAPTER 16

Italian ryegrass *(Lolium multiflorum)*

Morphology

ITALIAN ryegrass (Plate 9) is a tufted annual or short-lived perennial. Root system highly branched, dense, with fibrous adventitious root mass. Vegetative stems erect, often decumbent at the base. Vegetative tillers cylindrical. Leaf sheaths glossy green, red at base, becoming brown, persistent but not fibrous. Ligule short, auricles prominent. Leaf blades rolled in the sheath, opening flat without a prominent midrib, lax, slightly scabrid, strongly ribbed adaxially, glossy green abaxially. Culms slender, 2- to 4-noded. Inflorescence a simple spike of alternate sessile spikelets, edgeways on to the axis, spikelets long-awned, 3- to 10-flowered. Caryopsis enclosed in hardened palea and awned lemma.

The root system of Italian ryegrass is similar to that of perennial ryegrass both in mass and distribution in the soil. Vegetative stems are erect at first but are liable to become decumbent as the primary axes grow. The stems are fewer but more upright than in perennial ryegrass, since fewer basal tillers develop. Leaf blades range in size from 5 to 10 mm wide and up to 25 cm long. Culms reach a height of 90 cm.

There are no definitive morphological differences between cultivars (including colchicine-induced tetraploids), differences being mainly physiological.

The changes in the amount of leaf during growth have been studied by Minson *et al.* (1964) (Table 16.1).

Distribution

Lolium multiflorum is an introduced species (Clapham *et al.*, 1962) occurring in all vice-counties except three in Northern Ireland (Perring and Walters, 1962). It is widely cultivated and, as an escape, is frequently naturalized on roadsides, field margins and waste land (Hubbard, 1954; Clapham *et al.*, 1962). It is fairly common in the southwest, south, central and east parts of Britain, but is scattered and infrequent elsewhere, especially in Northern Ireland (Perring & Walters, 1962).

It is generally absent from land higher than 300 m and from wetter, acid uplands (Perring & Walters, 1962). The optimum soil pH for growth is 6·0 to 7·0, but it occurs over the range 5·0–8·0 (Spurway, 1941). It is commonest where the annual rainfall is less than 1 000 mm and only sparse where it exceeds 1 250 mm. It also tends to be commonest where the Meyer Precipitation–Saturation Deficit quotient is less than 200, though it has sometimes been found in areas where it is greater than 500 (Perring & Walters, 1962).

Compatibility

COMPATIBILITY WITH OTHER GRASSES

Italian ryegrass attains its full growth rate very quickly after emergence, a characteristic which would appear to make it particularly useful as a means

Table 16.1. Percentage leaf blade in the DM of S22 Italian ryegrass during the growth of the first crop

Date	% leaf blade
19 Apr.	71
3 May	53
27 May	25

Date of 50% ear emergence: 25 May.
From Minson *et al.* (1964)

of enhancing early production in leys otherwise comprised of grasses which develop relatively slowly. However, this capacity for rapid growth at early stages makes it very competitive against other species in mixtures and may delay their development to such an extent that they are unable to replace the Italian ryegrass satisfactorily when it subsequently dies out. Improvements in the species which have been achieved by plant breeders, mainly in terms of vigour, leafiness and longevity, aggravate difficulties in compatibility. In several experiments at the Grassland Research Institute, Hurley (1960, 1961, 1962), Italian ryegrass sown as a companion grass increased production in the seeding year from direct sowings of timothy, timothy plus meadow fescue, and perennial ryegrass sown at very low rates, all with white clover, with no reduction in total herbage production in the first harvest year (Grassland Research Institute, Unpublished data).

Lampeter (1960) calculated the suppressive effect of 5 grasses and two legumes on one another when grown in combinations of any two species. Over a period of two years, Italian ryegrass was more aggressive than meadow fescue, cocksfoot, timothy, red clover or lucerne, and only a little less so than perennial ryegrass. Where Italian ryegrass is to be used as the basic constituent of a seeds mixture, it is therefore relevant to consider the effects on its performance of the other species present. In an experiment at the Welsh Plant Breeding Station, Aberystwyth, pure stands of Italian ryegrass persisted well into the third harvest year, but the presence of another grass much reduced its persistence; in the presence of timothy, persistence was halved, but when perennial ryegrass was present the yield of the Italian ryegrass component was only 11% of that of a pure stand (Jones, 1958).

Level of nitrogen. In trials at the Welsh Plant Breeding Station, Jones (1958) found that Italian ryegrass was more competitive towards perennial ryegrass when fertilizer N was applied at 114–175 kg N/ha each year than when none was applied.

COMPATIBILITY WITH LEGUMES

Since red clover develops quickly and fits the pattern and duration of production of Italian ryegrass, the grass is often combined with it in mixtures designed for short-term use. In an experiment at the Grassland Research Institute, Hurley (1959) (Table 16.2), rates of 0·5 to 20 kg S22 Italian ryegrass seed/ha were combined with 18 to 0·5 kg of S123 red clover. The mixture

Table 16.2. Yield in the year of sowing from varying proportions of S123 red clover sown with S22 Italian ryegrass

High seed rate					Low seed rate				
Seeds mixture (kg/ha)		Yield of DM ('00 kg/ha)			Seeds mixture (kg/ha)		Yield of DM ('00 kg/ha)		
S22	S123	S22	S123	Total	S22	S123	S22	S123	Total
20	...	35·4	...	35·4	10	...	33·7	...	33·7
18	1	30·7	0·4	31·1	9	½	33·0	0·5	33·5
18	3	32·4	0·7	33·1	9	1½	33·2	1·3	34·5
18	6	31·0	2·1	33·1	9	3	29·8	2·2	32·0
18	18	33·0	4·8	37·8	9	9	30·3	4·9	35·2
6	18	32·3	7·9	40·2	3	9	27·7	11·4	39·1
3	18	27·6	16·2	43·8	1½	9	22·6	15·5	38·1
1	18	21·4	22·0	43·4	½	9	20·8	15·1	35·9
12	12	29·1	4·2	33·4	6	6	32·0	3·4	35·4
...	10	...	33·1	33·1	...	5	...	18·9	18·9

From Grassland Research Institute (1959)

yielded more than pure grass only when the Italian ryegrass was sown at less than 10 kg/ha and the red clover at more than 10 kg/ha. In another experiment to compare 5 legumes grown with Italian ryegrass, the highest gross yield in the first harvest year came from a mixture with lucerne, but less grass grew with lucerne than with red or white clover (1958) (Table 16.3).

Table 16.3. Effect of legumes on Italian ryegrass. Total yield of DM from 4 cuts in the first harvest year ('00 kg/ha)

	Italian ryegrass	Legume	Total
No legume	40	...	40
White clover	56	17	73
Broad red clover	52	44	96
Lucerne	36	63	99
Trefoil	50	31	81
Winter vetch	51	12	63

From Grassland Research Institute (1958)

Seed

CHARACTERS

The seed is lanceolate-oblong with a shallow groove along its ventral surface. Its length is 5 to 8 mm, with an average maximum width of 1·0 to 1·5 mm (Armstrong, 1948; Hubbard, 1954; Peto, 1961). The lemma is convex, usually smooth and is either blunt or has two minute points at the apex on either side of the base of a fine, straight awn. The awn is usually longer than the lemma (up to 10 mm in length) and may be shed when the seed is ripe. The palea is similar in length to the lemma. The rachilla is close to the palea, and may be flattened or almost cylindrical in cross-section (Armstrong, 1948; Hubbard, 1954).

The weight of seed varies between and within cultivars. Gerdes (1952) reported that 1 000-seed weight varied from 1·87 g to 2·26 g for seed of different national origins, and within the cultivar S22. Harkess (1965) found that the smallest seed averaged 1·8±0·2 mg (1·6–2·0 g per 1 000) and the largest 3·4±0·2 mg. Most cultivars in use are diploids, and the tetraploids recently introduced have larger seeds; in the tetraploid cultivar Tetrone, the smallest and largest seed respectively averaged 3·4 and 6·2 mg±0·2 mg (Harkess, 1965). A 1 000-seed weight of 3·91 g has been quoted for tetraploid seed (Twyford Seeds, 1964).

PRODUCTION

Crops intended for seed can be sown at 10–15 kg/ha up to the end of August without a cover crop, either broadcast or in drills up to 35 cm apart. An application of 100–155 kg N/ha in late February or early March is advised (National Institute of Agricultural Botany, 1959); autumn applications of N may not be effective (Schwanbom & Fröier, 1949), but the National Institute of Agricultural Botany (1959) recommends the application of NPK following trimming of the stubble immediately after harvest. Cutting or grazing in spring may be necessary to avoid an excess of straw at harvest;

Vyncke (1955) recommended a cut in April, and Evers and Sonneveld (1953) found such a cut plus 30 kg N/ha reduced straw production by 40% but did not reduce the yield of seed at harvest. Seed crops should be isolated by at least 180 m during anthesis (National Institute of Agricultural Botany, 1959). Principles and procedures for harvesting are well explained elsewhere (Ministry of Agriculture, Fisheries and Food, 1968).

Reported annual yields of seed have varied from 200 to 1 825 kg/ha, with the mean yield ranging from 575 to 1 100 kg, depending on the year (Lincolnshire Seed Growers' Association, 1965).

Yield

DRY MATTER

When sown as a pure crop, Italian ryegrass is mostly grown to provide grazing in early spring for dairy cows or ewes, after which it is often cut for silage. Its yields in early spring and in May or June are therefore of importance. The length of life of the crop varies with the cultivar used, since these differ greatly in persistency, overlapping in this respect the relatively short-lived Westerwolds ryegrass on the one hand, and the hybrids between Italian and perennial ryegrasses and the short-lived cultivars of perennial ryegrass on the other.

A yield of approximately 700 kg dry matter (DM)/ha was obtained in late April from 5 cultivars of Italian ryegrass, sown in the previous spring and given high rates of fertilizer (Hunt, 1962a). In the two succeeding years the same crops gave yields of 750 and 1 280 kg/ha in late April, although the total annual production fell from 15 000 kg/ha in the first harvest year to 12 000 and 9 000 kg/ha in the second and third years. In this trial there were considerable varietal differences and the New Zealand variety showed no decline in productivity from year to year. Jones (1953) found that the total annual yield of the cultivar S22 was similar for three years after sowing. Baker, Chard, Davies and Aldrich (1965) obtained yields of 1 500 kg DM/ha in late April 1962 from 7 cultivars of Italian ryegrass sown in late April 1961. However, after a very cold winter, yields in April from one-year-old crops of this species were, on average, only half this and stands of some cultivars had been almost eliminated.

Regarding the usefulness of Italian ryegrass as a crop for cutting for conservation, Wilman (1965) showed that, with heavy applications of fertilizer, 5 000 kg DM/ha could be produced by S22 in 6 weeks, with cutting starting any time between late March and late April. Corrall (1965) showed that S22 and Tetila increased in weight by about 7 000 kg DM/ha between mid-April and early June, by which date they were both at about 70% DM digestibility.

ENERGY AND PROTEIN

Attention was drawn by Hunt and Alexander (1961) to the frequently low crude protein (CP) content of Italian ryegrass. They found that the percentage of CP in the DM could be increased by up to 5 percentage units by raising the level of fertilizer input and that there were significant differences between varieties in CP concentration. In the same experiment, Hunt (1962a) demonstrated that there were significant differences between cultivars in the yield of CP over the 3 years of the experiment, but that the main effect was caused by the level of N input, the yield being 1 200 kg CP/ha at a low level and 1 900 kg at a high level. Increase in CP yield attributable to N manuring was thus due both to increased DM yield and to an increase in the percentage of CP in the DM.

Yields of 39·0 Mcal metabolizable energy/ha and 1 600 kg digestible CP/ha can be calculated from the data of Hunt (1962a) for Italian ryegrass in the first season (mean of 6 cultivars). Varietal differences in yield have been demonstrated, but with the exception of S22 (Minson *et al.*, 1964) there is no published information on the digestibility of regrowths. Under frequent cutting, the digestibility of the regrowth seldom declines below 70% digestibility and the N content rarely below 2·5%. Primary crop herbage harvested after the 70%-digestibility stage has a characteristically low protein content.

Response to management and environment

RESPONSE TO FERTILIZER

Nitrogen

Hunt (1962a, b) examined the response of 6 cultivars of Italian ryegrass to applied N. The grasses were cut 5 or 6 times in each of three years and were given 39 or 78 kg N/ha before each cut. Yield differences between cultivars were small, the range being about 10%, and there was no significant interaction between cultivar and rate of applied N. The mean yield of the cultivars at each rate of N and their N response are shown in Table 16.4. The decline

Table 16.4. Yield of DM and response to N of Italian ryegrass (average of 6 cultivars)

	Annual yield of DM		Response to N
	196 kg N/ha '00 kg/ha	392 kg N/ha '00 kg/ha	kg DM/kg N
1958	111·8	158·4	23·8
1959	101·5	131·5	15·3
1960*	75·4	91·2	8·6

* 20% more N in 1960

From Hunt (1962a)

in response from year to year was attributed to a progressively increasing deficiency of K.

Armitage and Templeman (1964) reported experiments in which N was applied to 2-year-old swards of Irish Italian ryegrass during 1956–9. The grass was cut 3 or 4 times each year and received 0, 420, 840 or 1 120 kg N/ha each year, divided for each cut. DM yield response to N was greater between 0 to 420 kg/ha than between 420 to 840 kg/ha. The responses to the successive rates of N were 15·7, 1·0 and −2·3 kg DM/kg N, respectively. The effect of 6 rates of applied N in the range 0 to 700 kg/ha on annual DM yields, examined in a further experiment by these authors, is shown in Table 16.5.

Wilman (1965) examined the response of S22 to N in spring, using a series of staggered growth periods. N was applied at 28, 84 or 140 kg/ha at the commencement of each period and yields of DM were estimated 2, 4 and 6 weeks later. The data demonstrated the increasing magnitude of the response to N with time after application. At N rates between 28 and 84 kg/ha, yield increases of 3·2, 15·4 and 28·6 kg DM/kg N were measured after 2, 4 and 6 weeks of growth, respectively (averages of two or three spring growth periods in 1961, 1963 and 1964). The comparable values for the yield increase to N rates between 28 and 140 kg N/ha were 1·4, 9·9 and 17·7 kg DM/kg N.

Table 16.5. Mean annual yield of DM and response to N of Italian ryegrass, 1960-61

N applied kg/ha	Yield of DM '00 kg/ha	Response to N* kg DM/kg N
0	61·5	...
117	78·7	14·7
233	103·6	21·3
350	111·5	6·8
466	117·1	4·8
700	120·3	2·7

* Response is calculated from the increase in yield above the next lowest level of N.

From Armitage and Templeman (1964)

In a two-year experiment at the Grassland Research Institute, Hurley, Cowling and Lockyer (1964) compared S22 Italian ryegrass with S24 perennial ryegrass. There were 5 or 6 cuts each year and N was given on 4 or 5 occasions at 22·5, 45 or 90 kg/ha. Yields and responses to N were almost identical for the two grasses. The response to N rates up to 200 kg/ha was linear, yields increasing by 22 kg DM/kg N. At rates between 200 and 400 kg N/ha, the response declined, yields increasing by 14·7 kg DM/kg N. Also at Hurley, a sward of S22 was given 0, 150, 300 or 600 kg N/ha, divided into

4 or 5 applications in each year (1963 and 1964). Response to N rates up to 300 kg/ha was almost linear, the increase in yield falling from 17·4 to 16·1 kg DM/kg N for the first two successive increments. Between the 300 and 600 kg N rates the increase fell to 4·1 kg DM/kg N.

Widdowson *et al.* (1962) studied the effect of 0, 125 and 250 kg N/ha, divided between three applications each year, on the yield of S22. The sward was cut three times each year. Response to the first 125 kg increment of N was equivalent to 29·2 and 46·2 kg DM/kg N in 1958 and 1959, respectively, and the response to the 125 and 250 kg rates/ha was 11·8 and 22·2 kg DM/kg N. In a further experiment, Widdowson *et al.* (1965) examined the response to N of two successively-sown swards of S22 given 0, 38, 76 or 114 kg N/ha for each of either two or three cuts annually. The increases in annual yield between successive increments of N averaged 34·8, 20·0 and 6·7 kg DM/kg N.

Phosphorus and potassium

The response of S22 Italian ryegrass to P and K was also measured in the latter experiment reported by Widdowson *et al.* (1965), but increases in yield due to either of these nutrients were small and not significant.

RESPONSE TO IRRIGATION

Little information is available on responses of Italian ryegrass to irrigation in Britain, although D'Aoust (1965) found that irrigation substantially increased its response to fertilizer N. In Germany, Klatt (1963) reported that 120 mm of irrigation increased the fresh matter yield from 71 400 to 90 000 kg/ha. Allowing for the usual fall in herbage DM content produced by irrigation, this indicates an irrigation efficiency of about 1–2 kg DM/m^3 water, which is typical for grasses.

RESPONSE TO TEMPERATURE

Italian ryegrass is reputed to start growth earlier in the year than some other grasses, although a comparison of the growth of S22 Italian ryegrass with that of S23 perennial ryegrass by Mudd (1965) showed less actual difference in early growth than would appear from the more erect habit of S22. Weihing (1963) found that the growth of an American race of Italian ryegrass was closely correlated with light energy receipt, but even more closely with the mean daily temperature between 7°C and 19°C. Nevertheless, Italian ryegrass may well be rather less sensitive to temperature than some other grasses. Ehara and Tanaka (1961) found that an increase in temperature in the range 15°–25°C decreased its growth slightly, whereas that of timothy was increased, and Ehara, Sasaki, Ikeda and Nata (1965) found the regrowth of Italian ryegrass to be the same at 30° and 20°C and only 50% greater than that at 10°C, whereas the growth rate of cocksfoot at 30°C was double that at 10°C. Beevers and Cooper (1964) showed that at 12°C Italian ryegrass grew twice as fast as at 25°C, although at an earlier growth stage this was not so, a maximum rate being then achieved in a 25°/12°C regime. They found no indication that Italian ryegrass was less sensitive to temperature than Irish perennial ryegrass. These results are to some extent conflicting, and clearly inadequate as a basis for understanding the temperature response of this species.

RESPONSE TO DEFOLIATION

S22 Italian ryegrass produced high yields under a rotational grazing management and when cut for conservation (Jones, 1953); regrowth was sufficiently rapid to allow 6 cuts for conservation during the season. Although considered to be a biennial (National Institute of Agricultural Botany, 1966a), yields have been maintained for three years under several defoliation regimes (Jones, 1953; Hunt, 1962b). Autumn defoliation has been considered particularly beneficial for this species (National Institute of Agricultural Botany, 1964).

Quality

CHEMICAL COMPOSITION

Italian ryegrass is relatively high in water-soluble carbohydrate compared with other grasses and tends to have a high Na content. Otherwise it is typical in chemical composition (see Chapter 13).

WATER CONTENT

Date of harvest has been found to influence the water content in successive periods of regrowth after cutting. Nicholson *et al.* (1953) took 7 successive harvests from the same Italian ryegrass swards between 30 May and 2 November when the herbage was between 8 and 10 inches high (20 to 25 cm). The water content fell progressively from 89·2% in May to 77·5% in November, even though the advance of the grass towards maturity was interrupted by repeated cutting.

Hunt and Alexander (1961) found that raising the amount of fertilizer N applied from 174 to 348 lb N/acre (from 191 to 383 kg N/ha) increased the percentage of water in Italian ryegrass. Values obtained at each of the 5 cuts taken are shown in Table 16.6.

NUTRITIVE VALUE (DRY MATTER DIGESTIBILITY)

Young leafy grass frequently cut

In swards of Italian ryegrass defoliated at 3- to 4-week intervals throughout the year, digestibility is similar to that of perennial ryegrass except that there

Table 16.6. Percentage of water in Italian ryegrass

Cut No.	High N (348 lb/ac)	Low N (174 lb/ac)
1	80	78
2	83·1	81·2
3	82·2	80·4
4	75·7	75·2
5	88·3	87·2

From Hunt and Alexander (1961)

may be a slight tendency for it to fall to lower levels in mid-summer, particularly under adverse conditions (e.g. drought). Its nutritive value under grazing conditions late in the season is excellent.

Uninterrupted primary growth

There are only small differences among varieties in heading dates in Italian ryegrass. Heading generally occurs in late May and digestibility remains at a satisfactory level until after heading. Minson *et al.* (1964) indicated a figure of about 75% DM digestibility in S22 at first ear emergence. Data from trials carried out by the National Institute of Agricultural Botany (1966a) indicate that S22, one of the latest-heading cultivars, is at 70% DM digestibility at 50% ear emergence or up to about 5 days later in late May or early June. The cultivars R.v.P. and Tetrone (tetraploid), which head about a week earlier, may be at 70% DM digestibility about 10 days after heading.

The digestibility curves for primary growth (Fig. 16.1, top) show that S22 has characteristics similar to S24 perennial ryegrass, although its heading date is about 2 weeks later. Late in the growth period, however, it maintains its digestibility better than S24, apparently due to a very high content of soluble carbohydrate (Table 16.7). DM yield during primary growth is similar to that of S24 (Fig. 16.1, middle) with a tendency to a slightly greater yield in S22. This, together with the level of digestibility, usually gives S22 some advantage over S24 in yield of digestible DM at the 70%-digestibility level.

Regrowths

Aftermath growth in the Italian ryegrasses after a conservation cut in early June is frequently of low digestibility owing to the tendency during regrowth to produce a large number of heads in second growth. In two months after cutting, digestibility may fall as low as 60% DM digestibility under some circumstances due to the stemmy nature of its growth. August regrowth also tends to be poor in quality, but growth later in autumn is of high digestibility.

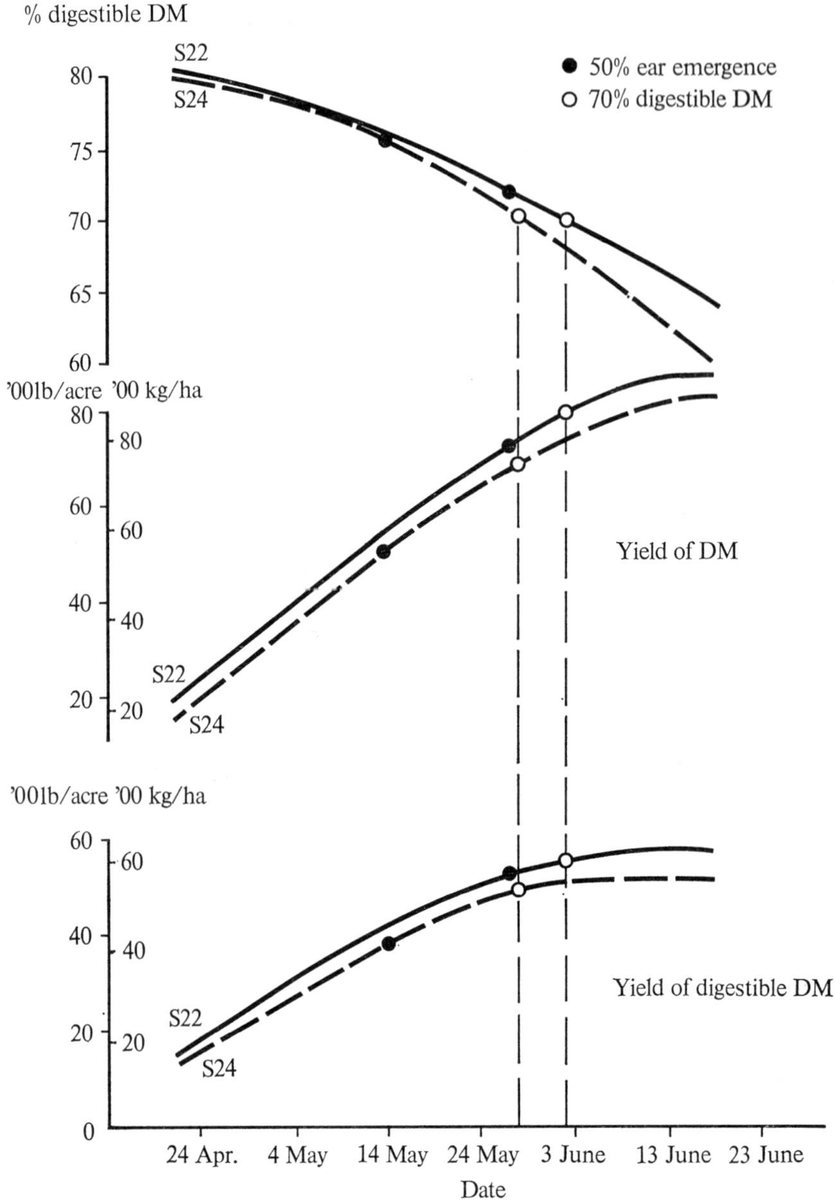

Fig. 16.1. Italian ryegrass S22. Seasonal changes in DM yield and digestibility. Values for S24 perennial ryegrass shown for comparison. (After Aldrich and Dent, 1967)

In general, therefore, Italian ryegrass in its first cropping year has some slight advantages over perennial ryegrass, but these may be offset in less favourable seasons by its stemmy, fibrous growth in mid-summer both in a frequently cut sward and in its aftermath growth following a conservation cut. The high digestibility of the species is associated with a very high soluble carbohydrate content (Table 16.7) and low protein content, particularly in extended primary growth.

Table 16.7. Percentage content on a DM basis of crude protein, crude fibre and soluble carbohydrate, and percentage digestible DM in Italian ryegrass

	Simulated grazing (weighted mean of 9 cuts)		Cut for hay (cut 10 days after 50% ear emergence)	
	S24	S22	S24	S22
Crude protein	19·0	17·5	8·5	6·5
Crude fibre	24·0	21·5	31·0	29·0
Soluble carbohydrate	15·0	19·5	20·0	28·0
Digestible DM	76·5	76·0	71·0	67·0

From National Institute of Agricultural Botany (1966b)

ANIMAL INTAKE CHARACTERISTICS

The New Zealand studies quoted in Chapter 15 suggest that the intake of Italian ryegrass by ruminants is greater than that of perennial ryegrass. Unpublished results at the Grassland Research Institute, Hurley (Thomson, 1963) have demonstrated that the intake of S22 Italian ryegrass is considerably greater than that of S48 timothy. Differences between two cultivars of Italian ryegrass have also been reported by Osbourn, Thomson and Terry (1966).

Longevity and persistency

Leaf longevity in this species appears to be similar to that in perennial ryegrass, but the information available is mainly inferred. In studies in New Zealand on an undefoliated sward, Hunt (1965) found that, in winter, Italian ryegrass maintained a constant three living leaves per tiller and produced one new leaf every 16–19 days.

Under similar conditions, Hunt and Brougham (1966) found that the number of leaves per tiller remained almost constant throughout the winter (May–September). Recalculation of their data on rate of leaf appearance indicates a range of 28 to 55 days in the life-span of individual leaves. They noted a fall in the amount of senescent material in a sward as many tillers approached the flowering stage; few new leaves were produced at that time, indicating that leaf longevity increases markedly when flowering takes place. Rates of leaf appearance and senescence are highest in summer; rates as high as 1·41 leaves per week were recorded in a controlled environment at 30°C by Edwards (1966).

Root longevity is less than in perennial ryegrass (Jacques & Schwass, 1956); rates of root death and renewal are high, and initiation of new roots begins later in the autumn than in perennial ryegrass. They attributed the decline in root and tiller numbers observed throughout the summer to rapid senescence of existing roots without root replacement, which made it impossible for the remaining roots to support all the tillers on the plant. Soper (1958) noted that the tendency of tillers to flower was greater than in perennial ryegrass, and that the production of a large number of inflorescences was associated with reduced longevity of the roots after the flower heads had died.

Italian ryegrass will flower in daylengths greater than 11 h, has no winter requirement for flowering (Cooper, 1951), and produces inflorescences throughout the summer, even when cut monthly (Cooper, 1950). This tendency to flower during much of the growing season adversely affects the potential longevity of tillers, since the florally-induced growing point not only loses the capacity to produce more leaves, but also is soon raised to a height at which it is destroyed if the sward is cut or grazed. Loss of tillers in this way is accentuated by management practices involving infrequent cutting.

Rate of tiller production is less than in perennial ryegrass, and most tillers are produced from aerial nodes (Cooper, 1951). Italian ryegrass usually behaves as a biennial (Cooper, 1951), but under rotational grazing is able to survive for extended periods if not sown with other grasses. For example, an S22/white clover sward which remained ryegrass-dominant for upwards of 5 years after sowing is mentioned by Charles (1964), who found that when perennial ryegrass was included in the seeds mixture, very few plants of Italian ryegrass remained after 3 years of grazing management.

Observations in the UK during the severe winter of 1962–63 indicated that Italian ryegrass is more susceptible than perennial ryegrass to winter killing (Miles & Williams, 1964); susceptibility is greater in spring-sown swards and in those over a year old than in those sown between mid-August and late September (Baker & David, 1963). Survival of plants exposed to drought and high temperature is reported to be significantly lower than in perennial ryegrass (Lucanus, Mitchell, Pritchard & Calder, 1960). Italian ryegrass is a species which produces a great deal of growth in the early spring and late autumn and only flourishes under conditions of high fertility (Stapledon *et al.*, 1927). Persistency may be greatly reduced if Italian ryegrass is undergrazed in the late autumn and carries a large amount of growth into the winter; this is especially true if winter conditions are severe. It has also been found that the greatest cumulative growth in March was obtained when swards of Italian ryegrass were harvested every month for the previous 6 months compared with similar swards which were harvested less frequently, the lowest total yield being the result of one cut only which was made in March (Stapledon *et al.*, 1927). Thus the autumn and winter management which has been found to encourage persistency during the first year also promotes the greatest total yield of DM.

A system of hard grazing when the crop is cleared very nearly to ground level, followed by a period of recovery, is one under which Italian ryegrass has been found to flourish (Stapledon *et al.*, 1927). There appear to be no data, however, which allow persistency under this rotational system of grazing to be compared with that under a system of continuous grazing.

References

ALDRICH, D. T. A.; DENT, J. W. 1967. The relationship between yield and digestibility in the primary growth of nine grass varieties. *J. natn. Inst. agric. Bot.* **11**, 104–13.

ARMITAGE, E. R.; TEMPLEMAN, W. G. 1964. Response of grassland to nitrogen fertilizer in the West of England. *J. Br. Grassld Soc.* **19**, 291–7.

ARMSTRONG, S. F. 1948. British grasses and their employment in agriculture. Cambridge: Univ. Press, 3rd edn.

BAKER, H. K.; CHARD, J. R. A.; DAVIES, J.; ALDRICH, D. T. A. 1965. A comparison of ryegrass varieties for early-bite production. *J. Br. Grassld Soc.* **20**, 151–5.

BAKER, H. K.; DAVID, G. L. 1963. Winter damage to grass. *Agriculture, Lond.* **70**, 380–2.

BEEVERS, L.; COOPER, J. P. 1964. Influence of temperature on growth and metabolism of ryegrass seedlings. 1. Seedling growth and yield components. *Crop Sci.* **4**, 139–43.

CHARLES, A. H. 1964. Differential survival of plant types in swards. *J. Br. Grassld Soc.* **19**, 198–204.

CLAPHAM, A. R.; TUTIN, T. G.; WARBURG, E. F. 1962. Flora of the British Isles. Cambridge: Univ. Press, 2nd edn.

COOPER, J. P. 1950. Day-length and head formation in the ryegrasses. *J. Br. Grassld Soc.* **5**, 105–12.

COOPER, J. P. 1951. Studies on growth and development in *Lolium*. 2. Pattern of bud development of the shoot apex and its ecological significance. *J. Ecol.* **39**, 228–70.

CORRALL, A. J. 1965. Development of first crop-yield and digestibility. *Internal Rep. 31 Grassld Res. Inst.*

COWLING, D. W.; LOCKYER, D. R. 1964. Nitrogenous fertilizers and the role of legumes. *Exps Prog. 16 Grassld Res. Inst. 1962–63* pp. 22–3.

D'AOUST, M. J. 1965. Some aspects of the interaction between nitrogen and water in the growth of grass swards. Ph.D. Thesis, Univ. Reading. (Cited in *Herb. Abstr.* **36**: 46)

EDWARDS, K. J. R. 1966. Developmental genetics of leaf formation in ryegrass. *Rep. Welsh Pl. Breed. Stn 1965* pp. 20–3.

EHARA, K.; SASAKI, T.; IKEDA, H.; NATA, Y. 1965. Physiological and ecological studies on the regrowth of herbage plants. *J. Jap. Soc. Grassld Sci.* **10**, 214–20.

EHARA, K.; TANAKA, S. 1961. Effect of temperature on the growth, behaviour and chemical composition of the warm- and cool-season grasses. *Proc. Crop Sci. Soc. Japan* **29**, 304–6.

EVERS, A.; SONNEVELD, A. 1953. Grass seed production trials. *Gestenc. Meded. 12 Cent. Inst. Landbouwk. Onderz. Wageningen.*

GERDES, G. 1952. [Studies on the absolute weight (1 000 seed weight) of the most important fodder grass species and their varieties in regard to the performance of grass breeding.] *Züchter* **22**, 353–66.

GRASSLAND RESEARCH INSTITUTE. 1958. *Exps Prog. 10 Grassld Res. Inst. 1956–7* p. 21.

GRASSLAND RESEARCH INSTITUTE. 1959. *Exps Prog. 11 Grassld Res. Inst. 1957–8* p. 23.

Grassland Research Institute. 1960. *Exps Prog. 12 Grassld Res. Inst. 1958–9* p. 16.

Grassland Research Institute. 1961. *Exps Prog. 13 Grassld Res. Inst. 1959–60* p. 14.

Grassland Research Institute. 1962. *Exps Prog. 14 Grassld Res. Inst. 1960–61* p. 16.

Harkess, R. D. 1965. The effect of seed size on early growth of diploid and tetraploid Italian ryegrass. *J. Br. Grassld Soc.* **20,** 190–3.

Hubbard, C. E. 1954. Grasses; a guide to their structure, identification, uses, and distribution in the British Isles. Harmondsworth, Middlesex: Penguin Books.

Hunt, I. V. 1962a. Productivity, persistence and response to nitrogen of Italian ryegrass varieties. *J. Br. Grassld Soc.* **17,** 125–9.

Hunt, I. V. 1962b. Comparing yield, persistence and nitrogen response of varieties of Italian and Westerwolds ryegrass 1957–1960. *Exp. Rec. 3 Grassld Husb. Dep. W. Scotl. agric. Coll.* pp. 31.

Hunt, I. V.; Alexander, R. H. 1961. The crude-protein content of Italian ryegrass. *J. Br. Grassld Soc.* **16,** 226–9.

Hunt, L. A. 1965. Some implications of death and decay in pasture production. *J. Br. Grassld Soc.* **20,** 27–31.

Hunt, L. A.; Brougham, R. W. 1966. Some aspects of growth in an undefoliated stand of Italian ryegrass. *Jnl appl. Ecol.* **3,** 21–8.

Jacques, W. A.; Schwass, R. H. 1956. Root development in some New Zealand pasture plants. 7. Seasonal root replacement in perennial ryegrass (*Lolium perenne*), Italian ryegrass (*L. multiflorum*) and tall fescue (*Festuca arundinacea*). *N.Z. Jl Sci. Technol. Sect. A* **37,** 570–83.

Jones, L. I. 1953. Preliminary note on S22 Italian ryegrass. *J. Br. Grassld Soc.* **8,** 345–8.

Jones, L. I. 1958. Grassland agronomy. *Rep. Welsh Pl. Breed. Stn 1950–56* pp. 97–9.

Klatt, F. 1963. [Results and conclusions from field irrigations 1962.] *Dt. Landwirt.* **14,** 227–30.

Lampeter, W. 1960. [Mutual influences of higher plants regarding growth of shoots and roots, mineral content and water consumption studied in some economically important forage plants.] *Wiss. Z. Karl-Marx-Univ. Lpz. 1959–60* **9,** 611–722.

Lincolnshire Seed Growers' Association. 1965. Report for 1964. Agric. House, Woodhall Spa, pp. 3–7.

Lucanus, R.; Mitchell, K. J.; Pritchard, G. G.; Calder, D. M. 1960. Factors influencing survival of strains of ryegrass during the summer. *N.Z. Jl agric. Res.* **3,** 185–93.

Miles, D. G.; Williams, I. G. 1964. Winter hardiness in pasture varieties. *Rep. Welsh Pl. Breed. Stn 1963* pp. 70–1.

Ministry of Agriculture, Fisheries and Food. 1968. Grass and clover crops for seed. *Bull. 204* London: H.M.S.O., pp. 42–60.

Minson, D. J.; Harris, C. E.; Raymond, W. F.; Milford, R. 1964. The digestibility and voluntary intake of S22 and H1 ryegrass, S170 tall fescue, S48 timothy, S215 meadow fescue and Germinal cocksfoot. *J. Br. Grassld Soc.* **19,** 298–305.

Mudd, C. H. 1965. A comparison between Italian ryegrass, perennial ryegrass and timothy under upland conditions. *Expl Husb.* **12,** 51–61.

National Institute of Agricultural Botany. 1959. Growing grasses for seed. *Herbage Seed Growers' Leaflet No. 2.*

National Institute of Agricultural Botany. 1964. Varieties of ryegrass. *Farmers' Leaflet No. 16.*

National Institute of Agricultural Botany. 1966a. Varieties of grasses. *Farmers' Leaflet No. 16.*

National Institute of Agricultural Botany. 1966b. Unpublished data.

Nicholson, H. H.; Firth, D. H.; Eden, A.; Alderman, G.; Baker, C. J. L.; Heimberg, M. 1953. The effect of ground water level upon productivity and composition of Fenland grass. 2. *J. agric. Sci., Camb.* **43,** 265.

Osbourn, D. F.; Thomson, D. J.; Terry, R. A. 1966. The relationship between voluntary intake and digestibility of forage crops, using sheep. *Proc. 10th int. Grassld Congr., Helsinki, 1966* pp. 363–7.

Perring, F. H.; Walters, S. M. 1962. Atlas of the British flora. London: Thomas Nelson and Sons Ltd, pp. 456.

PETO, D. H. 1961. Methods and problems of cleaning herbage seeds. *Rep. Lincs. Seed Growers' Assn 1960*. Agric. House, Woodhall Spa, pp. 19–21.

SCHWANBOM, N.; FRÖIER, K. 1949. Methods of herbage seed production. *J. Br. Grassld Soc.* **4,** 233–61.

SOPER, K. 1958. Effects of flowering on the root system and summer survival of ryegrass. *N.Z. Jl agric. Res.* **1,** 329–40.

SPURWAY, C. H. 1941. Soil reaction (pH) preferences of plants. *Spec. Bull. 306 Mich. agric. Exp. Stn* pp. 36.

STAPLEDON, R. G.; FAGAN, T. W.; EVANS, R. E.; MILTON, W. E. J. 1927. Italian ryegrass for winter and early spring keep: the effect of methods of grazing on productivity and palatability, and on the chemical and botanical composition of the herbage. *Bull. H5 Welsh Pl. Breed. Stn* pp. 5–41.

THOMSON, D. J. 1963. Grassland Research Institute, Hurley. Personal communication.

TWYFORD SEEDS. 1964. Tetraploid establishment. *Spring catalogue 1964* Twyford Seeds Ltd, Adderbury, Oxon., p. 9.

VYNCKE, A. 1955. [Cultivation of Italian ryegrass for seed.] *Revue Agric., Brux.* **8,** 1443–8.

WEIHING, R. M. 1963. Growth of ryegrass as influenced by temperature and solar radiation. *Agron. J.* **55,** 519–21.

WIDDOWSON, F. V.; PENNY, A.; WILLIAMS, R. J. B. 1962. An experiment comparing urea-formaldehyde fertilizer with 'nitro-chalk' for Italian ryegrass. *J. agric. Sci., Camb.* **59,** 263–8.

WIDDOWSON, F. V.; PENNY, A.; WILLIAMS, R. J. B. 1965. An experiment measuring effects of N, P and K fertilizers on yield and N, P and K contents of grass. *J. agric. Sci., Camb.* **64,** 93–100.

WILMAN, D. 1965. The effect of nitrogenous fertilizer on the rate of growth of Italian ryegrass. *J. Br. Grassld Soc.* **20,** 248–54.

CHAPTER 17

Westerwolds ryegrass

(Lolium multiflorum var. westerwoldicum)

Morphology

BOTANICALLY, Westerwolds ryegrass can be regarded as a group of ecotypes, the morphology of which is that of Italian ryegrass. Westerwolds ryegrass is regarded by some authorities as being a separate sub-species (*Lolium multiflorum* ssp. *gaudinii* (Parl. 1848) Sch. & Kell. 1914). Others regard it as an extreme 'form' of Italian ryegrass.

Distribution

No ecological data appear to be available.

Compatibility

Westerwolds ryegrass establishes faster than any other cultivated grass in Britain and comes into full production very soon after sowing. Also, as represented by the 'commercial' stocks of unimproved landraces, it is a true annual and the most short-lived grass. These characteristics make Westerwolds ryegrass suitable for use as a companion for species or mixtures which are slow developing in order to enhance production in the early stages. If sown in early spring, it is likely to be dead or nearly so by late summer, leaving time for the slower species to establish before winter.

In experiments at the Grassland Research Institute, Hurley, commercial stocks of Westerwolds ryegrass have been used successfully in this way. It has been grown with timothy/white clover (1961, 1962), timothy/meadow fescue/white clover (1960) and low seed rates of perennial ryegrass/white clover (1962). In the year of sowing, production was very much increased, with satisfactory suppression of weeds, compared with swards sown without Westerwolds ryegrass, and there was no disadvantage in the first harvest year.

It must be emphasized that the above qualities only apply to the unimproved landraces known usually as Dutch commercial Westerwolds ryegrass.

There are new cultivars available which more closely resemble Italian ryegrass in persistency and these present too long a period of competition to be suitable as companions for slow-establishing species. In experiments at the Grassland Research Institute, Hurley (1960, 1961), timothy developed better and gave greater yields in the first harvest year when grown with 'commercial' Westerwolds ryegrass than when a bred cultivar was used. The bred cultivar survived into the spring of the first harvest year.

Seed

CHARACTERS

There is no detailed description in the literature of the morphology of Westerwolds ryegrass seed, but it is probably very similar to Italian ryegrass. The 1 000-seed weight is in the range 1·82–2·27 g (Whyte *et al.*, 1959).

PRODUCTION

There appears to be little published advice and no experimental results on seed production from Westerwolds ryegrass. Sonneveld and Evers (1954) recommend that the crop be grown in the same way as a spring variety of cereals, with the application of 100–120 kg N/ha. Principles and procedures for harvesting are the same as for Italian ryegrass (Ministry of Agriculture, Fisheries & Food, 1968).

The average annual yield of seed has been reported to vary from 1 000 to 1 600 kg/ha in the Netherlands, with maximum yields of 1 800 kg (Sonneveld & Evers, 1954).

Yield

DRY MATTER

Westerwolds ryegrasses, selected over many years for rapid seedling growth, are chiefly noted for a rapid attainment of not very high yields shortly after sowing. For instance, in a trial sown in April, Westerwolds cultivars could be cut for hay in July and they also provided two subsequent cuts in the same year (National Institute of Agricultural Botany, 1961). There was little difference in the yield of hay between commercial and bred material. However, in monthly cuts beginning two months after sowing, the commercial varieties were at first the highest yielding but were soon surpassed by bred cultivars.

The capacity of Westerwolds ryegrasses to produce herbage in the spring of the year after sowing was examined by Baker *et al.* (1965). Dry matter (DM) yields between 1 000 and 1 500 kg/ha were obtained in late April. Frequent cutting in the year of sowing appeared to be necessary to ensure adequate survival of plants over winter, but the authors concluded that one-year-old Westerwolds swards should not be relied on for early-spring production.

ENERGY AND PROTEIN

Yields of energy from Westerwolds ryegrasses are generally lower than those from Italian and H.1 ryegrasses (Hunt, 1962) and yields of protein have also been found to be lower (Hunt, 1962).

Response to management and environment

RESPONSE TO FERTILIZER

There is only one report from the UK of an experiment dealing with the response of this grass to fertilizer. Barlow (1965) compared the N response of diploid and tetraploid forms with that of Italian ryegrass. The grasses were cut 4 times in each of two years and given from 125 to 500 kg N/ha in divided applications each year. The two forms of Westerwolds ryegrass were very similar, both giving a linear increase in yield up to 375 kg N/ha but no further increase in yield at the highest level of N. Over the linear part of the response curve, yield increased by 12·8 kg DM/kg N.

RESPONSE TO IRRIGATION

No results of experiments on the irrigation of Westerwolds ryegrass have been reported.

RESPONSE TO TEMPERATURE

Little is apparently known of the response of Westerwolds ryegrass to temperature, although it can be inferred from observations in the field and glasshouse that it is unlikely to differ much from that of other north-temperate grasses.

RESPONSE TO DEFOLIATION

Westerwolds ryegrasses are usually considered as annuals (National Institute of Agricultural Botany, 1964), but Hunt (1962) found yields were maintained for two years when the sward was cut 5 times during the season (4 to 8 weeks between cuts). Baker *et al.* (1965) considered that a very frequent cutting management in the year of sowing was necessary to maintain production in the following year. Bred cultivars are generally superior to commercial varieties after the initial cuts have been taken. Although the yields at the first cut for conservation were similar, bred cultivars yielded more aftermath 8 weeks later (National Institute of Agricultural Botany, 1961); also, bred cultivars subsequently outyielded commercial varieties under a monthly system of cutting.

Quality

CHEMICAL COMPOSITION

There appear to be no recent data on chemical composition, but Norman (1936) has reported values for some organic constituents.

WATER CONTENT

The water content of tetraploid cultivars of Westerwolds ryegrass has been shown to be higher than that of diploids; Glas (1961) found that they contained 1–2% more water than diploids. Korsten (1961) found that tetraploid cultivars gave higher yields of green herbage but lower yields of DM than diploid cultivars, the inference being that the percentage water content of the tetraploids was higher.

NUTRITIVE VALUE (DRY MATTER DIGESTIBILITY)

There is not much information available on the quality of Westerwolds ryegrass, but it is likely that frequently cut or grazed material would be similar to that of Italian ryegrass in digestibility.

In a spring sown trial at the National Institute of Agricultural Botany, Cambridge, varieties of Westerwolds ryegrass were at 70% DM digestibility in early July, about 10 days after ear emergence (Fig. 17.1). This behaviour is similar to that of early Italian ryegrass varieties, but yields at this stage of growth were much lower than those of Italian ryegrass (Unpublished data).

ANIMAL INTAKE CHARACTERISTICS

No evidence is available.

Longevity and persistency

No data are available on the longevity of leaves. Rate of leaf appearance is more rapid and rate of tiller production slower than in perennial ryegrass (Patel & Cooper, 1961). Westerwolds ryegrass is basically an annual species which can either be sown in the spring, or in the autumn for use in the following spring. If undefoliated during growth, the sward behaves as a cereal crop and does not recover after harvest; in summer and autumn, floral development will start at about the 6-leaf stage (Patel & Cooper, 1961) and nearly all tillers flower. Where there is soil-surface drought, the few remaining tillers cannot establish roots and the plant usually dies (Soper, 1958).

Intermittent defoliation close enough to the ground to prevent the flowering of individual tillers will induce the plants to persist after the normal time of flowering until the end of the current growing season. Hunt (1962) measured respective annual yields of about 14 000 and 11 000 kg DM/ha in 1958 and 1959 from Westerwolds ryegrass swards established in 1957, given a high level of N and cut 5 or 6 times per year. These swards, however, failed to

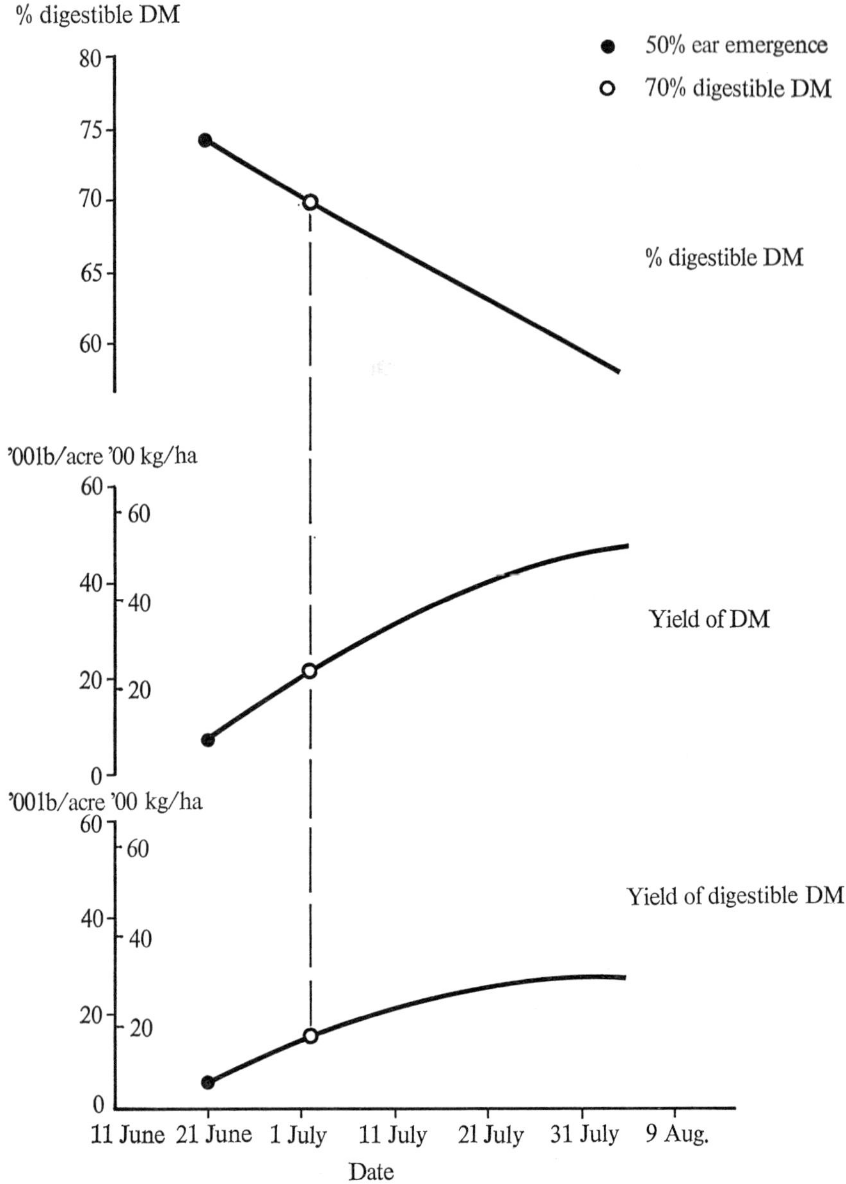

Fig. 17.1. Westerwolds ryegrass Million. Seasonal changes in DM yield and digestibility. (After Aldrich and Dent, 1967)

survive the 1959–60 winter. Westerwolds ryegrass thus needs careful management to enable it to persist for a whole season, and is not generally considered ideal as a grazing plant. Italian ryegrass is more persistent under cutting than Westerwolds ryegrass (Hunt, 1962; Barlow, 1965); the main distinction between them is that Westerwolds ryegrass is very short-lived and usually behaves as an annual under British conditions, though this may also be true of some Italian ryegrasses.

References

ALDRICH, D. T. A.; DENT, J. W. 1967. The relationship between yield and digestibility in the primary growth of nine grass varieties. *J. natn. Inst. agric. Bot.* **11,** 104–13.

BAKER, H. K.; CHARD, J. R. A.; DAVIES, J.; ALDRICH, D. T. A. 1965. A comparison of ryegrass varieties for early-bite production. *J. Br. Grassld Soc.* **20,** 151–5.

BARLOW, C. 1965. The use of special-purpose grasses for silage production. 1. A comparison of varieties. *J. agric. Sci., Camb.* **64,** 439–47.

GLAS, D. J. 1961. [Breeding tetraploid ryegrasses.] *Meded. ned. alg. KeurDienst LandbZaden Aardappelpootg.* **18,** 90–2.

GRASSLAND RESEARCH INSTITUTE. 1960. *Exps Prog. 12 Grassld Res. Inst. 1958–9* p. 16.

GRASSLAND RESEARCH INSTITUTE. 1961. *Exps Prog. 13 Grassld Res. Inst. 1959–60* p. 14.

GRASSLAND RESEARCH INSTITUTE. 1962. *Exps Prog. 14 Grassld Res. Inst. 1960–61* p. 16.

HUNT, I. V. 1962. Comparing yield, persistence and nitrogen response of varieties of Italian and Westerwolds ryegrass 1957–1960. *Exp. Rec. 3 Grassld Husb. Dep. W. Scotl. agric. Coll.* pp. 31.

KORSTEN, L. 1961. [CB tetraploid Westerwolds ryegrass.] *Meded. nat. coöp. Aan- en Verkoopv. Landb. Cent. Bur.* **45,** 86.

MINISTRY OF AGRICULTURE, FISHERIES AND FOOD. 1968. Grass and clover crops for seed. *Bull. 204* London: H.M.S.O., pp. 42–60.

NATIONAL INSTITUTE OF AGRICULTURAL BOTANY. 1961. Varieties of Italian ryegrass. *Rep. nat. Inst. agric. Bot. 1960* **41,** 39–40.

NATIONAL INSTITUTE OF AGRICULTURAL BOTANY. 1964. Varieties of ryegrass. *Farmers' Leaflet No. 16.*

NATIONAL INSTITUTE OF AGRICULTURAL BOTANY. Unpublished data.

NORMAN, A. G. 1936. The composition of forage crops. 1. Ryegrass (Western Wolths). *Biochem. J.* **30,** 1354–62.

PATEL, A. S.; COOPER, J. P. 1961. The influence of seasonal changes in light energy on leaf and tiller development in ryegrass, timothy and meadow fescue. *J. Br. Grassld Soc.* **19,** 299–308.

SONNEVELD, A.; EVERS, A. 1954. Some results of recent research on grass seed production in the Netherlands. *Proc. Europ. Grassld Conf. 1954* pp. 141–9.

SOPER, K. 1958. Effects of flowering on the root system and summer survival of ryegrass. *N.Z. Jl agric. Res.* **1,** 329–40.

WHYTE, R. O.; MOIR, T. R. G.; COOPER, J. P. 1959. Grasses in agriculture. *FAO Agric. Stud. No. 42.* Rome: FAO, pp. 416.

CHAPTER 18

Hybrid ryegrasses

(Lolium perenne × multiflorum)

Morphology

SEVERAL cultivars derived from crosses between perennial ryegrass and Italian ryegrass have been produced, notably Grasslands Manawa ryegrass (formerly H.1). These hybrids exhibit morphological characters intermediate between those of their parents; some cultivars breed true but others show segregation of parental characters. They are perennials, being longer-lived than Italian ryegrass but not as long-lived as the more persistent cultivars of perennial ryegrass. Perennial ryegrass also readily crosses with meadow fescue to form a natural sterile hybrid *Festulolium loliaceum* (Huds.) P. Fourn. which is highly competitive and vegetatively productive under sward conditions.

Distribution

No ecological data appear to be available on hybrid ryegrasses.

Compatibility

There is little or no published information for hybrid ryegrasses, but they are probably intermediate between perennial and Italian ryegrasses.

Seed

CHARACTERS

There is no detailed description of the seed of Grasslands Manawa ryegrass, but it has characters intermediate between those of perennial ryegrass and Italian ryegrass (Hubbard, 1954) and usually nearer those of Italian ryegrass. The seed is therefore lanceolate to ovate-oblong with a shallow groove along its ventral surface. Seed length is 5–8 mm. The lemma is blunt or slightly pointed at the apex, and may be awned or awnless (Hubbard,

1954), but usually has an awn (Hawkins & Clouting, 1964); the lemma is convex and usually smooth. The palea is similar in length to the lemma. The rachilla is close to the palea, and may be flattened or almost cylindrical in cross-section (cf. Armstrong, 1948; Hubbard, 1954).

The 1 000-seed weight, derived from data presented by Walters (1953), is in the range 2·12–2·17 g.

PRODUCTION

There appears to be no information specifically for Grasslands Manawa ryegrass, but the general advice given for the parent species (National Institute of Agricultural Botany, 1959) applies to seed production of the hybrid.

The average annual yield of seed has been reported to vary from 320 kg/ha (Portugal) and 400 kg (New Zealand) to 600–700 kg/ha (Germany) (FAO, 1961).

Yield

DRY MATTER

Williams (1947) described the origin of these grasses and gave their chief yield characteristics as more rapid growth in the year of sowing and in winter and early spring than the perennial parent. The hay yield was lower than that from perennial ryegrass but higher than from Italian ryegrass. However, consistently high yields in the three years after sowing have been reported by Williams, Stiles and Turner (1959 and 1960), 13 000, 12 000 and 12 000 kg dry matter (DM)/ha being recorded in three years under irrigation, and by Hunt (1962a), who obtained yields of 16 000, 14 000 and 10 000 kg DM/ha under a high level of N manuring.

Regarding early-spring production, Williams, Stiles and Turner (1960) showed the relative superiority of hybrid ryegrasses over an early cultivar of perennial ryegrass; they had attained a yield of 660 kg/ha by 1 April. Baker *et al.* (1965) showed that yields of this grass in late April were about 1 250 and 1 750 kg/ha, but a crop sown in the spring of one year was almost killed out during the following hard winter.

ENERGY AND PROTEIN

Yields and digestibility of hybrid ryegrasses tend to be lower than those of Italian ryegrass at comparable stages of growth. In the first harvest year, Hunt (1962b) recorded a DM yield of 16 000 kg/ha. On the basis of the limited information available (Minson *et al.*, 1964), the yield of metabolizable energy is lower than that of Italian ryegrass, but the N content of both primary and regrowth herbage is higher.

Hunt (1962a) found that yields of CP were dependent on the level of N manuring and were about 2 000 kg/ha where a high level of N was supplied.

Response to management and environment

RESPONSE TO FERTILIZER

Nitrogen

Although these grasses are frequently used in practice and are given large quantities of fertilizer N, there is little experimental information about their response to N. Hunt (1962b) compared H.1 (Grasslands Manawa) with a number of varieties of Italian ryegrass in a three-year experiment. The plots were cut 5 or 6 times each year and given either 39 or 78 kg N/ha before each cut. H.1 usually ranked first in total annual yield at each level of N and was significantly higher-yielding than most varieties of Italian ryegrass; response to N was, however, similar for all grasses. Yield was increased by 23·8, 18·0 and 9·0 kg DM/kg N in the three successive years of the experiment.

Phosphorus and potassium

There is no information from the UK about the response of these grasses to P or K.

RESPONSE TO IRRIGATION

Hybrid ryegrasses show the typical response to irrigation described in Chapter 14 (Williams *et al.*, 1959, 1960).

RESPONSE TO TEMPERATURE

Information on the effects of temperature is almost entirely derived from work by Mitchell (1954, 1956) on a commercial strain of New Zealand short-rotation ryegrass (Grasslands Manawa). In 1954, he compared its growth with that of S23 perennial ryegrass in temperature regimes of 21°/15°C and 13°/7°C and light intensities of 7 000 and 20 000 lux. An increase in light or in day and night temperature always increased growth, and short-rotation ryegrass invariably outyielded S23. In 1956 he reported an optimum temperature for growth of about 21°C in 12-h days, and reported with Lucanus (1962) an optimum similar in 16-h days but a degree or two lower in 8-h days. From these studies, and from others in collaboration with Lucanus (1960), he concluded that in short-rotation ryegrass, as in perennial ryegrass, cocksfoot and other grasses, day temperature was a more important determinant of yield than was night temperature.

RESPONSE TO DEFOLIATION

Hybrid ryegrasses include some selections similar to Italian ryegrass and others similar to perennial ryegrass (National Institute of Agricultural

Botany, 1966). Generally, hybrid ryegrasses produce high yields under a hay and aftermath system of management (Williams, 1947) and retain this capacity for high production when cut 4 times (Beddows, 1958) and 5 or 6 times during the growing season (Hunt, 1962a). More work on this group has been carried out in Australia and New Zealand, where it has performed well under grazing conditions (New Zealand Journal of Agriculture, 1945) and apparently combines some of the better characteristics of its parents (Walker, 1959).

Quality

CHEMICAL COMPOSITION

There is very little information on the chemical composition of the hybrid ryegrasses, but it is likely to be similar to that of perennial and Italian ryegrass.

WATER CONTENT

Hunt and Gardner (1956) grew H.1 (Grasslands Manawa) ryegrass with small amounts of Ladino clover or with S153 or Montgomery red clover. The seeds mixtures used consisted of 30 lb grass and 3 lb clover/ac (33·0 kg and 3·3 kg/ha, respectively). These three mixtures were harvested in June, August and October of 1951, and in May and August of 1952. Both the green and dry weight of herbage at each harvest are given in the above publication, so that percentage water content can be calculated for the first growth and two regrowths in 1951 and for the first growth and one regrowth in 1952. These values are shown in Table 18.1.

Table 18.1. Percentage water content of hybrid ryegrass in mixtures with clover

	Cutting date				
	1951			1952	
Herbage mixture	4 June	8 Aug.	11 Oct.	24 May	4 Aug.
H.1 + Ladino clover	75·9	72·7	83·6	82·1	77·1
H.1 + S123 red clover	77·4	72·8	83·3	80·1	79·5
H.1 + Montgomery red clover	79·4	77·4	82·0	82·4	78·1

In all instances in 1951, the herbage cut in August had the lowest water content and that cut in October the highest. In 1952, the herbage cut in August had a lower water content than that cut in May. This suggests a seasonal trend, with herbage which has grown early and late in the season having higher water contents than that grown in mid-season.

H

NUTRITIVE VALUE (DRY MATTER DIGESTIBILITY)

Young leafy grass frequently cut

As in both perennial and Italian ryegrass, the hybrid ryegrasses give material of excellent quality in swards defoliated frequently throughout the season.

Uninterrupted primary growth

Trials at the National Institute of Agricultural Botany, Cambridge (Unpublished) indicate that the two early cultivars from New Zealand, Grasslands Ariki and Grasslands Manawa, reach 70% DM digestibility about 10 days after 50% ear emergence. Vertas Long Rotation, which flowers about 10 days later, reaches this level of quality at about 5 days after ear emergence.

Regrowths

Flowering occurs particularly in the first regrowth and the resulting stemminess may cause low digestibility, particularly in hybrids which are nearer to the Italian type. Later regrowths are normally of good quality.

In all respects these hybrid ryegrasses are of a quality intermediate between that of perennial ryegrass and Italian ryegrass. Typical growth and digestibility curves may be deduced from those given for S24 perennial ryegrass and S22 Italian ryegrass in Fig. 16.1 (Chapter 16), while composition is indicated in Table 16.9 of the same chapter.

ANIMAL INTAKE CHARACTERISTICS

The differences between perennial and Italian ryegrasses as regards animal intake, referred to in Chapters 15 and 16, would seem to find expression within the hybrid ryegrasses in that some, e.g. Grasslands Ariki, resemble the perennial grasses and intake by animals is lower than with those resembling the Italian type, for example Grasslands Manawa.

Longevity and persistency

Hybrid ryegrasses are generally intermediate between Italian and perennial ryegrass and survive for an extended period under lenient grazing, but under continuous hard grazing are little better than annuals. The roots have been found to be short-lived (Soper, 1958). Eifrig (1968) reports that some cultivars respond to vernalization in the same way as Italian ryegrass, whereas others respond like perennial ryegrass. Since removal of tillers from the sward by their becoming reproductive is an important factor in the survival of ryegrass, this finding would suggest that the range of longevities to be found among cultivars of hybrid ryegrass may be quite considerable.

The reaction of hybrid ryegrasses to grazing is similar to that of Italian ryegrass, but the hybrids have a greater degree of persistency after the first season, inherited from the perennial parent. Managed in a similar way to that described for Italian ryegrass, the hybrids are therefore more desirable when grazing is to be carried beyond the second year after sowing. Except on soils of high fertility, the hybrids die out during the third year, but before that time their persistency has been found comparable to that of Irish perennial ryegrass. Given the same treatment, however, they are less persistent than S23 perennial ryegrass (Williams, 1947).

References

ARMSTRONG, S. F. 1948. British grasses and their employment in agriculture. Cambridge: Univ. Press, 3rd edn.

BAKER, H. K.; CHARD, J. R. A.; DAVIES, J.; ALDRICH, D. T. A. 1965. A comparison of ryegrass varieties for early-bite production. *J. Br. Grassld Soc.* **20,** 151–5.

BEDDOWS, A. R. 1958. Single plants as a basis of assessment in comparison with drills and plots. *Rep. Welsh Pl. Breed. Stn 1950–56* pp. 17–18.

EIFRIG, H. 1968. Laboratory methods for the identification of *Lolium* varieties. *Z. Acker- u. PflBau* **128,** 46–70.

F.A.O. 1961. Agricultural and horticultural seeds. *FAO agric. Stud. No. 55.* Rome: FAO.

HAWKINS, R. P.; CLOUTING, G. N. 1964. The classification of grass seed stocks. *J. natn. Inst. agric. Bot.* **10,** 104–9.

HUBBARD, C. E. 1954. Grasses; a guide to their structure, identification, uses, and distribution in the British Isles. Harmondsworth, Middlesex: Penguin Books.

HUNT, I. V. 1962a. Productivity, persistence and response to nitrogen of Italian ryegrass varieties. *J. Br. Grassld Soc.* **17,** 125–9.

HUNT, I. V. 1962b. Comparing yield, persistence and nitrogen response of varieties of Italian and Westerwolds ryegrass 1957–1960. *Exp. Rec. 3 Grassld Husb. Dep. W. Scotl. agric. Coll.* pp. 31.

HUNT, I. V.; GARDNER, A. L. 1956. Two-year leys for silage. *Res. Bull. 15 W. Scotl. agric. Coll.*

MINSON, D. J.; HARRIS, C. E.; RAYMOND, W. F.; MILFORD, R. 1964. The digestibility and voluntary intake of S22 and H.1 ryegrass, S170 tall fescue, S48 timothy, S215 meadow fescue and Germinal cocksfoot. *J. Br. Grassld Soc.* **19,** 298–305.

MITCHELL, K. J. 1954. Influence of light and temperature on the growth of ryegrass (*Lolium* spp.). 3. Pattern and rate of tissue formation. *Physiologia Pl.* **7,** 51–65.

MITCHELL, K. J. 1956. Growth of pasture species under controlled environments. 1. Growth at various levels of constant temperature. *N.Z. Jl Sci. Technol. Sect. A* **38,** 203–15.

MITCHELL, K. J.; LUCANUS, R. 1960. Growth of pasture species in controlled environment. 2. Growth at low temperature. *N.Z. Jl agric. Res.* **3,** 647–55.

MITCHELL, K. J.; LUCANUS, R. 1962. Growth of pasture species under controlled environment. 3. Growth at various levels of constant temperature with 8 and 16 hours of uniform light per day. *N.Z. Jl agric. Res.* **5,** 135–44.

NATIONAL INSTITUTE OF AGRICULTURAL BOTANY. 1959. Growing grasses for seed. *Herbage Seed Growers' Leaflet No. 2.*

NATIONAL INSTITUTE OF AGRICULTURAL BOTANY. Unpublished data.

NATIONAL INSTITUTE OF AGRICULTURAL BOTANY. 1966. Varieties of grasses. *Farmers' Leaflet No. 16.*

NEW ZEALAND JOURNAL OF AGRICULTURE. 1945. Short rotation ryegrass. Its breeding and characteristics. *N.Z. Jl Agric.* **71,** 465–70.

SOPER, K. 1958. Effects of flowering on the root system and summer survival of ryegrass. *N.Z. Jl agric. Res.* **1,** 329–40.

WALKER, M. H. 1959. H1 ryegrass production compared with that of Italian ryegrass at the New England Experimental Farm. *Agric. Gaz. N.S.W.* **70,** 528–31.

WALTERS, P. 1953. Re-seeding experiments: some factors affecting establishment and initial performance of grasses under hill conditions. *J. Br. Grassld Soc.* **8,** 71–90.

WILLIAMS, T. E. 1947. Short rotation ryegrass. *Agriculture, Lond.* **54,** 13–15.

WILLIAMS, T. E.; STILES, W.; TURNER, P. J. 1959. Grassland irrigation. *Exps Prog. 11 Grassld Res. Inst. 1957–8* pp. 56–8.

WILLIAMS, T. E.; STILES, W.; TURNER, P. J. 1960. Grassland irrigation. *Exps Prog. 12 Grassld Res. Inst. 1958–9* pp. 55–7.

CHAPTER 19

Timothy *(Phleum pratense)*

Morphology

TIMOTHY (Plate 10) is a procumbent to erect perennial. Root system fibrous, rather coarse adventitious roots. Vegetative stems erect to procumbent, stout, cylindrical, with a somewhat swollen basal node. Leaf sheath cylindrical, pale greyish-green, becoming fibrous, brown, persistent. Ligule long, white, membranous sheathing, no auricles. Leaf blade rolled in sheath, broad, smooth, flat, somewhat twisted as it emerges from the sheath; basal leaves shorter than upper leaves. Tillering intravaginal. Culms erect, stout, bulbous at base in summer; 3- to 6-noded, upper leaves on culm shorter than upper vegetative leaves. Inflorescence a spike-like panicle of more or less sessile 2-flowered spikelets with ciliate awned glumes. Caryopsis enclosed in thin lemma and palea.

Garwood (1967) found the following values in root distribution studies:

Table 19.1. Distribution of the root mass of S48 timothy in the soil profile

Horizon in.	Horizon cm	% of total root weight
0–2	0–5	79·9
2–4	5–10	8·3
4–6	10–15	4·4
6–12	15–30	4·7
12–18	30–45	1·7
18–24	45–60	0·7
24–30	60–75	0·3

From Garwood (1967)

The root mass of timothy was less than that of perennial ryegrass and was at a proportionately shallower depth.

In experiments by Ryle (1964) on the effect of changes in temperature and rate of N fertilizer application on the leaf size of S48 timothy under unheated glasshouse conditions, average sizes ranged from 2×60 mm for lower leaves to 7×150 mm for upper leaves, with considerable variation.

Under frequent defoliation much shorter leaves are produced on prostrate vegetative tillers than when plants are defoliated less frequently.

The variation in tiller size and density under monthly cutting has been studied by Anslow (Personal communication). Tiller size in S48 ranged from 2·5 to 7·4 g/1 000 tillers during the growing season, with a maximum in mid-June, and tiller population ranged from 40 to 100 tillers/dm^2 with a peak in April; Langer *et al.* (1964) recorded a similar pattern for S48 timothy. At the beginning of June, just before ear emergence, tiller buds are dormant for 2–3 weeks.

In April, the basal node of each tiller in which flower initiation has occurred swells up to form a basal bulb. New vegetative tillers can arise at the node above this and below the bulb if the culm is severed below the inflorescence.

Cultivars can be classified according to flowering date. Early-flowering cultivars are more erect and less densely tillered than late-flowering ones. Most cultivars of timothy are *Phleum pratense*, a hexaploid (2n = 42) but S50, which is regarded agronomically as timothy, is a cultivar of *Phleum bertolonii*, a diploid (2n = 14) (see Chapter 23). This species is morphologically similar to *Phleum pratense*, but is smaller and more variable, with forms ranging from stiffly erect types to prostrate, even stoloniferous types (e.g. S50).

Changes in the amount of leaf during development have been studied by Minson *et al.* (1964) (Table 19.2).

Table 19.2. Percentage leaf blade in the DM of S48 timothy during the growth of the primary crop

Date	Leaf blade %
23 May	57
1 June	46
9 June	56
15 June	32
29 June	32

Date of 50% ear emergence: 14 June
From Minson *et al.* (1964)

Distribution

Phleum pratense is a native species (Clapham *et al.*, 1962) occurring in all vice-counties. It is widely distributed in the east and west, becoming less frequent in north and north-west Scotland. In Ireland it is general in the north and east and less frequent in the south and west (Clapham *et al.*, 1962;

Perring & Walters, 1962). Its native habitat is probably restricted to the moist soils of water meadows and other low-lying grasslands (Hubbard, 1954), but it is also found in meadows, field margins, waste places, and roadsides (Hubbard, 1954; Clapham *et al.*, 1962), and now occurs in all but the highest parts of Scotland and Wales (Perring & Walters, 1962). Its distribution in Britain bears no relation to isotherms (Perring & Walters, 1962).

It grows best on moist, heavy soils (Hubbard, 1954) and its distribution is not related to chalk or limestone (Perring & Walters, 1962). The optimum pH range is 5·5–8·0 (Spurway, 1941). It is less common in areas with an annual rainfall in excess of 1 250 mm (Perring & Walters, 1962), even though it will tolerate a fairly high water table (Hubbard, 1954) and its roots will penetrate into high-moisture zones (Baumann & Klauss, 1955).

Early and late cultivars need a daylength of more than 12 h in order to flower, but early cultivars can produce flower primordia, but without stem elongation, in daylengths of 6–12 h (Tincker, 1925). Thus there are no day-length limits in Britain. Winter daylengths of 9–10 h plus a supplement of from 75–300 f. c. (750–3 000 lux) to a total of 12 h indicated that light intensity had no effect, but that longer duration allowed flowering (Sprague, 1948). Thus light intensity is also not limiting in Britain.

Compatibility

Timothy is not an aggressive grass. Anslow (1955) showed that in the early stages cocksfoot and perennial ryegrass were more inhibitory to grasses sown with them than was timothy. Davies (1960) stated that timothy does not compete well with ryegrass in a general-purpose mixture, but makes an excellent companion for meadow fescue and the legumes. Arens (1962) found that timothy was suppressed by perennial ryegrass, meadow fescue and cocksfoot. Similarly, Lampeter (1960) found that timothy was suppressed by meadow fescue, cocksfoot, perennial ryegrass and Italian ryegrass over the first two harvest years. He also found that timothy did suppress lucerne and red clover to some extent, but Williams (1950) found that timothy was not nearly so damaging to lucerne as was cocksfoot. However, in experiments reported by Green and Davis (1955), timothy at 6 kg/ha eventually became so dense that it smothered lucerne in wetter places. White clover establishes well in the presence of timothy, but under a cutting management it can become reduced in the second and subsequent harvest years to no more than will grow with perennial ryegrass or cocksfoot (see Tables 19.3 and 19.4).

Of the grasses that are commonly cultivated, timothy is among the slowest to establish and come into full production. Therefore a quick-growing but short-lived species is sometimes added to a direct sowing of timothy to provide early production with the minimum of depression to the development of the timothy. In an experiment at the Grassland Research Institute, Hurley

(1960), sowing timothy with 11 kg Italian ryegrass seed/ha increased yield in the year of sowing by 15% with little effect upon the first harvest year yield. However, the establishment of timothy was very much reduced. In another experiment in 1961, timothy developed better with broad red clover or

Table 19.3. Yield and percentage contribution of white clover grown with each of 5 grasses

	Perennial ryegrass S24	Perennial ryegrass S23	Timothy S48	Meadow fescue S215	Cocksfoot S37
		White clover contribution to yield, %			
1949					
Cut 1	22	20	49	49	19
2	33	29	45	31	18
3	9	6	9	7	2
1950					
Cut 1	8	14	8	13	9
2	20	15	8	16	10
3	19	9	9	13	4
4	14	6	6	7	2
5	5	...	4	10	1
		Total yield of white clover, '00 kg DM/ha			
1949	20·0	18·8	36·5	26·6	10·6
1950	15·6	11·1	8·3	12·8	7·1

From Grassland Research Institute (1952)

Table 19.4. Yield of white clover when grown with each of 5 grasses and cut 4 or 5 times each year

	Perennial ryegrass S24	Perennial ryegrass S23	Timothy S48	Meadow fescue S215	Cocksfoot S37
		Yield of white clover '00 kg DM/ha			
1960	11·0	9·0	11·5	10·4	6·3
1961	52·2	57·0	40·2	46·5	41·7
1962	31·3	33·7	24·0	31·2	26·9
Total	94·5	99·7	75·7	88·1	74·9

From Cowling and Lockyer (1965)

commercial Westerwolds ryegrass as a short-lived companion than with Italian ryegrass or 'improved' Westerwolds. In further work, a mixture of Westerwolds ryegrass (6·7 kg/ha) and broad red clover (4·5 kg/ha) was added to a sowing of timothy and white clover. The mixture proved very effective in controlling weeds and provided good production in the year of sowing. At the same time it was noted that the mixture was less harmful to the timothy than was pure Westerwolds ryegrass and suppressed white clover less than did pure red clover (Grassland Research Institute, 1961, 1962).

Seed

CHARACTERS

The seed is spherical-ovoid in shape, about 1·5–1·7 mm long and 0·8 mm wide. The lemma and palea are silvery-white, thin and membranous, and separate easily from the caryopsis (the skinned seed), which is a pale yellow-brown in colour and finely reticulated (Armstrong, 1948; Hubbard, 1954; Peto, 1961).

The 1000-seed weight varies greatly between and within cultivars. A range from 0·291 g (for S48) and 0·304 g (for S50) to 0·544 g (for Scandinavian) was found by Hunt (1954), whilst Gerdes (1952) reported variation from 0·315 g (for S48) 0·395 g (for S50) to 0·665 g (for Tammisto). These data also illustrate the variation within cultivars. Evans (1959) found a range from 0·23 g to 0·31 g in the 1000-seed weight of S48, and, in the same cultivar, Lambert (1967) found that the smallest and largest seeds in cleaned commercial samples were 0·10 mg and 0·60 mg, respectively; between these extremes the weight of individual seeds showed a normal distribution of values, with 23% of the total number of seeds weighing 0·30 mg$\pm$0·02 mg, which corresponds to the generally accepted 1000-seed weight of 0·28 g to 0·32 g for S48. Smaller seeds (0·05 mg per seed) in uncleaned samples were, however, found to be capable of germination (Lambert, 1967).

PRODUCTION

The seed should be sown in spring, usually not more than 1·5 cm deep in a fine tilth and the seed rate should be about 5 kg/ha in drills 45–60 cm apart (National Institute of Agricultural Botany, 1959; Lambert, 1964). Cover crops reduce yield (Roberts, 1958a). Application of about 90 kg N/ha, divided into equal doses after harvest and at the time of ear initiation, is effective (Lambert, 1963, 1967). The date of ear initiation varies from mid-April for early cultivars to mid-May for later ones (Lambert, 1966a). Post-harvest application of 30 kg P and 60 kg K/ha is recommended (National Institute of Agricultural Botany, 1959). Cutting or grazing in December or January is possible without reducing yield, but it is unlikely that defoliation will increase yield of seed (Evans, 1953, 1955; Roberts, 1958b; Lambert, 1966b). Crops should be isolated by at least 180 m during anthesis (National Institute of Agricultural Botany, 1959). Principles and procedures for harvesting are well explained elsewhere (Ministry of Agriculture, Fisheries & Food, 1968).

The annual yield of seed can vary from 12·5 to 875 kg/ha, with the mean yield ranging between 162 and 450 kg/ha, depending on variety and year (Lincs Seed Growers' Association, 1965).

Yield

DRY MATTER

The average annual dry matter (DM) production of 6 cultivars of timothy over three years was found to be 8 800 kg/ha by Hunt (1959) when they were grown at a high level of fertility and cut at monthly intervals. Very similar yields were recorded under similar cultural conditions by Anslow and Green (1967). S48 timothy gave an average yield of 7 600 kg/ha in an experiment in Wales (Henderson, Edwards & Hammerton, 1962), but higher yields were obtained by Williams *et al.* (1960) from S48 at the Grassland Research Institute, Hurley; they measured a yield of 10 800 kg/ha from an irrigated sward.

On upland peat, Hunt (1964) found much lower yields from timothy varieties. S50 best maintained the botanical composition as sown, but this only gave an average annual yield of 2 500 kg/ha, of which 64% was sown grass.

Studies on the seasonal distribution of the annual production of timothy generally show a reduced rate of production in midsummer. Hunt (1959) found that only about 7–8% of the production measured during the growing season of 6 months occurred in June. In another experiment, Hunt (1957) found that the growth of timothy was at a maximum from the middle to the end of May and that production thereafter until early July was almost negligible. Monthly cuts made by Williams *et al.* (1960) demonstrated lower yields on 20 July than a month earlier or a month later; the highest yield was obtained in May. Langer (1959b) also found that DM production was greatest in the early part of the season and that the onset of ear emergence had a pronounced effect on DM production.

ENERGY AND PROTEIN

Data from the literature are summarized in Table 19.5.

Table 19.5. Potential annual yields of energy and protein from timothy (DM basis)

	Frequent cutting	Infrequent cutting
DM digestibility (%)	75	63
Yield of digestible OM (kg/ha)	6 100	10 900
Yield of metabolizable energy (Mcal/ha)	22·4	32·0
Crude protein (%)	20·0	10·5
Yield of crude protein (kg/ha)	1 970	1 764
Yield of digestible crude protein (kg/ha)	1 330	1 050

Data compiled by D. J. Thomson (Grassland Research Institute, Hurley) from literature cited in the text. For conversion formulae see Chapter 5.

Response to management and environment

RESPONSE TO FERTILIZERS

Nitrogen

Holmes and MacLusky (1955) reported the responses to N of three varieties of timothy. The grasses were cut 3 to 6 times each year over a 4-year period and were given 0, 195 or 390 kg N/ha each year, divided into equal applications before each cut. S48 and Scotch gave similar yield responses, each increasing respectively by 21·9 and 23·8 kg DM/kg N when given 195 kg N/ha and by 13·6 and 12·7 kg DM/kg N for the additional N increment. Scotia gave a higher yield than S48 or Scotch where no N was applied, but its response to the first N increment, 13·9 kg DM/kg N, was lower than that of the other varieties. Scotia gave an increase of 12·7 kg DM/kg N for the second N increment.

Lambert (1962) examined the yield of S48 timothy under two cutting systems with either 58 or 175 kg N/ha given in split applications. Yields were much lower where 8 monthly cuts were taken than where there was a single cut at an advanced stage of development followed by 3 more cuts. Response to N was 20·1 kg DM/kg N in the former treatment, compared with 25·5 in the latter.

Cowling and Lockyer (1965) studied the response to N of S48 over a three-year period. The sward was given 0, 91, 182 or 364 kg N/ha each year divided between 4 or 5 applications, one before each cut. Yield and response to N were similar to those of the other commonly cultivated grasses. Yield increased by 29·5 kg DM/kg N up to 182 kg N/ha and by 15·7 kg DM/kg N between this level and the highest rate of N application.

In trials by Widdowson *et al.* (1963) the variety Scotch was given three levels of N (0, 37·7 or 75·4 kg N/ha for each cut), and harvested 7 times over a 2½-year period. Yield increased by 48·2 kg DM/kg N with the first increment of N and by a further 25·5 kg DM/kg N with the additional increment. Timothy was inferior to cocksfoot but superior to ryegrass and meadow fescue in response to N.

Phosphorus and potassium

There is no information from field trials about the response of timothy to P or K. In a sand-culture experiment, Langer (1959a) found that S48 responded to both P and K provided that the supply of N was high. He concluded from his study, a straightforward NPK factorial experiment, that "the response to any one nutrient tended to increase as the supply of one or both of the others was raised". This was particularly the case with N which, when present in only small amounts, prevented the expression of P and K effects.

RESPONSE TO IRRIGATION

Timothy shows the typical response to irrigation described in Chapter 14 (Williams *et al.*, 1959, 1960).

RESPONSE TO TEMPERATURE

Very little work has been done on the effect of temperature on the growth and yield of timothy, beyond simple observations that they are increased in winter by growing plants in a heated glasshouse. Sprague (1943) found a greater rate of growth at 21°/13°C than at 13°/4°C or 29°/21°C; Ehara and Tanaka (1961) found that dry weight was greater at 25°C than at 20°C, and least at 15°C. There is no reason to suppose that its response to temperature differs markedly from that of other north-temperate grasses.

RESPONSE TO DEFOLIATION

Anslow and Green (1967) showed that three varieties of timothy, one from each of the varietal groups (National Institute of Agricultural Botany, 1965), behaved similarly when cut at 3- to 4-week intervals. When cut monthly the dry matter yield was less than 50% of that under a hay and aftermath system of management (Lambert, 1962; Dent & Aldrich, 1963). In another experiment (Green *et al.*, 1960), reducing the interval between cuts from 30 to 10 days reduced yield by about 50%.

Regrowth of a timothy sward is generally slow (National Institute of Agricultural Botany, 1965), and was found to be particularly slow in the cultivar S48 following a cut for conservation in early June (Jewiss & Powell, 1966). Timothy has been considered to be more suited to hay and silage cropping (Davies, 1960), but the Aberystwyth cultivars can also perform well under grazing management (Welsh Plant Breeding Station, 1961).

Quality

CHEMICAL COMPOSITION

Timothy is relatively high in lignin content compared with other grasses and is inherently low in Na content. Otherwise it appears to be typical in chemical composition (see Chapter 13).

WATER CONTENT

A number of workers have studied the relationship between stage of growth and the water content of timothy. Waite and Sastry (1949) found that the water content of whole timothy (i.e. the above-ground parts) decreased during uninterrupted growth from 77% on 20 May to 46% on 25 September. In experiments in Norway, Hvidsten and Pedersen (1950) also found that water content fell with increasing stage of development when timothy which had not been previously cut in the current season was cut each week throughout the growing season in 5 successive years.

Kivimäe (1959), in Sweden, found quantitative relationships between the DM content of timothy and growth stage. The water content was at first low (69% on 8 May) and increased to a maximum at an early stage of stem elongation (79% on 1 June), after which it decreased fairly quickly to a minimum at seed maturity (53% on 27 July). The relationship was expressed by two linear regressions, one for the first 20 days of growth ($y = 29{\cdot}94 - 0{\cdot}254x$) and another for the 21st to the 73rd day of growth ($y = 14{\cdot}67 + 0{\cdot}481x$), where y = DM percentage and x = growth stage in days.

An effect of N fertilizer on the water content of timothy has been found by Ivins (1952), who observed that in stands of almost pure American and S48 timothy given either no fertilizer or 224 lb/ac (246 kg/ha), the application of N increased the percentage of water in the herbage.

NUTRITIVE VALUE (DRY MATTER DIGESTIBILITY)

Young leafy grass frequently cut

In swards of timothy frequently defoliated and kept in leafy condition, DM digestibility varies between about 72% and 82%, and in this respect under pasture conditions is almost equal to perennial ryegrass and is consistently better than cocksfoot. It shows a similar seasonal pattern to perennial ryegrass, digestibility being highest in the spring and lowest in midsummer, with intermediate values in autumn. Varietal differences are small under frequent cutting.

Uninterrupted primary growth

Timothy presents an entirely different pattern of digestibility in relation to ear emergence from that found in perennial ryegrass. It is in general a late-flowering species and at flowering its DM digestibility is already below 70%. At ear emergence the DM digestibility of an early cultivar such as S352 is about 65% to 70%, while that of a late cultivar such as S48 is 60% to 65% (Fig. 19.1, top). In some seasons there is a tendency for the rate of decline in digestibility to be slower after ear emergence, owing to the growth of new tillers. Minson *et al.* (1964) showed that the rate of decrease in digestibility of the primary growth of S48 timothy was about 0·4% per day. Compared with the later maturing pasture-type perennial ryegrass cultivar S23 (Fig. 19.1, top), it shows a lower digestibility throughout the growth period. Yield curves, which are close in April and May (Fig. 19.1, middle), diverge in favour of S23 at later stages, so that the latter shows a definite advantage in yield of digestible DM (Fig. 19.1, bottom).

Regrowths

The regrowth of timothy following a cut for conservation in late June is of

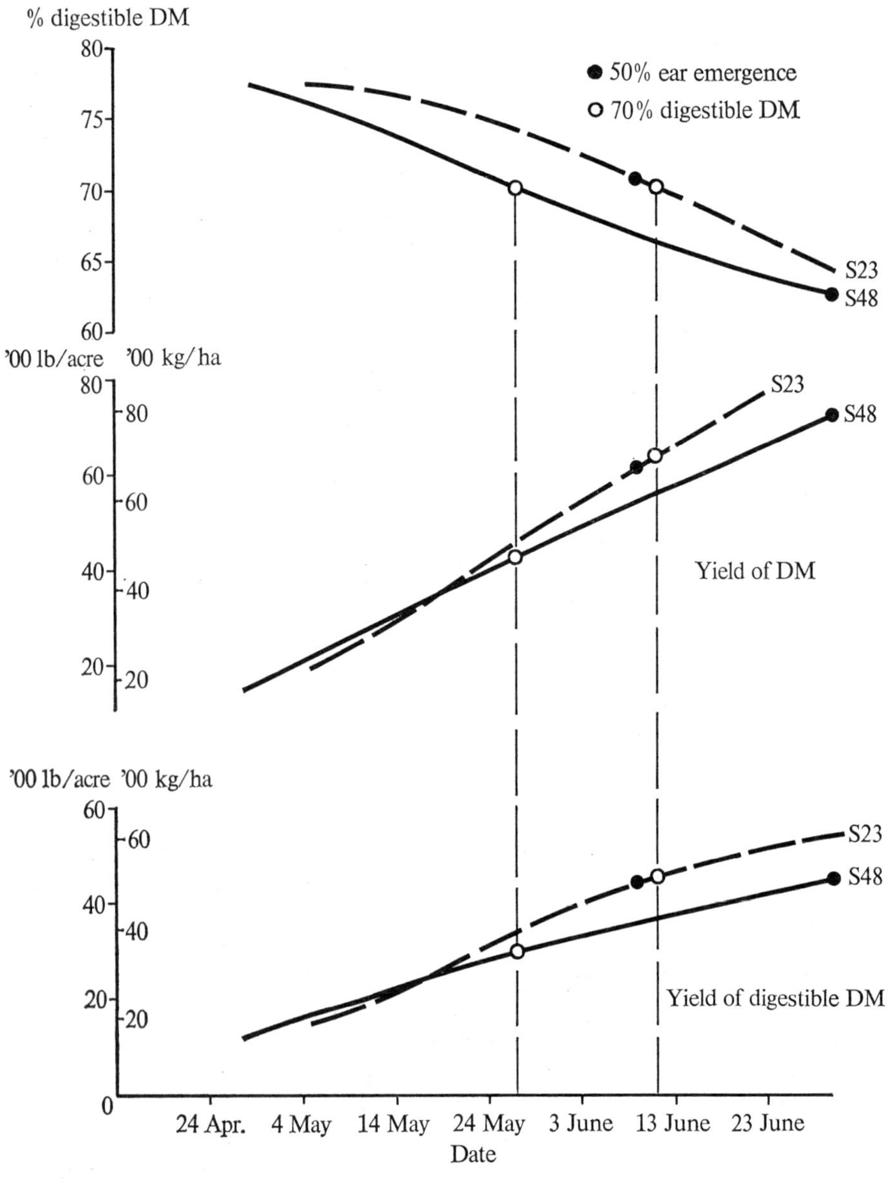

Fig. 19.1. Timothy S48. Seasonal changes in DM yield and digestibility. Values for S23 perennial ryegrass shown for comparison. (After Aldrich and Dent, 1967)

high quality, although like perennial ryegrass it produces inflorescences in an aftermath cut taken after a primary cut in June. After about 6 weeks its DM digestibility is usually about 75% and would be unlikely to drop below 70%

after two months. Late in the year, two-monthly regrowths produce material of about 75% DM digestibility.

In general, therefore, frequently cut or grazed timothy is similar in digestibility to perennial ryegrass, although it is considerably lower in soluble carbohydrate content (Table 19.6). The lower content of soluble carbohydrate is, however, partly offset by the higher protein content. In primary growth, digestibility of timothy, particularly that of the later cultivars, falls to low levels by the time of ear emergence.

Table 19.6. Percentage content (DM basis) of crude protein, crude fibre and soluble carbohydrate, and percentage digestible DM in S23 ryegrass and S48 timothy

	Simulated grazing (weighted mean of 9 cuts)		Cut for hay (cut 10 days after 50% ear emergence)	
	S23	S48	S23	S48
Crude protein	19·0	21·5	7·0	7·0
Crude fibre	23·0	23·0	31·5	34·5
Soluble carbohydrate	15·5	9·5	22·0	14·0
Digestible DM	76·5	75·0	67·5	63·0

From National Institute of Agricultural Botany (1966)

ANIMAL INTAKE CHARACTERISTICS

Minson *et al.* (1964) found the intake of S48 timothy to be considerably less than that of Germinal cocksfoot, H.1 (Grasslands Manawa) hybrid ryegrass, S22 Italian ryegrass and S215 meadow fescue, when compared at the same level of digestibility.

Longevity and persistency

Langer (1954) recorded rates of leaf appearance of one leaf every 13·5 days on vegetative tillers of timothy kept outdoors in winter, compared with one leaf every 9·3 days for plants of the same age transferred to a heated glasshouse at 50–60°F (10–15°C). Corresponding figures for number of living leaves per tiller were 4·6 and 6·1. This would give calculated leaf longevities of 62·1 and 56·7 days, respectively.

According to Evans (1927), the rate of leaf appearance under sward conditions in the USA ranges from an average of one leaf per 9·8 days in April–May to one leaf per 14 days in September–December, and the number of living leaves per tiller from 4 to 7 (exceptionally 8). During the growing season each leaf remains green for about two months or less.

Langer (1957) found averages of just over 5 leaves per tiller in early August and 3 per tiller in October. Average longevity of leaves on fertile tillers was found by Evans to be 50·3 days for flag leaves, 56 days for penultimate leaves and 58·3 days for third leaves; tip senescence began 12·8, 16·2 and 20·5 days before complete leaf death, and all three leaf types took approximately 8 days longer to die than they did to emerge.

Langer (1956) found that in uncut timothy grown outdoors in pots, the life span of tillers arising at any time of the year was not normally more than just over a year. The mean length of life of tillers on established plants exceeded 200 days for those originating between July and January and declined progressively to less than 100 days for those formed in mid-May. Relative death rates of tillers rose rapidly in the autumn but decreased during winter and early spring. Some tillers in timothy elongate in the year of formation but do not flower; these 'blind' tillers were found by Langer to survive for a shorter time than equivalent tillers which did not elongate until the following year. Tillers which survived longest were formed in July; 1·4% of these survived for 14 months. Jewiss (1966) stated that in timothy cut for hay, tillers which arise immediately after flowering in June and before September form the greater proportion of tillers in the sward for most of the year and ensure perennation of that sward; in timothy cut frequently, a small proportion of tillers survive beyond the time of inflorescence formation, but this is of little significance for perennation.

Investigations in the USA indicate that the roots of timothy are not perennial (Evans, 1927; Stuckey, 1941). Observations at the Grassland Research Institute, Hurley, tend to confirm this view, at least for roots formed on vegetative tillers. However, the basal internodes of tillers which have flowered persist for a time after they have lost their leaves, and there is evidence that their roots may continue to absorb nutrients for a considerable period of time. Tatarinova (1961) has demonstrated uptake of ^{32}P by roots attached to these stem bases 1, 2 and 4 years after flowering in established stands of timothy grown for seed production.

Mence (1964) found that timothy seed in soil responded to temperatures below 10°C by prolonged seed dormancy; this resulted in good survival-rates of young seedlings. Many seedlings and young plants die during the establishment of a sward. Charles (1961) found that only 7% of viable timothy seeds sown in March produced plants which survived their first winter. Langer *et al.* (1964) investigated the course of plant numbers per unit area in timothy swards cut three times per year or at monthly intervals. Of 390 plants present at establishment in October 1956, 116 survived until the first hay cut in May 1957, while in the frequently-cut treatment 244 plants were present at the same date. Plant numbers had fallen to 19 and 95, respectively, by October 1959.

There is little information on the ultimate longevity of timothy plants. Evans (1927) found no lack of vigour in plants 20 years old, but did note that the percentage of fertile tillers tended to decrease as swards aged. Timothy is relatively insensitive to management practices, and will withstand the coldest winters. Baker and David (1963) noted remarkably good survival of timothy in Britain after the very severe winter of 1962–3.

Many experiments have shown that the highest yields of timothy in the spring and summer were obtained after the sward had been allowed a long period of uninterrupted growth before harvesting. The greatest amount of tillering, however, resulted from intermittent and moderate grazing during the same period (Jones, 1939). Timothy has been found to persist in a grazed sward, making a satisfactory contribution to the botanical composition even after 7 years (Alder, 1966). This sward was grazed intermittently between April and July and again between November and February, but rested during the autumn (mid-July to late September). The ability of timothy to persist when grazed during the winter and early spring was also demonstrated in this work.

The species is tolerant of wet conditions during the winter and will persist in land which is poorly drained or liable to short periods of winter flooding. When grown and grazed on such land, timothy has been found to persist and to outyield swards of other species with which it was compared (Mudd & Mair, 1961).

References

ALDER, F. E. 1966. A summary of twelve years work on a farmlet type of experiment designed to study animal production on sown pastures. *A. Rep. Grassld Res. Inst. Hurley 1965* pp. 79–86.

ALDRICH, D. T. A.; DENT, J. W. 1967. The relationship between yield and digestibility in the primary growth of nine grass varieties. *J. natn. Inst. agric. Bot.* **11,** 104–13.

ANSLOW, R. C. 1955. The effects of competition on the shoot and root growth of three grass species. M.Sc. Thesis, Univ. Reading.

ANSLOW, R. C. Grassland Research Institute, Hurley. Personal communication.

ANSLOW, R. C.; GREEN, J. O. 1967. The seasonal growth of pasture grasses. *J. agric. Sci., Camb.* **68,** 109–22.

ARENS, R. 1962. The effect of rate of seeding on competition between species and the productivity of the first year's pasture. *Z. Acker- u. PflBau* **115,** 357–74.

ARMSTRONG, S. F. 1948. British grasses and their employment in agriculture. Cambridge: Univ. Press, 3rd edn.

BAKER, H. K.; DAVID, G. L. 1963. Winter damage to grass. *Agriculture, Lond.* **70,** 380–2.

BAUMANN, H.; KLAUSS, M. L. 1955. [Root development in an environment with a high water table.] *Z. Acker- u. PflBau* **99,** 410–26.

CHARLES, A. H. 1961. Differential survival of cultivars of *Lolium, Dactylis* and *Phleum. J. Br. Grassld Soc.* **16,** 69–75.

CLAPHAM, A. R.; TUTIN, T. G.; WARBURG, E. F. 1962. Flora of the British Isles. Cambridge: Univ. Press, 2nd edn.

COWLING, D. W.; LOCKYER, D. R. 1965. A comparison of the reaction of different grass species to fertilizer nitrogen and to growth in association with white clover. *J. Br. Grassld Soc.* **20,** 197–204.

DAVIES, I. 1960. Growth stage and related factors in grassland management. *Rep. Welsh Pl. Breed. Stn 1959* pp. 104–9.

DAVIES, W. 1960. The grass crop. London: E. & F. N. Spon Ltd, 2nd edn.

DENT, J. W.; ALDRICH, D. T. A. 1963. The interrelationships between heading date, yield, chemical composition and digestibility in varieties of perennial ryegrass, timothy, cocksfoot and meadow fescue. *J. natn Inst. agric. Bot.* **9,** 261–81.

EHARA, K.; TANAKA, S. 1961. Effect of temperature on the growth, behaviour and chemical composition of the warm- and cool-season grasses. *Proc. Crop Sci. Soc. Japan* **29,** 304–6.

EVANS, G. 1959. Seed rates of grasses for seed production. 1. Pasture varieties of ryegrass, cocksfoot, and timothy. *Emp. J. exp. Agric.* **27,** 291–9.

EVANS, M. W. 1927. The life history of timothy. *Bull. 1450 U.S. Dep. Agric.* pp. 56.

EVANS, T. A. 1953. Management and manuring for seed production in cocksfoot and timothy. *J. Br. Grassld Soc.* **8,** 245–59.

EVANS, T. A. 1955. Manuring and winter grazing for seed production in S48 timothy. *J. Br. Grassld Soc.* **10,** 254–62.

GARWOOD, E. A. 1967. Some effects of soil water conditions and soil temperature on the roots of grasses. 1. The effects of irrigation on the weight of root material under various swards. *J. Br. Grassld Soc.* **22,** 176–81.

GERDES, G. 1952. [Studies on the absolute weight (1000 seed weight) of the most important fodder grass species and their varieties in regard to the performance of grass breeding.] *Züchter* **22,** 353–66.

GRASSLAND RESEARCH INSTITUTE. 1952. *Exps Prog. 5 Grassld Res. Inst. 1952* pp. 23–31.

GRASSLAND RESEARCH INSTITUTE. 1960. *Exps Prog. 12 Grassld Res. Inst. 1958–9* p. 16.

GRASSLAND RESEARCH INSTITUTE. 1961. *Exps Prog. 13 Grassld Res. Inst. 1959–60* p. 14.

GRASSLAND RESEARCH INSTITUTE. 1962. *Exps Prog. 14 Grassld Res. Inst. 1960–61* p. 16.

GREEN, J. O.; DAVID, G. L.; EYLES, J. C.; ASTLE, W. R. 1960. Level and frequency of cutting: pure grass swards at two levels of nitrogen. *Exps Prog. 12 Grassld Res. Inst. 1958–9* pp. 19–21.

GREEN, J. O.; DAVIS, A. G. 1955. The seasonal use of lucerne/grass mixture for grazing. In: Grassland Research Institute. Memoir No. 1. Lucerne investigations, 1944–53, pp. 63–77.

HENDERSON, J. L.; EDWARDS, R. S.; HAMMERTON, J. L. 1962. The productivity of five grass species at six levels of compound fertilizer application. 1. Dry-matter production. *J. agric. Sci., Camb.* **59,** 5–11.

HOLMES, W.; MACLUSKY, D. S. 1955. The intensive production of herbage for crop-drying. *J. agric. Sci., Camb.* **46,** 267–86.

HUBBARD, C. E. 1954. Grasses; a guide to their structure, identification, uses, and distribution in the British Isles. Harmondsworth, Middlesex: Penguin Books.

HUNT, I. V. 1954. Seed establishment in the West of Scotland. *J. Br. Grassld Soc.* **9,** 85–98.

HUNT, I. V. 1957. Comparative productivity of 27 strains of grass in the West of Scotland. 4. Seasonal growth. *J. Br. Grassld Soc.* **12,** 127–37.

HUNT, I. V. 1959. Comparison of production from six cultivars of timothy 1954–1957. *Res. Bull. 27 W. Scotl. agric. Coll.* pp. 39.

HUNT, I. V. 1964. Comparative productivity of herbage varieties on upland peat. *J. Br. Grassld Soc.* **19,** 55–61.

HVIDSTEN, H.; PEDERSEN, E. 1950. Investigations on the content of dry matter, crude protein and carotene in meadow and pasture plants. *Forsk. Fors. Landbr.* **1,** 311–45.

IVINS, J. D. 1952. The relative palatibility of herbage plants. *J. Br. Grassld Soc.* **7,** 43–54.

JEWISS, O. R. 1966. Morphological and physiological aspects of growth of grasses during the vegetative phase. In: Milthorpe, F. L.; Ivins, J. D. [Eds] The growth of cereals and grasses. Proceedings of the twelfth Easter School in Agricultural Science, University of Nottingham, 1965. London: Butterworth and Co., pp. 39–54.

JEWISS, O. R.; POWELL, C. E. 1966. The growth of S48 timothy swards after cutting in relation to carbohydrate reserves and leaf area. *A. Rep. Grassld Res. Inst. Hurley 1965* pp. 67–72.

JONES, I. 1939. Pasture management and its effect on the sward. *Bull. H15 Welsh Pl. Breed. Stn* pp. 40–129.

KIVIMÄE, A. 1959. Chemical composition and digestibility of some grassland crops. *Acta Agric. scand.* Suppl. 5, 26–8.

LAMBERT, D. A. 1962. A study of growth in swards of timothy and meadow fescue. *J. agric. Sci., Camb.* **59,** 25–32.

LAMBERT, D. A. 1963. The effect of the rate and timing of nitrogen applications on the seed yield and components of yield in S48 timothy (*Phleum pratense L.*). *J. Br. Grassld Soc.* **18,** 154–7.

LAMBERT, D. A. 1964. The influence of density and nitrogen in seed production stands of S48 timothy (*Phleum pratense L.*) and S215 meadow fescue (*Festuca pratensis L.*). *J. agric. Sci., Camb.* **63,** 35–42.

LAMBERT, D. A. 1966a. Factors affecting seed yield. *Proc. Conf. Timothy Seed Production, W. Scotl. agric. Coll.* pp. 4–12.

LAMBERT, D. A. 1966b. The effect of cutting timothy (*Phleum pratense L.*) grown for production of seed. *J. Br. Grassld Soc.* **21,** 208–13.

LAMBERT, D. A. 1967. The effects of nitrogen and irrigation on timothy (*Phleum pratense*) grown for production of seed. 2. Reproductive growth and yield of seed. *J. agric. Sci., Camb.* **69,** 231–9.

LAMPETER, W. 1960. [Mutual influences of higher plants regarding growth of shoots and roots, mineral content and water consumption studied in some economically important forage plants.] *Wiss. Z. Karl-Marx-Univ. Lpz. 1959–60* **9,** 611–722.

LANGER, R. H. M. 1954. A study of leaf growth in timothy (*Phleum pratense*). *J. Br. Grassld Soc.* **9,** 275–84.

LANGER, R. H. M. 1956. Growth and nutrition of timothy (*Phleum pratense*). 1. The life history of individual tillers. *Ann. appl. Biol.* **44,** 166–87.

LANGER, R. H. M. 1957. Growth and nutrition of timothy (*Phleum pratense*). 2. Growth of the plant in relation to tiller development. *Ann. appl. Biol.* **45,** 528–41.

LANGER, R. H. M. 1959a. Growth and nutrition of timothy (*Phleum pratense L.*). 4. Effect of nitrogen, phosphorus and potassium on growth during the first year. *Ann. appl. Biol.* **47,** 211–21.

LANGER, R. H. M. 1959b. A study of growth in swards of timothy and meadow fescue. 2. The effects of cutting treatments. *J. agric. Sci., Camb.* **52,** 273–81.

LANGER, R. H. M.; RYLE, S. M.; JEWISS, O. R. 1964. The changing plant and tiller populations of timothy and meadow fescue swards. 1. Plant survival and pattern of tillering. *Jnl appl. Ecol.* **1,** 197–208.

LINCOLNSHIRE SEED GROWERS' ASSOCIATION. 1965. Report for 1964. Agric. House, Woodhall Spa, pp. 3–7.

MENCE, M. J. 1964. The differential mortality of *Trifolium repens* and *Phleum pratense* seedlings in relation to temperature. *Pl. Soil* **21,** 113–24.

MINISTRY OF AGRICULTURE, FISHERIES AND FOOD. 1968. Grass and clover crops for seed. *Bull. 204* London: H.M.S.O., pp. 42–60.

MINSON, D. J.; HARRIS, C. E.; RAYMOND, W. F.; MILFORD, R. 1964. The digestibility and voluntary intake of S22 and H.1 ryegrass, S170 tall fescue, S48 timothy, S215 meadow fescue and Germinal cocksfoot. *J. Br. Grassld Soc.* **19,** 298–305.

MUDD, C. H.; MAIR, R. B. 1961. Performance of eight seeds mixtures at Great House 1952–1958. *Expl Husb.* No. 6, 21–50.

NATIONAL INSTITUTE OF AGRICULTURAL BOTANY. 1959. Growing grasses for seed. *Herbage Seed Growers' Leaflet No. 2.*

NATIONAL INSTITUTE OF AGRICULTURAL BOTANY. 1965. Varieties of timothy, meadow fescue, and cocksfoot. *Farmers' Leaflet No. 21.*

NATIONAL INSTITUTE OF AGRICULTURAL BOTANY. 1966. Unpublished data.

PERRING, F. H.; WALTERS, S. M. 1962. Atlas of the British flora. London: Thomas Nelson and Sons Ltd, pp. 456.

PETO, D. H. 1961. Methods and problems of cleaning herbage seeds. *Rep. Lincs. Seed Growers' Assn 1960* Agric. House, Woodhall Spa, pp. 19–21.

ROBERTS, H. M. 1958a. The effect of a cover crop. *Rep. Welsh Pl. Breed. Stn 1950–56* pp. 147–8.

ROBERTS, H. M. 1958b. The effect of defoliation on the seed-producing capacity of bred strains of grasses. 1. Timothy and perennial ryegrass. *J. Br. Grassld Soc.* **13,** 255–61.

RYLE, G. J. A. 1964. A comparison of leaf and tiller growth in seven perennial grasses as influenced by nitrogen and temperature. *J. Br. Grassld Soc.* **19,** 281–90.

SPRAGUE, V. G. 1943. The effects of temperature and day-length on seedling emergence and early growth of several pasture species. *Proc. Soil Sci. Soc. Am.* **8,** 287–94.

SPRAGUE, V. G. 1948. The relation of supplementary light and soil fertility to heading in the greenhouse of several perennial forage grasses. *J. Am. Soc. Agron.* **40,** 144–54.

SPURWAY, C. H. 1941. Soil reaction (pH) preferences of plants. *Spec. Bull. 306 Mich. agric. Exp. Stn* pp. 36.

STUCKEY, I. H. 1941. Seasonal growth of grass roots. *Am. J. Bot.* **28,** 486–91.

TATARINOVA, N. K. 1961. [The longevity of the roots of meadow grasses.] *Bot. Zh. SSSR* **46,** 925–35.

TINCKER, M. A. H. 1925. The effect of length of day upon the growth and reproduction of some economic plants. *Ann. Bot.* **39,** 721–54.

WAITE, R.; SASTRY, K. N. S. 1949. The composition of timothy (*Phleum pratense*) and some other grasses during seasonal growth. *Emp. J. exp. Agric.* **17,** 179–87.

WELSH PLANT BREEDING STATION. 1961. The Aberystwyth varieties of grasses and clovers. *Leaflet Series No. 7.*

WIDDOWSON, F. V.; PENNY, A.; WILLIAMS, R. J. B. 1963. An experiment comparing responses to nitrogen fertilizer of four grass species. *Expl Husb.* No. 9, 28–36.

WILLIAMS, T. E.; STILES, W.; TURNER, P. J. 1959. *Exps Prog. 11 Grassld Res. Inst. 1957–8* p. 58.

WILLIAMS, T. E.; STILES, W.; TURNER, P. J. 1960. Grassland irrigation. *Exps Prog. 12 Grassld Res. Inst. 1958–9* pp. 55–7.

WILLIAMS, W. 1950. Experiments on the yield of lucerne strains as pure plots and on the behaviour of lucerne when sown in a mixture with various species of grasses. *J. Br. Grassld Soc.* **5,** 113–29.

CHAPTER 20

Cocksfoot *(Dactylis glomerata)*

Morphology

COCKSFOOT (Plate 11) is a perennial with a densely tufted habit. Root system thick, fibrous, highly-branched adventitious roots. Vegetative stems cylindrical, short. Leaf sheaths strongly laterally-compressed, keeled, entire at first, splitting later, more or less glabrous. Ligule long, white, folded, no auricles. Leaf blade folded on midrib but opening flat later, tapering to a fine point, with a single midrib furrow adaxially. Tillers intravaginal. Culms tall, cylindrical, 2- to 4-noded. Inflorescence a secund panicle, with the lower branches longer, and reflexed in fruit; spikelets 2- to 5-flowered, clustered at ends of panicle branches. Caryopsis slender, enclosed in hardened lemma and palea. Cocksfoot in the UK is tetraploid ($2n = 28$). Some Mediterranean forms are diploid ($2n = 14$) but are morphologically similar to the tetraploid.

There is little published information on the root distribution of cocksfoot. Garwood (Personal communication) has found the rooting depth and distribution to be similar to that of perennial ryegrass but the total root mass to be less.

In glasshouse studies Ryle (1964) found the leaves of cocksfoot to range from 4×50 mm to 8×150 mm in size. Anslow (Personal communication) recorded tiller weights of 2·0 to 10·0 g/1 000 tillers, with maxima in June and August, and tiller populations during the growing season ranging from 30 to 35 tillers/dm^2.

Cocksfoot cultivars can be conveniently classified according to flowering date. Early forms are more erect than late-flowering ones. In late-flowering cultivars, e.g. S143, peripheral vegetative tillers in a clump are prostrate and short. All cultivars and ecotypes will form tussocks under infrequent defoliation.

Changes in the proportion of leaf blade, sheath and stem during the development of the first crop have been studied by Terry and Tilley (1964) (Table 20.1).

Table 20.1. The proportion of leaf blade, sheath, stem and inflorescence in the DM of S37 cocksfoot during primary growth

Date	Percentage of dry weight		
	Leaf blade	Sheath	Stem + inflorescence + dead material
23 Apr.	67	21	12
8 May	60	27	13
29 May	27	19	54
9 June	25	15	60
2 July	20	13	67

Date of 50% ear emergence: 19 May

From Terry and Tilley (1964)

Distribution

Dactylis glomerata is a native species (Clapham *et al.*, 1962) occurring in all vice-counties of Britain. It is widespread and often very abundant (Hubbard, 1954) though less common in north and west Scotland and north-west Ireland (Perring & Walters, 1962). It occurs in hedgerows, meadows, rough grassland and, occasionally, in open woodland from 0 to 600 m above sea level, being absent only from the highest parts of Scotland (Hubbard, 1954; Beddows, 1959; Perring & Walters, 1962).

In the northern hemisphere it is normally found in areas bounded by the 10°C and 27°C isotherms (Beddows, 1959) and is therefore not limited by summer temperatures in Britain. A period of cold is necessary for floral initiation (Wycherly, 1952; Beddows, 1959), though late spring frosts can injure the developing inflorescence and cause "winter burn" of the leaves (Beddows, 1959). This might account for the observation that the species is less common in areas in Scotland outside the February 1°C isotherm (Perring & Walters, 1962). If the weekly mean temperature is less than 10°C, growth is very slow (Leasure, 1952).

Strong winds (hot, cold or salt laden) can cause damage analogous to winter burn (Beddows, 1959), though var. *collina* Schlechtd. grows in dry, exposed places (Hubbard, 1954).

It grows on all soils except the most acid (Beddows, 1959) but is often sparse on chalk and limestone (Perring & Walters, 1962). The optimum pH is 6·0–7·0, but a range from 5·5 to 8·0 is tolerated (Spurway, 1941; Beddows, 1959).

Cocksfoot tolerates high rainfall poorly and is less common in the wetter areas in northern Scotland and Ireland (Beddows, 1959; Perring & Walters, 1962); also, it will not tolerate a high water table (Baumann & Klauss, 1955; Beddows, 1959), under which conditions the forage becomes unpalatable even when young (Beddows, 1959).

The species is reputed to be drought resistant, though growth is certainly retarded by drought conditions (Beddows, 1959).

Both early- and late-flowering types need more than 12 h daylight for the initiation of the panicle primordia; daylength is thus not limiting in Britain. Very late types under 12-h days produce primordia which do not develop normally but revert to the vegetative state and the glumes develop as small leaves (Tincker, 1925). The work of Sprague (1948) suggests that light intensity is not a limiting factor in Britain.

Cocksfoot should be usable in all areas except those with a high water table or those liable to flooding.

Compatibility

Cocksfoot develops slowly from seed. Although the crop from an early spring sowing can attain a moderate level of production by the end of the summer, seedling growth is very much slower than that of the ryegrasses. During establishment cocksfoot is not aggressive towards other grasses, but becomes so as the sward develops. In two cutting experiments with sown swards, Anslow (1955) found that cocksfoot initially exhibited less aggression than ryegrass, but was more aggressive once established. Work on small plots under a cutting regime, reported by Cowling and Lockyer (1968), confirmed Anslow's findings and indicated that the increase in competitiveness by cocksfoot was accelerated by increasing the level of fertilizer N. Arens (1962) found a similar relationship when using combinations of cocksfoot with perennial ryegrass and meadow fescue under controlled grazing at a moderate level of manuring.

Legumes are generally suppressed by cocksfoot. In an experiment reported by Cowling and Lockyer (1965), less white clover grew with cocksfoot than with perennial ryegrass, timothy or meadow fescue. This result was consistent over three harvest years. The same relationship was found in another experiment at the Grassland Research Institute, Hurley (1952) when N was applied at 160–260 kg/ha, although little distinction could be seen between the species when no N was applied.

In many of the farm trials of lucerne/grass mixtures reported by Williams (1950), there was a reduction in the number of surviving lucerne plants when cocksfoot was present. Cocksfoot was more damaging to the lucerne than were perennial ryegrass, timothy or meadow fescue. Green and Davis (1955) showed that lucerne could be satisfactorily grown with cocksfoot in a mixture for grazing, but that 2–3 kg of grass seed/ha was ample, and it was necessary to control the grass by winter grazing or cutting.

Management can drastically affect the compatibility of cocksfoot with other grasses. Davies (1960) found that excessively hard grazing soon reduced the vigour of cocksfoot and killed it out, but that if grazing was too lenient it was apt to grow luxuriantly and gain complete mastery of the sward. The importance of winter grazing in controlling cocksfoot in a mixture with lucerne has already been mentioned. England (1965) compared the performance of cocksfoot and perennial ryegrass in pure swards with that in 50 : 50 mixtures of these species. When the swards were kept vegetative, ryegrass was more competitive than cocksfoot, but when they were allowed to pass out of the vegetative phase before the first cut, cocksfoot was competitively superior.

As in other species, improved varieties of cocksfoot are more aggressive to other grasses than are 'commercial' stocks and, similarly, cocksfoot is suppressed to a relatively greater extent by improved varieties of other grasses. At Aberystwyth (Jones, 1958) a 10-year-old mixture of Danish cocksfoot and Irish perennial ryegrass contained 10% cocksfoot in the herbage, while with S23 perennial ryegrass the cocksfoot content was negligible. On the other hand, S23 persisted better with Danish cocksfoot than with the cocksfoot cultivar S26. (This work was done when 'Danish' cocksfoot meant a commercial stock, and not a bred variety as it would mean today.)

Seed

SEED CHARACTERS

The seed is lanceolate to oblong, ranging from 4 to 9 mm but usually from 5 to 6 mm in length, and with an average width of 1·0 to 1·5 mm. It is distinctly keeled and triangular in cross-section (Armstrong, 1948; Hubbard, 1954; Peto, 1961). The lemma has a distinct keel, and forms two of the three sides of the seed; it may be blunt or pointed and is tipped with a rigid, often curved, awn up to 1·5 mm long. The keel is either fringed with hairs or is serrated. The palea is 4 to 6 mm long, thin and membranous. The rachilla is cylindrical, with an enlarged tip (Armstrong, 1948; Hubbard, 1954).

The 1 000-seed weight varies greatly between and within varieties. A range from 0·762 g (for Scotia) to 1·268 g (for S37) was found by Hunt (1954), and a similar variation was reported by Gerdes (1952). A large variation (0·75 g to 1·24 g) was also obtained within the cultivar S143 by Evans (1959), and Davies and Edwards (1953) found a range of 0·814 g to 0·917 g in S26.

PRODUCTION

The seed should be sown in spring at 4–10 kg/ha in drill rows 30–60 cm apart (Schwanbom & Fröier, 1949; National Institute of Agricultural Botany, 1959; Roberts, 1961; Lambert, 1963). Sowing in drill rows 30 cm apart and gapping to obtain plants on a 30-cm square pattern will increase yield by 30% compared with that from continuous drills (Lambert, 1963). Cover crops may reduce yield considerably (Evers & Sonneveld, 1953; Roberts, 1958). Application of about 200 kg N/ha appears desirable, with a small dose after harvest and the majority in spring (Lambert, 1963; 1965; 1966b). Application of 30 kg P and 60 kg K/ha after harvest is also recommended (National Institute of Agricultural Botany, 1959). Cutting or grazing in the December–March period may increase yield and will not reduce it appreciably (Green & Evans, 1956; 1957; Lambert, 1966a). Repeated grazing will reduce yield (Green & Evans, 1957), and grazing is more detrimental than cutting (Roberts, 1959). Crops should be isolated by at least 180 m during the period of anthesis (National Institute of Agricultural Botany, 1959). Principles and procedures for harvesting are well explained elsewhere (Ministry of Agriculture, Fisheries & Food, 1968).

The annual yield of seed can vary from 25 to 1 440 kg/ha, with the mean yield ranging between 390 and 815 kg, depending on variety and year (Lincs Seed Growers' Association, 1965).

Yield

DRY MATTER

Yields of 11 000 kg dry matter (DM)/ha have been recorded by Williams *et al.* (1960) from plots of cocksfoot irrigated and cut monthly. Hunt (1965) reported a yield of 9 000 kg DM/ha as the average of 12 varieties in three years under a high level of fertilizer input, more than twice the yield produced at a low level of input. Coppenet and Le Corre (1964) found that cocksfoot yielded about 8 000 kg DM/ha in Brittany.

Hunt (1965) found no great difference between the yield of different varieties in his experiment, although differences increased to give maximum variation in the third year. The only significant difference in the average yield of the three years was between the highest- and lowest-yielding varieties. However, Heddle (1965) found a range of over 1 000 kg DM/ha between the highest- and lowest-yielding varieties at a high level of fertilizer input. There was no close relationship between total yield and date of flowering of a variety. In this connection, Davis (1961) in Canada classified 40 varieties into 5 maturity classes, based on the date of ear emergence, and found that early maturity was correlated with a high total DM yield, but that the medium-maturity group gave the least variable seasonal distribution of DM.

Hunt (1964) found large differences between three varieties as regards their suitability for growth on upland peat, S143 maintaining a good botanical composition and giving an average annual yield of 3 500 kg DM/ha over 5 years.

In an examination of the seasonal production of cocksfoot by Anslow and Green (1967), the highest rate of production occurred in early May and there was a marked second peak in production in July/August in a year when either water or N had been deficient earlier in the season. Williams *et al.* (1960) also found that a cut on 25 May gave the highest yield of a succession of monthly cuts. Cuts in June, July and August gave values similar to each other. Hunt (1965) stated that differences between cultivars in the first cut of the year were largely balanced by compensatory differences in the second cut. Yields in June were lower than those in May.

ENERGY AND PROTEIN

The DM digestibility of cocksfoot can be maintained at about 74% by frequent cutting (Dent & Aldrich, 1963; Anslow & Green, 1967). Less frequent defoliation results in lower digestibility (68% or less) and yields of metabolizable energy in the region of 19·0 Mcal/ha (see Table 20.2). The yield of digestible crude protein under this type of management was 1 060 kg/ha in trials by Dent and Aldrich, 1963. Hunt (1965) and Dent and Aldrich (1963) have shown that varietal differences in digestibility have tended to be small.

Table 20.2. Potential annual yields of energy and protein from cocksfoot (DM basis)

	Frequent cutting	Infrequent cutting
DM digestibility (%)	74	68
Yield of digestible OM (kg/ha)	5 660	2 500–5 070
Yield of metabolizable energy (Mcal/ha)	21·2	18·8
Crude protein (%)	21·5	17–19
Yield of crude protein (kg/ha)	1 770	1 450
Yield of digestible crude protein (kg/ha)	1 360	514 – 1 060

Data compiled by D. J. Thomson (Grassland Research Institute, Hurley) from literature cited in the text. For conversion formulae see Chapter 5

A crude protein yield of 1 500 kg/ha was recorded by Hunt (1965) at a high level of fertilizer input, and Heddle (1965) found no great differences in crude protein between varieties under these conditions; such differences as existed were attributable to differences in DM yield and not to differences in crude protein percentage. Davis (1961) stated that very early and very late maturity classes generally had a high protein content throughout the year.

Response to management and environment

RESPONSE TO FERTILIZERS

Nitrogen

Holmes and MacLusky (1955) reported an experiment in which the response to N of a number of grasses, including three varieties of cocksfoot, was examined over a 4-year period. The grasses were cut 3–6 times each year and given 0, 195 or 390 kg N/ha divided between the cuts. There were small differences among the varieties in annual yield of DM; yield increased by

30·4, 23·8 and 27·3 kg DM/kg N for S26, Scotia and Danish, respectively, in response to the first N increment. The respective increases in yield with the final increment of N were 10·4, 13·5 and 12·7 kg DM/kg N.

The response of S37 to N can be calculated from data given by Herriott and Wells (1960) from trials in which the sward was cut 4 times each year over a 4-year period and was given 0 or 77 kg N/ha divided into two applications each year. Yield was increased by 19·0, 51·4 and 39·0 kg DM/kg N in the three successive years.

Davies *et al.* (1966) reported the response to N of the cultivars S37, Scotia, Pajbjerg II and CB cut initially after ear emergence and subsequently on three occasions and given 70 or 140 kg N/ha divided into 4 equal applications. There were no significant differences among the cultivars in yield, and the average response to N was 29·5 kg DM/kg N.

Widdowson *et al.* (1963) reported responses of 55·8 kg DM/kg N in the yield of S37 given a total of 264 kg N/ha in split applications over a 2·5-year period, the sward being harvested two or three times each year. The additional response to N where the application was doubled was 35·5 kg DM/kg N. In a further experiment, Widdowson *et al.* (1965) examined the response to N of the same cultivar over a two-year period. The grass was harvested 6 times during the experiment and was given 0, 38, 76 or 114 kg N/ha for each cut. Response declined with increasing level of N. With successive increments of N there were responses of 44·6 and 19·8 kg DM/kg N for the 38 and 76 kg rates and no further increase in yield for the 114 kg rate.

In an experiment comparing a number of grass species and varieties, cut 4 or 5 times each year, Cowling and Lockyer (1965) found S37 to give a linear response to N up to a level of 182 kg N/ha per year. Yield increased by 30·5 kg DM/kg N. The use of double this level of N produced a further response of 19·5 kg DM/kg N.

Cowling (1961) reported the response to N of S37 in two swards of three-year duration. The grass was harvested 4 or 5 times each year and given 0, 39, 117 or 234 kg N/ha divided into two applications each year. With successive increments of N, annual yield increased by 28·9, 34·1 and 26·0 kg DM/kg N, respectively.

Phosphorus and potassium

Widdowson *et al.* (1965) examined the effect of an increasing supply of P and K on the yield of S37 given various rates of N. The sward was sown following two successive sowings of Italian ryegrass which had received similar NPK treatments but which had shown no response to P or K. There were no significant increases in yield where up to 65 kg P/ha was given. Where N was applied at 152 to 456 kg/ha each year, the cocksfoot showed a response to increasing levels of K. Yield increased by 11·3 kg DM/kg K in the first year and by 19·4 kg DM/kg K in the second year when the supply of K was raised from nil to 31 kg/ha for each cut. There were no further significant increases in yield when the application of K was doubled. It is noteworthy that responses to K were usually found when the content of K in the DM was less than 2%. This agrees with findings at the Grassland Research Institute, Hurley, with ryegrass (Clement, 1967) and with the conclusions of McNaught (1958) in New Zealand.

RESPONSE TO IRRIGATION

Cocksfoot shows the typical response to irrigation described in Chapter 14. Although it is reputed to be particularly drought-resistant, this has not been confirmed in various experiments (Williams *et al.*, 1959, 1960; Penman, 1962; Reid & Castle, 1965) designed to study response to irrigation.

RESPONSE TO TEMPERATURE

Most of the controlled-environment work on cocksfoot has been carried out using a New Zealand commercial variety. Mitchell (1956) reported an optimum of 21°C in days with 12 h light, and Mitchell and Lucanus (1962) a similar optimum in 16-h days but a slightly lower one in 8-h days. As with ryegrass, day temperature is a more important determinant of yield than night temperature (Mitchell & Lucanus, 1960). American work agrees with this; Brown (1939) cites an optimum temperature of 21°C, and Sprague (1943) an optimum regime of 21°/13°C.

Davidson and Milthorpe (1965) found that small plants of cocksfoot grew slightly faster at 26°C than at 22°C, but very much faster than at 14°C. As the plants increased in size the effect of temperature was reduced, as has also been observed in tall fescue. In Japan, Ehara *et al.* (1965) compared the regrowth of cocksfoot and Italian ryegrass at temperatures of 10°, 20° and 30°C. Increasing temperature greatly increased the rate of regrowth, particularly of cocksfoot, but also the rate at which roots and stubble lost weight.

At Aberystwyth, ecotypes of cocksfoot and ryegrass have been shown to differ little in growth rate at high temperatures, but at 10°C, and even more at 5°C, material of Mediterranean origin expanded leaf more rapidly, and continental populations more slowly, than did maritime varieties (Cooper, 1964).

RESPONSE TO DEFOLIATION

Dent and Aldrich (1963) found little difference between the yields of 14 varieties of cocksfoot when cut 8 times during the growing season at approximately monthly intervals. Hunt (1965) obtained a similar result with 12 varieties cut 6 times at approximately monthly intervals; only extreme varieties differed significantly in yield.

In a comparison between monthly cutting and a hay-and-aftermath system of management, Dent and Aldrich (1963) included only the yield at the first of three aftermath cuts taken at intervals of 8 weeks. However, in contrast to three other species grown, the aggregate yield of cocksfoot with the monthly cutting system was only slightly less than that from the hay-and-aftermath system. This was apparently the result of large differences in yield between species when cut for hay. A more typical reduction in yield from increased frequency of cutting was found by Green *et al.* (1960), who recorded a 50% reduction in yield of dry matter between plots cut every 10 days compared with those cut every 30 days.

Satisfactory yields of cocksfoot are usually obtained when swards are grazed, so long as they are given some period of rest. Thus, recorded yields in one experiment were poor when defoliation was carried on throughout the season, but greatly increased when the sward was rested during the summer and autumn (Jones, 1939).

Quality

CHEMICAL COMPOSITION

Cocksfoot is relatively low in soluble carbohydrate and high in lignin content compared with other grasses and may be high in Na. Otherwise it is typical in chemical composition (see Chapter 13).

WATER CONTENT

Both stage of growth and amount of fertilizer N applied have been found to influence the water content of cocksfoot.

Johnston and Waite (1965) sampled first-growth S37 cocksfoot during May, June, July and August and found a progressive fall in water content during uninterrupted growth from 82·2% on 17 May to 56·5% on 15 August.

Ivins (1952) found that areas of Danish and S143 cocksfoot to which 224 lb nitrogenous fertilizer/ac (246 kg/ha) had been applied yielded herbage with a higher water content than similar areas which had received no N. Parish *et al.* (1955) measured the water content of S143 cocksfoot from areas which had received either no fertilizer N or 560 lb fertilizer N/ac (616 kg/ha). The areas were sampled at weekly intervals during uninterrupted growth, i.e. from areas not previously cut during that season. The application of fertilizer N increased the water content of herbage, but there was no indication that this was due to any delay in the onset of maturation. The results of this experiment are summarized in Table 20.3.

NUTRITIVE VALUE (DRY MATTER DIGESTIBILITY)

Young leafy grass frequently cut

In swards of cocksfoot frequently defoliated and kept in a leafy condition, DM digestibility values vary between about 68% and 79%. Under comparable conditions, cocksfoot varieties are about 4% lower in digestibility than perennial ryegrass, and although there is some variation between cocksfoot varieties, in trials at the National Institute of Agricultural Botany (Dent & Aldrich, 1968) the least digestible perennial ryegrass was found to be more digestible than the most digestible cocksfoot. The quality of cocksfoot as compared with other species is particularly poor in midsummer, when digestibility is low and fibre content high. However, Cooper *et al.* (1962) have suggested that there is considerable scope for improving the digestibility of cocksfoot by selection and breeding.

Uninterrupted primary growth

There have been a number of comparisons of the rate of fall in digestibility in the primary growth of S37 cocksfoot compared with that in S24 perennial ryegrass, which flowers on approximately the same date (Raymond & Terry, 1966). These workers have shown that S37 is about 5% lower in digestibility

Table 20.3. Percentage of water in S143 cocksfoot

Stage of growth, weeks	High N (616 kg/ha)	No N
1	81·9	78·9
2	83·1	79·3
3	82·7	79·9
4	83·9	82·5
5	79·5	77·0
6	74·0	73·8

From Parish *et al.* (1955)

than S24 up to ear emergence, but this difference increases at later growth stages (Fig. 20.1, top). The digestibility of S37 cocksfoot approaches 70% at the 50% ear-emergence date, whereas S24 perennial ryegrass makes two weeks' more growth before falling to this level. The latter usually gives higher yields of DM at any particular digestibility level, and a higher yield of digestible DM (Fig. 20.1). Terry and Tilley (1964) have shown that these differences between cocksfoot and ryegrass are due more to differences in digestibility of leaf blades and leaf sheaths than to differences in stem digestibility.

Although the range in flowering dates between early and late cocksfoot varieties is only about 2·5 weeks, the early, stemmy varieties are usually more digestible at 50% ear emergence than the late leafy ones. This has drawn attention to the fact that leafiness is not necessarily associated with high nutritive value, and it has been shown that up to ear emergence, the digestibility of the stem portion is equal to or slightly higher than that of the leaf blade (Terry & Tilley, 1964). There is a tendency also for the higher digestibility of the early cultivars such as Scotia to be associated with above-average soluble carbohydrate content, while a leafy pasture cultivar such as

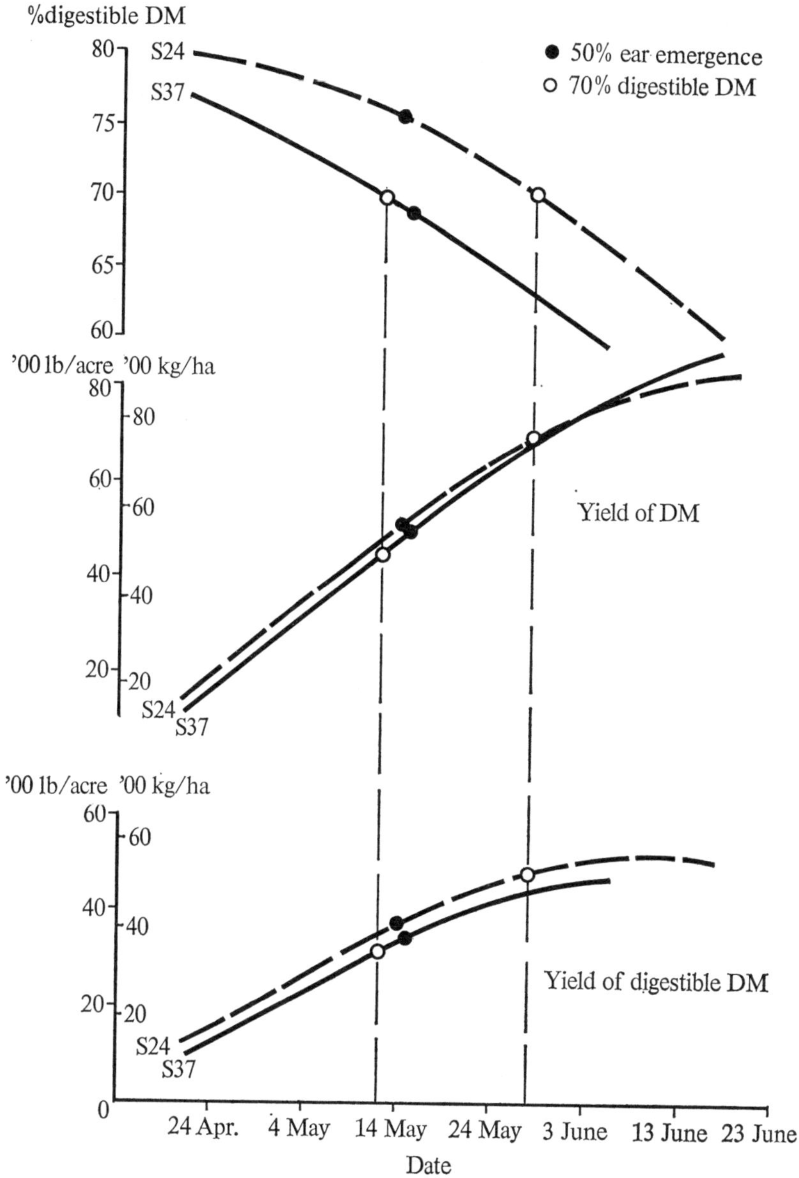

Fig. 20.1. Cocksfoot S37. Seasonal changes in DM yield and digestibility. Values for S24 perennial ryegrass shown for comparison. (After Aldrich and Dent, 1967)

S143 is below average in this respect. Grazing animals have shown preferences for the former, particularly early in the season (Bland & Dent, 1964).

Regrowths

Cocksfoot seldom produces flowering stems in the regrowth after a primary cut in late May or early June, but Minson *et al.* (1960) found that S37 cocksfoot regrowths were less digestible than regrowths of S24 perennial ryegrass at both the one-month and the two-month stage. After one month the approximate values for DM digestibility were 70% and 75%, and after two months 65% and 70%. Variety trials reported by Dent and Aldrich (1968) have shown that mid- and late-season regrowths of cocksfoot are of low digestibility.

Cocksfoot is less digestible throughout the season than perennial ryegrass; at the same time it is higher in protein content, lower in soluble carbohydrate content and higher in fibre content (Dent & Aldrich, 1966). Typical analyses of S24 ryegrass and S37 cocksfoot under grazing management and for a hay cut taken 10 days after 50% ear emergence are given in Table 20.4. These cultivars are directly comparable as they have similar heading dates.

Table 20.4. Percentage content (DM basis) of crude protein, crude fibre and soluble carbohydrate, and percentage digestible DM

	Simulated grazing (weighted mean of 9 cuts)		Cut for hay (cut 10 days after 50% ear emergence)	
	S24	S37	S24	S37
Crude protein	19·0	20·0	8·5	10·5
Crude fibre	24·0	24·5	31·0	32·5
Soluble carbohydrate	15·0	9·0	20·0	8·5
Digestible DM	76·5	72·5	71·0	64·0

From National Institute of Agricultural Botany (1966)

ANIMAL INTAKE CHARACTERISTICS

Greenhalgh (1966) found that grazing dairy cows voluntarily consumed more organic matter from a sward of S24 perennial ryegrass than from one of S37 cocksfoot. However, the ryegrass was some 5 to 6 percentage units higher in digestibility than the cocksfoot, and it is possible that this accounted for the observed difference in intake.

Longevity and persistency

The total length of life of leaves of cocksfoot does not appear to have been studied. Data on the life-span of leaves between full extension and complete senescence have been given by Taylor and Templeton (1966) for cocksfoot in the year after sowing. Longevity increased from 18–22 days for leaves fully expanded in March, to 36 days for those becoming fully expanded in June, then declined to 24–28 days when observations were discontinued in

September. The sixth leaf below the flag leaf lived for 19 days after full expansion; subsequent leaves on flowering tillers progressively increased in longevity to 38–39 days for the last three leaves which emerged.

Treharne and Taylor (1967) found that leaves used for the measurement of photosynthesis lived for about 36 days after complete expansion. Rate of photosynthesis in 1 000 f.c. light (10 000 lux) at 29°C remained fairly steady until the leaf began to senesce visibly. In further trials (Taylor *et al.*, 1968), the period between full expansion and senescence was 27 days at 16 140 lux and 30·8 days at 10 760 lux. The number of living leaves per vegetative tiller was found by Lambert (1967) to increase steadily from 1·76 in February to 3·13 in early July. Other measurements (Bean, 1961) indicate a range of 2·5–3·5 leaves per tiller during most of the growing season. The tillers are thought to be annual (Beddows, 1959) but their longevity has apparently not been measured. Lambert (1967) found that many rooted tillers died between late May and mid-June in spaced plants grown for seed production, though the number of rooted tillers per plant remained almost constant between February and early July. Seminal roots of cocksfoot are short-lived, adventitious roots are perennial (Stuckey, 1941), but there are no data on longevity of individual roots.

Mortality of cocksfoot during the establishment phase is high (Charles, 1961). Longevity of undefoliated plants may be lower than that of mown or grazed ones. Luff (1965) found that the pattern of leaf production and decay in cocksfoot tussocks remained similar to that of grazed plants for about 7 years, after which time the accumulation of dead leaves began to choke new growth. Tussocks had a life of at least 9 years, but in old tussocks new growth took place only on the windward side.

Beddows (1959) noted that plants grown for seed production were fairly vigorous after 11 years, and was of the opinion that grazed plants of cocksfoot in 'fattening pastures' were almost certainly old. Cocksfoot is able to resist grazing damage to some extent by the production of tillers in the horizontal as well as in the vertical plane. However, frequent defoliation has been shown to reduce the total yield (Jones, 1939) and, as in perennial ryegrass, the ability of the species to resist grazing may mean that there is a reduction in the proportion of the sward which is taken by the animal (Kydd, 1966).

The greatest yield of cocksfoot has been recorded after a period of two years during which the sward was neither grazed nor mown, and the greatest number of tillers also developed under this treatment (Jones, 1939). Under conditions of low fertility, however, such neglected use of the sward has led first to an increase in the proportion of cocksfoot, followed by its rapid replacement by Yorkshire fog (Kydd, 1964). Again, when grazed at intervals from April to January and also in March, the yield of S143 was very much reduced in the third year after sowing, and it died out in the fourth year. S37, however, was found to persist for 9 years in a mixture which was grazed from April to July and then rested until November before being grazed again (Alder, 1966). Persistence of cocksfoot is adversely affected by grazing in early autumn, winter and early spring (Jones, 1933; Beddows, 1959) and by damage from trampling (Beddows, 1959; Edmond, 1964). Cocksfoot showed good survival during the 1962–3 winter in Britain, but subsequent spring growth lacked vigour (Baker & David, 1963).

References

ALDER, F. E. 1966. A summary of twelve years' work on a farmlet type of experiment designed to study animal production on sown pastures. *A. Rep. Grassld Res. Inst. Hurley 1965* pp. 79–86.

ALDRICH, D. T. A.; DENT, J. W. 1967. The relationship between yield and digestibility in the primary growth of nine grass varieties. *J. natn. Inst. agric. Bot.* **11,** 104–13.

ANSLOW, R. C. 1955. The effects of competition on the shoot and root growth of three grass species. M.Sc. Thesis, Univ. Reading.

ANSLOW, R. C. Grassland Research Institute, Hurley. Personal communication.

ANSLOW, R. C.; GREEN, J. O. 1967. The seasonal growth of pasture grasses. *J. agric. Sci., Camb.* **68,** 109–22.

ARENS, R. 1962. The effect of rate of seeding on competition between species and the productivity of the first year's pasture. *Z. Acker- u. PflBau* **115,** 357–74.

ARMSTRONG, S. F. 1948. British grasses and their employment in agriculture. Cambridge: Univ. Press, 3rd edn.

BAKER, H. K.; DAVID, G. L. 1963. Winter damage to grass. *Agriculture, Lond.* **70,** 380–2.

BAUMANN, H.; KLAUSS, M. L. 1955. [Root development in an environment with a high water table.] *Z. Acker- u. PflBau* **99,** 410–26.

BEAN, E. W. 1961. The effect of nitrogen and light intensity on the growth of grass swards. Ph.D. Thesis, Univ. Reading, pp. 132.

BEDDOWS, A. R. 1959. Biological flora of the British Isles. *Dactylis glomerata* L. *J. Ecol.* **47,** 223–39.

BLAND, B. F.; DENT, J. W. 1964. Animal preference in relation to the chemical composition and digestibility of varieties of cocksfoot. *J. Br. Grassld Soc.* **19,** 306–15.

BROWN, E. M. 1939. Some effects of temperature on the growth and chemical composition of certain pasture grasses. *Res. Bull. 299 Univ. Mo. agric. Exp. Stn* pp. 1–76.

CHARLES, A. H. 1961. Differential survival of cultivars of *Lolium*, *Dactylis* and *Phleum*. *J. Br. Grassld Soc.* **16,** 69–75.

CLAPHAM, A. R.; TUTIN, T. G.; WARBURG, E. F. 1962. Flora of the British Isles. Cambridge: Univ. Press, 2nd edn.

CLEMENT, C. R. 1967. Potassium for the grass crop, its supply from soil and fertilisers. *A. Rep. Grassld Res. Inst. Hurley 1966* pp. 80–5.

COOPER, J. P. 1964. Climatic variation in forage grasses. 1. Leaf development in climatic races of *Lolium* and *Dactylis*. *Jnl appl. Ecol.* **1,** 45–61.

COOPER, J. P.; TILLEY, J. M. A.; RAYMOND, W. F.; TERRY, R. A. 1962. Selection for digestibility in herbage grasses. *Nature, Lond.* **195,** 1276–7.

COPPENET, M.; LE CORRE, L. 1964. Grass variety trial in Quimper, Brittany. *Fourrages* **17,** 81–97.

COWLING, D. W. 1961. The effect of white clover and nitrogenous fertilizer on the production of a sward. *J. Br. Grassld Soc.* **16,** 281–90.

COWLING, D. W.; LOCKYER, D. R. 1965. A comparison of the reaction of different grass species to fertilizer nitrogen and to growth in association with white clover. *J. Br. Grassld Soc.* **20,** 197–204.

COWLING, D. W.; LOCKYER, D. R. 1968. A comparison of the yield of three grass species at various levels of nitrogenous fertilizer sown alone or in a mixture. *J. agric. Sci., Camb.* **71,** 127–36.

DAVIDSON, J. L.; MILTHORPE, F. L. 1965. The effect of temperature on the growth of cocksfoot (*Dactylis glomerata*). *Ann. Bot.* **29,** 407–17.

DAVIES, H.; THOMAS, B. V.; ALDRICH, D. T. A. 1966. The effects of different levels of nitrogen fertilizer on the yields of varieties of perennial ryegrass and cocksfoot. *J. Br. Grassld Soc.* **21,** 245–9.

DAVIES, J.; EDWARDS, G. H. A. 1953. The influence of nitrogen on the yield of a cocksfoot seed crop. *J. Br. Grassld Soc.* **8,** 261–6.

DAVIES, W. 1960. The grass crop. London: E. & F. N. Spon Ltd, 2nd edn.

DAVIS, W. E. P. 1961. An evaluation of varietal characteristics of orchard-grass (*Dactylis glomerata,* L.) subjected to a silage-pasture type of utilization. *Can. J. Pl. Sci.* **41,** 653–63.

DENT, J. W.; ALDRICH, D. T. A. 1963. The inter-relationships between heading date, yield, chemical composition and digestibility in varieties of perennial ryegrass, timothy, cocksfoot and meadow fescue. *J. natn. Inst. agric. Bot.* **9,** 261–81.

DENT, J. W.; ALDRICH, D. T. A. 1966. The *in vitro* digestibility of herbage species and varieties and its relationship with cutting treatment, stage of growth and chemical composition. *Proc. 10th int. Grassld Congr., Helsinki, 1966* pp. 419–24.

DENT, J. W.; ALDRICH, D. T. A. 1968. Systematic testing of quality in grass varieties. 2. The effect of cutting dates, seasons and environments. *J. Br. Grassld Soc.* **23,** 13–19.

EDMOND, D. B. 1964. Some effects of sheep treading on the growth of 10 pasture species. *N.Z. Jl agric. Res.* **7,** 1–16.

EHARA, K.; SASAKI, T.; IKEDA, H.; NATA, Y. 1965. Physiological and ecological studies on the regrowth of herbage plants. *J. Jap. Soc. Grassld Sci.* **10,** 214–20.

ENGLAND, F. J. W. 1965. Interactions in mixtures of herbage grasses. *Rec. Scot. Pl. Breed. Stn 1965* pp. 125–49.

EVANS, G. 1959. Seed rates of grasses for seed production. 1. Pasture varieties of ryegrass, cocksfoot, and timothy. *Emp. J. exp. Agric.* **27,** 291–9.

EVERS, A.; SONNEVELD, A. 1953. Grass seed production trials. *Gestenc. Meded. 12 Cent. Inst. Landbouwk. Onderz., Wageningen.*

GARWOOD, E. A. Grassland Research Institute, Hurley. Personal communication.

GERDES, G. 1952. [Studies on the absolute weight (1000 seed weight) of the most important fodder grass species and their varieties in regard to the performance of grass breeding.] *Züchter* **12,** 353–66.

GRASSLAND RESEARCH INSTITUTE. 1952. *Exps Prog. 5 Grassld Res. Inst. 1952* pp. 23–31.

GREEN, J. O.; CORRALL, A. J.; TAYLER, J. C. 1960. Level and frequency of cutting: pure grass swards at two levels of nitrogen. *Exps Prog. 12 Grassld Res. Inst. 1958–9* pp. 19–21.

GREEN, J. O.; DAVIS, A. G. 1955. The seasonal use of lucerne/grass mixture for grazing. In: Grassland Research Institute. Memoir No. 1. Lucerne investigations, 1944–53, pp. 63–77.

GREEN, J. O.; EVANS, T. A. 1956. Manuring and grazing for seed production in S37 cocksfoot and S215 meadow fescue. *J. Br. Grassld Soc.* **11,** 165–73.

GREEN, J. O.; EVANS, T. A. 1957. Grazing management for seed production in leafy strains of grasses. *J. Br. Grassld Soc.* **12,** 4–9.

GREENHALGH, J. F. D. 1966. Studies of herbage consumption and milk production in grazing dairy cows. *Proc. 10th int. Grassld Congr., Helsinki, 1966* pp. 351–5.

HEDDLE, R. G. 1965. Varieties of cocksfoot 1957–1960. Varieties of Italian ryegrass 1961–62. *Tech. Bull. 25 Edinb. Sch. Agric.* pp. 28.

HERRIOTT, J. B. D.; WELLS, D. 1960. Clover nitrogen and sward productivity. *J. Br. Grassld Soc.* **15,** 63–9.

HOLMES, W.; MACLUSKY, D. S. 1955. The intensive production of herbage for crop-drying. *J. agric. Sci., Camb.* **46,** 267–86.

HUBBARD, C. E. 1954. Grasses; a guide to their structure, identification, uses, and distribution in the British Isles. Harmondsworth, Middlesex: Penguin Books.

HUNT, I. V. 1954. Seed establishment in the West of Scotland. *J. Br. Grassld Soc.* **9,** 85–98.

HUNT, I. V. 1964. Comparative productivity of herbage varieties on upland peat. *J. Br. Grassld Soc.* **19,** 55–61.

HUNT, I. V. 1965. Comparisons of production from twelve varieties of cocksfoot. *Res. Bull. 36 W. Scotl. agric. Coll.* pp. 30.

IVINS, J. D. 1952. The relative palatability of herbage plants. *J. Br. Grassld Soc.* **7,** 43–54.

JOHNSTON, M. J.; WAITE, R. 1965. Studies in the lignification of grasses. 1. Perennial ryegrass (S24) and cocksfoot (S37). *J. agric. Sci., Camb.* **64,** 211–19.

JONES, I. 1939. Pasture management and its effect on the sward. *Bull. H15 Welsh Pl. Breed. Stn* pp. 40–129.

JONES, L. I. 1958. Grassland agronomy. *Rep. Welsh Pl. Breed. Stn 1950–56* pp. 97–9.

JONES, M. G. 1933. Grassland management and its influence on the sward. 3. The management of a 'grassy' sward and its effects. *Emp. J. exp. Agric.* **1,** 223–34.

KYDD, D. D. 1964. The effect of different systems of cattle grazing on the botanical composition of permanent downland pasture. *J. Ecol.* **52,** 139–49.

KYDD, D. D. 1966. The effect of intensive sheep stocking over a five year period on the development and production of the sward. 3. Ecological discussion. *J. Br. Grassld Soc.* **21,** 298–302.

LAMBERT, D. A. 1963. The influence of density and nitrogen in seed production stands of S37 cocksfoot (*Dactylis glomerata L.*). *J. agric. Sci., Camb.* **61,** 361–73.

LAMBERT, D. A. 1965. The effect of high levels of nitrogen on seed yield of cocksfoot (*Dactylis glomerata* L.). *J. Br. Grassld Soc.* **20,** 164–7.

LAMBERT, D. A. 1966a. The effect of relating applications of nitrogen to the stage of development of cocksfoot (*Dactylis glomerata L.*) grown for production of seed. *J. Br. Grassld Soc.* **21,** 127–34.

LAMBERT, D. A. 1966b. The effect of cutting cocksfoot (*Dactylis glomerata L.*) grown for production of seed. *J. Br. Grassld Soc.* **21,** 200–7.

LAMBERT, D. A. 1967. Cocksfoot (*Dactylis glomerata*) grown for production of seed. *Ann. Bot.* **31,** 371–83.

LEASURE, J. K. 1952. Growth patterns of mixtures of orchard grass and tall fescue with Ladino white clover in relation to temperature. *Proc. Ass. sth. agric. Wkrs* **49,** 177–8.

LINCOLNSHIRE SEED GROWERS' ASSOCIATION. 1965. Report for 1964. Agric. House, Woodhall Spa, pp. 3–7.

LUFF, M. L. 1965. The morphology and microclimate of *Dactylis glomerata* tussocks. *J. Ecol.* **53,** 771–87.

MCNAUGHT, K. J. 1958. Potassium deficiency in pastures. 1. Potassium content of legumes and grasses. *N.Z. Jl agric. Res.* **1,** 148–81.

MINISTRY OF AGRICULTURE, FISHERIES AND FOOD. 1968. Grass and clover crops for seed. *Bull. 204* London: H.M.S.O., pp. 42–60.

MINSON, D. J.; RAYMOND, W. F.; HARRIS, C. E. 1960. Studies in the digestibility of herbage. 8. Digestibility of S37 cocksfoot, S23 ryegrass and S24 ryegrass. *J. Br. Grassld Soc.* **15,** 174–80.

MITCHELL, K. J. 1956. Growth of pasture species under controlled environments. 1. Growth at various levels of constant temperature. *N.Z. Jl Sci. Technol. Sect. A* **38,** 203–15.

MITCHELL, K. J.; LUCANUS, R. 1960. Growth of pasture species in controlled environment. 2. Growth at low temperature. *N.Z. Jl agric. Res.* **3,** 647–55.

MITCHELL, K. J.; LUCANUS, R. 1962. Growth of pasture species under controlled environment. 3. Growth at various levels of constant temperature with 8 and 16 hours of uniform light per day. *N.Z. Jl agric. Res.* **5,** 135–44.

NATIONAL INSTITUTE OF AGRICULTURAL BOTANY. 1959. Growing grasses for seed. *Herbage Seed Growers' Leaflet No. 2.*

NATIONAL INSTITUTE OF AGRICULTURAL BOTANY. 1966. Unpublished data.

PARISH, D. H.; COEY, W. E.; ROBINSON, K. L. 1955. The chemical composition of cocksfoot S143 at successive stages of growth as influenced by fertilizer treatment. *Res. exp. Rec. Minist. Agric. Nth. Ire. 1952–3* **3,** 46–54.

PENMAN, H. L. 1962. Woburn irrigation 1951–59. Results for grass. *J. agric. Sci., Camb.* **58,** 349–64.

PERRING, F. H.; WALTERS, S. M. 1962. Atlas of the British flora. London: Thomas Nelson and Sons Ltd, pp. 456.

PETO, D. H. 1961. Methods and problems of cleaning herbage seeds. *Rep. Lincs. Seed Growers' Assn 1960.* Agric. House, Woodhall Spa, pp. 19–21.

RAYMOND, W. F.; TERRY, R. A. 1966. Studies of digestibility by an *in vitro* method. *Outl. Agric.* **5**, 60–8.

REID, D.; CASTLE, M. E. 1965. The response of grass-clover and pure-grass leys to irrigation and fertilizer nitrogen treatment. 1. Irrigation effects. *J. agric. Sci., Camb.* **64**, 185–94.

ROBERTS, H. M. 1958. The effect of a cover crop. *Rep. Welsh Pl. Breed. Stn 1950–56* pp. 147–8.

ROBERTS, H. M. 1959. The effect of defoliation on the seed-producing capacity of bred strains of grasses. 2. Cocksfoot. *J. Br. Grassld Soc.* **14**, 58–64.

ROBERTS, H. M. 1961. The effect of drill width on the seed productivity of leafy varieties of three grass species. *J. Br. Grassld Soc.* **16**, 37–42.

RYLE, G. J. A. 1964. A comparison of leaf and tiller growth in seven perennial grasses as influenced by nitrogen and temperature. *J. Br. Grassld Soc.* **19**, 281–90.

SCHWANBOM, N.; FRÖIER, K. 1949. Methods of herbage seed production. *J. Br. Grassld Soc.* **4**, 233–61.

SPRAGUE, V. G. 1943. The effects of temperature and day-length on seedling emergence and early growth of several pasture species. *Proc. Soil Sci. Soc. Am.* **8**, 287–94.

SPRAGUE, V. G. 1948. The relation of supplementary light and soil fertility to heading in the greenhouse of several perennial forage grasses. *J. Am. Soc. Agron.* **40**, 144–54.

SPURWAY, C. H. 1941. Soil reaction (pH) preferences of plants. *Spec. Bull. 306 Mich. agric. Exp. Stn* pp. 36.

STUCKEY, I. H. 1941. Seasonal growth of grass roots. *Am. J. Bot.* **28**, 486–91.

TAYLOR, T. H.; COOPER, J. P.; TREHARNE, K. J. 1968. Growth response of orchardgrass (*Dactylis glomerata* L.) to different light and temperature environments. 1. Leaf development and senescence. *Crop Sci.* **8**, 437–40.

TAYLOR, T. H.; TEMPLETON, W. C. 1966. Tiller and leaf behaviour of orchardgrass (*Dactylis glomerata* L.) in a broadcast planting. *Agron. J.* **58**, 189–92.

TERRY, R. A.; TILLEY, J. M. A. 1964. The digestibility of leaves and stems of perennial ryegrass, cocksfoot, timothy, tall fescue, lucerne and sainfoin, as measured by an *in vitro* procedure. *J. Br. Grassld Soc.* **19**, 363–72.

TINCKER, M. A. H. 1925. The effect of length of day upon the growth and reproduction of some economic plants. *Ann. Bot.* **39**, 721–54.

TREHARNE, K. H.; TAYLOR, T. H. 1967. Changes in photosynthetic activity with age of leaf. *Rep. Welsh Pl. Breed. Stn 1966* p. 27.

WIDDOWSON, F. V.; PENNY, A.; WILLIAMS, R. J. B. 1963. An experiment comparing responses to nitrogen fertilizer of four grass species. *Expl Husb.* No. 9, 28–36.

WIDDOWSON, F. V.; PENNY, A.; WILLIAMS, R. J. B. 1965. An experiment measuring effects of N, P and K fertilizers on yield and N, P and K contents of grass. *J. agric. Sci., Camb.* **64**, 93–100.

WILLIAMS, T. E.; STILES, W.; TURNER, P. J. 1959. *Exps Prog. 11 Grassld Res. Inst. 1957–8* p. 58.

WILLIAMS, T. E.; STILES, W.; TURNER, P. J. 1960. Grassland irrigation. *Exps Prog. 12 Grassld Res. Inst. 1958–9* pp. 55–7.

WILLIAMS, W. 1950. Experiments on the yield of lucerne strains as pure plots and on the behaviour of lucerne when sown in a mixture with various species of grasses. *J. Br. Grassld Soc.* **5**, 113–29.

WYCHERLY, P. R. 1952. Temperature and photoperiod in relation to flowering in three perennial grass species. *Meded. LandbHogesch. Wageningen* **52**, 75–92.

CHAPTER 21

Meadow fescue *(Festuca pratensis)*

Morphology

MEADOW fescue (Plate 12) is an erect, loosely-tufted perennial, very similar morphologically to Italian ryegrass. Vegetative stems erect, often decumbent at the base. Leaf sheaths red at base, glabrous, becoming brown, papery, persistent. Ligule short, auricles prominent, glabrous. Leaf blades rolled in sheath, opening flat, without a marked midrib; strongly ribbed adaxially, glossy abaxially. Culms slender, 2- to 4-noded, producing vegetative tillers from the first or even the second node above the ground. Inflorescence a little-branched, secund panicle of terete 5- to 14-flowered spikelets. Caryopsis broadly oblong, enclosed in hardened lemma and palea. In the vegetative phase meadow fescue has a somewhat less lax, more erect habit than Italian ryegrass and is a slightly darker green.

There are almost no published data on the root distribution of meadow fescue, but what evidence there is indicates that the root distribution and bulk are similar to the other important herbage grasses.

Langer *et al.* (1964) studied changes in the tiller population of S215 meadow fescue and found that it produced fewer tillers than S48 timothy; in a fully established sward they recorded 15–80 tillers/dm^2, with tiller density greater under more frequent defoliation. Leaf sizes ranging from 4×100 mm to 8×240 mm were recorded by Ryle (1964) in plants grown in a glasshouse. Early-flowering cultivars are more erect and less densely tillered than late-flowering ones.

Changes in the amount of leaf blade during the development of the first crop have been studied by Minson *et al.* (1964) (Table 21.1).

Distribution

Festuca pratensis is a native species (Clapham *et al.*, 1962), occurring in all but 6 vice-counties of Britain (Perring & Walters, 1962). It is common in south Scotland and England, except in the south-west where it is somewhat less frequent. It is considerably less frequent in Wales and north Scotland.

In Ireland it is common in the north, becoming rare in the south (Hubbard, 1954; Clapham *et al.*, 1962; Perring & Walters, 1962).

It occurs in meadows and grassy places, old pastures and roadsides (Hubbard, 1954; Clapham *et al.*, 1962) and is often abundant in water meadows and low-lying grassland (Hubbard, 1954). It is generally absent above 750 m and much less common above 300 m (Perring & Walters, 1962). Distribution in Britain bears no relation to isotherms (Perring & Walters, 1962). It occurs on a wide range of soils, but especially on rich, heavy loams (Hubbard, 1954). It grows best on soils with pH values in the range 4·5–7·0 (Spurway, 1941). It is much less common in areas where the annual rainfall is above 1 000 mm (Perring & Walters, 1962) although it is quite tolerant of high water tables, is often abundant in water meadows, and its roots will grow vigorously in substrates with a high water content (Hubbard, 1954; Baumann & Klauss, 1955).

Table 21.1. Percentage leaf blade in the DM of S215 meadow fescue during primary growth

Date	Leaf blade %
6 May	58
17 May	36
27 May	15
9 June	23
16 June	26

Date of 50% ear emergence: 18 May
From Minson *et al.* (1964)

Compatibility

Meadow fescue has the reputation of being compatible with other grasses and with legumes. The combination of meadow fescue with timothy and white clover forms the basis of many sowings where ease of management takes priority over maximum herbage yield (which would require the use of ryegrass).

A number of workers have assessed the relative competitive ability of meadow fescue with other grasses. Crocker and Martin (1964) found, 9 months after sowing, that the size of meadow fescue plants growing with perennial ryegrass was only half that of meadow fescue plants in a pure stand. Under grazing, Arens (1962) found that in the first year meadow fescue was less competitive than perennial ryegrass but more so than cocksfoot. However, he found that in later years cocksfoot became dominant. Lampeter (1960) calculated 'suppression factors' for the effect of one species on another over two harvest years under cutting management. Meadow fescue was suppressed by Italian and perennial ryegrass but was aggressive towards timothy, red clover and lucerne. Here again, cocksfoot was suppressed by meadow fescue only in the early stages.

Although meadow fescue has been reported above as being suppressed by perennial ryegrass, when established the two grasses are alike in their compatibility with white clover. From the results of an experiment at the Grassland Research Institute, Hurley (1952), and later work reported by Cowling and Lockyer (1965), it seems that after the initial stages equal amounts of white clover grow with meadow fescue and perennial ryegrass. Meadow fescue was consistently more compatible with white clover than was cocksfoot. More white clover grew with timothy than with meadow fescue at first, but as the timothy developed it became more aggressive and the balance changed.

Meadow fescue has been recommended as a companion grass for lucerne. Green and Davis (1955) concluded from a series of experiments that it was more suitable than timothy or cocksfoot, offering less competition to the lucerne and yet controlling weeds effectively. In a summary of trials on farms in Wales, Williams (1950) stated that meadow fescue was less damaging to lucerne than was cocksfoot.

Seed

CHARACTERS

The seed is lanceolate-oblong with a shallow groove along its ventral surface. Seed length is 5–7 mm, and the average maximum width 1·0–1·5 mm (Armstrong, 1948; Hubbard, 1954; Peto, 1961). The lemma is pointed, convex, awnless and smooth or slightly rough near the tip; the palea is similar in length to the lemma. The rachilla is cylindrical and stands out from the palea (Armstrong, 1948; Hubbard, 1954).

The 1 000-seed weight varies between and within cultivars. Gerdes (1952) reported values to range from 1·72 g for S53 to 2·38 g for a German variety. Within S53, 1 000-seed weights of 2·12 g (Evans, 1962), 1·81 g (Hunt, 1954), 1·72 g (Gerdes, 1952) and 1·04 g (Lewis, 1966) have been recorded. Similarly, in S215 a range from 1·56 g (Lewis, 1966), 1·6 g to 2·1 g (Green, 1964), and 2·04 g (Hunt, 1954) to 2·12 g (Gerdes, 1952) is apparent.

PRODUCTION

The seed should be sown in spring at 10–15 kg/ha in drills not more than 35 cm apart or alternatively broadcast (Lambert, 1964; National Institute of Agricultural Botany, 1959; Roberts, 1961); yield may be reduced considerably by a cover crop (Roberts, 1958). The maximum amount of fertilizer required appears to be 100–120 kg N/ha applied in two equal doses in autumn and spring, but half this amount may give comparable yields (Green & Evans,

1956; Lambert, 1964; Roberts, 1961). Application of 30 kg P and 60 kg K/ha after harvest is recommended (National Institute of Agricultural Botany, 1959). Grazing in December will probably increase yield (Green & Evans, 1956; 1957); repeated grazing will reduce yield from drills but may be beneficial in old broadcast stands (Green & Evans, 1957). Crops should be isolated by at least 180 m during anthesis (National Institute of Agricultural Botany, 1959). Principles and procedures for harvesting are well explained elsewhere (Ministry of Agriculture, Fisheries & Food, 1968).

The annual yield of seed can vary from 12·5 to 1 200 kg/ha, with the mean yield ranging between 140 and 675 kg/ha, depending on the cultivar and year (Lincolnshire Seed Growers' Association, 1965).

Yield

DRY MATTER

Dent and Aldrich (1963) examined the growth of 11 varieties of meadow fescue at Cambridge and at Trawscoed, Wales. Under a system of monthly cuts, they found that there was no clear relationship between the mean date of ear emergence of a variety and its annual yield. The average annual yield of dry matter (DM) under this system of cutting was about 7 500 kg/ha and the difference in yield between sites was slight. When cut for hay with a second cut 8 weeks later, the annual yield was higher, averaging 10 400 kg/ha at the two sites. In experiments in Brittany, Coppenet and Le Corre (1964) obtained annual yields of about 8 000 kg DM/ha when the crop was cut to simulate grazing.

In a comparison of the cultivars S215 and S53 by Hunt (1961), their annual yields were 9 500 and 9 000 kg/ha, respectively, under a sequence of monthly cuts and at a high level of fertility. In further trials on upland peat (Hunt, 1964), the two cultivars respectively yielded 2 600 and 2 900 kg/ha, the slight superiority of S53 being due primarily to its better persistency over the 5-year period of the trial and a smaller proportion of weeds in the sward. The average annual production of S215 grown in Wales by Henderson, Edwards and Hammerton (1962) was 6 800 kg/ha over a three-year period when crops were cut two or three times per year.

Regarding the seasonal distribution of DM yield, Hunt (1957) gave data showing that under a monthly cutting regime the heaviest yields were obtained in early June. There was no evidence in this species of a second production peak in late summer.

ENERGY AND PROTEIN

The annual production of crude protein by S215 and S53 in trials reported by Hunt (1961) was 1 650 and 1 600 kg/ha, respectively. Data for energy production, derived from the literature, are summarized in Table 21.2.

Table 21.2. Potential annual yields of energy and protein from meadow fescue (DM basis)

	Frequent cutting	Infrequent cutting
DM digestibility (%)	76	67
Yield of digestible OM (kg/ha)	7 100	6 460
Yield of metabolizable energy (Mcal/ha)	23·0	23·8
Crude protein (%)	18·0	12·0
Yield of crude protein (kg/ha)	1 620	1 250
Yield of digestible crude protein (kg/ha)	1 190	790

Data compiled by D. J. Thomson (Grassland Research Institute, Hurley) from literature cited in the text. For conversion formulae see Chapter 5.

Response to management and environment

RESPONSE TO FERTILIZER

Nitrogen

The response to N of S53 meadow fescue cut 3–6 times each year over a 4-year period and given 0, 195, or 390 kg N/ha each year, divided between the cuts, was reported by Holmes and MacLusky (1955). Annual yield increased by 18·1 kg DM/kg N with the first increment of N and by a further 15·1 kg DM/kg N with the second increment. Meadow fescue gave a yield similar to that of the other commonly-used grasses when no N was given. Its response to N was similar to that of perennial ryegrass but inferior to that of cocksfoot and timothy.

Lambert (1962) examined the response to N of S215 meadow fescue either cut 8 times at monthly intervals or cut first at an advanced stage of development and then three times subsequently. The grass was given 58 or 175 kg N/ha in split applications. Yield and N response were both reduced under the increased frequency of defoliation. The increase in yield was 20·4 kg DM/kg N in the frequently-cut treatment and 31·7 kg DM/kg N in that less frequently cut.

Cowling and Lockyer (1965) examined the response to N of a number of grasses, including S215 meadow fescue. In each of three years the sward was cut 4 or 5 times and was given 0, 91, 182 or 364 kg N/ha. Yield increased by 25·0 kg DM/kg N where up to 182 kg N/ha were given, and by a further 19·7 kg DM/kg N when this N level was doubled.

Widdowson *et al.* (1963) studied the response to N of S215 meadow fescue over a 2·5-year period. The sward was cut on 7 occasions over the whole period and was given 0, 37·7 or 75·4 kg N/ha before each cut. The response in yield was 41·7 kg DM/kg N with the first increment of N, with an additional increase of 22·3 kg DM/kg N when the quantity of N given was doubled.

Phosphorus and potassium

There is no information about the response to P and K of meadow fescue in field trials in Britain.

RESPONSE TO IRRIGATION

In the only reported experiments in Britain on the irrigation of meadow fescue, only small responses were found (Castle & Reid, 1960; Long & French, 1967). So far as it goes, this evidence suggests that meadow fescue shows the typical irrigation response described in Chapter 14.

RESPONSE TO TEMPERATURE

As with cocksfoot and timothy, Sprague (1943) found meadow fescue to yield more at 21°/13°C day/night temperatures than at higher or lower temperatures. Again, there is no reason to suppose that its response to temperature would differ markedly from that of other north-temperate grasses. Observations in the field and on its growth in heated and unheated glasshouses support this view.

RESPONSE TO DEFOLIATION

In trials with S215 meadow fescue, Lambert (1962) obtained a high yield from a hay cut followed by two aftermath cuts approximately 12 and 24 weeks later, but obtained only about half of this yield from cuts taken at intervals of 4 weeks. In a similar comparison of managements, but including the yield of only the first aftermath cut taken 8 weeks after the cut for hay, Dent and Aldrich (1963) obtained similar results at one experimental site. At the other site, differences in yield between managements were much less evident because the hay yields were much lower.

Quality

CHEMICAL COMPOSITION

Meadow fescue has an inherently low N content but otherwise appears to be typical in chemical composition (see Chapter 13).

WATER CONTENT

Stählin and Daniel (1965) measured the carotene, chlorophyll, protein and DM contents and leaf colour of meadow fescue at three stages of maturity. Water content was found to fall with advancing maturity, the values at heading, inflorescence emergence and full flowering being 83, 80 and 75%, respectively. A relationship was also found between water content and leaf colour. At each growth stage light green leaves had a lower water content than dark green leaves.

Ivins (1952) found that an area of S215 meadow fescue to which nitrogenous fertilizer at 224 lb/ac (246 kg/ha) had been applied yielded herbage with a higher percentage water content than a similar area given no fertilizer.

NUTRITIVE VALUE (DRY MATTER DIGESTIBILITY)

Young leafy grass frequently cut

In swards of meadow fescue defoliated throughout the year at intervals of 3–4 weeks, digestibility is high and very close to that of perennial ryegrass. As in perennial ryegrass, digestibility is lowest in midsummer, but when cut frequently it rarely falls below 70% (DM basis) even at this time. Varietal differences in digestibility are negligible.

Uninterrupted primary growth

There is less evidence in meadow fescue concerning the rate of decline in digestibility of primary growth than in some other species. Minson *et al.* (1964) have plotted a curve for S215 which indicates that it is similar to S24 perennial ryegrass. There is not a wide range of heading dates in meadow fescue, the difference between S215 and S53 being only about 10 days. S215 is at 70% DM digestibility at about 50% ear emergence to 5 days later, while S53, as a representative of the late-flowering group, is at this level of digestibility at about ear emergence, or a few days before.

The digestibility curve for primary growth of S215 (Fig. 21.1, top) is very close to that of S24, but the mean heading date of S215 is about 8 days later. At a comparable stage of growth, therefore, it is lower in digestibility. Its yield curve (Fig. 21.1, middle) is also similar but a little lower than that of S24, resulting in a lower yield of digestible DM.

Regrowths

Meadow fescue does not produce heads in aftermath growth after a primary cut taken in June. Growth for 6 weeks will give a DM digestibility of about 75%, and this will rarely fall below 70% in two months. High quality is maintained in later growth periods in the autumn.

Meadow fescue, therefore, when frequently cut or grazed gives herbage of high digestibility similar to that of perennial ryegrass and timothy. This digestibility level is associated with a slightly higher protein content than that of perennial ryegrass but a lower content of soluble carbohydrate (Table 21.3). The latter, however, is not so low as in frequently cut timothy or cocksfoot. In primary growth the meadow fescues are at about 70% DM digestibility at 50% ear emergence but generally give a lower yield of digestible DM at a given growth stage than perennial ryegrass.

ANIMAL INTAKE CHARACTERISTICS

There is insufficient evidence to allow a definite conclusion, but the studies of Minson *et al.* (1964) showed no differences between the intake of S215 meadow fescue, of cocksfoot varieties and of S22 Italian ryegrass.

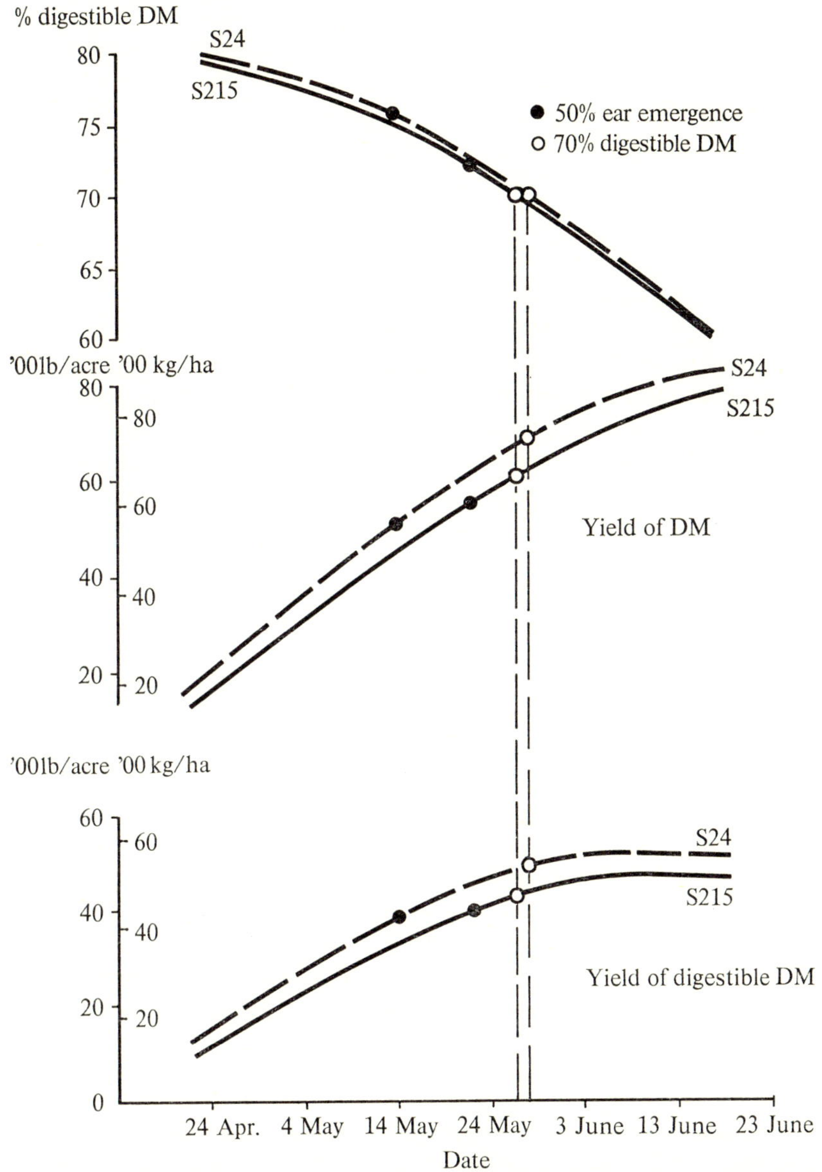

Fig. 21.1. Meadow fescue S215. Seasonal changes in DM yield and digestibility. Values for S24 perennial ryegrass shown for comparison. (After Aldrich and Dent, 1967)

Table 21.3. Percentage content (DM basis) of crude protein, crude fibre and soluble carbohydrate, and percentage digestible DM

	Simulated grazing (weighted mean of 9 cuts)		Cut for hay (cut 10 days after 50% ear emergence)	
	S24	S215	S24	S215
Crude protein	19·0	20·0	8·5	9·0
Crude fibre	24·0	23·0	31·0	35·0
Soluble carbohydrate	15·0	12·0	20·0	12·0
Digestible DM	76·5	76·5	71·0	67·0

From National Institute of Agricultural Botany (1966)

Longevity and persistency

No data are available on leaf longevity. The number of living leaves per tiller is about three in April to May (Langer, 1959), and fluctuates between 1·5 and 4 throughout the year, with low numbers recorded in June and again in early August (when cut monthly) or in early October (when cut infrequently) (Lambert, 1962).

Jewiss *et al.* (Unpublished data) found that in infrequently cut meadow fescue, between 10 and 50% of older tillers could survive beyond the first period of flowering and then follow a similar pattern of tiller death in the next year. Large numbers of new tillers formed between June and August died in the months up to October; tiller deaths thereafter were low until April. A high proportion (57 to 89%) of tillers which were classified as dead after the hay cut had produced reproductive shoots. Frequently-cut meadow fescue had a higher number of tillers per unit area; there was proportionately much less tiller death in summer and from March onwards than under infrequent cutting, but actual tiller death was numerically very similar in both treatments. Under frequent cutting, a larger percentage of tillers survived the flowering period and continued in the sward for a second and sometimes a third year.

Roots of meadow fescue are reported to be annual (Stuckey, 1941), but uptake of ^{32}P by roots two and three years old has been demonstrated in plants grown for seed production (Tatarinova, 1961); weak activity was found in roots 5 years old.

Langer *et al.* (1964) investigated the course of plant numbers per unit area in meadow fescue swards cut three times per year or at monthly intervals. Of 175 plants present at establishment in October 1956, 60 survived until the first hay cut in May 1957, while in the frequently cut treatment 114 plants were present at the same date. By October 1959, plant numbers had fallen to 30 and 50, respectively.

Some evidence of changes in meadow fescue plants with age was presented by Ermakova (1968), who observed this species in flood meadow swards cut once per year. Number of fertile tillers as a percentage of total tillers fell from 18–20% for young plants to 11·5–14% for older plants; some plants 6–11 years old did not flower. Plants 3 years old were the most vigorous. After 3 to 6 years the plants tended to fragment into clones which persisted for a further 3 to 5 years.

Meadow fescue is persistent under systems of moderate grazing but reacts unfavourably both to heavy grazing and to understocking. Established in a pure stand, meadow fescue was found not to persist well in swards which were either ungrazed for two years or in those which were heavily grazed by sheep, yield and tiller number being greatest in swards infrequently mown or lightly grazed (Jones, 1939). When sown in a mixture with tall fescue, however, a period of rest during the autumn to provide for late autumn and winter grazing favoured the persistence of the tall fescue which in later years dominated the sward (Alder & Redford, 1958; Baker *et al.*, 1965).

S215 meadow fescue has been found to persist when grazed by dairy cows on heavy, poorly-drained soils of the Pennines. Under these conditions, when grown with cocksfoot and timothy in a mixed sward it dominated the botanical composition (area cover) after 6 years. The yield of these swards, however, was less than that of a pure stand of S48 timothy which was grown under the same conditions (Mudd & Mair, 1961). It has been found for many species that the system of management which encourages the greatest tiller production may influence persistence, but is not conducive to the greatest total DM yield, though in some cases both aspects of production go together (Jones, 1939). The fact that meadow fescue is winter hardy and grows early in the spring, combined with the ability to persist under systems of moderate grazing, may outweigh in importance some shortfall in total DM yield when this species is compared with others. During the severe winter of 1962–3 in the UK, meadow fescue generally overwintered fairly well (Baker & David, 1963), and no winter killing was recorded in any cultivar of this species at the Welsh Plant Breeding Station, Aberystwyth (Miles & Williams, 1964).

References

ALDER, F. E.; REDFORD, R. A. 1958. Further observations on grassland management for meat production. *J. Br. Grassld Soc.* **13,** 239–46.

ALDRICH, D. T. A.; DENT, J. W. 1967. The relationship between yield and digestibility in the primary growth of nine grass varieties. *J. natn. Inst. agric. Bot.* **11,** 104–13.

ARENS, R. 1962. The effect of rate of seeding on competition between species and the productivity of the first year's pasture. *Z. Acker- u. PflBau* **115,** 357–74.

ARMSTRONG, S. F. 1948. British grasses and their employment in agriculture. Cambridge: Univ. Press, 3rd edn.

BAKER, H. K.; CHARD, J. R. A.; HUGHES, W. E. 1965. A comparison of cocksfoot and tall fescue-dominant swards for out-of-season production. *J. Br. Grassld Soc.* **20,** 84–90.

BAKER, H. K.; DAVID, G. L. 1963. Winter damage to grass. *Agriculture, Lond.* **70,** 380–2.

BAUMANN, H.; KLAUSS, M. L. 1955. [Root development in an environment with a high water table.] *Z. Acker- u. PflBau* **99,** 410–26.

CASTLE, M. E.; REID, D. 1960. Irrigation of grassland in south-west Scotland and its influence on the utilisation of fertilizer nitrogen. *Proc. 8th int. Grassld Congr., Reading, 1960* pp. 146–50.

CLAPHAM, A. R.; TUTIN, T. G.; WARBURG, E. F. 1962. Flora of the British Isles. Cambridge: Univ. Press, 2nd edn.

COPPENET, M.; LE CORRE, L. 1964. Grass variety trial in Quimper, Brittany. *Fourrages* **17,** 81–97.

COWLING, D. W.; LOCKYER, D. R. 1965. A comparison of the reaction of different grass species to fertilizer nitrogen and to growth in association with white clover. *J. Br. Grassld Soc.* **20,** 197–204.

CROCKER, R. L.; MARTIN, P. M. 1964. Competition between perennial ryegrass and meadow fescue under field-plot conditions. *J. Br. Grassld Soc.* **19,** 27–9.

DENT, J. W.; ALDRICH, D. T. A. 1963. The inter-relationships between heading date, yield, chemical composition and digestibility in varieties of perennial ryegrass, timothy, cocksfoot and meadow fescue. *J. natn. Inst. agric. Bot.* **9,** 261–81.

ERMAKOVA, I. M. 1968. [On age conditions of fescue (*Festuca pratensis* Huds.).] *Nauch. Dokl. vŷssh. Shk.* (*biol. Nauki*) No. 7, 70–7.

EVANS, G. 1962. Seed rates of grasses for seed production. 2. Bred varieties of tall fescue, meadow fescue, cocksfoot (hay type), and Phalaris hybrid. *Emp. J. exp. Agric.* **30,** 181–91.

GERDES, G. 1952. [Studies on the absolute weight (1 000 seed weight) of the most important fodder grass species and their varieties in regard to the performance of grass breeding.] *Züchter* **22,** 353–66.

GRASSLAND RESEARCH INSTITUTE. 1952. *Exps Prog. 5 Grassld Res. Inst. 1952* pp. 23–31.

GREEN, J. 1964. Harvesting relationship of moisture, maturity and weight of herbage seeds. *Rep. Lincs. Seed Growers' Assn 1963.* Agriculture House, Woodhall Spa, pp. 13–17.

GREEN, J. O.; DAVIS, A. G. 1955. The seasonal use of lucerne/grass mixture for grazing. In: Grassland Research Institute. Memoir No. 1. Lucerne investigations, 1944–53, pp. 63–77.

GREEN, J. O.; EVANS, T. A. 1956. Manuring and grazing for seed production in S37 cocksfoot and S215 meadow fescue. *J. Br. Grassld Soc.* **11,** 165–73.

GREEN, J. O.; EVANS, T. A. 1957. Grazing management for seed production in leafy strains of grasses. *J. Br. Grassld Soc.* **12,** 4–9.

HENDERSON, J. L.; EDWARDS, R. S.; HAMMERTON, J. L. 1962. The productivity of five grass species at six levels of compound fertilizer application. 1. Dry-matter production. *J. agric. Sci., Camb.* **59,** 5–11.

HOLMES, W.; MACLUSKY, D. S. 1955. The intensive production of herbage for crop-drying. *J. agric. Sci., Camb.* **46,** 267–86.

HUBBARD, C. E. 1954. Grasses. Harmondsworth, Middlesex: Penguin Books.

HUNT, I. V. 1954. Seed establishment in the West of Scotland. *J. Br. Grassld Soc.* **9,** 85–98.

HUNT, I. V. 1957. Comparative productivity of 27 strains of grass in the West of Scotland. 4. Seasonal growth. *J. Br. Grassld Soc.* **12,** 127–37.

HUNT, I. V. 1961. Comparison of production from three varieties of tall fescue and three varieties of meadow fescue. *Res. Bull 29. W. Scotl. agric. Coll.* pp. 43.

HUNT, I. V. 1964. Comparative productivity of herbage varieties on upland peat. *J. Br. Grassld Soc.* **19,** 55–61.

IVINS, J. D. 1952. The relative palatability of herbage plants. *J. Br. Grassld Soc.* **7,** 43–54.

JEWISS, O. R.; RYLE, S. M.; LANGER, R. H. M. The changing plant and tiller populations of timothy and meadow fescue swards. 2. Tiller longevity. (Unpublished)

JONES, I. 1939. Pasture management and its effect on the sward. *Bull. H15 Welsh Pl. Breed. Stn* pp. 40–129.

LAMBERT, D. A. 1962. A study of growth in swards of timothy and meadow fescue. 3. The effect of two levels of nitrogen under two cutting treatments. *J. agric. Sci., Camb.* **59,** 25–32.

LAMBERT, D. A. 1964. The influence of density and nitrogen in seed production stands of S48 timothy (*Phleum pratense L.*) and S215 meadow fescue (*Festuca pratensis L.*). *J. agric. Sci., Camb.* **63,** 35–42.

LAMPETER, W. 1960. [Mutual influences of higher plants regarding growth of shoots and roots, mineral content and water consumption studied in some economically important forage plants.] *Wiss. Z. Karl-Marx-Univ. Lpz. 1959–60* **9**, 611–722.

LANGER, R. H. M. 1959. A study of growth in swards of timothy and meadow fescue. 2. The effects of cutting treatments. *J. agric. Sci., Camb.* **52**, 273–81.

LANGER, R. H. M.; RYLE, S. M.; JEWISS, O. R. 1964. The changing plant and tiller populations of timothy and meadow fescue swards. 1. Plant survival and pattern of tillering. *Jnl appl. Ecol.* **1**, 197–208.

LEWIS, J. 1966. The relationship between seed yield and associated characters in meadow fescue (*Festuca pratensis*). *J. agric. Sci., Camb.* **67**, 243–8.

LINCOLNSHIRE SEED GROWERS' ASSOCIATION. 1965. Report for 1964. Agriculture House, Woodhill Spa, pp. 3–7.

LONG, I. F.; FRENCH, B. K. 1967. Measurement of soil moisture in the field by neutron moderation. *J. Soil Sci.* **18**, 149–66.

MILES, D. G.; WILLIAMS, I. G. 1964. Winter hardiness in pasture varieties. *Rep. Welsh Pl. Breed. Stn 1963* pp. 70–1.

MINISTRY OF AGRICULTURE, FISHERIES AND FOOD. 1968. Grass and clover crops for seed. *Bull. 204* London: H.M.S.O., pp. 42–60.

MINSON, D. J.; HARRIS, C. E.; RAYMOND, W. F.; MILFORD, R. 1964. The digestibility and voluntary intake of S22 and H.1 ryegrass, S170 tall fescue, S48 timothy, S215 meadow fescue and Germinal cocksfoot. *J. Br. Grassld Soc.* **19**, 298–305.

MUDD, C. H.; MAIR, R. B. 1961. Performance of eight seeds mixtures at Great House 1952–1958. *Expl Husb.* No. 6, 21–50.

NATIONAL INSTITUTE OF AGRICULTURAL BOTANY. 1959. Growing grasses for seed. *Herbage Seed Growers' Leaflet No. 2.*

NATIONAL INSTITUTE OF AGRICULTURAL BOTANY. 1966. Unpublished data.

PERRING, F. H.; WALTERS, S. M. 1962. Atlas of the British flora. London: Thomas Nelson and Sons Ltd, pp. 456.

PETO, D. H. 1961. Methods and problems of cleaning herbage seeds. *Rep. Lincs. Seed Growers' Assn 1960* Agricultural House, Woodhall Spa, pp. 19–21.

ROBERTS, H. M. 1958. The effect of a cover crop. *Rep. Welsh Pl. Breed. Stn 1950–56* pp. 147–8.

ROBERTS, H. M. 1961. The effect of drill width on the seed productivity of leafy varieties of three grass species. *J. Br. Grassld Soc.* **16**, 37–42.

RYLE, G. J. A. 1964. A comparison of leaf and tiller growth in seven perennial grasses as influenced by nitrogen and temperature. *J. Br. Grassld Soc.* **19**, 281–90.

SPRAGUE, V. G. 1943. The effects of temperature and day-length on seedling emergence and early growth of several pasture species. *Proc. Soil Sci. Soc. Am.* **8**, 287–94.

SPURWAY, C. H. 1941. Soil reaction (pH) preferences of plants. *Spec. Bull. 306 Mich. agric. Exp. Stn* pp. 36.

STÄHLIN, A.; DANIEL, P. 1965. [Leaf colour and chemical composition of grass varieties.] *Z. Acker- u. PflBau* **121**, 106–15.

STUCKEY, I. H. 1941. Seasonal growth of grass roots. *Am. J. Bot.* **28**, 486–91.

TATARINOVA, N. K. 1961. [The longevity of the roots of meadow grasses.] *Bot. Zh. SSSR* **46**, 925–35.

WIDDOWSON, F. V.; PENNY, A.; WILLIAMS, R. J. B. 1963. An experiment comparing responses to nitrogen fertilizer of four grass species. *Expl Husb.* No. 9, 28–36.

WILLIAMS, W. 1950. Experiments on the yield of lucerne strains as pure plots and on the behaviour of lucerne when sown in a mixture with various species of grasses. *J. Br. Grassld Soc.* **5**, 113–29.

CHAPTER 22

Tall fescue *(Festuca arundinacea)*

Morphology

TALL fescue (Plate 13) is a stout, glabrous perennial. Root system fibrous, brown, highly-branched adventitious roots. Vegetative stems short, more or less erect, slender to stout, tillering only near the base, intravaginally. Leaf sheaths dark green, glabrous, slightly ribbed becoming fibrous, whitish, persistent. Ligule short, auricles prominent, ciliate on margin. Leaf blades broad, stiff, dark green, glossy abaxially, very strongly ribbed adaxially, tapering to a fine point. Culms stout, tall, tillering from the first or even second aerial node. Inflorescence a much-branched panicle of minutely-awned 5- to 6-flowered spikelets. Caryopsis enclosed in persistent lemma and palea.

There are no published data on the root distribution of tall fescue grown in Britain. There is no evidence to suggest, however, that the root system differs substantially from that of the other important herbage grasses. The plant is large, coarse and forms tussocks readily unless frequently defoliated. Ryle (1964) recorded leaf sizes of 4×80 mm to 9×400 mm in the cultivar S170 grown in a glasshouse. The two tall fescue cultivars in this study, S170 and an Algerian ecotype, had larger upper leaves than any other species in the trial. Similarly, fewer tillers developed at the upper nodes than on the other species, including perennial ryegrass.

S170 tall fescue is finer-leaved and generally less coarse than other commercial cultivars, e.g. Alta and Kentucky 31 from North America.

Changes in the proportion of leaf during the flowering period have been studied by Terry and Tilley (1964) (see Table 22.1).

Distribution

Festuca arundinacea is native to Britain (Clapham *et al.*, 1962) where it occurs in all but three vice-counties (Perring & Walters, 1962). It is commonest in south England, less frequent in Wales and very uncommon in north Scotland. It is of scattered occurrence in Ireland and is absent from the

Table 22.1. Proportion of leaf blade, sheath, stem and inflorescence in the DM of S170 tall fescue primary growth

	Percentage of dry weight		
Date	Leaf blade	Sheath	Stem + inflorescence
12 Apr.	72	26	2
19 „	56	33	11
26 „	62	22	16
3 May	38	32	30
10 „	40	24	36
16 „	37	19	44
24 „	28	20	53
1 June	20	18	63
7 „	17	18	65

Date of 50% ear emergence: 3 May

From Terry and Tilley (1964)

south-west (Perring & Walters, 1962). It is generally absent at altitudes above 300 m and its distribution in Britain bears no relation to isotherms (Perring & Walters, 1962). In Tennessee it has been reported to grow whenever the weekly mean temperature is above 4·5°C (Leasure, 1952), though growth is reduced at temperatures below 18°C (Cowan, 1956). It was, however, never completely dormant, even at temperatures as low as 2°C (Leasure, 1952).

It grows on a variety of soils, but different strains occupy distinct habitats, the taller, stouter kinds occurring on heavy soils in low-lying meadows and by rivers and streams, the shorter types on drier calcareous or sandy soils in rough-hill and downs grassland (Hubbard, 1954). It grows best on soils with pH between 6·5 and 8·0 (Spurway, 1941), with a lower limit of 4·7 (Rampton, 1950) and an upper limit of 9·5 (Cowan, 1956).

It is most common in areas with an annual rainfall of less than 750 mm, and is much less common where rainfall exceeds 1 000 mm (Perring & Walters, 1962). The taller types tolerate a high water table and the roots grow well in wet soil (Hubbard, 1954; Baumann & Klauss, 1955). It is especially tolerant of wet soil in winter (Cowan, 1956).

Compatibility

There is very little published information relating to the compatibility of tall fescue with other grasses and legumes. It is slow to establish from seed, which implies that it would be susceptible to aggression from the ryegrasses during this phase. Harkess (1966) studied the development of seedlings of S24 perennial ryegrass and S170 tall fescue sown at various seed rates. During the first 4 months, both species grew better when grown amongst tall fescue seedlings than when grown amongst those of perennial ryegrass.

Once tall fescue is established it grows strongly and is capable of spreading by means of limited rhizome development. In early work with commercial varieties at Aberystwyth, Davies (1928) sowed mixtures of timothy with each of 7 companion grasses. The establishment of timothy sown with tall fescue was high three months after sowing, but at 15 months it had dropped to the level of that growing with perennial ryegrass.

Seed

CHARACTERS

The seed is lanceolate to lanceolate-oblong, with a shallow groove along the ventral surface; seed length is 5–9 mm. The lemma varies from pointed to blunt, convex, and may either be awnless or have the middle nerve of the lemma continued as a fine rough awn 1–4 mm long. The palea is similar in length to the lemma, and is acutely pointed. The rachilla is cylindrical and stands out from the palea (Armstrong, 1948; Hubbard, 1954).

The 1 000-seed weight has been recorded as 2·22 g for unnamed commercial seed (Hunt, 1954) and 2·96 g for S170 (Evans, 1962).

PRODUCTION

The seed should be sown in spring in drills 45–75 cm apart at not more than 5 kg/ha; if satisfactory establishment can be obtained from lower seed rates yields of seed will probably be increased (Green & Evans, 1957; Evans, 1962; National Institute of Agricultural Botany, 1959). Cover crops may reduce yields considerably (Roberts, 1958). No critical trial to examine N requirements has been performed, but an annual application of 80–120 kg N/ha, divided between autumn and spring, appears satisfactory (Green & Evans, 1957; National Institute of Agricultural Botany, 1959; Evans, 1962). Application of 30 kg P and 60 kg K/ha after harvest is recommended (National Institute of Agricultural Botany, 1959). Grazing will reduce seed yield (Green & Evans, 1957). Crops should be isolated by at least 180 m during anthesis (National Institute of Agricultural Botany, 1959). Principles and procedures for harvesting have been well explained elsewhere (Ministry of Agriculture, Fisheries & Food, 1968).

The annual yield of seed can vary from 25 to 940 kg/ha, with the mean yield ranging between 325 and 657 kg/ha, depending on the year (Lincs Seed Growers' Association, 1965).

Yield

DRY MATTER

Tall fescue has generally been found high-yielding; for instance, Coppenet and Le Corre (1964) measured annual yields of about 11 000 kg dry matter (DM)/ha in Brittany when this species was cut to simulate grazing, and Hunt (1961) found that S170, the best of three cultivars tested, produced 11 000 kg DM/ha at a high level of manuring. Baker *et al.* (1965) recorded annual yields of 10 600 kg DM/ha from swards dominated by S170. On upland peat,

S170 produced an average yield of 3 500 kg/ha over a 5-year period (Hunt, 1964).

In experiments, Frame and Hunt (1964) found no effect of seed rates ranging from 11 to 55 kg/ha on the total herbage yield in either of the two subsequent years; yields were 9 600 kg/ha in the first year and 7 900 kg/ha in the second. There were only slight effects of seed rate on the contribution of tall fescue to total herbage.

Considerable interest has been centred on the usefulness of tall fescue for early spring production. Hunt (1961) mentioned that the characteristic features of the cultivar S170 were its marked earliness and uniform mid-season production, and Baker *et al.* (1965) measured an average yield at the first cut in spring of 1 700 kg DM/ha from 17 experimental sites in three years. Differences have been found between genotypes of tall fescue with regard to the distribution of their DM yield. Thus Fyfe and Rogers (1965) found that, although S170 produced a higher annual yield than synthetic cultivars bred from material of Mediterranean origin, seasonal growth differences resulted in the latter growing more vigorously than S170 in autumn; all the grasses tested were grown in mixtures with lucerne. Chatterjee (1961) also found that S170 produced a greater annual yield (11 000 kg/ha) than did Mediterranean ecotypes (8 800 kg/ha), but that winter production of the former (220 kg/ha) was lower than that of the latter (820 kg/ha).

ENERGY AND PROTEIN

Hunt (1961) found that the average percentage crude protein content in the herbage of S170 tall fescue at a high level of fertilizer input was 17% and that the annual crude protein yield was 1900 kg/ha. Data for yields of energy and protein, derived from the literature, are summarized in Table 22.2.

Table 22.2. Potential annual yields of energy and protein from tall fescue (DM basis)

	Frequent cutting
DM digestibility (%)	73
Yield of digestible OM (kg/ha)	7 420
Yield of metabolizable energy (Mcal/ha)	28·0
Crude protein (%)	17·0
Yield of crude protein (kg/ha)	1 900
Yield of digestible crude protein (kh/ga)	1 350

Data compiled by D. J. Thomson (Grassland Research Institute, Hurley) from literature cited in the text. For conversion formulae see Chapter 5

Response to management and environment

RESPONSE TO FERTILIZERS

There is no information on the response of tall fescue to N, P or K from field trials in Britain. However, there is no reason to believe that its response to N is much different from that of the other commonly cultivated species. In the USA, for example, Wagner (1954) found responses comparable to those obtained with cocksfoot.

RESPONSE TO IRRIGATION

No results of experiments in Britain on the irrigation of tall fescue have been reported. In Montana, Stitt (1955) found the yield of irrigated Alta tall fescue was near to the average of several grass species, while in Alabama, where it is a cool-season crop, Bennett and Doss (1963) found that it gave a greater yield when irrigated to field capacity from a 50-mm soil-moisture deficit than from one of 25 mm or 60 mm.

RESPONSE TO TEMPERATURE

Tall fescue grows more during the low temperatures of winter than many other north-temperate grasses. Leasure (1953) reported that it continued to grow at a mean ambient temperature of 1°C and grew rapidly at temperatures above 4°C, whereas cocksfoot was dormant at 4°C and only grew rapidly above 10°C. At the Grassland Research Institute, Hurley, S170 tall fescue yielded twice as much herbage as S23 perennial ryegrass between October and December (Green & Evans, 1957). Yet ecotypes from North Africa grow even better than S170 during winter in Britain. Robson and Jewiss (1968) showed that this superior growth resulted from a different response to temperature; in regimes of 15°/5°C and 10°/0°C the types hardly differed, but in 5°/−3°C those from North Africa greatly outyielded S170.

Controlled-environment studies (Robson & Jewiss, 1968) show that S170 grows most rapidly when the day temperature is in the region of 23°C and when the night temperature, which is a less important determinant of growth, is the same or slightly lower. The growth rate of small plants at or near optimum temperatures is more than twice that at 10°C; as with cocksfoot, larger plants are less sensitive to temperature.

RESPONSE TO DEFOLIATION

S170 tall fescue produced high yields when a hay cut was taken and the aftermath was grazed (Baker *et al.*, 1965). When grown with other species, the contribution by S170 to the total yield increased from year to year both when rotationally grazed and when cut for hay and the aftermath grazed (Jones & Charles, 1961).

Quality

CHEMICAL COMPOSITION

There is a lack of information on the organic constituents of tall fescue but as regards inorganic constituents it appears to be of average composition (see Chapter 13).

WATER CONTENT

Recalculation of data of Hunt and Gardner (1956) on fresh and dry weights of herbage harvested in June, August and October 1951 and May and August 1952 from a stand of tall fescue containing small amounts of Montgomery red clover, showed that the water contents of the herbage on these 5 occasions were 81·3, 81·5, 80·2, 79·5 and 76·8%, respectively.

NUTRITIVE VALUE (DRY MATTER DIGESTIBILITY)

Young leafy grass frequently cut

Tall fescue is a rather coarse-leaved grass, and its digestibility is generally lower than that of the other common species except cocksfoot. In swards defoliated at 3- to 4-week intervals throughout the year, DM digestibility may range from 65% to 80%. The lowest figure in midsummer, however, may vary with weather conditions but, generally, as with cocksfoot, fibre content is high and digestibility is low at this time.

Uninterrupted primary growth

Evidence concerning the rate of decline in digestibility in primary growth is lacking in this species. Minson *et al.* (1964) plotted a curve for S170 which indicated that its DM digestibility at first ear emergence was about 72%. Recent data (Green & Corrall, Unpublished) show a digestibility curve similar to that of S24 perennial ryegrass, which has a similar heading date, but S170 is about 2% lower in digestibility. Data from trials at the National Institute of Agricultural Botany, however, indicate a difference of 5% between the two varieties 10 days after 50% ear emergence (Table 22.3). Other results also showed that early tall fescues such as McGillsmith's Early or the French cultivar Manade are at 70% DM digestibility at about 5–10 days after 50% ear emergence, while S170, which heads about 10 days later, is at 70% digestibility for about 5 days after 50% ear emergence. The yield curve of S170 is similar to and a little lower than that of S24 except in early spring (Fig. 22.1, middle). Its lower digestibility, however, results in a greater disparity between the species in yield of digestible DM.

Regrowths

There is little evidence concerning the digestibility of individual aftermath growths of tall fescue. Results from trials conducted by the National Institute of Agricultural Botany (Unpublished) indicate, however, that aftermath growth may be coarse and fibrous and of low digestibility, particularly under dry conditions in midsummer.

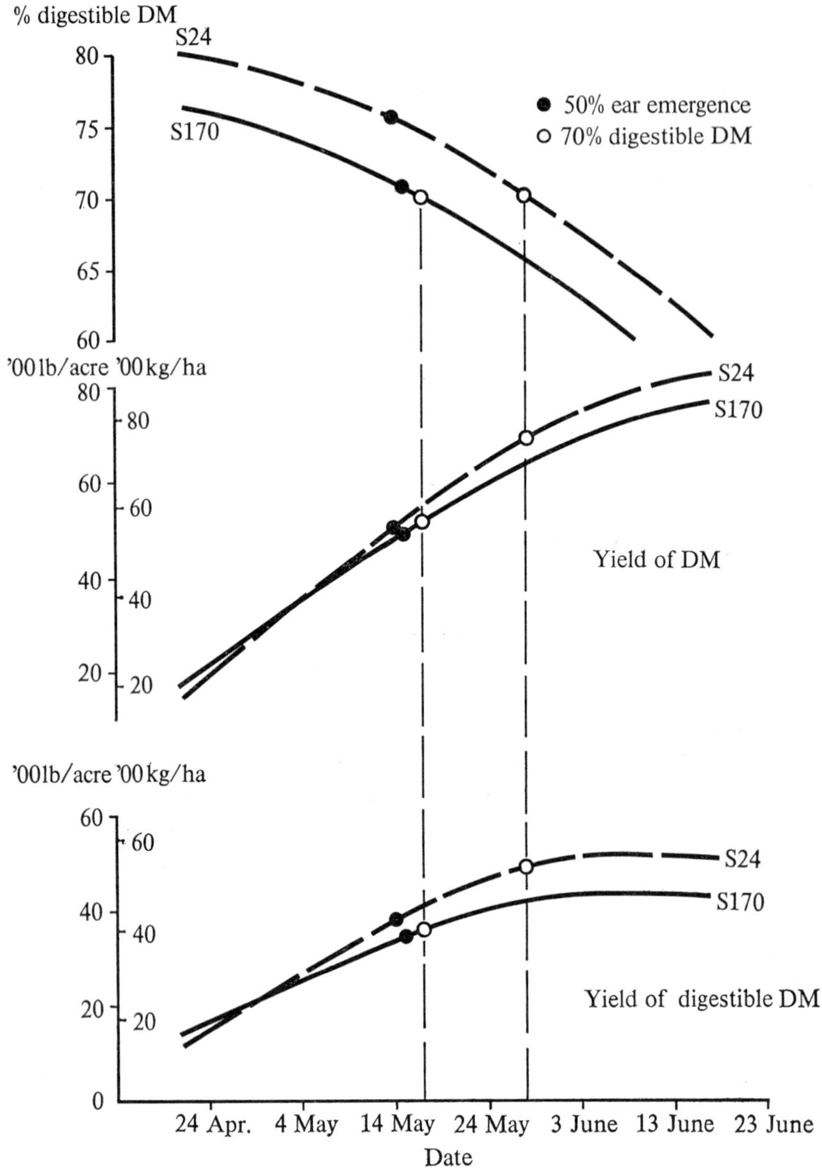

Fig. 22.1. Tall fescue S170. Seasonal changes in DM yield and digestibility. Values for perennial ryegrass S24 shown for comparison. (After Aldrich and Dent, 1967)

Tall fescue is coarse and rather unpalatable; its digestibility, however, is generally higher than that of cocksfoot, being associated with a higher content of soluble carbohydrate, while its fibre content is not as high as the appearance and physical nature of the plant might suggest (see Table 22.3).

Table 22.3. Percentage content (DM basis) of crude protein, crude fibre and soluble carbohydrate, and percentage digestible DM in S24 ryegrass and S170 tall fescue

	Simulated grazing (weighted mean of 9 cuts)		Cut for hay (cut 10 days after 50% ear emergence)	
	S24	S170	S24	S170
Crude protein	19·0	21·0	8·5	10·0
Crude fibre	24·0	23·0	31·0	31·0
Soluble carbohydrate	15·0	12·0	20·0	13·0
Digestible DM	76·5	74·0	71·0	66·0

From National Institute of Agricultural Botany (1966)

ANIMAL INTAKE CHARACTERISTICS

S170 tall fescue was examined for intake by Minson *et al.* (1964); in one year its intake by sheep appeared normal, but in a second year very low voluntary intakes were recorded on the highly digestible early spring growth. Van Soest (1965), working in the USA, found a negative relationship between intake and digestibility for tall fescue.

Longevity and persistency

Leaf longevity of tall fescue has been little studied. The length of time taken from complete expansion of the lamina until photosynthesis declined to zero was found by Woledge (Unpublished) to range from 30 to 60 days, depending on temperature. In a 12-h day at 1500 f.c. and 20°/15°C, the fifth seedling leaf of S170 lived for 36 days after full expansion (Jewiss & Woledge, 1967), but for a large part of this experiment the leaves consisted of a dead terminal portion and an apparently healthy green basal part.

In studies on winter growth of S170, Chatterjee (1961) found that the number of living leaves per tiller ranged from 1·72 in January to 2·99 in early May; from his data on rate of leaf appearance it can be calculated that leaves lived for about 90 days in winter and about 60 days in spring.

Robson (1968) followed the life history of tillers of tall fescue from sowing in April 1962 until July 1964. During the trial, 50% of tillers died without having produced an ear. Those which elongated in autumn but produced no inflorescence usually failed to survive the winter. The main period of tiller mortality coincided approximately with the peak period of flowering during April to May, while rate of tiller production was highest immediately following flowering. Some tillers died within days or weeks of being produced.

A second peak of tiller mortality occurred in September; all but 8% of tillers that died between 5 September and 10 October 1963 had arisen during the preceding 10 weeks. Of tillers present in July 1964, 65% had arisen in 1964, 34% in 1963 and less than 1% in 1962; 700 days was the greatest tiller age recorded. Median age of tillers exceeded 300 days for tillers produced during late May and June and fell below 200 days for tillers produced between October and December.

No information is available on longevity of roots or of whole plants. Matches (1966) is of the opinion that in the USA, grazing to a 3-in. (7·5-cm) stubble height should ensure continuous survival of individual plants. Indigenous tall fescue is extremely winter-hardy in the UK, but Mediterranean ecotypes are not (Baker & David, 1963). At the Grassland Research Institute, Hurley, spaced plants of S170 all survived the hard winter of 1962–3, but only 4 plants out of 25 of Mediterranean introductions did so (Jewiss & Robson, 1964).

Tall fescue is depressed by heavy grazing and, during the first two years of establishment, has shown increased persistence when allowed to grow without interruption. Persistence of S170 tall fescue was quite satisfactory when the sward was grazed intermittently until mid-July or the first week of August and then rested until November. After November the grown material can be used as winter foggage. Under these conditions tall fescue in a mixed sward with meadow fescue persisted for 9 years, at the end of which time it dominated the sward (Alder & Redford, 1958; Baker *et al.*, 1965).

The persistency and total production of tall fescue have been compared with those of S143 cocksfoot when both were given the same treatment and rested in the autumn for the production of winter keep. Both species persisted satisfactorily into the sixth year from the time of sowing and there was no difference between their total annual yield of DM; however, a greater proportion of the fescue was produced in the early spring (Baker *et al.*, 1965).

References

ALDER, F. E.; REDFORD, R. A. 1958. Further observations on grassland management for meat production. *J. Br. Grassld Soc.* **13**, 239–46.

ALDRICH, D. T. A.; DENT, J. W. 1967. The relationship between yield and digestibility in the primary growth of nine grass varieties. *J. natn. Inst. agric. Bot.* **11**, 104–13.

ARMSTRONG, S. F. 1948. British grasses and their employment in agriculture. Cambridge: Univ. Press, 3rd edn.

BAKER, H. K.; CHARD, J. R. A.; HUGHES, W. E. 1965. A comparison of cocksfoot and tall fescue-dominant swards for out-of-season production. *J. Br. Grassld Soc.* **20**, 84–90.

BAKER, H. K.; DAVID, G. L. 1963. Winter damage to grass. *Agriculture, Lond.* **70**, 380–2.

BAUMANN, H.; KLAUSS, M. L. 1955. [Root development in an environment with a high water table.] *Z. Acker- u. PflBau* **99**, 410–25.

BENNETT, O. L.; DOSS, B. D. 1963. Effects of soil moisture regime on yield and evapotranspiration from cool-season perennial forage species. *Agron. J.* **55**, 275–8.

CHATTERJEE, B. N. 1961. Analysis of ecotypic differences in tall fescue (*Festuca arundinacea* Schreb.) and their agronomic significance. Ph.D. Thesis, Univ. Reading, pp. 233.

CLAPHAM, A. R.; TUTIN, T. G.; WARBURG, E. F. 1962. Flora of the British Isles. Cambridge: Univ. Press, 2nd edn.

COPPENET, M.; LE CORRE, L. 1964. Grass variety trial in Quimper, Brittany. *Fourrages* **17,** 81–97.

COWAN, J. R. 1956. Tall fescue. *Adv. Agron.* **8,** 283–320.

DAVIES, W. 1928. The factor of competition between one species and another in seeds mixtures. *Bull. H8 Welsh Pl. Breed. Stn.*

EVANS, G. 1962. Seed rates of grasses for seed production. 2. Bred varieties of tall fescue, meadow fescue, cocksfoot (hay type), and Phalaris hybrid. *Emp. J. exp. Agric.* **30,** 181–91.

FRAME, J.; HUNT, I. V. 1964. The effect of companion grass and seed rate on the productivity of a tall-fescue sward. *J. Br. Grassld Soc.* **19,** 330–5.

FYFE, J. L.; ROGERS, H. H. 1965. The effect of varying variety and spacing on yields and composition of mixtures of lucerne and tall fescue. *J. agric. Sci., Camb.* **64,** 351–9.

GREEN, J. O.; CORRALL, A. J. 1967. Grassland Research Institute, Hurley. Unpublished data.

GREEN, J. O.; EVANS, T. A. 1957. Grazing management for seed production in leafy strains of grasses. *J. Br. Grassld Soc.* **12,** 4–9.

HARKESS, R. D. 1966. The assessment of inter- and intra-specific competition by the wire grid technique. *Proc. 9th int. Grassld Congr., São Paulo, 1965* pp. 417–19.

HUBBARD, C. E. 1954. Grasses. Harmondsworth, Middlesex: Penguin Books.

HUNT, I. V. 1954. Seed establishment in the West of Scotland. *J. Br. Grassld Soc.* **9,** 85–98.

HUNT, I. V. 1961. Comparison of production from three varieties of tall fescue and three varieties of meadow fescue. *Res. Bull. 29 W. Scotl. agric. Coll.* pp. 43.

HUNT, I. V. 1964. Comparative productivity of herbage varieties on upland peat. *J. Br. Grassld Soc.* **19,** 55–61.

HUNT, I. V.; GARDNER, A. L. 1956. Two year leys for silage. *Res. Bull. 15 W. Scotl. agric. Coll.* pp. 32.

JEWISS, O. R.; ROBSON, M. J. 1964. Studies on the growth of ecotypes of tall fescue. *Exps Prog. 16 Grassld Res. Inst. 1962–3* pp. 32–4.

JEWISS, O. R.; WOLEDGE, J. 1967. The effect of age on the rate of apparent photosynthesis in leaves of tall fescue (*Festuca arundinacea* Schreb.). *Ann. Bot.* **31,** 661–71.

JONES, L. I.; CHARLES, A. H. 1961. Observations on S170 tall fescue. *Rep. Welsh Pl. Breed. Stn 1960* pp. 55–9.

LEASURE, J. K. 1952. Growth patterns of mixtures of orchard grass and tall fescue with Ladino white clover in relation to temperature. *Proc. Ass. sth. agric. Wkrs* **49,** 177–8.

LEASURE, J. K. 1953. Some factors affecting grass-legume competition in swards. *Diss. Abstr.* **13,** 941.

LINCOLNSHIRE SEED GROWERS' ASSOCIATION. 1965. Report for 1964. Agriculture House, Woodhall Spa, pp. 3–7.

MATCHES, A. G. 1966. Influence of intact tillers and height of stubble on growth responses of tall fescue (*Festuca arundinacea* Schreb.). *Crop Sci.* **6,** 484–7.

MINISTRY OF AGRICULTURE, FISHERIES AND FOOD. 1968. Grass and clover crops for seed. *Bull. 204* London: H.M.S.O., pp. 42–60.

MINSON, D. J.; HARRIS, C. E.; RAYMOND, W. F.; MILFORD, R. 1964. The digestibility and voluntary intake of S22 and H.1 ryegrass, S170 tall fescue, S48 timothy, S215 meadow fescue and Germinal cocksfoot. *J. Br. Grassld Soc.* **19,** 298–305.

NATIONAL INSTITUTE OF AGRICULTURAL BOTANY. 1959. Growing grasses for seed. *Herbage Seed Growers' Leaflet No. 2.*

NATIONAL INSTITUTE OF AGRICULTURAL BOTANY. 1966. Unpublished data.

PERRING. F. H.; WALTERS, S. M. 1962. Atlas of the British flora. London: Thomas Nelson and Sons Ltd, pp. 456.

RAMPTON, H. H. 1950. Alta fescue for high yielding pastures. *What's new Crops Soils* **2,** 18–19 and 25.

ROBERTS, H. M. 1958. The effect of a cover crop. *Rep. Welsh Pl. Breed. Stn 1950–56* pp. 147–8.

ROBSON, M. J. 1968. The changing tiller population of spaced plants of S170 tall fescue (*Festuca arundinacea*). *Jnl appl. Ecol.* **5,** 575–90.

ROBSON, M. J.; JEWISS, O. R. 1968. A comparison of British and North African varieties of tall fescue (*Festuca arundinacea*). 3. Effect of light, temperature and daylength on relative growth rate and its components. *Jnl appl. Ecol.* **5,** 191–204.

RYLE, G. J. A. 1964. A comparison of leaf and tiller growth in seven perennial grasses as influenced by nitrogen and temperature. *J. Br. Grassld Soc.* **19,** 281–90.

SOEST, P. J. VAN. 1965. Symposium on factors influencing the voluntary intake of herbage by ruminants: voluntary intake in relation to chemical composition and digestibility. *J. Anim. Sci.* **24,** 834–43.

SPURWAY, C. H. 1941. Soil reaction (pH) preferences of plants. *Spec. Bull. 306 Mich. agric. Exp. Stn* pp. 36.

STITT, R. E. 1955. Yields of grass species and varieties under irrigation. *Bull. 508 Montana agric. Exp. Stn.*

TERRY, R. A.; TILLEY, J. M. A. 1964. The digestibility of leaves and stems of perennial ryegrass, cocksfoot, timothy, tall fescue, lucerne and sainfoin, as measured by an *in vitro* procedure. *J. Br. Grassld Soc.* **19,** 363–72.

WAGNER, R. G. 1954. Influence of legume and fertilizer nitrogen on forage production and botanical composition. *Agron. J.* **46,** 167–71.

WOLEDGE, J. Grassland Research Institute. Unpublished data.

CHAPTER 23

Other grasses

Introduction

THE grasses so far described in this book are those currently regarded as of most value to British agriculture, but there are at least another 150 other species which occur in Britain. These show a wide diversity of growth habit and nutrient requirement, and many are very variable in morphology. Many are restricted to specialized habitats and are unlikely ever to be of more than botanical interest. Some, however, have definite if not always well defined agricultural value and may be important, for example, in upland areas or as minor components in seeds mixtures. Some, notably *Agrostis* and *Festuca* species, may be important for lawns and sports turf and other specialized uses.

Because of their agricultural potential, a few species have been exploited commercially and have been the subject of experimentation, but outside these, the amount of information available is insufficient to enable a clear distinction to be drawn between those grasses with potential as crop plants and those of no practical value. An exhaustive description of all species is beyond the scope of this book, and attention has therefore been confined to describing the more agriculturally useful ones and to examining the reasons why they are not classed as 'important' in British agriculture.

The species described in this chapter include the following: *Agrostis gigantea* (redtop bent), *A. stolonifera* (creeping bent), *A. tenuis* (common bent), *Alopecurus pratensis* (meadow foxtail), *Anthoxanthum odoratum* (sweet vernal-grass), *Arrhenatherum elatius* (tall oatgrass), *Cynosurus cristatus* (crested dogstail), *Festuca rubra* (red fescue and Chewings fescue), *Holcus lanatus* (Yorkshire fog), *Phleum bertolonii* (smaller cat's tail), *Poa pratensis* (smooth-stalked meadowgrass), *Poa trivialis* (rough-stalked meadowgrass) and *Trisetum flavescens* (yellow oatgrass). Of these grasses, all but sweet vernal-grass, Yorkshire fog and smaller cat's tail have been classed as probably among the world's principal sown pasture grasses by Hartley and Williams (1956).

In accordance with the objectives of this book, the above species include only those used in artificially sown grasslands, for which reason such species as *Festuca ovina* have been omitted, despite their importance in, for example, upland grazings. Other species have been omitted because they are found only in such natural habitats as hedgerows, woodlands, sand dunes, etc.

The above species are classed as 'secondary' in their importance to British agriculture, and are considered to be of less value, chiefly because their yield under average conditions of soil fertility, moisture regime and climate is lower, or they are of poorer quality than the 'better' grasses. Under special or extreme conditions, however, these grasses may be of greater value than the 'better' grasses, and so it is probably no paradox that a conscious effort is made to encourage the growth on hill soils of species actively regarded as weeds under fertile lowland conditions.

Within each species to be described there are usually many strains differing in character according to their place of origin, and other differences arise from the selection of breeding lines, culminating sometimes in recognized cultivars. It is not possible, therefore, to make any statements which apply satisfactorily to a species as a whole, and the data refer only to the particular samples tested.

Agrostis species

Bent (Welsh Plant Breeding Station, 1932; Jones, 1967)

It is necessary to deal with *Agrostis* species in general, as there is an appreciable quantity of published data for which the species was not determined or was not stated.

The term 'bent' is taken to imply any or all *Agrostis* species, usually termed *A. tenuis*, *A. stolonifera*, *A. canina* and *A. gigantea*, although other nomenclature may be used (Armstrong, 1945). Generally speaking, however, literature referring to 'bent' in British pastures probably implies *A. tenuis* in areas of low soil fertility and *A. stolonifera* or *A. stolonifera* × *tenuis* in high-fertility areas. In very poor areas it may include both *A. canina* and *A. tenuis* (Bradshaw, 1968).

The bent grasses are low-quality grasses occurring so frequently that in England, for example, they dominate some 80% of lowland permanent pasture (Baker, 1963). Opinions regarding their value vary; some consider them comparatively useless (Lynn & Lynn, 1924), and they are often regarded as weeds (Jones, 1933c) or indeed, with *Poa trivialis*, as the major weeds of sown grassland (Baker, 1963). In contrast to this, they are considered of value for hill conditions (Davies, 1940), although growth in upland peat is sometimes poor (Hughes & Nicholson, 1961b).

Bent grasses are relatively aggressive in permanent pastures (Davies, 1928; Jones, 1933b) and have been described as the greatest of all competitors against ryegrass (Stapledon, 1939). They do not require a high level of soil fertility (Stapledon, 1939). *Agrostis*/*Deschampsia flexuosa* communities may occur on soils of pH 4·6, in contrast to the higher pH of 5·9 under which *Festuca rubra/Trifolium repens* occur (Welsh Plant Breeding Station, 1965).

Compared with other naturally occurring species in upland areas, bent grasses are relatively acceptable to stock (Milton, 1933, 1953; Welsh Plant Breeding Station, 1960); acceptability increases with liming (Lynn & Lynn, 1924). Native hill *Agrostis* has been found to contain 17·5% crude protein (CP), 14·0% water-soluble carbohydrate and to have an *in vitro* digestibility of 71·2% compared with values of 20·5%, 18·2% and 80·6%, respectively, for S23 perennial ryegrass in fertilized, reseeded hill areas (Welsh Plant Breeding Station, 1964). It yields best under close grazing (Stapledon, 1939; Jones, 1939) and on peat its yield has been known to surpass that of perennial ryegrass (West of Scotland Agricultural College, 1957).

Undesirable attributes of *Agrostis* species are their ability to carry infection of *Ophiobolus graminis* ('take-all' of cereals) (Walker, 1945; Brooks, 1965), and a tendency for the foliage to become 'winter-burned' (Welsh Plant Breeding Station, 1958a,b, 1960). The general impression given by these records is that bent grasses are disfavoured, but that little evidence is cited in support of this opinion.

Agrostis gigantea Roth.

Redtop bent (Hartley & Williams, 1956)
Black bent (Hubbard, 1954)
Red top (Hubbard, 1954)
Fiorin (Whyte *et al.*, 1959)

A loosely-tufted perennial with tough, creeping rhizomes; chromosome number 2n = 42. It is widely distributed in lowland districts of the British Isles, and sometimes called *A. alba* (Hartley & Williams, 1956) although this is thought to be incorrect (Hubbard, 1954). It differs from *A. tenuis* by its longer ligules and larger leaf-blades, and from *A. stolonifera* by its rhizomes and very loose panicles (Hubbard, 1954).

Its use in Britain is very limited (Hubbard, 1954). An estimate of the demand for its seed around the year 1948 was that it was less than 0·5% of that of the ryegrasses (Evans, 1948). Indeed, it can become a serious weed of arable land and its cultivation may therefore be inadvisable (Hubbard, 1954). It has been regarded as one of the major factors limiting the intensive production of wheat and barley (Weed Research Organization, 1967). The species is included in mixtures for wet permanent pasture in Europe (Whyte *et al.*, 1959).

It would appear that there is not enough precise information to judge whether this species could be valuable in agriculture in Britain.

Agrostis stolonifera L.

Creeping bent (Hubbard, 1954)

Fiorin (Hubbard, 1954)
White bent (Hubbard, 1954)
Sometimes wrongly called *A. alba* L. (Hubbard, 1954). A tufted perennial, spreading by leafy stolons and forming a close turf; chromosome number 2n = 28. It is widely distributed and abundant in poorer British grasslands (Hubbard, 1954). In Britain it is frequently regarded as a weed (Hubbard, 1954; Whyte *et al.*, 1959) which can invade swards of sown grasses (Kydd, 1966).

An undesirable aspect of the presence of creeping bent in cultivated land is the ease with which it can carry infection of *Ophiobolus graminis avenae* (Brooks, 1965).

The species is leafy and acceptable to stock when young. It is said to make a large proportion of its growth in late autumn (Whyte *et al.*, 1959), but there is little detailed information on yield to support this view. In trials on fertile soil, the total dry matter (DM) yield from cuts from April to September was 8 170 kg/ha, not significantly different from rather poor swards of S24 perennial ryegrass or swards of S48 timothy (Corrall, 1966b). A further trial showed that the growth before a first cut taken between late April and early June, gave a yield as high as that of S48 timothy, and only slightly less than that of S24 perennial ryegrass (Corrall, 1966c). The grass has been considered well worth encouraging by Armstrong (1945). It has been shown that *A. stolonifera* gives a greater yield response to fertilizer N, P and Ca than does *A. tenuis*, and it is concluded that the former is agronomically superior (Bradshaw *et al.*, 1958, 1960, 1964; Bradshaw, 1968).

Agrostis tenuis Sibth.

Common bent (Armstrong, 1945; Hubbard, 1954)
Fine bent-grass (Armstrong, 1945)
Brown top (Welsh Plant Breeding Station, 1932; Hubbard, 1954)
Colonial bent (Armstrong, 1945; Whyte *et al.*, 1959)
Browntop bent (Hartley & Williams, 1956)
Browntop (Armstrong, 1945)
A tufted perennial, spreading by short rhizomes and stolons, forming a loose or dense turf. Chromosome number 2n = 28 (Hubbard, 1954). This grass is abundant and widely distributed in Britain, and large areas of permanent grassland, particularly on poorer soils, are dominated by it (Hubbard, 1954).

Agrostis tenuis is considered to be of relatively low agricultural value in Britain (Armstrong, 1945; Henderson *et al.*, 1962a) except on poorer soils, especially those of hill land (Hubbard, 1954). Because its productivity is low, it is generally regarded as a weed (Whyte *et al.*, 1959), and it is used as a grassland weed for the purpose of testing herbicides (Weed Research Organization, 1965). It is therefore not surprising that the annual seed requirement in about 1948 was estimated to be less than 0·5% of that of the ryegrasses (Evans, 1948).

The value of *Agrostis tenuis* with *Festuca rubra* in hill areas has long been recognized. Sheep mostly frequent areas under these grasses (Fenton, 1952), and the stocking rate which can be tolerated on a hill farm is greatly influenced by the proportion of *Agrostis*/fescue in the pasture (Hunter, 1958). Attention has therefore been given to the reseeding of hill areas with *Agrostis tenuis* and it is thought that selected strains of this species may be the only ones of real value for very poor or dry hill soils (Hubbard, 1954).

It is often slow to establish after reseeding, but eventually forms a dense sward that is resistant to invasion and inhibits the growth of clover and other sown species (Welsh Plant Breeding Station, 1962). The yield in mixtures with white clover on mineralized upland peat is poor, less than that of red fescue and other grasses (Hughes & Nicholson, 1962; Hunt, 1964a) and may be rather greater (Hunt, 1964a) or lower (Hughes & Nicholson, 1961a) than that of perennial ryegrass, depending on the circumstances and the variety used. It is apparently not well adapted to areas with a high water table (Welsh Plant Breeding Station, 1964). The yield and winter-greenness of spaced plants are inferior to those of red fescue (Hill Farming Research Organization, 1965).

In lowland situations there have been several cases of *Agrostis tenuis*-dominated swards outyielding perennial ryegrass where grasses were sown either alone or in mixtures with white clover, and other instances where the reverse was true (Stapledon & Milton, 1932). Typical yield figures at various levels of fertilizer input are given in Table 23.1. These comparisons were made

Table 23.1. DM yields of *Agrostis tenuis* compared with those of perennial ryegrass under the same conditions (kg/ha)

Agrostis tenuis	Perennial ryegrass Yield	Variety	Reference from:
19 398	19 599	S101*	Henderson *et al.* (1962a)
9 439	13 530	S24	Cowling and Lockyer (1965)
7 074	6 491	Irish	
7 074	9 338	S23	Cowling and Lockyer (1963)
5 246	5 213	Irish*	
5 246	6 053	S23*	
7 590	8 240	S24	Corrall (1966a)
9 439	13 519	S24	
3 823	2 522	S23*	Grassland Research Institute (1961)
3 823	3 610	S24*	

* Sown with white clover; yield includes grass and legume

using a species which had not been bred for high yield. It is therefore not surprising to find that *Agrostis tenuis* is relatively inefficient in terms of DM production per unit of N applied (Grassland Research Institute, 1966; Cowling & Lockyer, 1967).

It has been considered that the seasonality of growth in *Agrostis tenuis* is unsatisfactory and in need of improvement early and late in the year (Welsh Plant Breeding Station, 1962). A higher proportion of its annual yield may occur in the second and third cuts of each year (Henderson *et al.*, 1962a) than in other grasses. This is supported by other data which have shown that in comparison with S24 perennial ryegrass, *Agrostis tenuis* makes relatively less growth in April and early May than later in the year (Corrall, 1966b). This is likely to be due to the relatively low rate of development of leaves which is known to occur under some circumstances (Grassland Research Institute, 1967), rather than to a low number of tillers per unit of ground area (Troughton, 1961).

Other undesirable attributes of *Agrostis tenuis* are low acceptability to stock (Hubbard, 1954), although there is little analytical evidence for or against this. Data reported by ap Griffith and Cooper (1959) showed that the normal-acid fibre content of the species cut in late May was 35–38 % compared with about 31 % in S23 perennial ryegrass and 36–39 % in S24 perennial ryegrass cut at the same time. It is slightly susceptible to snow mould (*Fusarium nivale*) in hill pastures, and is an efficient carrier of *Ophiobolus graminis avenae* (Brooks, 1965).

It has an I content intermediate between that of cocksfoot and perennial ryegrass (Butler & Johnson, 1957), and a Mo content more than twice that of Italian ryegrass (Fleming, 1965).

Alopecurus pratensis L.

Meadow foxtail (Welsh Plant Breeding Station, 1932; Armstrong, 1945; Whyte *et al.*, 1959)
Meadow fox-tail (Hubbard, 1954)
Common fox-tail (Hubbard, 1954)

A tufted perennial. Chromosome number 2n = 28. Occurs throughout Britain. In addition to indigenous strains, numerous introduced varieties have now been established in this country (Hubbard, 1954).

Meadow foxtail is valuable for grazing and mowing (Hubbard, 1954; Whyte *et al.*, 1959) on deep moist soils (Whyte *et al.*, 1959) and rich, fertile, heavy soils (Moore, 1966).

It occurs naturally on moist soils, where productivity can be high (Watson & More, 1962). It was observed that in the wet season of 1954, its spring growth and annual yield were superior to those of perennial ryegrasses (Welsh Plant Breeding Station, 1958a). It is suitable for growing under irrigation (Whyte *et al.*, 1959) and is reported to grow well under moderate shade (Watson & More, 1962).

Meadow foxtail has the added advantage of being resistant to cold (Hubbard, 1954). It suffers less loss by senescence during the winter than do perennial or Italian ryegrass (Welsh Plant Breeding Station, 1958a), and it is conspicuously winter-green (Fagan, 1932). It does not readily carry *Ophiobolus graminis* infection (Brooks, 1965).

The species gives a high yield (Stapledon, 1924; Hubbard, 1954) of good quality herbage (Fagan & Jones, 1924), typical data being given in Table 23.2. Yields are, however, lower than those from perennial or Italian ryegrass, as shown also by Fagan (1932). The mean DM content of herbage cut at various times of the year is relatively high, figures reported by Milton (1935) being 27·6% compared with 21·3% for Italian ryegrass.

Growth begins very early in the year (Watson & More, 1962) and it is one of the first grasses to flower. The cultivar S55 produced by the Welsh Plant Breeding Station has been examined particularly for spring and midsummer production (Welsh Plant Breeding Station, 1959); the spring yield is high (Stapledon & Milton, 1932), although it is reported to be only 70% of that of Italian ryegrass (Welsh Plant Breeding Station, 1958a). Other evidence suggests that the late autumn and early spring yields may be higher than those of Italian or perennial ryegrasses (Fagan, 1932).

Meadow foxtail has been described as a 'leafy' species (Stapledon & Milton, 1932), although this varies widely, some types being succulent, while others are relatively stemmy and of little agricultural value. The cultivar S55 is very leafy (Hubbard, 1954).

The species is reported to have a high acceptability and nutritive value to livestock (Fagan & Jones, 1924; Fagan, 1932; Moore, 1966); it is less readily eaten than timothy, and more (Stapledon & Milton, 1932) or less (Reid, 1951) readily eaten than perennial ryegrass. In pastures it has been shown to have a CP yield similar to that of cocksfoot and perennial ryegrass, a figure of about 54 g/yd^2 (65 g/m^2) being reported by Fagan and Jones (1924). It has the lowest fibre content of all grasses, according to Armstrong (1945).

It produces a hay of high quality and with a higher CP content than that of most other species; also, its aftermath growth is good (Fagan & Jones, 1924; Whyte *et al.*, 1959).

Meadow foxtail has several disadvantages. It is difficult to establish (Watson & More, 1962; Moore, 1966) and it generally lacks aggressiveness under competition (Davies, 1928). Its performance is poor in upland areas (Davies, 1940; Hughes & Nicholson, 1961a,b), probably due to its preference for fertile soil. The grass was formerly used to a considerable extent in seeds mixtures, but some authors do not now recommend this (Moore, 1966) and it has been largely replaced by timothy. One important reason for the change was that merchants had difficulty in supplying good meadow foxtail seed, mainly because of damage caused by certain species of gall midge (Barnes, 1930; Hubbard, 1954; Watson & More, 1962; Moore, 1966).

Anthoxanthum odoratum L.

Sweet vernal-grass (Hubbard, 1954)

Table 23.2. DM yield of *Alopecurus pratensis* compared with that of other grasses (kg/ha)

	Alopecurus pratensis	Perennial ryegrass	Italian ryegrass	Timothy	Reference from:
Yield 28 Aug. (sowing year)	1 693	2 253	2 836	1 356	Stapledon and Davies (1930)
Yield 12 May (1st harvest year)	2 930	3 310	...	2 430	Corrall (1966c)
Yield 1st week Apr.	1 193	1 444	1 695	...	Welsh Plant Breeding Station (1958a)
Total annual yield					
cut monthly	7 680	8 240	...	8 400	Corrall (1966b)
under grazing	9 642	10 936	10 145	...	Welsh Plant Breeding Station (1958a)
Mean total annual yield over 3 years—range between varieties grazed	5 538–10 414	7 978–9 581	6 046–6 599	4 413–9 715	Stapledon and Milton (1932)

Sweet vernal (Stapledon & Davies, 1930; Welsh Plant Breeding Station, 1932; Jones, 1967)

Sweet-scented vernal grass (Armstrong, 1945)

A tufted perennial. Chromosome number 2n = 20. This is an early-flowering grass, strongly scented with coumarin, occurring in Britain in a wide variety of habitats and abundant in poorer grasslands (Hubbard, 1954). It has been regarded as useless (Lynn & Lynn, 1924) or has been considered a weed (Jones, 1933c) and cannot be recommended (Armstrong, 1945).

Although it is thought to be relatively unpalatable to stock (Stapledon & Davies, 1930; Hubbard, 1954), there are reports that it is readily grazed when present in unimproved grassland, and is more readily accepted than *Festuca ovina* and other hill species (Milton, 1933; Milton, 1953). It has been reported to be grazed to a greater extent than tall fescue (Stapledon & Davies, 1930). Its acceptability to stock is probably not affected by the presence of coumarin (Welsh Plant Breeding Station, 1959). It has a high proportion of stem to leaf (Hubbard, 1954).

Sweet vernal-grass is a low-yielding species (Stapledon, 1924). For example, a cut in August gave a DM yield of 510 lb/acre (572 kg/ha) compared with 2 010 lb/acre (2 253 kg/ha) from perennial ryegrass (Stapledon & Davies, 1930). It is sensitive to defoliation and does best in a 'hay-and-aftermath' system of harvesting, from which a DM yield slightly larger than that of timothy has been recorded (Stapledon & Davies, 1930). Other reports also show that it does better mown than grazed (Jones, 1939). In many ways its behaviour has been found similar to that of Yorkshire fog (Thomas, 1936).

Sweet vernal-grass is frequently present in hill grassland (Hubbard, 1954). It may be highly (Grant & Hunter, 1968) or moderately (Hill Farming Research Organization, 1965) winter-green. It was at one time included in seeds mixtures, but this is not now generally recommended (Hubbard, 1954).

Arrhenatherum elatius (L.) J. & C. Presl.

Tall oatgrass

Tall oat-grass (Armstrong, 1945; Hubbard, 1954)

False oat(-)grass (Armstrong, 1945; Hartley & Williams, 1956)

Tall oat (Welsh Plant Breeding Station, 1932)

= *A. avenaceum* Beauv. (Pfitzenmeyer, 1962)

A tall, usually erect, tussock-forming perennial. Bulb formation can occur (Jenkin, 1931), bulb size being very variable. Chromosome number 2n = 28. It occurs widely in Britain and is abundant in the lowlands on soils of moderate to high fertility (Pfitzenmeyer, 1962).

Although tall oatgrass was once included in seeds mixtures (Stapledon & Davies, 1928b; Hubbard, 1954), there appears to be no particular place for it in Britain at present (Watson & More, 1962); indeed the form known as 'onion couch' can be a troublesome weed (Hubbard, 1954; Woodford & Evans, 1965).

The species is used as a hay grass in France and is highly productive under hot, dry conditions. It will not tolerate close grazing (Armstrong, 1945; Hubbard, 1954; Whyte *et al.*, 1959; Watson & More, 1962) or close cutting, owing to the relatively small number of basal axillary buds available for tiller formation (Pfitzenmeyer, 1962). Similarly, it cannot withstand excessive trampling (Whyte *et al.*, 1959; Watson & More, 1962; Pfitzenmeyer, 1962) and a further disadvantage is that it is a very good carrier of *Ophiobolus graminis avenae* infection (Brooks, 1965).

Tall oatgrass is, however, being evaluated for pasture and for hay production (Welsh Plant Breeding Station, 1959). It is known that it can establish rapidly when sown at 40 lb/acre (45 kg/ha) with white clover (Welsh Plant Breeding Station, 1959) and it has been described as aggressive in the seedling stage and in aftermath (Davies, 1928). Nevertheless, instances have been reported of difficulty in establishing it as a pure sward (Stapledon & Davies, 1930).

It is high yielding (Stapledon, 1924); under cutting it has been reported to yield 8 320 kg DM/ha per year compared with 8 240 kg/ha from S24 perennial ryegrass and 8 400 kg/ha from S48 timothy (Corrall, 1966b), although some workers have stated that it grows only for a short period and yields little bulk compared with ryegrass (Thomas & Davies, 1964). This is supported by yields obtained by repeated mowing of swards (Stapledon & Davies, 1930). Growth is, however, rapid and the product very leafy (Hubbard, 1954). When cut in late May at an *in vitro* DM digestibility of about 70%, its yield of digestible DM (4 100 kg/ha) has been intermediate between that of S24 and S48 (Corrall, 1966c).

Tall oatgrass has been shown to be a valuable source of carotene, producing 31 mg/100 g DM, which is greater than the concentration in ryegrass (Moon, 1939). It is one of the best drought-resisting British grasses (Armstrong, 1945).

Cynosurus cristatus L.

Crested dogstail (Welsh Plant Breeding Station, 1932; Hartley & Williams, 1956)

Crested dog's(-)tail (Armstrong, 1945; Whyte *et al.*, 1959)

A tufted perennial. Chromosome number 2n = 28. Widely distributed throughout Britain and very common in rough grasslands (Hubbard, 1954) and in permanent pasture on agricultural soils (Lodge, 1959).

Crested dogstail has been considered a desirable grass for Britain (Lynn & Lynn, 1924; Jones, 1933c; Hubbard, 1954; Moore, 1966), though of minor importance (Evans, 1948). It is described as a 'bottom' grass (Davies 1939; Scottish Agricultural Industries, 1967), particularly suited to grazing by sheep (Hubbard, 1954; Whyte *et al.*, 1959; Watson & More, 1962; Scottish

Agricultural Industries, 1967), and has been included in seeds mixtures intended for grazing (Stapledon & Davies, 1928b, 1930). There is some evidence that it performs best with only light defoliation (Jones, 1939). It is recommended for poor soils (Hubbard, 1954; Watson & More, 1962; Spedding, 1965; Moore, 1966; Scottish Agricultural Industries, 1967).

It is used as an ingredient in seeds mixtures (Stapledon, 1939; Milton, 1939; Heddle, 1947; Hubbard, 1954; Whyte *et al.*, 1959; Hunt, 1964b; Spedding, 1965), in permanent pastures (Whyte *et al.*, 1959; Watson & More, 1962; Moore, 1966), often on dry soils (Bridge & Price, 1938; Bond, 1938; Whyte *et al.*, 1959; Moore, 1966), since it can withstand drought (Hubbard, 1954; Jones, 1967). It establishes slowly (Armstrong, 1945; Moore, 1966), its first-year yield being low compared with that of perennial ryegrass (Stapledon & Davies, 1930). Although it is a prolific self-seeder (Watson & More, 1962; Moore, 1966), it is not aggressive (Davies, 1928), does not seem to persist well and does not appear to resist the ingress of weeds (Jones, 1933a; Ivins, 1952).

Crested dogstail is often used in hill areas (Bridge & Price, 1938; Whyte *et al.*, 1959; Moore, 1966; Scottish Agricultural Industries, 1967), where its establishment is reported to be better than that of other grasses (Jones, 1967). Other work has shown its performance to be variable in uplands, having been reported to be good by Davies (1940), occasionally so by Hughes and Nicholson (1961b), intermediate by Hunt, (1964a) and poor by Hughes and Nicholson (1961a, 1962). It withstands cold and is winter-green (Hubbard, 1954; Watson & More, 1962; Moore, 1966; Jones 1967).

Its yield is often low (Stapledon, 1924; Davies, 1939; Hubbard, 1954; Whyte *et al.*, 1959; Watson & More, 1962; Jones, 1967), and usually below that of perennial ryegrass (Fagan & Jones, 1924). For example, in trials reported by Stapledon and Davies (1930), repeated mowing of 'pure' swards gave an annual DM yield of 4 160 lb/acre (4 663 kg/ha) compared with 5 610 lb/acre (6 289 kg/ha) from perennial ryegrass, although at certain times of the year its yield exceeded that of perennial ryegrass. The yield of mixtures with white clover has been variously reported to be good (Stapledon & Davies, 1928a) or poor compared with mixtures of perennial ryegrass and white clover (North of Scotland College of Agriculture, 1954).

Crested dogstail has been reported to produce most of its yield late in the year (Moore, 1966). Although the proportion of unpalatable stems is high (Armstrong, 1945; Hubbard, 1954; Watson & More, 1962; Moore, 1966; Jones, 1967), the leaf material always appears to be very palatable to stock (Milton, 1933; Armstrong, 1945) despite a CP content lower than that of perennial ryegrass (Ashton & Morgan, 1953). The species is so poor in minerals that its presence was considered by Thomas *et al.* (1948) to reduce the overall 'mineral-producing efficiency' of a sward, although this is not apparent in other work on trace elements (Fleming, 1965).

The Cu content is very low, but higher than in red fescue (Thomas *et al.*, 1948). Its P concentration is only moderate (Thomas *et al.*, 1948; Ashton & Morgan, 1953), but its concentration of Zn and Mo is relatively high (Fleming, 1965). Mg (Ashton & Morgan, 1953) and Mn (Thomas *et al.*, 1952) contents are intermediate between those of ryegrass and cocksfoot. The Cl content is equal to or greater than that of Italian ryegrass and certain other grasses (Fleming, 1965).

The annual seed requirement has in some years been estimated to be less than 2% of that of the ryegrasses (Evans, 1948).

Festuca rubra L.

Red fescue (Armstrong, 1945)

F. rubra L. subsp. *rubra.* Red fescue (Hubbard, 1954); creeping fescue (Hubbard, 1954); red fescue (creeping) (Whyte *et al.*, 1959)

F. rubra L. subsp. *commutata* Gaud. Chewings fescue (Hubbard, 1954)

A perennial, widely distributed in Britain. Chromosome number 2n = 42 or 56.

In Britain, red fescue is of limited agricultural use, although it is valuable on poor upland soils and in areas which tend to dry out (Hubbard, 1954; Whyte *et al.*, 1959; Spedding, 1965), in exposed pastures (Spedding, 1965) and in poultry runs (Bridge & Price, 1938). It is regarded as a weed (Jones, 1933c; Weed Research Organization, 1967) in lowland pastures and is used as a test species for herbicide evaluation (Weed Research Organization, 1965).

Some forms of this grass, notably Chewings fescue, are valuable lawn grasses and appreciable quantities of seed are marketed for this purpose (Evans, 1948). The cultivar S59 is the principal source of seed for agricultural use in Britain.

Growth of red fescue is sometimes poor in the first year (Stapledon & Davies, 1930; Thomas, 1936), perhaps because it is not aggressive (Davies, 1928), although it eventually spreads to form a dense sward having a large number of small tillers (Troughton, 1961).

It is sometimes low-yielding (Stapledon, 1924; Stapledon & Davies, 1928a; Whyte *et al.*, 1959). For example, Green and Evans (1957) reported an October DM yield of 710 lb/acre (796 kg/ha) compared with 1 120 lb/acre (1 256 kg/ha) from S23 perennial ryegrass, and in April the yields were 250 and 360 lb/acre (280 kg/ha and 404 kg/ha) respectively.

In other instances, red fescue has given relatively high yields (Fagan, 1932), e.g. a total annual DM yield of 9 170 kg/ha compared with 8 240 kg/ha for S24 perennial ryegrass and 8 400 kg/ha for S48 timothy (Corrall, 1966b); 640 lb/acre (717 kg/ha) of DM in December compared with 450 lb/acre (504 kg/ha) from S23 perennial ryegrass (Green & Evans, 1957). Yields in spring may also be high (Stapledon & Milton, 1932; Corrall, 1966c), and late autumn and early spring yields have been reported to be greater than those from Italian and perennial ryegrasses (Fagan, 1932).

Red fescue performs particularly well on reseeded hill soils or peat (Davies, 1940; West of Scotland Agricultural College, 1958, 1960; Hughes & Nicholson, 1961a; Hill Farming Research Organization 1962). In these circumstances, usually in mixtures with white clover, it has regularly outyielded perennial ryegrass (West of Scotland Agricultural College, 1957; Hughes & Nicholson, 1961, 1962; Hunt, 1964a,b; Welsh Plant Breeding Station, 1966). There have been, for example, spring cuts giving a DM yield of 530 lb/acre (594 kg/ha) compared with 240 lb/acre (269 kg/ha) from S23 and 410 lb/acre (460 kg/ha) from S24 perennial ryegrass. There is, however, one recorded case of S23 perennial ryegrass being more vigorous than S59 red fescue, irrespective of soil type or altitude (Welsh Plant Breeding Station, 1965). Compared with other hill species, red fescue may give a conspicuously high yield, for example when cut in June (Grant & Hunter, 1968).

Part of its value in hill situations is its high potential for winter survival (Welsh Plant Breeding Station, 1964) associated with winter-greenness (Hubbard, 1954; Welsh Plant Breeding Station, 1958b, 1963; Thomas & Davies, 1964; Hill Farming Research Organization, 1965; Grant & Hunter, 1968), although, as might be expected, yields are less than at lower altitudes (West of Scotland Agricultural College, 1964). Its value is reduced by a tendency to form seed heads in early summer (Hill Farming Research Organization, 1962).

Except when young (Armstrong, 1945) red fescue is less palatable to livestock than other hill and lowland species (Stapledon & Milton, 1932; Welsh Plant Breeding Station, 1932; Ivins & Shaw, 1950; Reid, 1951), although more so than *Nardus*, *Molinia* and tall heather (Welsh Plant Breeding Station, 1960). This low palatability is due to a high fibre content (Griffith & Cooper, 1959).

The herbage is of low quality compared with that of perennial ryegrass (Welsh Plant Breeding Station, 1964); for example, the CP content is only about two thirds that of the latter. It has a high DM content (28·9 % reported by Milton, 1935), but is one of the poorest sources of major and trace elements (Thomas *et al.*, 1948) and is low in Na (Welsh Plant Breeding Station, 1967; Thomas *et al.*, 1952), Ca and Mg (Thomas *et al.*, 1952). It has been considered that a sward dominated by red fescue would be incapable of meeting the mineral needs of livestock (Thomas *et al.*, 1948). The fact that the growth rates of lambs on this species have sometimes been much lower than those on perennial ryegrass (Spedding *et al.*, 1966) probably lends some support to this view. Red fescue is, however, high in Co (Thomas *et al.*, 1952). There are reasonable concentrations of Fe and Mn (Thomas *et al.*, 1948) in this species and it is one of the few in which the Mn content has been found to vary with maturity (Thomas *et al.*, 1952). In judging the value of analytical data it should, however, be borne in mind that there is wide genetic variation within the species (Smith, 1965; Ratcliffe, 1965). This may perhaps explain the diversity in the recorded level of Cu (Thomas *et al.*, 1948; Thomas *et al.*, 1952).

Red fescue can withstand close grazing (Whyte *et al.*, 1959), which also reduces its tendency to become unpalatable to stock, although it sometimes does better under lenient than under severe grazing (Jones, 1939). In hill areas, sheep tend to frequent vegetational types which contain a high proportion of this grass, and the stocking rate which can be tolerated is greatly influenced by the presence of *Agrostis*/fescue pastures.

The species is fairly resistant to *Fusarium nivale* but can carry *Ophiobolus graminis* infection (Brooks, 1965).

Holcus lanatus L.

Yorkshire fog (Armstrong, 1945; Hubbard, 1954)

Meadow soft-grass (Armstrong, 1945)

A tufted perennial. Chromosome number 2n = 14. It is a very common grass occurring throughout Britain, particularly in pastures, rough grassland and on wasteland (Hubbard, 1954; Beddows, 1961). It is often regarded as comparatively useless agriculturally (Lynn & Lynn, 1924) or as a weed (Jones, 1933c; Evans, 1964; Weed Research Organization, 1965) and to constitute a problem in permanent pastures in England and Wales (Weed Research Organization, 1967).

Serious attempts have been made to utilize the species both in lowland and in hill situations (West of Scotland Agricultural College, 1957, 1958) and improved selections have been produced (New Zealand Journal of Agriculture, 1963). It may be aggressive in pastures (Davies, 1928; Jones, 1967) and needs to be carefully controlled by management (Thomas, 1936).

For reseeding hill areas, Yorkshire fog has been one of the most successful grasses (Davies, 1940; West of Scotland Agricultural College, 1960; Hughes & Nicholson, 1961a; Hill Farming Research Organization, 1962). Initial establishment may be rapid (Jones, 1967), after which it may yield more than S23 perennial ryegrass (Hughes & Nicholson, 1961b; Hunt, 1964a).

There is much variation in performance between different types of Yorkshire fog (Jones, 1967) and it has been undergoing selection as a pioneer grass for poorer soils (New Zealand Journal of Agriculture, 1963). It is less winter-green than red fescue (Hill Farming Research Organization, 1965) and may become badly winter-burned.

Its palatability has been reported to vary from high to low (Stapledon & Milton, 1932; Milton 1933; New Zealand Journal of Agriculture, 1963). It is usually eaten readily in the spring (Milton, 1953; Jones, 1967), but later becomes stemmy (Stapledon & Milton, 1932) and unacceptable to livestock

(Jones, 1967). It does better under lenient than under hard grazing (Jones, 1939).

The yield at a first cut taken before the end of June was reported by Corrall (1966c) to be similar to that of S48 timothy. For example, a cut on 2 June gave 6 180 kg DM/ha (4 430 kg digestible DM/ha) for *Holcus* and 5 660 kg DM/ha (3 950 kg digestible DM/ha) for S48.

The species probably has a relatively low DM content compared with other grasses (Hunt, 1964a) but it is capable of absorbing large quantities of most mineral elements (Fleming, 1965). The mineral content of the hay compares favourably with that of the more desirable grasses (Ashton & Morgan, 1953; Beck, 1962), although I (Butler & Johnson, 1957), Mg and Co (Thomas *et al.*, 1948) contents are relatively low. It has been shown to have relatively high Cl (1·0% in the DM) and Mo contents (Fleming, 1965).

A disadvantage to its use in pasture is the prevalence of infections by rust fungi (*Puccinia* spp.) (New Zealand Journal of Agriculture, 1963) and its ability to carry *Ophiobolus graminis* infection (Walker, 1945).

Phleum bertolonii D.C.

Smaller cat's tail (Hubbard, 1954)

A perennial consisting of a number of strains, some prostrate, spreading by stolons, and others erect. It has usually been regarded as a variety of *Phleum pratense*. Chromosome number 2n = 14 (Hubbard, 1954). The species is distributed throughout Britain, particularly in old pastures.

Strains have been selected for their leafy growth, high yield and acceptability to livestock. On account of their persistence they are particularly suitable for long-term pastures (Hubbard, 1954; Whyte *et al.*, 1959). A notable British cultivar is S50, on which most of the following statements are based.

The species has performed well in reseeded hill areas (Hughes & Nicholson, 1961a, 1962), where the yield may be similar to (Welsh Plant Breeding Station, 1966), or even surpass that of perennial ryegrass (Hughes & Nicholson, 1961b). For example, a spring cut yielded 740 lb DM/acre (830 kg/ha) compared with 240 lb/acre (269 kg/ha) from S23 and 410 lb/acre (460 kg/ha) from S24 perennial ryegrass (Welsh Plant Breeding Station, 1963).

It is usually grown in mixtures with white clover which, under lowland conditions, may (Welsh Plant Breeding Station, 1958) or may not (Hunt, 1956c) be suppressed by competition with it. The DM and CP yield as hay and aftermath of such mixtures has been concluded to be generally low compared with that of certain perennial ryegrass varieties and with that of several other species (Hunt, 1956c).

Herbage quality is often low compared with perennial ryegrass, especially in midsummer (Welsh Plant Breeding Station, 1964). The content of normal-acid fibre in the spring is similar to that in S24 perennial ryegrass (Griffith & Cooper, 1959).

The growth rate of S50 is less in mid-season than that of S23 perennial ryegrass (Hunt, 1957); however, it is winter-green (Welsh Plant Breeding Station, 1958a, b) and has been described as having a particularly good potential for winter production. It is resistant to rusts (*Puccinia* spp.) and is useful in orchards and on land liable to flooding (Welsh Plant Breeding Station, 1958b), although it is reported to occur more frequently in drier situations than *Phleum pratense* (Armstrong, 1945).

Poa pratensis L.

Smooth(-)stalked meadow(-)grass (Armstrong, 1945; Whyte *et al.*, 1959)
Meadow grass (Hubbard, 1954)

A very variable perennial with creeping, slender rhizomes. Chromosome number 2n = 50 to 124. It is widespread in Britain (Hubbard, 1954; Spedding, 1965), being common in old meadows and pastures.

Smooth-stalked meadowgrass has limited agricultural use, and is thought to be of little agricultural value (Hubbard, 1954). It is sometimes considered to be a useful component of permanent pasture (Lynn & Lynn, 1924; Scottish Agricultural Industries, 1967) and it is useful in poultry runs (Robinson, 1936). Despite this diversity of opinion, it appears to have certain qualities which suggest that it could be a profitable subject for improvement (Scottish Plant Breeding Station, 1967). Its rhizomatous nature may make it difficult to eradicate and it may therefore be a potential arable weed (Watson & More, 1962).

Smooth-stalked meadowgrass is usually sown in mixtures (Spedding, 1965) for permanent pastures (Whyte *et al.*, 1959; Moore, 1966) and, although it is described by some as non-aggressive (Davies, 1928) and to take two years to establish (Whyte *et al.*, 1959), it sometimes has a tendency to dominate the sward (Moore, 1966). It is fairly drought-resistant (Whyte *et al.*, 1959; Watson & More, 1962) and is useful on dry soils (Armstrong, 1945; Hubbard, 1954).

Smooth-stalked meadowgrass exists in a wide variety of races and strains, many of which are apomictic, and this may account for its tolerance of a wide variety of soil conditions. Although it responds to fertility and is a species of rich soils (Moore 1966), it also does well on infertile soil (Scottish Agricultural Industries, 1967). It is often found in areas of high soil N status (Welsh Plant Breeding Station, 1962).

Described as a cool-season grass, most of its growth occurs in the spring; summer production is poor (Whyte *et al.*, 1959) and it may be summer-dormant (Watson & More, 1962).

The view that it may be relatively low-yielding (Stapledon, 1924) is not supported by trials in which the cultivar S63 gave a total annual DM yield of 7 750 kg/ha, compared with 8 240 kg/ha from S24 perennial ryegrass and 8 400 kg/ha from S48 timothy (Corrall, 1966b). A first-growth cut in late

June (Corrall, 1966c) gave a yield of (*in vitro*) digestible DM of 5 090 kg/ha, compared with 5 310 kg/ha from S24 and 5 150 kg/ha from S48. In another experiment, an annual DM yield of 6 800 lb/acre (7 623 kg/ha) was obtained, compared with 7 400 lb/acre (8 295 kg/ha) from S24 perennial ryegrass (Allen, 1966).

Smooth-stalked meadowgrass is a bottom grass. It is palatable, resistant to close grazing and trampling, resistant to cold, is winter-green and may make an appreciable amount of growth in milder winters (Whyte *et al.*, 1959; Moore, 1966; Scottish Agricultural Industries, 1967).

It has been tested for hill areas with variable results. In some trials early growth has been poor and yields lower than those of S23 perennial ryegrass (Welsh Plant Breeding Station, 1962). In others its performance has been moderate to poor (Hughes & Nicholson, 1961a,b, 1962), and in others good enough for it to be regarded as one of the best grasses for reseeding hill land (West of Scotland Agricultural College, 1957, 1958, 1960, 1964; Hunt, 1964).

Smooth-stalked meadowgrass is considered to be highly nutritious to livestock (Moore, 1966). It has been recorded as having a high Cu content of about 21 ppm. (Beeson *et al.*, 1947), a CP content similar to that of perennial ryegrass, a Ca content greater than and a P content less than that of perennial ryegrass (Ashton & Morgan, 1953).

Smooth-stalked meadowgrass is considered inferior to rough-stalked meadowgrass (Moore, 1966). In Britain, less seed is used than of the latter (Evans, 1948).

Poa trivialis L.

Rough(-)stalked meadow(-)grass (Armstrong, 1945; Whyte *et al.*, 1959; Allen, 1966)

Rough meadow(-)grass (Hubbard, 1954; Hartley & Williams, 1956; Allen, 1966)

A loosely tufted perennial with creeping stolons. Chromosome number 2n = 14 (Hubbard, 1954). Widely distributed in Britain and common in lowland pastures.

Rough-stalked meadowgrass is considered a desirable grass (Armstrong, 1945; Hubbard, 1954), superior to smooth-stalked meadowgrass (Moore, 1966). It is accepted as being useful (Lynn & Lynn, 1924; Jones, 1933c) in mixtures for long-term pastures (Hubbard, 1954; Whyte *et al.*, 1959; Moore, 1966) but it is also considered to be a weed of arable land (Watson & More, 1962; Allen, 1966) and can invade sown pastures (Baker, 1963; Allen, 1966; Kydd, 1966). Although it can thus be regarded as a weed of pastures (Weed Research Organization, 1965, 1967; Allen, 1966), it has been strongly recommended for inclusion in seeds mixtures (Evans, 1964) and there are many records of its having been sown as such (Jones, 1933b; Milton, 1939; Heddle,

1947; Bennetts, 1958; Hunt, 1964b), particularly on better soils (Stapledon & Davies, 1928b; Laird, 1949). Generally speaking, seeds mixtures before 1930 included 1 lb of rough-stalked meadowgrass/acre (1·1 kg/ha) (Davies, 1958), but the practice has declined (Allen, 1966). Only a relatively small amount of seed is used in Britain (Evans, 1948) and the case for its use is now rather weak (Watson & More, 1962).

Some workers consider the grass persistent (Stapledon & Davies, 1928a; Robinson, 1933), to have a tendency to dominate the sward (Moore, 1966) and to hinder the development of more valuable plants (Hubbard, 1954). Alternatively, it has been reported to be difficult to establish as a pure sward (Stapledon & Davies, 1930) and it has been reported as lacking persistence (Ivins, 1952) so that its frequency fell to less than 30% of the sward (Jones, 1933a), in keeping with the view that it is generally non-aggressive (Davies, 1928). Similarly, it does not inhibit the growth of white clover when the two are grown together (Hunt, 1956c).

Rough-stalked meadowgrass is a low-yielding grass (Stapledon, 1924; Stapledon & Davies, 1928a, 1930; Davies, 1939; Cowlishaw *et al.*, 1956; Hunt, 1956b). A total annual DM yield of 6 060 kg/ha has been reported by Corrall (1966b), compared with 8 240 kg/ha from S24 perennial ryegrass and 8 400 kg/ha from S48 timothy. Most of this growth occurred in spring and early summer. Yield has been alleged to decline rapidly later in the year (Hunt, 1957; Grassland Research Institute, 1964; Corrall, 1966a, b), although other work suggests that the production curve is similar to that of perennial ryegrass, after allowing for the fact that the latter has a higher growth rate (Anslow & Green, 1967). Other work (Corrall, 1966a) has shown an annual DM yield of 5 800 lb/acre (6 502 kg/ha) compared with 7 400 lb/acre (8 295 kg/ha) from S24 perennial ryegrass.

Mixtures of the species with white clover have given the following yields (3-year totals) compared with those of perennial ryegrass (Hunt, 1956a).

Grass in mixture with white clover	DM lb/ac (kg/ha)	CP lb/ac (kg/ha)	Starch equivalent lb/ac (kg/ha)
S24 perennial ryegrass	19 130 (21 445)	2 470 (2 769)	10 170 (11 401)
S23 perennial ryegrass	14 910 (16 714)	2 000 (2 242)	7 940 (8 901)
Rough-stalked meadow-grass	14 560 (16 322)	2 310 (2 590)	8 180 (9 170)

The species grows best on rich, heavy, moist soils (Whyte *et al.*, 1959; Watson & More, 1962; Allen, 1966; Moore, 1966; Scottish Agricultural

Industries, 1967). Its performance in reseeded hill land varies from poor to good (Hughes & Nicholson, 1961a, b, 1962). It is fairly frost-resistant, gives an appreciable amount of early spring growth and is improved by irrigation (Allen, 1966). It is susceptible to infection by *Fusarium nivale* (Hill Farming Research Organization, 1959). Rough-stalked meadowgrass is palatable to livestock (Robinson, 1933; Ivins & Shaw, 1950; Hubbard, 1954; Whyte *et al.*, 1959; Moore, 1966), though perhaps eaten less readily than timothy (Milton, 1934). It has higher Cl, Cu, Mo and Zn contents than has Italian ryegrass (Fleming, 1965) and relatively low I (Butler & Johnson, 1957) and Mn (Fleming, 1965) contents compared with perennial ryegrass. The species is described as of high nutritive value (Moore, 1966; Allen, 1966). A comprehensive review of the literature on rough-stalked meadowgrass is available (Allen, 1966).

Trisetum flavescens (L.) Beauv.

Yellow oatgrass (Hartley & Williams, 1956)
Yellow oat(-)grass (Hubbard, 1954)
Golden oat-grass (Hubbard, 1954)

A loosely tufted perennial. Chromosome number 2n = 28. Common in England and Wales, but infrequent in Scotland and Ireland. Abundant in poorer grassland (Hubbard, 1954).

Yellow oatgrass seed is expensive and scarce (Whyte *et al.*, 1959) and is not now included in seeds mixtures used in Britain (Hubbard, 1954), although it is not without desirable qualities (Lynn & Lynn, 1924; Hubbard, 1954).

It is fairly drought-resistant and tolerates a range of soils (Hubbard, 1954). It is readily consumed by sheep and cattle (Hubbard, 1954; Whyte *et al.*, 1959) and its stemmy character is offset by the unusual characteristic that sheep eat as much stem as leaf (Stapledon & Milton, 1932), although the fibre content may be relatively high (Fagan, 1932). Pastures are reported to produce less CP per unit area than those of perennial ryegrass (Fagan & Jones, 1924). It produces hay of good quality and with a high CP content (Fagan & Jones, 1924; Fagan, 1932). There appears to be relatively little published information on the performance of yellow oatgrass in Britain.

Productivity has been reported as low (Whyte *et al.*, 1959) and moderate (Stapledon, 1924). A single cut taken in the year of sowing gave a DM yield about 41 % of that of Italian ryegrass (Stapledon & Davies, 1930). In the first harvest year it was reported to have given a DM yield of 7 010 lb/acre (7 858 kg/ha), compared with 6 210 lb/acre (6 961 kg/ha) from perennial ryegrass (Stapledon & Milton, 1932); production declined in subsequent years, so that the total yield for three years was greater in the latter species, i.e. 25 640 lb/acre (28 742 kg/ha) compared with 13 670 lb/acre (15 324 kg/ha) from yellow oatgrass. Fagan (1932) also reported a decline in yield in the years after sowing due to its inability to withstand hard grazing.

Conclusions

It is difficult to summarize the available information in any way which would show which species are most useful to British agriculture. The value of each grass will depend on the situation for which it is considered. Some species excel on poor soils, particularly in hill areas, where they tend to be superior to perennial ryegrass.

In considering their use on fertile lowland soils, the following facts are worth noting:

Agrostis gigantea is difficult to eradicate, and there is little information to suggest that it has good qualities. *Agrostis stolonifera* is also difficult to eradicate, but has a promising yield and further work might show it to have value. *Agrostis tenuis* gives a reasonably high yield but may have to be rejected because of its low palatability to livestock.

Alopecurus pratensis appears to be a desirable grass, although it is difficult to establish and seed is in short supply. It may have value for irrigated land.

There is little information on *Anthoxanthum odoratum*. Its yield and palatability to livestock are both low.

Arrhenatherum elatius in the bulbous form is probably difficult to eradicate, but can give a high yield of good quality, although this is achieved only with certain methods of harvesting, owing to its intolerance of close defoliation. It is drought-resistant.

Cynosurus cristatus is difficult to maintain as a close sward, becomes stemmy, is low in minerals and has a low yield.

Festuca rubra has a low yield of herbage which is of low mineral content and not very acceptable to livestock.

There is little information on *Holcus lanatus*, which has herbage of relatively high mineral content, but which tends to become unpalatable to livestock as it ages.

There is little information on *Phleum bertolonii*. The herbage is of low quality. Winter growth is good.

Poa pratensis and *Poa trivialis* are acceptable and nutritious to livestock. The former is drought resistant and the latter has a good mineral content and although its yield is low when grown alone, the yield of mixtures with white clover is reasonably high. Both species have the disadvantage of a poor summer yield and are difficult to eradicate.

There is little information on *Trisetum flavescens*. Its seed is expensive and scarce.

Some 130 British grasses have not been discussed individually in this book because there is no evidence that they have sufficient importance as pasture species to warrant their inclusion.

Thus the number of British grasses which are cultivated is low. This situation is paralleled by that in the world as a whole, where 40 out of the 10 000 or so known grass species probably account for 99% of the sown grass pasture, excluding species used for range reseeding or soil conservation (Hartley & Williams, 1956).

References

ALLEN, G. P. 1966. The botany, ecology, agronomy and control of *Poa trivialis* L.—rough stalked meadow-grass. *Tech. Rep.* 6 *A.R.C. Weed Res. Org., Kidlington* pp. 23.

ANSLOW, R. C.; GREEN, J. O. 1967. The seasonal growth of pasture grasses. *J. agric. Sci., Camb.* **68,** 100–22.

ARMSTRONG, S. F. 1945. British grasses and their employment in agriculture. Cambridge: Univ. Press.

ASHTON, W. M.; MORGAN, D. E. 1953. The chemical composition and nutritive value of Welsh hay in comparison with other British hays. Part 2. *J. Br. Grassld Soc.* **8,** 131–47.

BAKER, H. K. 1963. A survey of British grasslands. *Proc. 6th Br. Weed Control Conf., 1962* pp. 23–30.

BARNES, H. F. 1930. On the biology of the gall-midges (Cecidomyidae) attacking meadow foxtail grass (*Alopecurus pratensis*), including the description of one new species. *Ann. appl. Biol.* **17,** 339–66.

BECK, A. B. 1962. The levels of copper, molybdenum and inorganic sulphate in some Western Australian pastures—a contribution to the study of copper-deficiency diseases in ruminants. *Aust. J. exp. Agric. Anim. Husb.* **2,** 40–5.

BEDDOWS, A. R. 1961. *Holcus lanatus* L. Biological flora of the British Isles. *J. Ecol.* **49,** 421–30.

BEESON, K. C.; GRAY, L.; ADAMS, M. B. 1947. The absorption of mineral elements by forage plants. 1. The phosphorus, cobalt, manganese, and copper content of some common grasses. *J. Am. Soc. Agron.* **39,** 356–62.

BENNETTS, J. 1958. Grassland management at Scobbiscombe. *J. Br. Grassld Soc.* **13,** 118–20.

BOND, J. R. 1938. The care of grass runs. *Sci. Poult. Breed. Reg.* pp. 81–5.

BRADSHAW, A. D. 1968. Hartley Botantical Laboratories, Liverpool Univ. Personal communication.

BRADSHAW, A. D.; CHADWICK, M. J.; JOWETT, D.; LODGE, R. W.; SNAYDON, R. W. 1960. Experimental investigations into the mineral nutrition of several grass species. Part 3. Phosphate level. *J. Ecol.* **48,** 631–7.

BRADSHAW, A. D.; CHADWICK, M. J.; JOWETT, D.; SNAYDON, R. W. 1964. Experimental investigations into the mineral nutrition of several grass species. Part 4. Nitrogen level. *J. Ecol.* **52,** 665–76.

BRADSHAW, A. D.; LODGE, R. W.; JOWETT, J.; CHADWICK, M. J. 1958. Experimental investigations into the mineral nutrition of several grass species. Part 1. Calcium level. *J. Ecol.* **46,** 749–57.

BRIDGE, R. D. H.; PRICE, W. T. 1938. Poultry husbandry in the West. *J. Bath W. sth. Cties' Ass.* **13,** 68–86.

BROOKS, D. H. 1965. Wild and cultivated grasses as carriers of the take-all fungus (*Ophiobolus graminis*). *Ann. appl. Biol.* **55,** 307–16.

BUTLER, G. W.; JOHNSON, J. M. 1957. Factors influencing the iodine content of pasture herbage. *Nature, Lond.* **179,** 216–17.

CORRALL, A. J. 1966a. Data on Exp. H.373, Grassland Research Institute, Hurley. Personal communication.

CORRALL, A. J. 1966b. Annual Summary Exp. H.674, Grassland Research Institute, Hurley. Personal communication.

CORRALL, A. J. 1966c. Annual Summary Exp. H.676, Grassland Research Institute, Hurley. Personal communication.

COWLING, D. W.; LOCKYER, D. R. 1963. Nitrogenous fertiliser and the role of legumes. *Exps Prog. 15 Grassld Res. Inst. 1961–2* pp. 16–19.

COWLING, D. W.; LOCKYER, D. R. 1965. A comparison of the reaction of different grass species to fertilizer nitrogen and to growth in association with white clover. 1. Yield of dry matter. *J. Br. Grassld Soc.* **20,** 197–204.

COWLING, D. W.; LOCKYER, D. R. 1967. A comparison of the reaction of different grass species to fertilizer nitrogen and to growth in association with white clover. 2. Yield of nitrogen. *J. Br. Grassld Soc.* **22,** 53–61.

COWLISHAW, S. J.; EYLES, D. E.; ASTBURY, J. G. 1956. The effect of two rates of stocking on contrasting swards for poultry. *J. Br. Grassld Soc.* **11,** 155–61.

DAVIES, W. 1928. The factor of competition between one species and another in seeds mixtures. *Bull. H8 Welsh Pl. Breed. Stn* pp. 82–149.

DAVIES, W. 1939. Temporary leys: comparison of station-bred and commercial grasses used in simple mixtures. *Bull. H15 Welsh Pl. Breed Stn* pp. 1–24.

DAVIES, W. 1940. The yield and other characteristics of grasses and clovers when sown in simple mixtures or on exposed hill at an elevation of 1300 ft. *Welsh J. Agric.* **16,** 175–205.

DAVIES, W. 1958. Grassland developments in the United Kingdom in the last twenty five years. *Emp. J. exp. Agric.* **26,** 195–207.

EVANS, G. 1948. Needs and supplies of seeds of adapted strains. *J. Br. Grassld Soc.* **3,** 35–42.

EVANS, S. 1964. The herbicidal control of broad-leaved and grass weeds in established grassland. *J. Br. Grassld Soc.* **19,** 205–11.

FAGAN, T. W. 1932. The influence of nitrogenous manures on the chemical composition of the produce of individual grasses as pasture, hay and aftermath. *Bull. H13 Welsh Pl. Breed. Stn* pp. 80–92.

FAGAN, T. W.; JONES, H. T. 1924. The nutritive value of grasses as shown by their chemical composition. *Bull. H3 Welsh Pl. Breed. Stn* pp. 85–130.

FENTON, E. 1952. The influence of man and animals on the vegetation of certain hill grazings in South East Scotland. 2. *Bull. 5 Edinb. E. Scotl. Coll. Agric. Tech.*

FLEMING, G. A. 1965. Trace elements in plants with particular reference to pasture species. *Outl. Agric.* **4,** 270–85.

GRANT, S. A.; HUNTER, R. F. 1968. Variation in yield, maturity type, winter greenness and sensitivity to cutting of hill grass species. *J. Br. Grassld Soc.* **23,** 149–55.

GRASSLAND RESEARCH INSTITUTE. 1961. *Exps Prog. 13 Grassld Res. Inst. 1959–60.*

GRASSLAND RESEARCH INSTITUTE. 1964. *Exps Prog. 16 Grassld Res. Inst. 1962–63.*

GRASSLAND RESEARCH INSTITUTE. 1966. *A. Rep. Grassld Res. Inst. Hurley 1965.*

GRASSLAND RESEARCH INSTITUTE. 1967. *A. Rep. Grassld Res. Inst. Hurley 1966.*

GREEN, J. O.; EVANS, T. A. 1957. Grazing management for seed production in leafy strains of grasses. *J. Br. Grassld Soc.* **12,** 4–9.

GRIFFITH, G. AP; COOPER, J. P. 1959. The normal-acid fibre content of species and varieties of herbage grasses. *J. Br. Grassld Soc.* **14,** 272–6.

HARTLEY, W.; WILLIAMS, R. J. 1956. Centres of distribution of cultivated pasture grasses and their significance for plant introduction. *Proc. 7th int. Grassld Congr., Palmerston North, 1956* pp. 190–201.

HEDDLE, R. G. 1947. Increased production from reseeded grassland. *J. Br. Grassld Soc.* **2,** 83–92.

HENDERSON, J. L.; EDWARDS, R. S.; HAMMERTON, J. L. 1962a. The productivity of five grass species at six levels of compound fertilizer application. 1. Dry matter production. *J. agric. Sci., Camb.* **59,** 5–11.

HENDERSON, J. L.; EDWARDS, R. S.; HAMMERTON, J. L. 1962b. The productivity of five grass species at six levels of compound fertilizer application. 2. Crude protein production. *J. agric. Sci., Camb.* **59,** 199–206.

HILL FARMING RESEARCH ORGANIZATION. 1959. *Rep. Hill Fmg Res. Org. 1954–58.*

HILL FARMING RESEARCH ORGANIZATION. 1962. *Rep. Hill Fmg Res. Org. 1958–61.*

HILL FARMING RESEARCH ORGANIZATION. 1965. *Rep. Hill Fmg Res. Org. 1961–4.*

HUBBARD, C. E. 1954. Grasses. Harmondsworth, Middlesex: Penguin Books Ltd.

HUGHES, R.; NICHOLSON, I. A. 1961a. Comparisons of grass varieties for surface seeding upland pasture types. 1. *Molinia* and *Nardus* pastures. *J. Br. Grassld Soc.* **16,** 210–21.

HUGHES, R.; NICHOLSON, I. A. 1961b. Comparisons of grass varieties for surface seeding upland pasture types. 2. Deep peat. *J. Br. Grassld Soc.* **16,** 314–22.

HUGHES, R.; NICHOLSON, I. A. 1962. Comparisons of grass varieties for surface seeding upland pasture types. 3. *Calluna* heath. 4. General discussion. *J. Br. Grassld Soc.* **17,** 74–84.

HUNT, I. V. 1956a. Comparative productivity of 27 strains of grass in the West of Scotland. Part 1. Total production for 3 years. *J. Br. Grassld Soc.* **11,** 49–55.

HUNT, I. V. 1956b. Comparative productivity of 27 strains of grasses in the West of Scotland. Part 2. Annual production during three years. *J. Br. Grassld Soc.* **11,** 174–81.

HUNT, I. V. 1956c. Comparative productivity of 27 strains of grasses in the West of Scotland. Part 3. First year hay and aftermath. *J. Br. Grassld Soc.* **11,** 207–15.

HUNT, I. V. 1957. Comparative productivity of 27 strains of grass in the West of Scotland. Part 4. Seasonal Growth. *J. Br. Grassld Soc.* **12,** 127–37.

HUNT, I. V. 1964a. Comparative productivity of herbage varieties on upland peat. *J .Br. Grassld Soc.* **19,** 55–61.

HUNT, I. V. 1964b. Seed mixtures for hill farms. *Advisory Leaflet 44 W. Scotl. Agric. Coll.*

HUNTER, R. F. 1958. The direction and problems of research in hill pasture improvement. *J. Br. Grassld Soc.* **13,** 121–5.

IVINS, J. D. 1952. The relative palatability of herbage plants. *J. Br. Grassld Soc.* **7,** 43–54.

IVINS, J. D.; SHAW, R. B. 1950. Pastures for poultry. *Agriculture, Lond.* **57,** 86–90.

JENKIN, T. J. 1931. Swollen stem internodes and other characters in *Arrhenatherum* Beauv. *Bull. H12 Welsh Pl. Breed. Stn* pp. 126–47.

JONES, L. I. 1939. Pasture management and its effect on the sward. *Bull. H15 Welsh Pl. Breed. Stn* pp. 40–129.

JONES, L. I. 1967. Studies on hill land in Wales. *Tech. Bull. 2 Welsh Pl. Breed. Stn* pp. 1–179.

JONES, M. G. 1933a. Grassland management and its influence on the sward. Part 1. Factors influencing the growth of pasture plants. *Emp. J. exp. Agric.* **1,** 43–57.

JONES, M. G. 1933b. Grassland management and its influence on the sward. Part 2. The management of a clovery sward and its effects. *Emp. J. exp. Agric.* **1,** 122–7.

JONES, M. G. 1933c. Grassland management and its influence on the sward. Part 4. The management of poor pastures. Part 5. Edaphic and biotic influences on pastures. *Emp. J. exp. Agric.* **1,** 361–7.

KYDD, D. D. 1966. The effect of intensive sheep stocking over a five year period on the development and production of the sward. *J. Br. Grassld Soc.* **21,** 284–302.

LAIRD, R. 1949. The grassland of the West of Scotland. *J. Br. Grassld Soc.* **4,** 85–94.

LODGE, R. W. 1959. *Cynosurus cristatus.* [Biological Flora of the British Isles.] *J. Ecol.* **47,** 511–18.

LYNN, J. C.; LYNN, D. A. 1924. Observations on the improvement of poor pasture in the West Riding of Yorkshire. *Ann. appl. Biol.* **11,** 135–52.

MILTON, W. E. J. 1933. The palatability of the self-establishing species contributing to different types of grassland. *Emp. J. exp. Agric.* **1,** 347–60.

MILTON, W. E. J. 1934. The relative palatability of seed-mixtures and a study of the influence of fertilisers on natural hill pastures. *Emp. J. exp. Agric.* **2,** 51–64.

MILTON, W. E. J. 1935. The dry-matter content of certain grass and clover species. *Emp. J. exp. Agric.* **3,** 197–202.

MILTON, W. E. J. 1939. An experiment on the yield and persistency of strains of grass and clover species grown in mixtures, and an experiment on the blending of species in simple mixtures. *Bull. H15 Welsh Pl. Breed. Stn* pp. 25–39.

MILTON, W. E. J. 1953. The palatability of herbage on undeveloped grasslands in West-Central Wales. *Emp. J. exp. Agric.* **21,** 116–22.

MOON, F. E. 1939. The carotene-contents of some grass and clover species with a note on pasture weeds. *Emp. J. exp. Agric.* **7,** 235–43.

MOORE, I. 1966. Grass and grasslands. London: Collins, pp. 189.

NEW ZEALAND JOURNAL OF AGRICULTURE. 1963. Improved selections of Yorkshire Fog. *N.Z. Jl Agric.* **105,** 455.

NORTH OF SCOTLAND COLLEGE OF AGRICULTURE. 1954. *Cal. Sch. Agric., Aberdeen 1954–5* p. 166.

PFITZENMEYER, C. D. C. 1962. *Arrhenatherum elatius.* [Biological Flora of the British Isles.] *J. Ecol.* **50,** 235–45.

RATCLIFFE, D. 1965. Aspects of the mineral nutrition of hill pasture grasses. *Rec. Scot. Pl. Breed. Stn* pp. 151–62.

REID, D. 1951. A quantitative method for determining palatability of pasture plants. *J. Br. Grassld Soc.* **6,** 187–95.

ROBINSON, D. H. 1933. The formation of grass turf for poultry. *J. Minist. Agric.* **40,** 510–15.

ROBINSON, D. H. 1936. The establishment and management of poultry turf. *Harper Adams Util. Poult. J.* **21,** 144–6.

SCOTTISH AGRICULTURAL INDUSTRIES FARM SEEDS CATALOGUE. 1967.

SCOTTISH PLANT BREEDING STATION. 1967. 46th Annual report, 1966–67.

SMITH, A. 1965. The assessment of patterns of variation in *Festuca rubra* L. in relation to environmental gradients. *Rec. Scot. Pl. Breed. Stn* pp. 163–95.

SPEDDING, C. R. W. 1965. Sheep production and grazing management. London: Baillière Tindall and Cox.

SPEDDING, C. R. W.; LARGE, R. V.; KYDD, D. D. 1966. The evaluation of herbage species by grazing animals. *Proc. 10th int. Grassld Congr., Helsinki, 1966* pp. 479–83.

STAPLEDON, R. G. 1924. The seasonal productivity of herbage grasses with particular reference to the influence of different systems of cutting on indigenous and non-indigenous strains respectively. *Bull. H3 Welsh Pl. Breed. Stn* pp. 5–84.

STAPLEDON, R. G. 1939. The establishment and maintenance of temporary leys. *Bull. H15 Welsh Pl. Breed. Stn* pp. 130–50.

STAPLEDON, R. G.; DAVIES, W. 1928a. Seeds mixture experiments with special reference to the influence of environmental factors. *Bull. H8 Welsh Pl. Breed Stn* pp. 7–81.

STAPLEDON, R. G.; DAVIES, W. 1928b. Sensible seeds mixtures. *Bull. H8 Welsh Pl. Breed. Stn* pp. 150–62.

STAPLEDON, R. G.; DAVIES, W. 1930. Experiments to test the yield and other properties of various species and strains of herbage plants under different methods of management. *Bull. H10 Welsh Pl. Breed. Stn* pp. 5–41.

STAPLEDON, R. G.; MILTON, W. E. J. 1932. Yield, palatability and other studies on strains of various grass species. *Bull. H13 Welsh Pl. Breed. Stn* pp. 1–79.

THOMAS, B.; THOMPSON, A. 1948. The ash content of some grasses and herbs on the palace leas hay plots at Cockle Park. *Emp. J. exp. Agric.* **16,** 221–30.

THOMAS, B.; THOMPSON, A.; OYENUGA, V. A.; ARMSTRONG, R. H. 1952. The ash constituents of some herbage plants at different stages of maturity. *Emp. J. exp. Agric.* **20,** 10–22.

THOMAS, J. O.; DAVIES, L. J. 1964. Common British grasses and legumes. London: Longmans, Green and Co. Ltd, 4th edn, pp. 122.

THOMAS, M. T. 1936. The introduction and maintenance of nutritious and palatable species and strains. *Bull. H14 Welsh Pl. Breed. Stn* pp. 4–57.

TROUGHTON, A. 1961. The effect on the soil of the components and management of a ley. *J. Br. Grassld Soc.* **16,** 1–5.

WALKER, A. G. 1945. Rhizomatous grass weeds and *Ophiobolus graminis* Sacc. *Ann. appl. Biol.* **32,** 177–8.

WATSON, J. A. S.; MORE, J. A. 1962. Agriculture, the science and practice of farming. Edinburgh and London: Oliver and Boyd, 11th edn, pp. 836.

WEED RESEARCH ORGANISATION. 1965. *Rep. Weed Res. Orgn 1960–64.*

WEED RESEARCH ORGANISATION. 1967. *Rep. Weed Res. Orgn 1965–66.*

WELSH PLANT BREEDING STATION. 1932. Trials with pedigree strains of herbage grasses. *Bull. H13 Welsh Pl. Breed. Stn.*

WELSH PLANT BREEDING STATION. 1958a. *Rep. Welsh Pl. Breed. Stn 1950–56.*

WELSH PLANT BREEDING STATION. 1958b. The Aberystwyth strains of grasses and clovers. *Welsh Pl. Breed. Stn Leaflet Series S 6* pp. 1–18.

WELSH PLANT BREEDING STATION. 1959. *Rep. Welsh Pl. Breed. Stn 1956–58.*

WELSH PLANT BREEDING STATION. 1960. *Rep. Welsh Pl. Breed. Stn 1959.*

WELSH PLANT BREEDING STATION. 1962. *Rep. Welsh Pl. Breed. Stn 1961.*

WELSH PLANT BREEDING STATION. 1963. *Rep. Welsh Pl. Breed. Stn 1962.*

WELSH PLANT BREEDING STATION. 1964. *Rep. Welsh Pl. Breed. Stn 1963.*

WELSH PLANT BREEDING STATION. 1965. *Rep. Welsh Pl. Breed. Stn 1964.*

WELSH PLANT BREEDING STATION. 1966. *Rep. Welsh Pl. Breed. Stn 1965.*
WELSH PLANT BREEDING STATION. 1967. *Rep. Welsh Pl. Breed. Stn 1966.*
WEST OF SCOTLAND AGRICULTURAL COLLEGE. 1957. *Rep. W. Scotl. agric. Coll.*
WEST OF SCOTLAND AGRICULTURAL COLLEGE. 1958. *Rep. W. Scotl. agric. Coll.*
WEST OF SCOTLAND AGRICULTURAL COLLEGE. 1960. *Rep. W. Scotl. agric. Coll.*
WEST OF SCOTLAND AGRICULTURAL COLLEGE. 1964. *Rep. W. Scotl. agric. Coll.*
WHYTE, R. O.; MOIR, T. R. G.; COOPER, J. P. 1959. Grasses in agriculture. *FAO Agric. Stud. No. 42* Rome: FAO, pp. 416.
WOODFORD, E. K.; EVANS, S. A. 1965. Weed Control Handbook. Oxford: Blackwell Scientific Publications, 4th edn, pp. 445.

PART III

THE LEGUMES

CHAPTER 24

Origins of Cultivated Herbage Legumes

HERBAGE legumes were derived primarily from the eastern Mediterranean and adjoining countries, whence they have spread gradually over a long period of time into other temperate parts of the world, initially by natural means and more recently by deliberate introduction. One of the earliest references, *c.* 480 B.C., to the use of a legume as fodder mentions the use of lucerne by the Medes and Persians as feed for war horses. The beneficial effects of legumes on both the soil and other crops have also been appreciated for a long time, but it was not until the 1880's that it was realized that this was due to fixation of atmospheric N by associated root nodule bacteria.

The spread of legumes from their centres of origin was followed by their differentiation and adaptation to a range of new environments. This process is more complex in herbage legumes than in herbage grasses, since success with the former depends on co-operation with two unrelated organisms—bees for pollinating self-sterile flowers, and bacteria for N fixation. Adaptation of the plant must therefore move in step with that of these two organisms.

Wild white clover became adapted in Britain as a small-leaved, low-growing perennial, contrasting with some of the populations from nearer the Mediterranean region which have very large leaves and are not strongly perennial. However, during the 18th and 19th centuries, practically all white clover seed sown in Britain was imported from the Netherlands and neighbouring countries. This so-called Dutch white clover had leaves of intermediate size and was demonstrated by Gilchrist (1909) to be agronomically inferior to the indigenous white clover, since it disappeared from pastures within two or three years from sowing, whereas the latter type persisted almost indefinitely. Improvement of white clover in Britain consisted in maintaining and propagating the best local ecotypes, such as Kent, and in selecting from them improved cultivars, such as Aberystwyth S184. At the same time, larger-leaved perennial types, e.g. Aberystwyth S100, were selected from plants collected over a wide range of geographic areas but with similar environments. In new cultivars currently being produced, the winter-hardiness of established

British cultivars is being combined with the larger leaf size and late-season productivity of Mediterranean types.

Red clover also became adapted in Britain as a small-leaved type, which because of its thin wiry stems has played only a small part in breeding work. Most of the improvement in Britain has been obtained from derivatives of the large-leaved cultivated red clover imported from the Netherlands after the middle of the 17th century. Three main types have evolved through inter-crossing and adaptation in response to specific farming systems. In the traditional 4-course arable areas, broad red clover (double cut) met the need for one-year leys from which the first cut was taken for hay and the aftermath for seed. The late-flowering prostrate type (e.g. Montgomery) evolved in areas where the farming system required leys of longer duration and which were sometimes used for grazing as well as for conservation; here, seed was harvested after 3 or 4 years. Finally, single-cut cultivars, which are intermediate, constitute the third type.

There are detailed records of the spread of lucerne from the southeast Caucasus–Caspian Sea area, across Europe to reach England in 1630. This species differs from clovers in that no distinctive locally adapted cultivars have arisen in Britain, probably because of its failure to evolve an efficient pollination mechanism in the colder British climate. The best-adapted type of lucerne (Flamande) for Britain has originated from an area in northern France where the climate is similar. The cultivar du Puits is probably the best known and has proved very satisfactory, in marked contrast to Provence lucerne from the Mediterranean area.

The continental origin and history of sainfoin appears to be similar to that of lucerne, although its cultivation was largely restricted to limestone soils. There has been no problem with seed production, and two distinct types are present in Britain: giant sainfoin, which is seldom left down for more than a year, and common sainfoin which is more persistent and can last for 4–5 years.

Reference

GILCHRIST, D. A. 1909. Trials of wild white clover. *J. Bd Agric.* **16,** 713–18.

CHAPTER 25

Morphology

THE external form of a plant is to some degree determined by its environment. Classical systems of taxonomy imply a standard basic morphology for any one group of plants. The degree of relationship in any such group depends on whether a species, genus, family or order is being considered. Within any one of these groups, there is a range of variation in the characteristics determining that grouping. It is the 'norm' of this range for the family Papilionaceae of the order Leguminosae that will be considered here. Variations from the 'norm', which imply different types of agricultural use of these plants, will be discussed later in the chapters on individual species and in relation to the reaction of the plant to variations in the environment.

It is possible to distinguish a number of stages in the annual growth cycle from the seedling to the flowering plant; for the purpose of this chapter, it is convenient to consider three such stages. Although these stages may be only tenuously related to the developmental morphology of the plant, they serve as useful growth phases for descriptive morphology.

Seed and seedling

Seeds of legumes are usually round or kidney-shaped with a smooth testa; the hilum is normally prominent (Robinson, 1947). Both colour and size of seed vary between genera and species but can be used to identify varieties within a species (Hector, 1936). In the seed of common forage legumes there is no endosperm outside the embryo. The seed consists of two fleshy cotyledons containing starch and protein and enclosing a rudimentary plumule. The radicle is frequently distinguishable as a 'hump' in the seed. In most species, a proportion of the seed have testas which are impermeable to water, the so-called 'hard seeds' (Robinson, 1947). It is possible to scarify the seeds mechanically or to rupture the testa by chemical means and thus initiate imbibition of water and germination.

Germination in the fodder legumes is usually epigeal (Hector, 1936), the cotyledons forming the first photosynthetic leaves. The part of the plant between the cotyledons and the radicle, the hypocotyl, can vary in length depending on the environment at germination and the depth at which the seed is sown. The first foliage leaf is simple, round, and has a long petiole. The second and subsequent leaves are usually of the adult form, but in some species (e.g. sainfoin) the 'juvenile' form, or some modification of it, may persist to the third leaf (Hector, 1936). Lateral roots appear on the radicle at an early stage and the plant is then able to live independently without further recourse to supplies in the cotyledons (Thomas & Davies, 1964).

The vegetative plant

LEAVES

The leaves of legumes are arranged alternately on the stem. Apart from the first, and occasionally the second and third, the leaves are compound, consisting usually of three leaflets; bifoliate and multipinnate forms also occur in some species. Leaflet shape varies greatly between genera, from ovate or almost circular in white clover (Erith, 1924) to elliptical in sainfoin (Robinson, 1947). The margins of leaves may be entire or conspicuously toothed. In the clovers, distinctive pale 'horse-shoe', 'V-shaped' or 'band' markings occur. These marks are unique for any one species (Erith, 1924). The terminal leaflet may be represented by a tendril in some species. At the base of each leaf petiole is a pair of leaf-like appendages, the stipules. The colour, shape and the presence or absence of hairs on the stipules are frequently used as diagnostic features of a variety or a species (Robinson, 1947).

ROOT SYSTEM

All legumes typically have a well developed tap-root. Lateral branches develop soon after germination and produce numerous secondary branches which again branch profusely. The depth of the tap-root system varies greatly between species but, although roots have been recorded to a depth of more than 30 ft (9 m) in lucerne, the greater part of the root system is in the top 1-ft (30-cm) layer of soil (Bolton, 1962). In some legume species the depth of rooting determines to a great extent their resistance to drought. A secondary root system develops in those legumes which form stolons (e.g. white clover). Adventitious roots arise at the distal nodes of a stolon when moisture conditions in the soil or in the microclimate around the plant are suitable (Sanderson, 1966). These adventitious roots grow downwards for up to 6 in. (15 cm) before forming second- and third-order lateral branches; the structural similarity to the tap-root system is thus maintained.

The feature that has made legumes so valuable agriculturally is the presence of nodules on their roots. This symbiotic relationship of the legume with *Rhizobium* (see Chapters 26 and 27) makes the plant independent of a mineral-N supply from the soil. When parts of a legume root or shoot decay, or when the whole plant is ploughed in, the N-enriched residues improve the N status of the soil and benefit concurrently growing or subsequent crops.

GROWTH HABIT

The growth form of any species of forage legume determines whether it is used primarily for grazing or cutting. There are essentially two growth habits, creeping and erect. However, some types or varieties of a species which is typically erect may exhibit a semi-prostrate growth habit (Bolton, 1962).

The creeping growth habit is exemplified by white clover (Erith, 1924; Sanderson, 1966). A short stem bears alternately arranged leaves. The buds in the axils of these leaves, apart from two or three near the apex, develop to form stolons which, because of the short internodes on the main stem, radiate in all directions in a horizontal plane. The stolons bear leaves in the axils of which the buds form either inflorescences or secondary stolons. Because all the meristematic activity of the shoot system is at or near ground level, legumes with this growth habit are suited especially to grazing.

The other extreme of growth form is represented by lucerne. In its first year of growth, this erect perennial has its meristematic apex well above ground level. The first stem arises from the plumule between the two cotyledons and subsequent stems develop in the axils of the cotyledons or of the lower leaves. As these stems grow, they become woody at the base and, at ground level, a crown of tissue is formed from which buds develop to give successive crops, usually 3 or 4 per season (Bolton, 1962). Grazing in the first year of growth is likely to be detrimental to a stand of plants of this type, but later growth can be grazed without damage to the meristematic region in the crown.

The reproductive plant

The floral structure of forage legumes and, in particular, the structure of the pistil, is the definitive feature of the order Leguminosae (Hector, 1936). The inflorescence is, however, of less importance as a determinant of yield than in the grasses. Leguminous flowers are distinctive, brightly coloured and adapted specifically for insect pollination. They are typically arranged in a raceme which may be terminal, as in red clover (Thomas & Davies, 1964), or more usually axillary, as in white clover (Erith, 1924; Sanderson, 1966). The typical flower is zygomorphic and consists of 5 fused sepals, 5 petals, 10 stamens and one carpel.

CALYX

The fused sepals form a tube at the base of the flower, the 5 teeth indicating its pentamerous structure. In some species the sepals may be arranged as two groups—one of three and one of two.

COROLLA

The conspicuous corolla is composed of 5 petals—the posterior one (the standard) being the largest; the two lateral petals (wings) are free but frequently 'keyed' to the keel so that movement of the wing petals is transferred to the keel. The two anterior petals are fused to form a keel, inside which are the androecium and gynaeceum.

ANDROECIUM

The filaments of the 10 stamens are either fused for most of their length to form a staminal column or tube (monadelphous forms), or this fusion includes only 9 of the staminal filaments leaving one free (diadelphous forms). When nectaries are present, they are situated at the base of the stamens.

GYNAECEUM

The single carpel consists of a superior ovary with a single loculus containing one or more ovules, a style which varies in length with species and ends in a stigma which projects well beyond the anthers. The stigma may be covered with a protective membrane which is ruptured when the flower is 'tripped' by pressure on the wing petals. The ovules are borne alternately on each side of the central suture.

FRUIT

The fruit is typically a pod or legume, varying in shape and size according to species and usually dehiscent, splitting along both dorsal and ventral sutures, or by a transverse split (pyxidium) as in red clover. Sainfoin produces an indehiscent, single-seeded fruit and lucerne has indehiscent, several-seeded pods.

References

BOLTON, J. L. 1962. Alfalfa. London: Leonard Hill (Books) Ltd, pp. 474.

ERITH, A. G. 1924. White clover: a monograph. London: Duckworth and Co., pp. 150.

HECTOR, J. M. 1936. Introduction to the botany of field crops. Vol. II. Non-cereals. Johannesburg: Central News Agency Ltd, pp. 479–1127.

ROBINSON, D. H. 1947. Leguminous forage plants. London: Edward Arnold & Co., 2nd edn, pp. 119.

SANDERSON, J. F. 1966. Aspects of the growth of white clover with particular reference to the effects of defoliation and shade. Ph.D. Thesis, Univ. Reading, pp. 117.

THOMAS, J. O.; DAVIES, J. L. 1964. Common British grasses and legumes. London: Longmans, 4th edn pp. 122.

CHAPTER 26

Principles of Establishment

Germination

THE environmental conditions which favour the germination of legumes are similar to those described for grasses (Chapter 8). Compared with that of grasses, the germination of clovers is less affected by low temperature, similar numbers of seed germinating from summer and winter sowings (Stapledon & Wheeler, 1948).

The testa of a high proportion of seeds of legume species is covered by a cuticle which is impermeable to water. During threshing and hulling, the testa of most seeds sustains damage so that the seed can imbibe water and germination follows. Commercial legume seed contains up to 15% hard seeds (those in which the testa is intact), the proportion being greatest in the smaller-seeded species and varieties. In lucerne the majority of hard seeds imbibe water within one year from sowing, but only about one-third of hard clover seed will germinate within this period (Mercer, 1948). The chance of a seedling from a late-germinating seed surviving in an established grass/clover sward is low, so that the value of hard seeds is limited in practice.

Symbiosis with nodule bacteria

Leguminous plants form a symbiotic association with bacteria of the genus *Rhizobium*. These bacteria, which are able to fix N from the atmosphere, supply their host legume with nitrogenous compounds, the legume providing the bacteria with carbohydrates. Legumes can therefore be independent of the soil for their N supply. Successful symbiosis is, however, only possible with species and strains of *Rhizobium* appropriate to the legume species sown.

The natural geographic distribution of *Rhizobium* spp. usually corresponds with that of their legume hosts. While these bacteria are able to live independently as normal components of the soil microflora, in the absence of a legume host their number in the soil declines at a roughly exponential rate which is accelerated by desiccation, high temperature and soil acidity (Nutman, 1965). The introduction of a legume strongly stimulates the bacteria and they rapidly increase in number. Less than 100 bacteria in the rhizosphere are sufficient to initiate infection in a legume plant, but once the first nodule is formed, progressively more bacteria are required for the formation of each succeeding nodule (Nutman, 1965).

It is usually assumed that British soils contain the appropriate strains of *Rhizobium* for the legumes commonly sown, with the exception of lucerne. To ensure effective nodulation in lucerne, artificial inoculation is usually necessary. This can be conveniently done by dressing the seed with a pure culture of the appropriate species of *Rhizobium* in a skim-milk medium. The treated seed should be protected from direct sunlight and sown as soon as possible after inoculation. The bacteria introduced on the seed are able to migrate along the root as rapidly as the root grows (Nutman, 1965), and the first infection occurs 3–4 days after germination.

Emergence

Legume seeds have no endosperm, the food reserves being stored within the cotyledons. In the forage legumes commonly sown in Britain, the cotyledons and the plumular bud are carried to the soil surface by the elongation of the hypocotyl. After emergence, the cotyledons soon unfold and become the initial photosynthetic organs of the seedling. The larger-seeded species are able to emerge from a greater depth than the smaller-seeded ones. Similarly, within a species, seed size determines the depth from which emergence can take place.

In the annual subterranean clover it has been shown (Black, 1956) that seed weight also determines the initial area of the cotyledons, and that seeds of similar weight, when sown at various depths, produce seedlings with cotyledons of different weight but of similar area. Since seedling growth depends directly on cotyledonary area, sowing depth has little effect on the subsequent growth of emerged seedlings. Similar data on other legumes are not available, but it is reasonable to postulate that in species having epigeal cotyledons, growth is likely to be proportional to seed weight during the early post-emergence period, regardless of sowing depth. Removal of varying amounts of cotyledon has been shown to reduce greatly the subsequent growth of lucerne seedlings, an effect which persisted during the 5 months that the plants were under observation (Bignoli, 1950). In a competitive situation, however, seedlings from deep-sown seed will be at a disadvantage, since emergence is delayed as sowing depth increases (Moore, 1943).

In a favourable environment, the interval from sowing to seedling emergence in legumes is 5–7 days, a slightly shorter period than that for the associated grasses with which they are usually sown. Under unfavourable soil conditions, legume emergence is delayed to a lesser extent than that of the grasses (Stapledon & Wheeler, 1948). Though germination and emergence are little affected by low temperature, legume seedlings from autumn-sown seed seldom survive until the spring. Mortality is mainly due to attack by *Pythium* and other fungi which cause a sudden wilting (damping-off) (Martin, 1949).

Viable seed can be invaded by *Pythium* spp. before the radicle breaks through the testa, or the elongated hypocotyl can be invaded just before the seedling emerges from the soil, or near the soil surface after emergence (Jackobs, 1947). Further seedling development is prevented in each case. Invasion of the seed by this fungus before radicle emergence is commonest in seeds which are immature or have a damaged testa, whereas seedling invasion is more dependent on environmental factors, as it increases under high temperature and humidity. Mortality from invasion by *Pythium* spp. is usually high following low-temperature damage to seedling tissue (Martin, 1949). Of the legumes, red clover is the most susceptible to winter injury, and lucerne is more resistant than white clover (Martin, 1949). *Pythium* spp. have proved to be more pathogenic to lucerne on soil with high pH than to lucerne on soil with low pH (Grandfield *et al.*, 1935).

Growth and development

Root development in legume seedlings is rapid. In white clover, three longitudinal rows of lateral roots develop before the plumule appears from between the cotyledonary petioles (Erith, 1924). In all legumes, repeated branching of the primary root gives rise to a tap-root system.

The plumule develops relatively slowly. Leaves are produced from crowded nodes on the main stem. The first leaf to appear is unifoliate in all species, with subsequent leaves being trifoliate or pinnate according to species. In clovers, leaves appear at a rate similar to that in grasses. Successive leaves are larger and petioles longer, but stem formation in clovers proceeds much more slowly than does tillering in grasses. In white clover, stem elongation does not commence until 7 leaves have appeared on the main stem, which was 9 weeks after seedling emergence at 12°C in trials reported by Mitchell (1956).

Internode elongation in white clover results in the formation of stolons, while in red clover, lucerne and sainfoin, semi-erect or erect stems are formed.

As a group, therefore, legumes germinate and emerge from the soil in a shorter period than the grasses, but their early post-emergence development is slower.

References

BIGNOLI, D. P. 1950. The effect of early defoliation on lucerne seedlings. *J. Br. Grassld Soc.* **5**, 281–6.

BLACK, J. N. 1956. The influence of seed size and depth of sowing on pre-emergence and early vegetative growth of subterranean clover. *Aust. J. agric. Res.* **7**, 98–109.

ERITH, A. G. 1924. White clover: a monograph. London: Duckworth and Co., pp. 150.

GRANDFIELD, C. O.; LEFEBVRE, C. L.; METZGER, W. H. 1935. Relation between fallowing and the damping-off of alfalfa seedlings. *J. Am. Soc. Agron.* **27,** 800–6.

JACKOBS, J. A. 1947. Factors affecting seed rotting caused by *Pythium* spp. in sweetclover with preliminary tests in alfalfa and red clover. *J. Am. Soc. Agron.* **39,** 702–18.

MARTIN, B. F. 1949. The establishment of autumn sown legumes. *J. Br. Grassld Soc.* **4,** 47–55.

MERCER, S. P. 1948. Farm and garden seeds. London: Crosby Lockwood and Son Ltd.

MITCHELL, K. J. 1956. Growth of pasture species. 3. White clover, subterranean clover and *Lotus major*. *N.Z. Jl Sci. Technol. Sect. A* **37,** 395–413.

MOORE, R. P. 1943. Seedling emergence of small-seeded legumes and grasses. *J. Am. Soc. Agron.* **35,** 370–81.

NUTMAN, P. S. 1965. The relation between nodule bacteria and the legume host in the rhizosphere and in the process of infection. In: Baker, K. F.; Snyder, W. C. [Eds] Ecology of soil-borne pathogens. London: John Murray, pp. 231–47.

STAPLEDON, R. G.; WHEELER, D. E. 1948. An experiment to test the influence of date and depth of sowing on four of the chief grasses and on white clover. *J. Br. Grassld Soc.* **3,** 263–72.

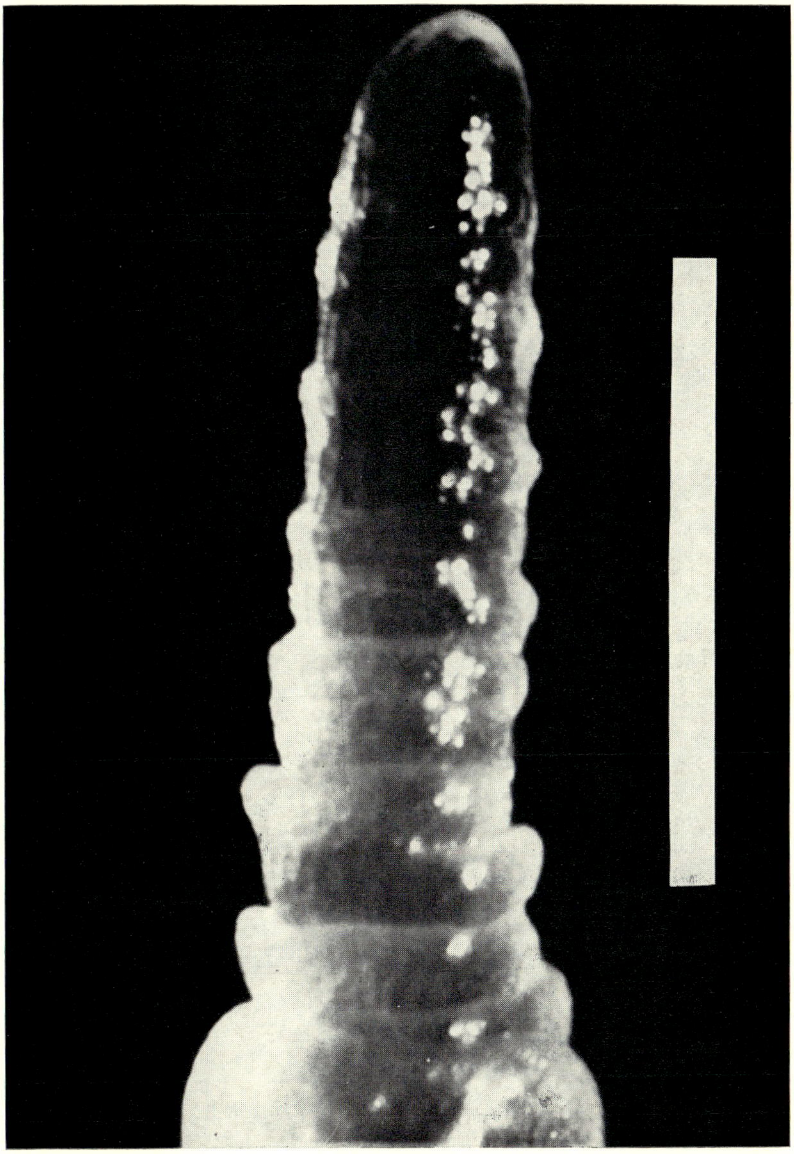

Plate 1. S24 *Lolium perenne*: shoot apex at transition to reproduction (early double-ridge stage). Scale = 1 mm.

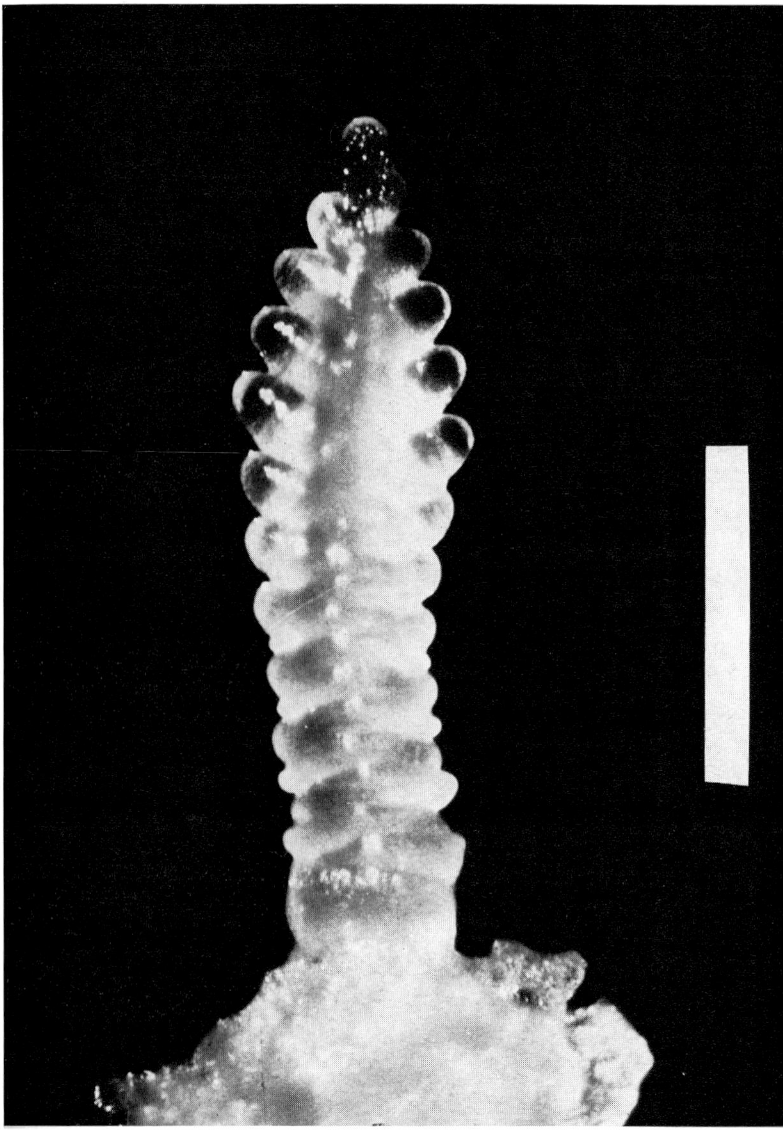

Plate 2. S24 *Lolium perenne*: reproductive shoot apex showing double ridges near the base and inflorescence branch primordia at the tip. Scale = 1mm.

Plate 3. S24 *Lolium perenne*: reproductive shoot apex showing fifteen young inflorescence branches, each with developing floret primordia. Scale = 1mm.

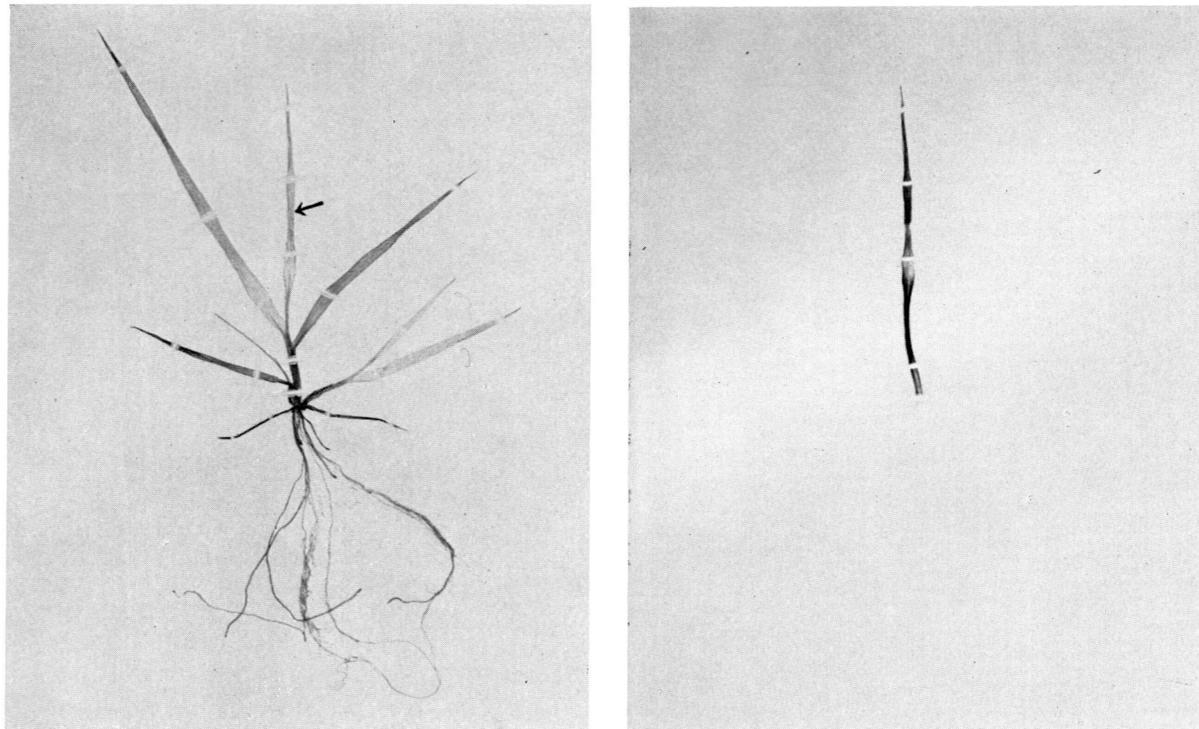

Plate 4. Distribution of labelled assimilates in timothy 24 h after feeding an expanding leaf (arrowed). Darkened areas in right-hand illustration indicate radioactivity and show that there was no movement of assimilate out of the treated leaf. (Williams, 1964)

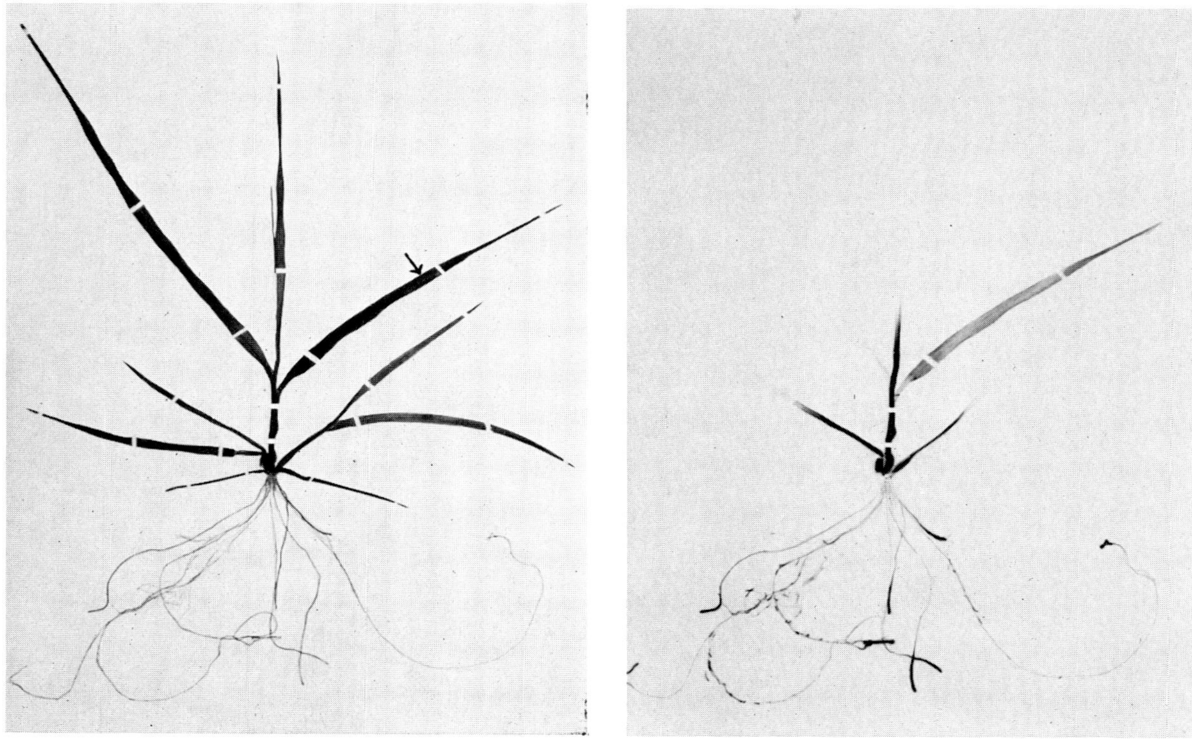

Plate 5. Distribution of labelled assimilates in timothy 24 h after feeding the most recently fully expanded leaf (arrowed). Assimilates moved to the growing leaf, tillers and roots. (Williams, 1964)

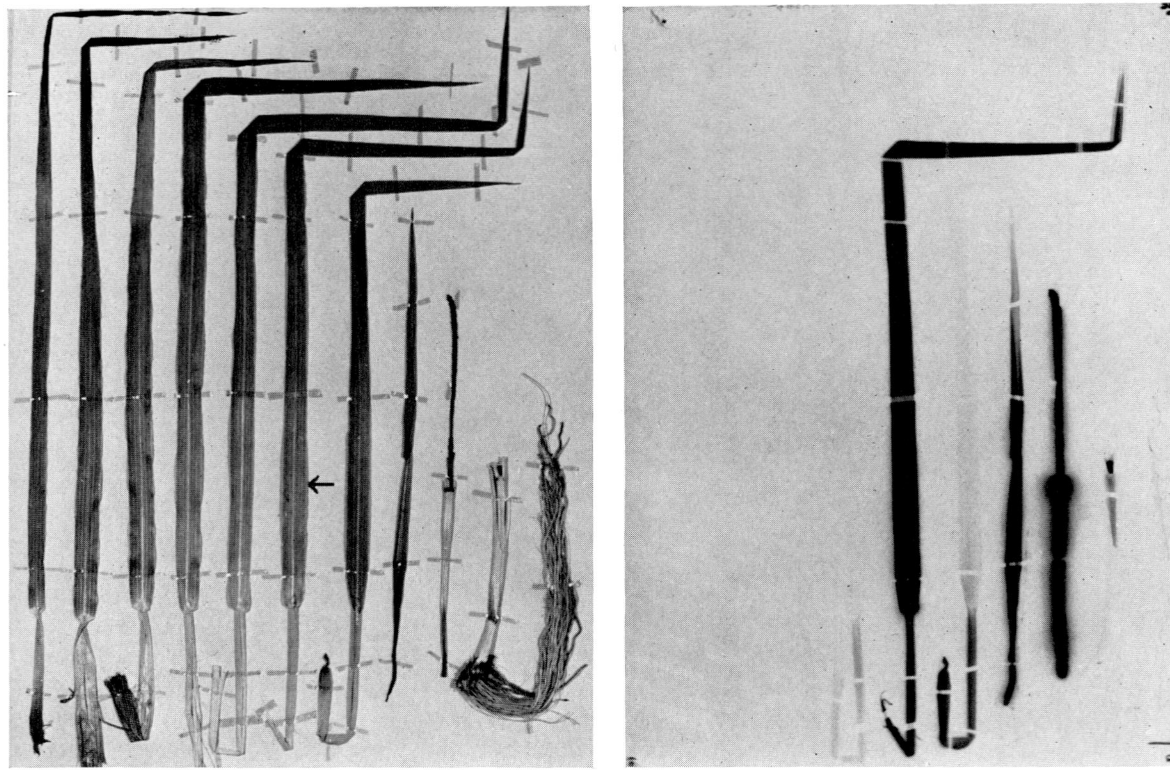

Plate 6. Distribution of labelled assimilates in a reproductive shoot of timothy 24 h after feeding labelled CO_2 to the most recently expanded leaf (arrowed). Darkened areas in right-hand illustration indicate radioactivity and show movement of assimilate to the younger leaves and stem apex. The oldest leaf is that furthest to the left. (Williams, Unpublished)

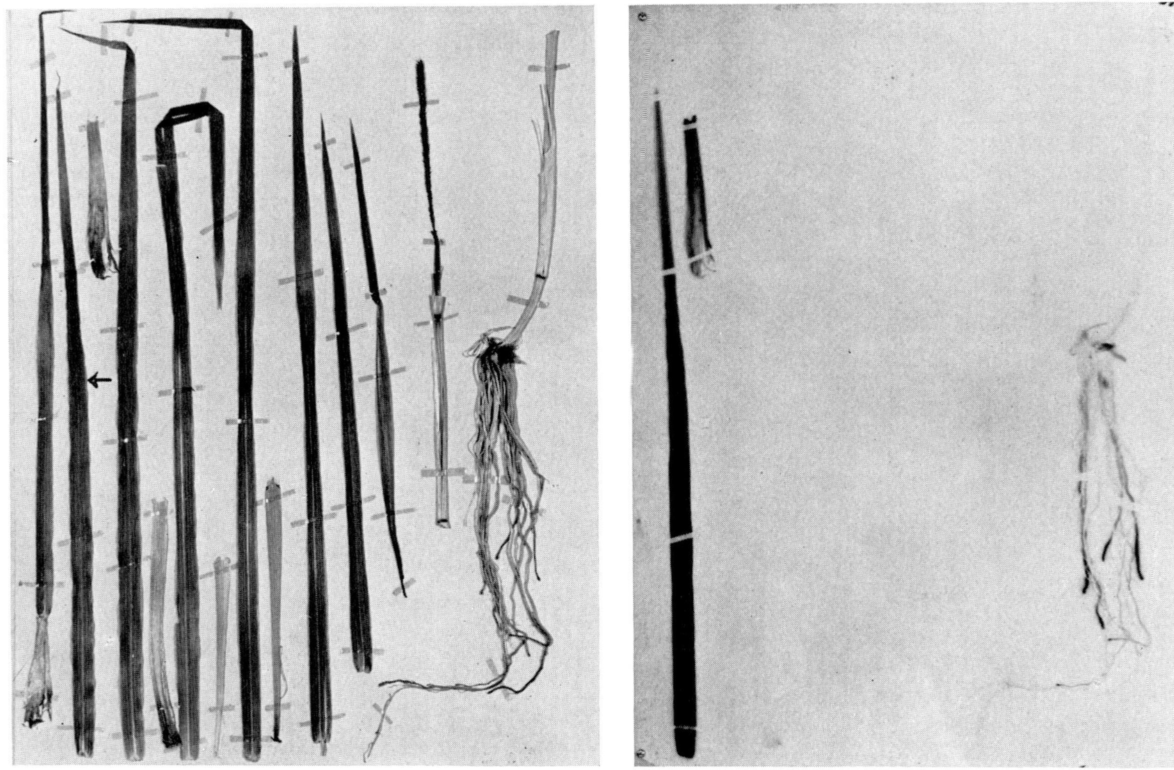

Plate 7. Distribution of labelled assimilates in a reproductive shoot of timothy 24 h after feeding labelled CO_2 to the oldest green leaf (arrowed). Movement of assimilate is principally to the roots. (Williams, Unpublished)

9 May 23 May 7 June

Plate 8. Perennial ryegrass

9 May

23 May

7 June

Plate 9. Italian ryegrass

Plate 10. Timothy

Plate 11. Cocksfoot

9 May

23 May

7 June

Plate 12. Meadow fescue

9 May

23 May

7 June

Plate 13. Tall fescue

White clover

Red clover
Plate 14

Lucerne

Sainfoin
Plate 15

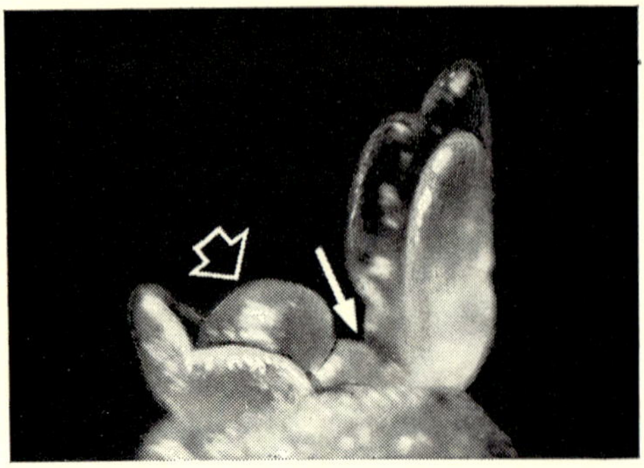

A

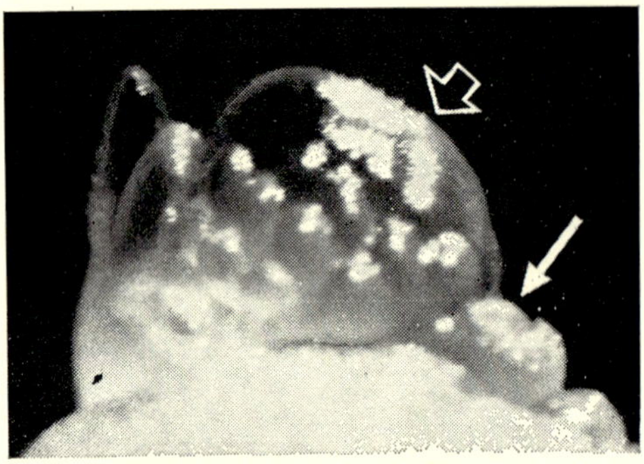

B

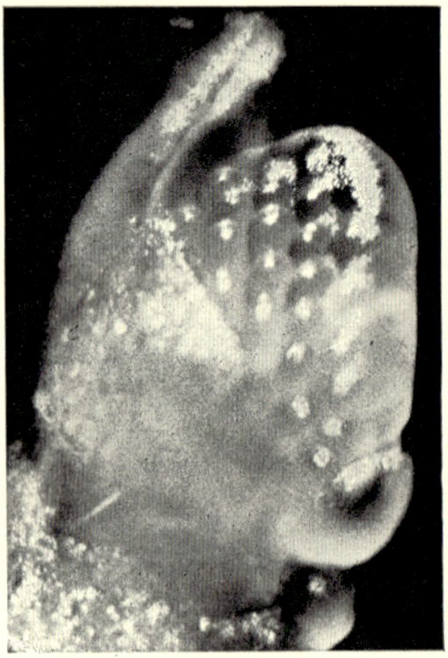

C

Plate 16

A. Red clover inflorescence primordium (below broad arrow) in the axil of the youngest leaf primordium. The stipule in front of the dome (below white arrow) has been removed.

B. Red clover inflorescence bulge (below broad arrow) with the first appearance of floret primordia. Note the large size of the inflorescence compared with the dome (below white arrow).

C. Red clover inflorescence bulge after absorption of the dome with most of the floret primordia formed. Figures from Aitken (1964), reproduced by permission of CSIRO Melbourne, Australia.

CHAPTER 27

Vegetative Growth

THOUGH herbage legumes are subject to the same environmental factors as the grasses, there are striking differences between the two groups of plants in their response to these factors. The most important difference arises from the ability of legumes to make use of atmospheric N, in contrast to the grasses, which are completely dependent on N provided by the soil or by fertilizer. Herbage legumes do not generally require fertilizer N, but need adequate P, K and Ca (see Chapter 32). In general, they prefer moderately acid or neutral soils, are intolerant of shade, and are at their best under conditions of high light intensity.

Nitrogen fixation

N fixation can only take place when there is effective symbiosis with root-nodule bacteria of the genus *Rhizobium*. A *Rhizobium* strain may produce nodules on one group of related legumes, but not usually on legumes of other groups. This has led to the designation of cross-inoculation groups and is the basis of a widely accepted species classification. Thus, isolates of *Rhizobium trifolii* from white clover will induce nodulation on most other *Trifolium* spp. but not, for example, on lucerne, which requires *R. meliloti* for symbiosis. Strain-specific groups, e.g. *Lotus* species, are restricted in the range of organisms able to nodulate them, and these organisms are sometimes lumped together under the term 'soybean-cowpea miscellany' (Kleczkowska *et al.*, 1968). It has become clear that infection specificity is not an adequate basis for classifying *Rhizobium* in general, although it is reasonably satisfactory in the case of bacteria of the legumes native to temperate regions.

Within given cross-inoculation groups, strains of bacteria vary in their symbiotic capacity and can be classified as completely effective, intermediate or ineffective. Nodules produced by the last type are generally small, very numerous, whitish, and the host plant remains small, yellow and stunted in the absence of available soil N. Fully effective strains produce nodules which are large, relatively few in number, and reddish in colour, while the host plant has dark green foliage and grows vigorously.

Numerous surveys in Britain have shown that the proportion of ineffective strains is greater in poor, acid soils at high altitudes (Jones & Davies, 1966). Although it is not normally considered necessary to inoculate clover seeds, good responses to inoculation have been obtained in white clover on acid peat soils. Lucerne seed, on the other hand, is almost always inoculated, particularly when it is to be sown on areas where lucerne has not been grown before.

The amount of N fixed by legumes can be considerable, as shown by the figures given in Table 27.1. The data for species are not comparable because of the presence or absence of a companion grass and differences in utilization practice. Lucerne maintained a high N yield over 4 years, but N yields from red clover declined as the stand thinned out. Comparable pure stands of white clover have yielded 280–300 kg N/ha per year (Unpublished data).

Table 27.1. Nitrogen yield* from lucerne and clover (kg/ha)

Harvest year	Lucerne alone	Red clover + grass	White clover + S24 ryegrass
1	290	249	120
2	331	175	168
3	319	103	195
4	243	109	221
Average	296	159	176

* This includes the annual soil contribution of N amounting to 22–34 kg/ha per year

From Davies (1964)

Types of vegetative growth

Two main types can be distinguished: (i) the white clover type, and (ii) the red clover or lucerne type.

WHITE CLOVER TYPE

In white clover the primary stem is short (25–75 mm) and its growth is limited. The lower axillary buds on this stem develop into stolons which spread radially over the soil surface. Leaves are borne at the nodes and flowers or branch stolons may arise from the leaf axils. The primary root is a tap-root, but its activity soon diminishes, and it has usually died within 18 months (Tesar & Ahlgren, 1950). Adventitious roots are produced at the nodes, and these enable the new shoots to become independent units. Most of these roots are confined to the top 6-in. (15-cm) layer of soil.

In New Zealand, the duration of individual leaves from bud to senescence is 40 days under cool, moist conditions, but is much shorter under conditions of high temperature and drought. New leaves are formed throughout the

growing season and a maximum leaf area index (LAI) of 5·5 is reached 20 days after defoliation (Brougham, 1958). The leaf area is made up of: bud leaves, 0·5 units; fully expanded synthesizing leaves, 3–3·5 units; axillary leaves, 0·5 units; leaves in various stages of senescence, 1–1·5 units.

In New Zealand (40° S. lat.) at about noon in midsummer, the LAI at which 95% of the incident light is intercepted is 3·5 (Brougham, 1958).

The type of growth found in white clover is adapted to withstand frequent defoliation and the indeterminate growth of the branching stolons enables the plant to spread and 'fill' the sward.

RED CLOVER TYPE

In red clover or lucerne, the plant possesses a well-marked crown and does not spread. Hence, when the plant dies it leaves an empty space, and although the surrounding plants can compensate for this by increased aerial growth, there is a limit to this process and the stand eventually thins out. In spring, growth takes the form of a lengthening of the overwintered shoots or buds in preparation for flowering. Active internodes increase in length, new leaves are produced and lower leaves senesce. In lucerne the proportion of leaf to stem decreases with age from 70% in April to 35% in late June (Terry & Tilley, 1964). The lucerne shoot is indeterminate, with the racemes borne in the leaf axils. In red clover, the production of internodes continues until terminated by an inflorescence at the apex. Later-flowering types have 6–13 internodes, compared with 5–9 in the earlier-flowering broad red clover (Williams, 1927).

When red clover and lucerne crops are cut, regrowth occurs from the crown, from new buds, or from buds which have already begun to elongate. This regrowth takes the same course as the spring growth, until daylength becomes insufficient and the plant goes into the rosette condition in preparation for overwintering.

Both red clover and lucerne have a well-developed tap-root system which may penetrate the soil to a great depth. The root serves also as a storage organ into which reserves are accumulated to enable the plant to overwinter. Cutting must therefore be timed so as to allow a build-up of reserve carbohydrates before the onset of winter (Smith, 1964).

Factors affecting growth

These can be conveniently divided into (a) those that can be easily manipulated, and (b) those that cannot be altered in farm practice.

CONTROLLABLE FACTORS

Fertilizer application, and to some extent irrigation, fall into the first group. A considerable amount of empirical knowledge on the effect of various mineral elements is available, but data are insufficient to enable response curves to be constructed (Cooke, 1960). In addition, the method of crop utilization, the different amounts of nutrients returned under cutting or grazing systems, and the limitations of soil-analysis data complicate the issue.

Herbage legumes vary in their sensitivity to soil pH as shown in Table 27.2. Wild white clover tolerates more acid conditions than the other legumes, but below the quoted values growth of all legumes is increasingly poor. While pH may have no effect in itself, soils with extreme pH values have other limitations due to excess or deficiency of various elements, e.g. low pH may be associated with Ca deficiency and excess of soluble Mn and Al (Russell, 1966).

Table 27.2. Optimum range of soil pH values for herbage legumes

White clover*	5·6–7·0
White clover (Ladino)	5·6–7·0
Red clover*	6·0–7·5
Alsike clover	5·5–7·5
Lucerne	6·2–7·8

The plants listed grow fairly well at one half pH unit or more, and those marked * at one pH unit below the optimum.

After Ignatieff and Page (1958)

There are a few experiments which have shown the importance of K fertilization for lucerne (Shotton & James, 1960) and red clover (I. G. Williams, Personal communication). Jewiss (1966) observed that P promoted the growth of white clover, but found no yield increment in pot experiments with increased rates of the other major elements K and Mg.

A useful indication of the amounts of various elements needed can be obtained from the amounts removed by a high yielding crop (Table 27.3).

Table 27.3. Weight of nutrients removed in crops of clover and lucerne (kg/ha)

	Yield (metric tons/ha)	P	K	Ca kg/ha	Mg	S
Lucerne	6·7[1]	39	151	136	17	18
	11·2[2]	27	309	101	19	...
Red clover	4·5[1]	22	78	68	17	7
	11·2[2]	20	288	102	21	...
White clover	6·7[2]	19	201	78	12	...

[1] After Ignatieff and Page (1958)
[2] After Davies, ap Griffith and Ellington (1966 and Unpublished)
Based on notional yields

The data refer to the top growth only and, because a proportion of the elements are retained in the roots, quantities in excess of those removed in the harvested crop must be supplied. The value of such calculations depends in part on the efficiency of uptake, which may be as low as 5–20% for P.

UNCONTROLLABLE FACTORS

Little can be done to change temperature or light conditions, but adequate understanding of their effects is necessary for the optimum exploitation of legumes as crops. These factors affect the sowing date, season of growth, and time of harvesting or grazing.

In temperature-response curves constructed by Mitchell (1956) white clover, large birdsfoot trefoil (*Lotus uliginosus* = *L. major*), subterranean clover and perennial ryegrass showed maximum growth rates at 23·9°, 24·4°, 18·9° and 20°C, respectively (Fig. 27.1).

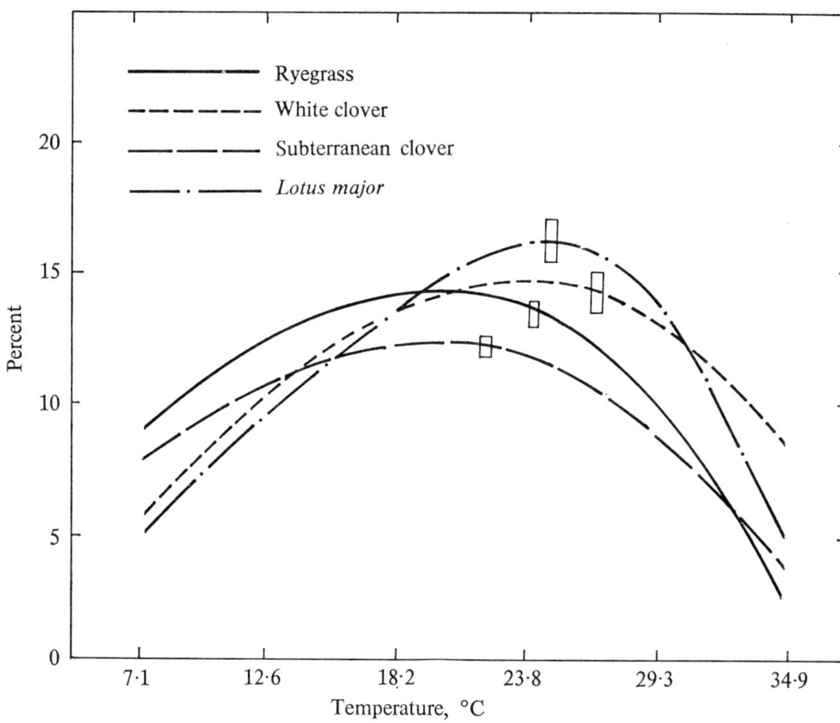

Fig. 27.1. Daily increase in dry weight of the total shoot. (From Mitchell, 1956)

The work of Nielsen *et al.* (1960) on lucerne, and of Gibson (1963) on subterranean clover, showed the marked influence of the temperature of the rooting medium. Symbiotic N fixation in clover was reduced at temperatures below 22°C, and at 5°C was only 10–17% of that at 18°C. At 30°C there was a marked reduction in N fixation by some host/*Rhizobium* strain combinations.

Gist and Mott (1957) grew seedlings of lucerne, red clover and birdsfoot trefoil at 15·6°, 21·2°, 26·7° and 32·2°C and found that with increasing temperature, growth of both tops and roots was reduced, the roots being more severely affected than the tops.

The total quantity of light reaching the plant cannot be increased, but the management practices adopted should aim as far as possible to maintain LAI values at an optimum for growth rate (Brougham, 1958). Blackman's early work (1938) indicated the deleterious effect of shading on white clover, and Black (1957) has published a generalized diagram to illustrate the response of legumes to shading (Fig. 27.2). Blackman and Black (1959) have

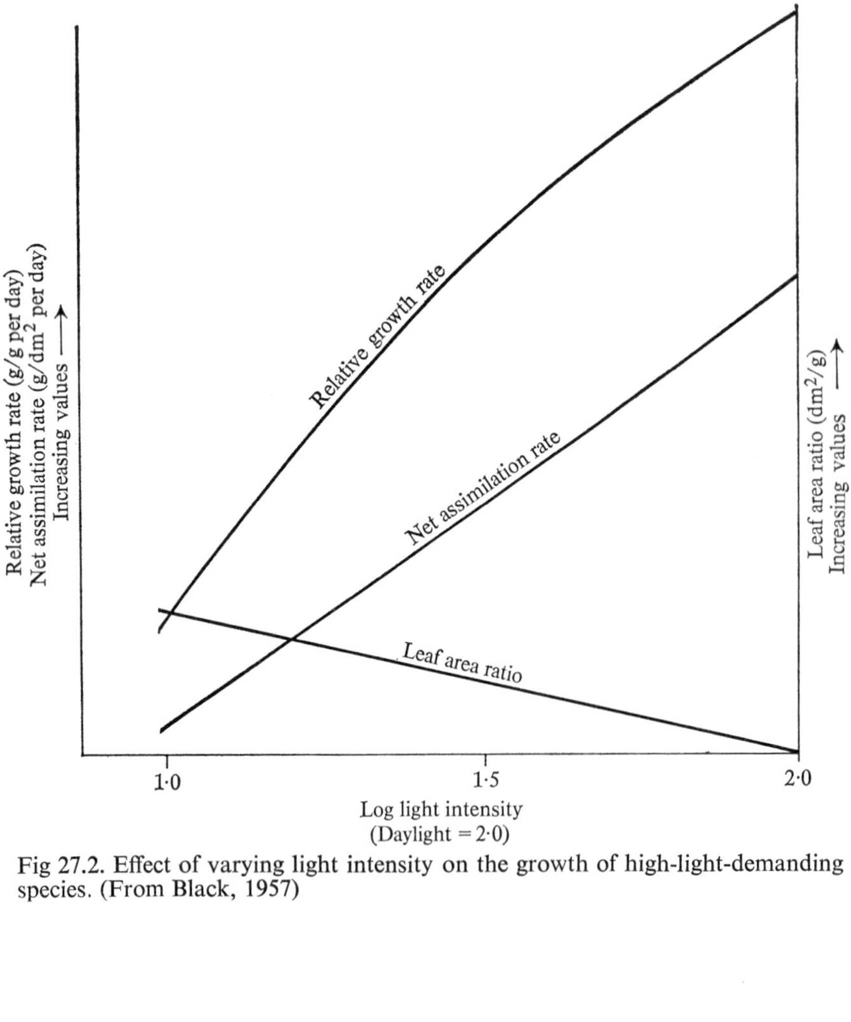

Fig 27.2. Effect of varying light intensity on the growth of high-light-demanding species. (From Black, 1957)

also estimated the light intensities necessary to obtain maximum growth rates of spaced seedlings of various herbage species (Table 27.4). These estimates

Table 27.4. Optimum light level at which relative growth rate is at its maximum (daylight = 1·0)

Lucerne	2·5
Red clover	1·0
White clover	1·8, 0·8
Alsike clover	1·5
Italian ryegrass	0·7

After Blackman & Black (1959)

may not strictly apply to closed swards, which might well be able to utilize much higher light intensities. These data show that herbage legumes generally are sun-loving species, and shading is detrimental to their growth. Jewiss and Sanderson (1963), working with S100 white clover, also found curvilinear relationships for both leaf area ratio and net assimilation rate when both were plotted against the log of light intensity (Fig. 27.3); these relationships were strikingly similar to those found for Ladino white clover (Blackman & Black, 1959).

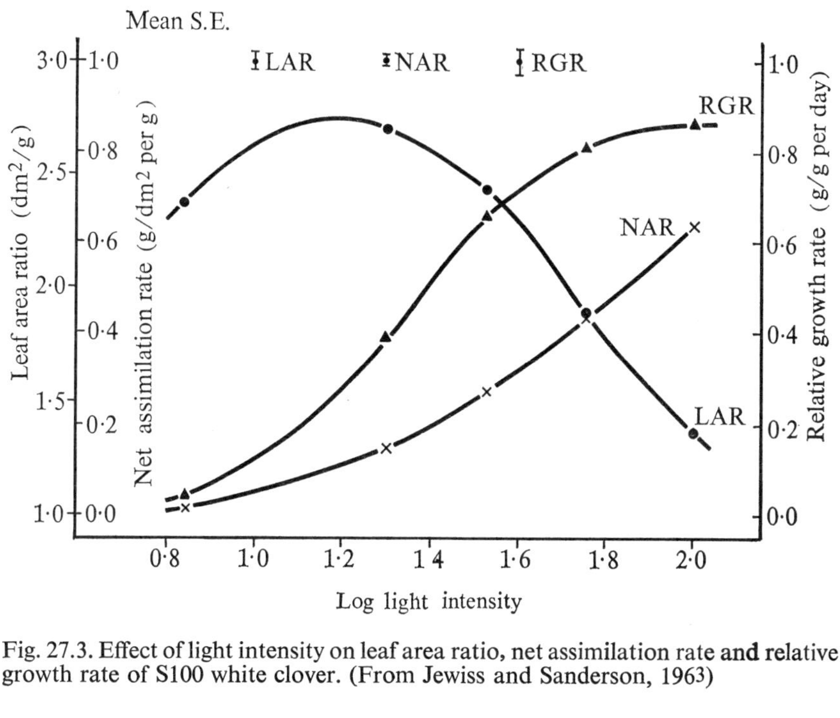

Fig. 27.3. Effect of light intensity on leaf area ratio, net assimilation rate and relative growth rate of S100 white clover. (From Jewiss and Sanderson, 1963)

Top growth and root growth are interdependent and both are affected by cutting and shading. Thus, Butler, Greenwood and Soper (1959) found that shading led to a marked loss of roots and nodules in white clover, with little subsequent regrowth. Under repeated defoliation, however, this loss was more than counterbalanced by new growth of roots and nodules. Regrowth of red clover and birdsfoot trefoil treated similarly was, however, much less rapid.

References

BLACK, J. N. 1957. The influence of varying light intensity on the growth of herbage plants. *Herb. Abstr.* **27**, 89–98.

BLACKMAN, G. E. 1938. The interaction of light intensity and nitrogen supply on the growth and metabolism of grasses and clover (*Trifolium repens*). 1. The effect of light intensity and nitrogen supply on the clover content of a sward. *Ann. Bot.* **2**, 257–80.

BLACKMAN, G. E.; BLACK, J. N. 1959. Physiological and ecological studies in the analyses of plant environment. 11. A further assessment of the influence of shading on the growth of different species in the vegetative phase. *Ann. Bot.* **23**, 51–63.

BROUGHAM, R. W. 1958. Leaf development in swards of white clover (*Trifolium repens* L.). *N.Z. Jl agric. Res.* **1**, 707–18.

BUTLER, G. W.; GREENWOOD, R. M.; SOPER, K. 1959. Effects of shading and defoliation on the turnover of root and nodule tissue of plants of *Trifolium repens*, *Trifolium pratense* and *Lotus uliginosus*. *N.Z. Jl agric. Res.* **2**, 415–26.

COOKE, G. W. 1960. Optimum fertilizer dressings for agricultural crops. *Jl R. agric. Soc.* **121**, 21–33.

DAVIES, W. E. 1964. A reappraisal of the value of herbage legumes. *Rep. Welsh Pl. Breed. Stn 1963* pp. 126–34.

DAVIES, W. E.; GRIFFITH, G. AP; ELLINGTON, A. 1966. The assessment of herbage legume varieties. 2. *In vitro* digestibility, water soluble carbohydrate, crude protein and mineral content of primary growth of clover and lucerne. *J. agric. Sci., Camb.* **66**, 351–7.

GIBSON, A. H. 1963. Physical environment and symbiotic nitrogen fixation. 1. The effect of root temperature on recently nodulated *Trifolium subterraneum* L. plants. *Aust. J. biol. Sci.* **16**, 28–42.

GIST, R. G.; MOTT, G. O. 1957. Some effects of light intensity, temperature and soil moisture on the growth of alfalfa, red clover and birdsfoot trefoil. *Agron. J.* **49**, 33–6.

IGNATIEFF, V.; PAGE, H. J. [Eds] 1958. Efficient use of fertilizers. *Agric. Studies. 43 FAO* Rome: FAO, pp. 355.

JEWISS, O. R. 1966. Effects of mineral nutrition. *A. Rep. Grassld Res. Inst. Hurley 1965* p. 31.

JEWISS, O. R.; SANDERSON, J. F. 1963. Effects of shading. *Exps Prog. 15 Grassld Res. Inst. 1961–2* pp. 27–8.

JONES, D. G.; DAVIES, W. E. 1966. The improvement of symbiotic effectiveness in white clover (*Trifolium repens*). *Proc. 10th int. Grassld Congr., Helsinki, 1966* pp. 648–51.

KLECZKOWSKA, J.; NUTMAN, P. S.; SKINNER, P. A.; VINCENT, J. M. 1968. The identification and classification of *Rhizobium*. In: Gibbs, B. M.; Skinner, F. A. [Eds] Identification methods for microbiologists. Vol. B. London, New York: Academic Press, pp. 51–65.

MITCHELL, K. J. 1956. Growth of pasture species under controlled environment. 1. Growth at various levels of constant temperature. *N.Z. Jl Sci. Technol. Sect. A* **38**, 203–16.

NIELSEN, K. F.; HALSTEAD, R. L.; MACLEAN, A. J.; HOLMES, R. M.; BOURGET, S. J. 1960. Effect of soil temperature on the growth and chemical composition of lucerne. *Proc. 8th int. Grassld Congr., Reading, 1960* pp. 287–92.

RUSSELL, E. W. 1966. Soil conditions and plant growth. London: Longmans, 9th edn, pp. 523.

SHOTTON, F. E.; JAMES, P. J. 1960. Potash manuring of lucerne. Trials in north-west Norfolk (1953–56). *Jl R. agric. Soc.* **121**, 34–45.

SMITH, D. 1964. Winter injury and the survival of forage plants. *Herb. Abstr.* **34**, 203–9.
TERRY, R. A.; TILLEY, J. M. A. 1964. The digestibility of the leaves and stems of perennial ryegrass, cocksfoot, timothy, tall fescue, lucerne and sainfoin, as measured by an *in vitro* procedure. *J. Br. Grassld Soc.* **19**, 363–72.
TESAR, M. B.; AHLGREN, H. L. 1950. Effect of height and frequency of cutting on the productivity and survival of ladino clover (*Trifolium repens* L.). *Agron. J.* **42**, 230–5.
WILLIAMS, R. D. 1927. Red clover investigations, 1919–1926. *Bull. H7 Welsh Pl. Breed. Stn* p. 18.

CHAPTER 28

Reproduction

REPRODUCTION in legumes is mainly by seeds, though species with stoloniferous habit, such as white clover, are also able to root at the nodes and so increase vegetatively.

The first visible sign of inflorescence development is the appearance of a bulge in the axil of the uppermost leaf initial (Aitken, 1960, 1964) (Plate 16A). In species such as red clover which have terminal inflorescences, this bulge develops into an inflorescence, absorbing the apical dome in the process, and the leaf initial develops into a leaf (Plate 16 B, C). In species which have axillary inflorescences (e.g. white clover) similar bulges appear also in the axils of subsequent leaves and develop into inflorescences; the leaf initials develop into leaves and the stem between the leaves expands to form elongated internodes.

Environmental control of flowering

The initiation and subsequent development of flowers are dependent on favourable conditions of light and temperature, neither of which is amenable to control in the field; however, flowering can be delayed by cutting or grazing. Some relevant data have been assembled in Table 28.1.

Light and temperature hardly ever vary independently under natural conditions and their effects interact. In most of the investigations reported, these two factors will therefore be considered jointly.

The herbage legumes appear to be basically quantitative long-day plants, i.e. they generally do not flower at all unless the daylength reaches a certain minimum, and produce more flowers to the extent that the daylength is increased above the threshold value. However, Thomas (1961) found some clones of New Zealand certified white clover to be short-long-day plants, as they flowered in long days only if first exposed to short days. Bawolski (1966) claimed that sainfoin showed a similar response.

Most investigators have reported that the legume species studied have no critical temperature requirements; however, exposure to relatively low

Table 28.1 Effect of cutting on date of flowering under field conditions

White clover mean for cv. S100, S184 and Kent		Red clover cv. English Broad		Lucerne cv. Eynsford	
Cutting date	Flowering date*	Cutting date	Flowering date**	Cutting date	Flowering date*
uncut	29 June 1958	Oct. 1922	20 July 1923	uncut	22 July 1956
15 May 1958	2 July 1958	2 Apr. 1923	23 July 1923	14 May 1956	15 Aug. 1956
9 June 1958	16 July 1958	13 June 1923	4 Aug. 1923	uncut	20 July 1957
uncut	22 June 1959			3 May 1957	5 Aug. 1957
7 May 1959	23 June 1959			uncut	18 Aug. 1958
1 June 1959	7 July 1959			22 May 1958	9 Sept. 1958
uncut	27 June 1960				
9 May 1960	28 June 1960				
2 June 1960	12 July 1960				

* Date when 100% of plants in flower. ** Date of full bloom.
White clover: from Zaleski (1961). Red clover: from Williams (1927). Lucerne: from Zaleski (1959)

temperatures during the winter often causes earlier flower induction, though flower development is retarded by low temperature (Jewiss, 1960, 1961, 1962; Haggar, 1961; Weller, 1964). As reported by Cooper (1960), Stoddart found an obligatory winter requirement in the white clover cultivars S100 and S184, but not in Ladino clover, while Dutch and New Zealand white clover each contained some plants which responded to low temperature and others which did not. Zaleski (1964) found a range of responses to temperature in the cultivar S184, and Beatty and Gardner (1961) recorded an interaction between temperature and light in clones of Ladino and common white clover from Arizona.

The critical daylength for wild white clover was found by Haggar and Holmes (1963) to be approximately 13·5 h. That for 4 German varieties of the same species tested by Weller (1964) was less than 13 h. Results of studies by Beatty and Gardner (1961) showed a range in daylength response, with Ladino bearing more flowers at 15 or more h, to a clone of common white clover which flowered under a 14-h daylength, but produced progressively fewer flowers as the daylength increased.

Red clover, which also exhibits a long-day response, has a critical daylength of 13–15 h or more (Ludwig, Barrales & Steppler, 1953; Gorman, 1955; Aitken, 1964; Weller, 1964). It appears to have no absolute low-temperature requirement, though Aitken (1964) reported that high temperatures combined with short days strongly retarded flowering in the variety Cowgrass and appeared to prevent its initiation in the variety Montgomery.

Lucerne is also a long-day species, but Weller (1964) found that it reacted less strongly than the other legumes he investigated. Under Weller's conditions, the critical daylength for lucerne was less than 13 h, as for white clover, whereas Van Dobben (1958) concluded that lucerne had a shorter critical daylength than white clover and that low temperature during winter was necessary to induce flowering. Davies (1959), however, found that low temperatures applied during germination had no consistent effect.

Pollination and seed-setting

The structure of the Papilionaceous flower has been described in Chapter 25. All the herbage legumes have diadelphous flowers with nectar, and are generally cross-pollinated by insects, though subterranean clover and some exotic species of *Trifolium* produce cleistogamous flowers that are of necessity self-pollinated. There is a complete range from self-fertility to self-sterility, the species here considered in detail being largely self-sterile (Robinson, 1947; Clapham, Tutin & Warburg, 1952), though there may be considerable variability between individual plants within a species.

The anthers dehisce inwards while the flowers are in bud, shedding pollen within the staminal column, but self-pollination is prevented because the style either projects beyond the anthers (as in the clovers and sainfoin) or is protected by a membrane. Once the flower opens, it can be "tripped" by any force which tends to separate the top edges of the keel (which is keyed to the wings) as, for example, by an insect alighting on the wings in search of nectar or pollen. During this process, pollen is deposited on the stigma by the insect, which also removes some pollen from the flower.

In the clovers and sainfoin, the flower is reset and can be tripped again, though one pollination is normally sufficient to achieve fertilization. White clover and sainfoin are normally pollinated by honey bees visiting them for nectar, or by bumble bees, but red clover has very long corolla tubes, so that only long-tongued bumble bees (e.g. *Bombus agrorum* and *B. hortorum*) are able to reach the nectaries inside the base of the staminal column. However, honey bees do visit the flowers of red clover for pollen and, according to Free (1965), pollen gatherers visit more flowers per head and are more efficient pollinators than are nectar gatherers.

In lucerne, self-pollination is prevented by a membrane which covers the stigma and which must be ruptured before the stigma becomes receptive. The tripping mechanism is explosive and cannot be reset (Armstrong & White, 1935); if pollination does not occur at the time of tripping, it is thus unlikely to occur subsequently. Bumble bees have been claimed to be better pollinators than honey bees, though the latter are effective if they are collecting pollen, which they cannot do without tripping the flower. Nectar-collecting honey bees generally develop the habit of approaching the flower from the side and thus extract nectar without tripping it. Even so, accidental tripping often occurs, and Zaleski (1959) concluded that honey bees were mainly responsible for seed setting.

The amount of pollination and subsequent seed-setting is influenced by various factors including the amount of nectar available, the proximity of other preferred plant species, abundance of bees, and their activity, which depends largely on the weather (Hawkins, 1961, 1965). Cutting and grazing red clover until early June delay flowering to a time when bumble bees are more abundant and active (Williams, 1927; Stoddart, 1960). Competition for pollinators occurs between leguminous species, and Zaleski (1959) recorded that bees often ignored lucerne and favoured red clover.

Reproduction, yield and persistence

Legume flowers are less important as yield constituents than are those of the grasses. In species with erect habit, however, the highest yield is obtained at flowering, at which time the leafy stems are at their longest. In prostrate forms, only the peduncles become erect, and vegetative growth may appear to be unaffected by flowering. Sanderson (1966) found that the cultivar S100 of white clover continued to produce branches throughout the year, with removal of inflorescences having no significant effect on branching or yield, but Erith (1924) found that branching in wild white clover ceased while inflorescences were being produced; Chow (1967) explained this by showing that as each node produces only one axillary bud, this can only grow into either a secondary stolon or an inflorescence. Chow also implied a negative correlation between flowering and persistence, agreeing in this with Hollowell (1966).

Red clover is normally a short-lived perennial, but it can be kept alive longer when photoperiods are too short to induce flowering, while late-flowering varieties are on the whole the most persistent.

References

AITKEN, Y. 1960. Axial origin of the terminal inflorescence in the genus *Trifolium*. *Nature, Lond.* **187,** 622–3.

AITKEN, Y. 1964. Flower initiation in pasture legumes. 4. Flower initiation in *Trifolium pratense* L. *Aust. J. agric. Res.* **15,** 21–36.

ARMSTRONG, J. M.; WHITE, W. J. 1935. Factors influencing seed-setting in alfalfa. *J. agric. Sci., Camb.* **25,** 161–79.

BAWOLSKI, S. 1966. Investigations on the development of sainfoin (*Onobrychis viciaefolia* Scop.). *Pam. Puław.* No. 23, 97–132.

BEATTY, D. W.; GARDNER, F. P. 1961. Effect of photoperiod and temperature on the flowering of white clover *Trifolium repens* L. *Crop Sci.* **1,** 323–6.

CHOW, H.-S. 1967. The structure and growth habit of white clover in relation to its persistence. *Diss. Abstr.* **27,** 3369 B.

CLAPHAM, A. R.; TUTIN, T. G.; WARBURG, E. F. 1962. Flora of the British Isles. Cambridge: Univ. Press, 2nd edn, pp. 1269.

COOPER, J. P. 1960. The use of controlled life-cycles in the forage grasses and legumes. *Herb. Abstr.* **30,** 71–9.

DAVIES, W. E. 1959. Vernalization and time of flowering. *Rep. Welsh Pl. Breed. Stn 1958* pp. 33–4.

DOBBEN, W. H. VAN. 1958. [The influence of the light period on growth and development of clover and lucerne.] *Jaarb. Inst. biol. scheik. Onderz. LandbGewass. 1958* pp. 35–43.

ERITH, A. G. 1924. White clover: a monograph. London: Duckworth and Co., pp. 150.

FREE, J. B. 1965. The ability of bumble bees and honey bees to pollinate red clover. *J. appl. Ecol.* **2,** 289–94.

GORMAN, L. W. 1955. Effect of photoperiod on varieties of red clover (*Trifolium pratense* L.). *N.Z. Jl Sci. Technol. Sect. A* **37,** 40–54.

HAGGAR, R. J. 1961. Flower initiation in Kent wild white clover (*Trifolium repens* L.) under controlled environment conditions. *Nature, Lond.* **191,** 1120–1.

HAGGAR, R. J.; HOLMES, W. 1963. Kent wild white clover. Ashford, Kent: Wye College Dep. Agric., pp. 40.

HAWKINS, R. P. 1961. Observations on the pollination of red clover by bees. 1. The yield of seed in relation to the numbers and kinds of pollinators. *Ann. appl. Biol.* **49,** 55–65.

HAWKINS, R. P. 1965. Factors affecting the yield of seed produced by different varieties of red clover. *J. agric. Sci., Camb.* **65,** 245–53.

HOLLOWELL, E. A. 1966. White clover *Trifolium repens* L., annual or perennial? *Proc. 10th int. Grassld Congr., Helsinki, 1966* pp. 184–7.

JEWISS, O. R. 1960. The growth and flowering behaviour of S100 white clover. *Exps Prog. 13 Grassld Res. Inst. 1959–60* p. 91.

JEWISS, O. R. 1961. Flowering studies. *Exps Prog. 14 Grassld Res. Inst. 1960–1* pp. 26–7.

JEWISS, O. R. 1962. Flowering studies. *Exps Prog. 15 Grassld Res. Inst. 1961–2* pp. 28–9.

LUDWIG, R. A.; BARRALES, H. G.; STEPPLER, H. 1953. Studies on the effect of light on the growth and development of red clover. *Can. J. agric. Sci.* **33,** 274–87.

ROBINSON, D. H. 1947. Leguminous forage plants. London: Edward Arnold and Co., 2nd edn, pp. 119.

SANDERSON, J. F. 1966. Aspects of the growth of white clover with particular reference to the effects of defoliation and shade. Ph.D. Thesis, Univ. Reading, pp. 117.

STODDART, J. L. 1960. Seed yield of late-flowering red clover. *Rep. Welsh Pl. Breed. Stn 1959* pp. 59–60.

THOMAS, R. G. 1961. Flower initiation in *Trifolium repens* L.: a short-long-day plant. *Nature, Lond.* **190,** 1130–1.

WELLER, R. 1964. [The influence of photoperiodism and vernalization on the readiness to flower of several lucerne varieties and various species of clover and their connection with seed production under changed climatic conditions.] *Z. Acker- u. PflBau* **119,** 241–62.

WILLIAMS, R. D. 1927. Studies concerning the pollination, fertilization and breeding of red clover. *Bull. H4 Welsh Pl. Breed. Stn.*

ZALESKI, A. 1959. Technique and methods of lucerne seed production. *J. Br. Grassld Soc.* **14,** 277–84.

ZALESKI, A. 1961. White clover investigations. 1. Effect of seed rates and cutting treatments on flower formation and yield of seed. *J. agric. Sci., Camb.* **57,** 199–212.

ZALESKI, A. 1964. Effect of density of plant population, photoperiod, temperature and light intensity on inflorescence formation in white clover. *J. Br. Grassld Soc.* **19,** 237–47.

CHAPTER 29

Perenniality and Senescence

THE 4 species of forage legumes most widely grown in the United Kingdom—white clover, red clover, lucerne and sainfoin—are generally regarded as perennials. The term 'perennial' is defined by Clapham, Tutin and Warburg (1962) as meaning 'living for more than two years and usually flowering each year'. In red clover (Smith, 1963), white clover (Gibson, 1957; Crowder & Craigmiles, 1960; Hollowell, 1966) and sainfoin (Thomson, 1938), profusion of flowering in the year of sowing is associated with lack of ability in some plants to survive the winter; these individuals thus behave as annuals. Reasons for this are not known, though genetic factors may be involved, and the possibility that small differences in the micro-environment may play some part in differential survival should not be excluded.

As plant cells age, they become progressively less active and eventually die. Many plant species grow and develop in such a way that the senescence and death of some parts of the plant does not result in the death of the whole plant. Such mechanisms are present in all the above-mentioned species of legumes. Inherently long-lived herbaceous dicotyledons such as lucerne exhibit well-developed secondary thickening of their underground parts (Seth & Dexter, 1958), which may show a different pattern in hardy and non-hardy varieties; sainfoin probably also has this property. Continued renewal of the conducting tissues which transport water and nutrients upward and assimilates downward may enable the plants to survive longer than the active life-span of any given cell. White clover survives by growing horizontally and rooting at the nodes at a rate which outpaces the rate of senescence of the older parts of the plant, except in winter, when the rate of senescence exceeds the growth rate (Sanderson, 1966; Hollowell, 1966). Red clover survives for a few seasons by pulling down its stem bases (crowns) into the soil, but this appears to be of limited survival value, since the root system gradually declines in vigour and eventually disintegrates (Terekhova, 1956; Krylova, 1965).

White clover overwinters in a relatively unchanged form, except that in winter it does not flower and has fewer and smaller leaves than in other seasons. Red clover, sainfoin and lucerne survive the winter by forming a rootstock with buds at soil level, the above-ground shoots dying back. Survival of the growing points is assisted by the ability of the roots to contract and pull the crown of the plant below the soil surface. Reserve substances are stored in the crown tissues and provide substrates for regrowth in spring and after cutting, though the nature and fate of these reserves is still somewhat obscure. Mortality in these three species is usually highest in winter. Factors involved in the winter survival of legumes have been reviewed by Smith (1964).

Longevity of leaves has been adequately studied only in white clover. Leaves survive for about 40 days in full light in summer (Brougham, 1958) and live for a slightly shorter time in plants subjected to cutting than in uncut plants (Sanderson, 1966). Low light intensity and low temperature increase leaf longevity considerably; Beinhart (1962), however, working with rooted cuttings, found that each unit of leaf area contributed the same net amount of dry matter increment to the plant during its life-span, regardless of light intensity. This might suggest the existence of a feed-back mechanism controlling leaf senescence in white clover.

Little work has been done on fundamental aspects of growth in lucerne, red clover and sainfoin, probably because they are rather difficult plants to work with. It is not even known with certainty what type of tissues (root, stem, hypocotyl) are involved in the regeneration of roots and shoots from crowns and rootstocks of these legumes, though adventitious roots and shoots are known to occur in some atypical lucernes not grown in Britain. Data are lacking on the course of photosynthesis in leaves, the fate of translocated assimilates, leaf senescence patterns, and the interrelationships between these factors; on the relationship between age and functioning of crown tissues; and on the effect of root contraction on the activity and longevity of the various parts of the root system. More studies on the effects of environment, inter-plant competition and the role of disease are needed. Data exist on the longevity of the various parts of the white-clover plant and of higher-order roots of red clover, but the underlying mechanisms governing the life span of these legumes are not understood.

Data on longevity of individual species are to be found in Chapters 33–36.

References

BEINHART, G. 1962. Effects of temperature and light intensity on CO_2 uptake, respiration and growth of white clover. *Pl. Physiol., Lancaster* **37**, 709–15.

BROUGHAM, R. W. 1958. Leaf development in swards of white clover. *N.Z. Jl agric. Res.* **1**, 707–18.

CLAPHAM, A. R.; TUTIN, T. G.; WARBURG, E. F. 1962. Flora of the British Isles. Cambridge: Univ. Press, 2nd edn, pp. 1269.

CROWDER, L. V.; CRAIGMILES, J. P. 1960. The effect of soil temperature, soil moisture and flowering on the persistency and forage production of white clover stands. *Agron. J.* **52,** 382–5.

GIBSON, P. B. 1957. Effect of flowering on the persistence of white clover. *Agron. J.* **49,** 213–15.

HOLLOWELL, E. A. 1966. White clover *Trifolium repens* L., annual or perennial? *Proc. 10th int. Grassld Congr., Helsinki, 1966* pp. 184–7.

KRYLOVA, N. P. 1965. [On the expression of the senile period in the life cycle of the clovers *Trifolium pratense* L. and *T. hybridum* L.] *Bot. Zh. SSSR* **50,** 692–4.

SANDERSON, J. F. 1966. Aspects of the growth of white clover with particular reference to the effects of defoliation and shade. Ph.D. Thesis, Univ. Reading, pp. 117.

SETH, J.; DEXTER, S. T. 1958. Root anatomy and growth habit of some alfalfa varieties in relation to wilt resistance and winter-hardiness. *Agron. J.* **50,** 141–4.

SMITH, D. 1963. Reliability of flowering as an indicator of winter survival in red clover. *Can. J. Pl. Sci.* **43,** 386–9.

SMITH, D. 1964. Winter injury and the survival of forage plants. *Herb. Abstr.* **34,** 203–9.

TEREKHOVA, A. F. 1956. [Study of the root system in red clover.] *Bot. Zh. SSSR* **41,** 553–8.

THOMSON, J. R. 1938. The development of sainfoin in its seeding year. *Ann. appl. Biol.* **25,** 457–71.

CHAPTER 30

Pests and Diseases

Introduction

Red clover populations and yields are often greatly reduced by pests and diseases, especially stem eelworm (*Ditylenchus dipsaci*) and clover rot (*Sclerotinia trifoliorum*) and lucerne yields by stem eelworm and Verticillium wilt (*Verticillium albo-atrum*). Between them, these three organisms limit any major increase in the acreage of red clover and lucerne in Britain.

Many plants of red clover, lucerne and also sainfoin are killed during the autumn and winter. The crown tissue of these plants is sometimes injured by weather and by defoliation, and occasionally this can lead to damage by pests and disease. The microclimate around the crowns is often very humid and conducive to disease, and this is probably one reason why the timing of defoliations in the autumn can be critical for the survival of these crops. An important problem is to define the part played by pests and diseases in this winter-killing.

With legumes, the death of a single mature plant can leave a large gap in the sward in which weeds are likely to become established; moreover, the interval between defoliations is often long, and this allows such weeds to set seed. When a legume is grown with grass, disorders in the legume often make it extra difficult to prevent the grass becoming dominant.

Although white clover harbours many of the pests and diseases affecting other legumes, it is generally less severely damaged. Moreover, unlike the others, white clover can regenerate from widely ranging stolons and thus recover from attack (Vanderveken, 1966).

Like the grasses, herbage legumes are perennial hosts of certain pests and diseases which attack other crops in summer. Of particular note are the virus diseases, which often appear to be of little account in the herbage legumes, but are very damaging when carried to certain field or fruit crops. For example, the winter hosts for the bean leaf roll virus and its aphid vectors are lucerne and white clover, and the host for pea mosaic virus and its aphids

Table 30.1. Pests of herbage legumes in Britain

Pest	Crop	Prevalence	Damage	Notes and control measures*	References
Stem eelworm, *Ditylenchus dipsaci* (Kühn) Filipjev.	Red clover and kidney vetch severely attacked by one biological race, lucerne by another and white clover by another. Alsike and crimson clovers, trefoil and sainfoin only lightly attacked.	Common on heavy soils. In dry areas it fluctuates widely from year to year. Of 1349 red clover crops examined during 1953–5 in England, 271 contained stem eelworm. In a survey of 408 fields containing lucerne crops during 1958–60, in England and Wales, up to 10% contained stem eelworm.	Important, especially in broad red clover and lucerne. Seedlings and established plants are destroyed. Of the 1 349 red clover crops, in 92 stem eelworm caused a loss of more than 5% of the stand.	See text.	A.L.† 409 Goodey (1922) Lester & Large (1958) Roberts & Large (1963)
Clover seed weevils (adults and larvae): *Apion aestivum* Germ. *A. apricans* Herbst. *A. dichroum* Bedel.	Red and white clovers.	Common.	Adults feed on the leaves, larvae feed in the flower-head. Seed yields of single-cut and late-flowering red clover are reduced by $\frac{1}{2}$ cwt or mor eper acre (> 62 kg/ha).	Control: For red clover seed crops, isolate from hay crops and hay stacks and defoliate red clover not grown for seed before a quarter of the heads are in full flower. Apply DDT. Limited experience has suggested that fenitrothion or malathion give control. Best applied before flower buds appear, i.e. within 2 weeks of taking the first cut.	Freeman (1967) Jenkins (1929) Jones (1950) Legowski (1956) Legowski (1967)
Pea and bean weevils (adults and larvae): *Sitona* spp. especially *S. lineatus* (L.).	All clovers, trefoil, lucerne, sainfoin.	Common in areas where pulses are grown.	Young clovers and lucerne seedlings are damaged.	Attacks occur in spring and autumn. Undersown crops are attacked when the cereal is harvested. Control: In poor growing conditions spray the edges of the field with DDT when damage is first seen.	A.L. 61 Jackson (1921)
Clover leaf weevils (adults and larvae): *Hypera variabilis* (Herbst.) *H. nigrirostris* (F.).	All clovers, trefoil, lucerne, sainfoin.	Fairly common.	Young leaves and flower buds are shredded and holed. Yields of seed are reduced.	Control: Locate seed crops away from those of the previous season and cut crops not grown for seed before the flowers open. Apply DDT before flower buds appear, i.e. within 2 weeks after the first cut.	
Gall midges (larvae): *Bremiola onobrychidis* (Bremi.), *Campylomyza ormerodi* (Kieff.), *Contarinia* spp., *Dasyneura* spp., *Jaapiella medicaginis* (Rübs.), *Tricholaba barnesi* Milne.	Red and white clovers, lucerne, sainfoin, trefoil.	Common.	Yields of red clover seed are sometimes reduced by more than 15%.	Seed-borne. Control: Plough deeply in autumn to bury overwintering larvae; practise good crop rotations; sow early; give adequate isolation; burn infested plant material; cut first growths and other crops not grown for seed before flowers open. Apply DDT. The best times for application are not fully known, but for the midges which attack the heads, apply before flower buds appear.	Barnes (1946) Metcalfe (1933) Milne (1960) Milne (1962)

Pest	Crop	Prevalence	Damage	Notes and control measures*	References
Aphids: *Acyrthosiphon pisum* (Harris) s. lat. *Aphis craccivora* Koch. *Aphis fabae* Scop. *Megoura viciae* Bukt.	All legumes.	Common.	Yields are reduced.	Damage is caused by feeding and because photosynthesis is interfered with. Each of these aphids may transmit a number of viruses. In particular, *A. fabae* transmits bean yellow mosaic virus to red clover and *A. pisum* lucerne mosaic virus. Control: Apply demeton-S-methyl or dimethoate or certain other systemic organophosphorus pesticides. Avoid applying chemicals during main flush of flowering because of danger to bees. Danger to bees is probably best avoided by using *granular* formulations of such pesticides, for example phorate or disulfoton.	A.L. 54 Free, Needham, Racey & Stevenson (1967)
Slugs: *Agriolimax reticulatus* (Müll) *Arion hortensis* (Fer).	All clovers, lucerne. Large-leaved Ladino-type white clovers are very susceptible	Common on heavy soils.	Young plants are destroyed. Young foliage and buds are selectively eaten.	Damage is underground or at the soil surface and mainly occurs at night in the spring and autumn, and during mild periods in winter. Control: Remove trash, especially before ploughing for new sowing, so that soil surface becomes dry. Cultivate thoroughly before sowing. Use methiocarb, or bran bait with metaldehyde, with or without addition of Paris green. Best applied in warm, humid, but not rainy, weather.	A.L. 115 Stephenson (1968)
Leatherjackets, *Tipula* spp. especially *T. paludosa* Meig.	All clovers, lucerne.	Common, especially in cool wet regions and under wet soil conditions.	Young plants are destroyed. In older swards spring growth is reduced. Swards become grass-dominant.	Control: Cultivate thoroughly before sowing; harrow and roll after brairding to kill the larvae; re-sow in summer rather than spring. Apply γ-BHC, or fenitrothion is sometimes effective. Or, except in rainy weather, bran bait with Paris green, which also controls slugs.	A.L. 179 Cameron (1940) White & French (1968)
Wood pigeons, *Columba palumbus* (L.).	All clovers, lucerne, sainfoin.	Occasional. Important in E. and S.E. England.	Buds and young shoots are selectively eaten. Damage done to red clover in winter can halve the winter yield, but there is little reduction in summer yield.	Clovers form an important part of the diet of pigeons in March and autumn.	A.L. 165 Murton (1965) Murton *et al.* (1966)
Wireworms, *Agriotes* spp.	Lucerne.	Occasional.	Seedlings are destroyed.	Control: When sowing, increase the seed rate; cultivate thoroughly and manure well. Apply γ-BHC as seed-dressing and later.	A.L. 199 *Bull* 128 *Min. Agric. Fish. Fd.*

* A warning on the use of pesticides, especially DDT, is given in Chapter 12, p. 91.

† A.L. = Advisory Leaflet of the Ministry of Agriculture, Fisheries and Food

is red clover. In summer these viruses are carried to and fro between herbage crops and field beans and peas (Quantz, 1961).

Tables 30.1 and 30.2 list the important pests and diseases of herbage legumes in their estimated decreasing order of importance, with notes on their distribution, effects on yield, control, and literature references. Further information on the importance of individual organisms in Britain and their chemical control is given by Gair and Roberts (1969). Bulletin No. 154 of the Ministry of Agriculture, Fisheries and Food (1966) contains a chapter on herbage pests and diseases and their control. Moore (1959) has described the distribution and prevalence of diseases, and life histories and other details of the pests concerned are given by Jones and Jones (1964) and Edwards and Heath (1964), and of the diseases by Sampson and Western (1954). Notes on seed-borne diseases are given by Noble and Richardson (1968), and Gibbs (1964) has reviewed the literature on the virus diseases. Disorders of lucerne are discussed in Advisory Leaflet No. 67 published by the Ministry of Agriculture, Fisheries and Food and disorders of herbage and seed crops are discussed in bulletins issued by the Welsh Plant Breeding Station (1967) and by the Ministry of Agriculture, Fisheries and Food (1968). Advice on the resistance of particular varieties of herbage legumes to pests and diseases has been published by the National Institute of Agricultural Botany (1968). Useful accounts of the disorders of herbage legumes in France have been published by Courtillot (1961) and by the Institut Technique des Céréales et des Fourrages (1964), and illustrations of the disorders have been issued by the Institut National de la Recherche Agronomique (1961). Disorders occurring in Denmark are well illustrated by Gram, Bovien and Stapel (1969). Close (1967) discusses diseases of lucerne in New Zealand, many of which also occur in Britain.

A warning about the use of the persistent organochlorine pesticides such as DDT is given in Chapter 12.

Pests and diseases of new sowings

Damping-off diseases, such as *Pythium*, are probably highly destructive to emerging legume seedlings.

An early attack by adults of the pea and bean weevils (*Sitona* spp.) can destroy sowings of lucerne or clover, but this pest is easily controlled by DDT. In the chemical control of weevils it is often only economically justifiable to treat the edges of fields.

If legumes are sown within a year of ploughing old grass, leatherjackets, slugs or wireworms can severely reduce the numbers of seedlings. These pests are often especially troublesome where so-called 'chemical ploughing' techniques employing paraquat are used. Cabbage thrips (*Thrips angusticeps*) sometimes damage the leaves of young clovers or lucerne. These also can be controlled by DDT.

Attacks by clover stem eelworm can occur on isolated seedlings, denoting an infestation derived from the seed. Subsequently, the areas of attacked plants increase and patches of plants are killed out.

In the autumn of the sowing year, clover rot can attack any of the herbage legumes. On a badly infected soil, red clover and occasionally trefoil (*Medicago lupulina*) and sainfoin, fail to survive the first winter.

Red clover and especially lucerne are subject to a number of pests and diseases which carry-over in the soil or stubble; therefore, direct re-seeding is inadvisable. However, if this is practised, care should be taken to kill all the plants in the old sward, as these may provide a nucleus for virus infection.

Pests and diseases of established crops

STEM EELWORM

Stem eelworm can seriously damage red clover, wild white clover and lucerne. Different biological races of stem eelworm have been recognized, but they are very heterogeneous and the hosts of a particular race cannot be closely defined (Hesling, 1966).

There are cultivars with some degree of resistance in the broad, single-cut and late-flowering types of red clover (Fiddian & Aldrich, 1964; National Institute of Agricultural Botany, Farmers Leaflet No. 4). The most resistant of the established varieties is the late-flowering red clover cultivar S123 (National Institute of Agricultural Botany, 1968). In broad red clover, the aim of the plant breeder has been to combine high yields with improved resistance to stem eelworm and also to clover rot. A diploid and a tetraploid cultivar, selected on these lines by the Plant Breeding Institute, Cambridge, are under test in national trials. Both are highly resistant to stem eelworm, and the tetraploid has outyielded the local diploid variety, Essex broad red.

In lucerne cultivars being bred in Hungary and Sweden, the aim is to combine resistance to stem eelworm with resistance to the *Verticillium* and *Fusarium* wilts (Bingefors, 1969).

Stem eelworm is soil-borne and seed-borne and is liable to be spread by any farm operations involving the movement of infested plant material or soil, as well as by rain, floodwater or wind. In severe attacks the plants are killed in patches which extend through the crop. The symptoms in clover have been described by Goodey (1922) and in the Ministry of Agriculture, Fisheries and Food Advisory Leaflet No. 409, and Brown (1957 a, b) has described those in lucerne.

It is especially important to try to prevent the introduction and spread of stem eelworm on heavy soils because it persists in these soils even where the legume host is grown infrequently. A break of 8 years between hosts of the same race of eelworm may be necessary to avert damage on heavy soils, whereas 2 years may be sufficient on light, sandy soils. Brown (1957 a, b,

Table 30.2. Diseases of herbage legumes in Britain

Disease	Crop	Prevalence	Damage	Notes and control measures	References
FUNGI					
Rot of clover, *Sclerotinia trifoliorum* Erikss.	All herbage legumes, especially red clover.	Common. In broad or late-flowering clovers, rot was present in 854 out of 1 349 crops examined during 1953–5 in England. It was most common when 4- to 5-year rotations were used (in Yorkshire, Midlands, E. and S.E. regions). Of 51 crops examined in Wales, none was affected. Seldom found in Ireland.	Of the 1 349 red clover crops examined, rot caused a loss of over 5% of the stand in 242, and in 114 of these, the loss was over 24%. Causes death in white clover and loss of stems in young crops of lucerne. At Hurley has halved the population of Giant sainfoin in the first winter.	See text.	A.L.† 266 Lester & Large (1958)
Verticillium wilt, *V. albo-atrum* Reinke & Berth. Different physiologic races exist.	Lucerne.	Common, especially where much lucerne is grown. In a survey of England and Wales in 1958–60, wilt was present in 50 out of 183 crops in their first harvest year.	Crops often rendered worthless after 2–3 years. In the survey, 40% of crops in their 4th year suffered severe reduction in yield.	See text.	Roberts & Large (1963)
Leaf spot: *Pseudopeziza trifolii* (Biv.-Bern. Ex Fr.) Fuckel. *P. medicaginis* (Lib.) Sacc.	Red, white, alsike and crimson clovers, lucerne, trefoil.	Common, except in N. and N.E. Scotland. Present on most lucerne crops.	Leaf losses can be serious in red clover and lucerne.	Most severe in the autumn. To minimize loss of leaf, and re-infection from fallen leaves, take the second cut early. *Pseudopeziza* has caused oestrogenic activity in lucerne (see text).	Loper & Hanson (1964) Schmiedeknecht (1967)
Black stem of lucerne, *Ascochyta imperfecta* Peck.	Lucerne, red and white clovers, trefoil, black medick.	Fairly common on lucerne in S. half of England. In Scotland mainly confined to S.E. Common on clovers in Ireland.	Destroys leaves and stems.	Seed-borne. Most severe in the spring and autumn; damage reduced by taking the first cut of lucerne early. Control: Apply mercurial seed-dressing.	Toovey, Waterston & Brooks (1936)
Scorch of red clover, *Kabatiella caulivora* (Kirchn.) Karak.	Red and white clovers.	Common on red clovers, especially broad red.	Causes losses of foliage and can cause total loss of seed.	Seed-borne. Carried over from year to year on plants and plant debris.	Sampson (1928)
Anthracnose, *Colletotrichum trifolii* Bain & Essary.	Lucerne, red clover.	Fairly common on lucerne. Anthracnose was present on about 10% of the lucerne crops surveyed in England and Wales in 1958–60. Not common on red clover.	Destroys young stems.	Probably seed-borne.	Roberts & Large (1963) Sampson (1928)

Disease	Crop	Prevalence	Damage	Notes and control measures	References
Powdery mildew of clover, *Erysiphe polygoni* D.C. ex Mérat. Different physiologic races exist.	Red and crimson clovers, sainfoin.	Fairly common on red clover and sainfoin.	Reduces yields of herbage and seed.	Most severe in the autumn.	
Crown wart of lucerne, *Physoderma alfalfae* (Magn.) Karling.	Lucerne.	Fairly common on wet soils, especially in S.E. England.	Reduces yields.	Soil-borne and very persistent. Control: Improve the drainage; use long rotations.	
Downy mildew, *Peronospora trifoliorum* de Bary. There are different physiologic races.	Red, white and alsike clovers, trefoil lucerne.	Fairly common.	Reduces yields.	Is most in evidence before the first cut and again in the autumn.	
Violet root rot, *Helicobasidium purpureum* Pat.	Red, white and alsike clovers, trefoil, lucerne.	Fairly common on sandy, well-aerated soils; not in N. of UK.	Seldom serious.	Soil-borne and persistent, but spreads slowly. Control: Improve the soil nutrient status; use long rotations.	A.L.3 46
Black blotch, *Cymadothea trifolii* Wolf.	Red, white and crimson clovers.	Fairly common.	Seldom serious. Damaging to white clover in Ireland.		
VIRUSES					
Clover phyllody.	Red, white, alsike and crimson clovers. In white clover, more common in S100 and Kersey than in wild types. Not common in red clover.	Common. In seed crops, during 1959–62 in England and Wales, it was found in 47% of crops of S100 and Kersey white clovers. Incidence increases with age of sward and is greatest in pure stands.	Increases winter-killing. The survey showed that in S100 it caused 11% reduction in seed yield in a quarter of second-year crops. In an experiment with half the number of plants infected, the yield of white clover herbage was reduced by 45%, that of seed by 30%. The yield of the companion grass was also reduced. The next year the healthy clover grew well and yields recovered.	See text.	Carr (1965) Carr & Catherall (1964) Carr & Large (1963)
Tomato stolbur.	White clover.	Common. In a survey of 23 seed crops in E. Anglia, it was absent from only 1.	Plants are destroyed. Yields of seed are reduced.	Shows 'red leaf' symptoms. See text.	Carr (1960) Davies (1964)

† A.L. = Advisory Leaflet of the Ministry of Agriculture, Fisheries and Food

1965) discusses this and other measures for control. The production of healthy seed is very important, because the pest multiplies so quickly that even a small incidence of seed infestation can lead to serious damage to the crop. For seed crops, the original seed should be derived from healthy crops and sown on land where eelworm is not known to occur. The crop should be inspected for eelworm during the spring. Because eelworms are not present under the seed coat and occur only in seed debris (Palo, 1962), efficient cleaning of the harvested seed helps considerably to reduce the risk of infestation, but seed known to be infested should also be fumigated commercially with methyl bromide. In a field experiment using lucerne seed contaminated with stem eelworm, treatment of the seed with the systemic insecticide phorate controlled the pest (Gostick, 1963).

CLOVER ROT

Clover rot is a serious disease of red clover, and trefoil can also be badly damaged. There is some evidence of the existence of different physiologic races of the fungus, as red clover is sometimes more severely damaged than trefoil growing in the same field, and vice versa. Damage to alsike clover and white clover is less severe. Lucerne and sainfoin are often attacked in their first year, lucerne generally suffering only a temporary growth check, but sainfoin can be killed out.

A few cultivars of red clover and white clover have some resistance to the disease. In red clover, the late-flowering cultivar Merkur is more resistant than the established broad red cultivars. In the newer broad red clovers, especially the tetraploid ones introduced from the Netherlands and Hungary, there is resistance to clover rot. A few diploid varieties, for example Kuhn, also show resistance. The tetraploid variety from Cambridge, now under test and already mentioned in connection with stem eelworm, is at present the only variety to combine resistance to clover rot with resistance to stem eelworm.

In white clover the variety Pajbjerg Milka has shown resistance to clover rot in trials at Cambridge (Aldrich & Doling, 1967).

Clover rot is both soil- and seed-borne. Resting bodies (sclerotia) of the fungus are formed in the crown of the host. Infection occurs in autumn from wind-borne spores produced by the sclerotia. Patches of infection develop and the spores infect neighbouring crops. Therefore, new sowings should not be made near infected crops. As a control measure Behringer (1963) advocates defoliating and rolling in the autumn to reduce the friability of the soil. This may also help by exposing the plant crown to drying conditions and low temperatures; for the same reason N fertilizer should be withheld late in the season.

Much care should be taken with produce from an infected crop, because sclerotia can be disseminated in clover hay and seed-sized ones can contaminate seed lots. Sclerotia have been reported to remain viable for up to $6\frac{1}{2}$ years in the soil in the absence of a host plant (Pape, 1937); this suggests that there should be a break of about this duration between susceptible crops. Where rot is already prevalent a break of 8 to 12 years is advisable.

VERTICILLIUM WILT

Wilt is a serious disease of lucerne and is spread by the same agencies and practices as stem eelworm. The two are often found together and the control measures are similar for both. Verticillium wilt is also serious on potatoes, hops and other crops. Isolates of the fungus from these crops are not normally pathogenic to lucerne, but those from lucerne can attack potatoes (Heale & Isaac, 1963). A Swedish variety of lucerne, Vertus, has recently shown valuable resistance to Verticillium wilt in trials at Cambridge.

Sainfoin and the clovers are attacked by the related species *V. dahliae* Kleb. (Roberts, 1963) and so occasionally is lucerne, but as the damage is less serious, the following observations refer only to *V. albo-atrum* in lucerne.

Symptoms appear on isolated plants after the first cut, after which the fungus spreads rapidly. Infection of 1% of the stems in one year can lead to 25% plant mortality in the next year, and the crop is often worthless within 2 or 3 years.

The fungus sporulates freely on the lower stems and the spores infect the ends of severed stems. The fungus spreads in the vascular tissues throughout the stems, leaves and roots. Spread to other crops of lucerne occurs mainly via infected plant debris from which spores are subsequently disseminated. Infective plant debris can be spread by wind, by cutting machinery, by livestock and by footwear, the spread often being from an old, heavily infected crop, to a neighbouring young crop (Isaac, 1957, 1959). New sowings should obviously be isolated from old, infected stands and healthy crops should be cut before infected ones. Contractor's machinery can carry infection from one farm to another. In the soil, wilt is largely confined to the decaying tissues of diseased plants, further spread occurring when healthy roots grow into contact with such tissues (Sewell, 1959; Heale & Isaac, 1963).

As with stem eelworm and clover rot, wilt is soil-borne and seed-borne. In the soil it can survive on lucerne residues, perhaps for as long as 10 years. A very long rotation is therefore required for its control and care has to be taken to eliminate volunteer lucerne plants in other crops. Potatoes may carry over the disease if they are grown after diseased lucerne or on a field adjacent to diseased lucerne. A number of common weeds also appear to be carriers of the disease (Heale & Isaac, 1963). Experiments with the hop race of Verticillium wilt suggest that it can be eradicated from the soil after only 4 years in the absence of dicotyledonous plants as, for example, by keeping grass swards weed-free by the use of selective herbicides (Sewell & Wilson, 1966).

Wilt is frequently introduced with seed. Conidia of the fungus on the surface of the seed can remain viable for up to 7 months and fragments of lucerne plants in seed lots can carry resting mycelia which can remain viable for up to 13 months (Isaac & Heale, 1961). It is therefore important to use only seed from healthy crops. It is possible to obtain healthy seed from the USA because the disease is not present there on lucerne; however, much of the seed used in Britain is imported from France, where Verticillium wilt is widely distributed. It is necessary therefore to ensure that seed is efficiently cleaned and, where there is a risk of infection, to treat it against Verticillium wilt. Fungicidal seed-dressings, such as thiram, are effective against the seed-borne propagules and do not interfere with the inoculation of the seed with *Rhizobium* (Isaac & Heale, 1961). The same authors have shown that wilt in lucerne seed can be controlled by fumigation with formaldehyde and certain other fumigants, though not with methyl bromide under the conditions of their tests. The development of an inexpensive process for disinfecting lucerne seed to give simultaneous kill of both stem eelworm and Verticillium wilt would be very desirable.

There are a number of other fungal diseases attacking these crops, the most injurious of which are listed in Table 30.2.

BACTERIAL WILT OF LUCERNE

Outbreaks of bacterial wilt (*Corynebacterium insidiosum*), a highly contagious and extremely virulent disease, occurred at two locations in East Anglia in 1964 and 1965 (Plant Breeding Institute, 1965, 1966). Precautions were taken to eradicate the disease and no further outbreaks have since been reported in Britain.

As it may not be possible to prevent further introductions with imported seed despite precautions (see section on Seed Health below), the Plant Breeding Institute, Cambridge, have incorporated resistance to this disease in their new selections of lucerne. These have yet to be tested on a field scale.

VIRUS DISEASES

Most virus diseases are capable of transmission by sap inoculation which can occur as a result of trampling or cutting the crop, and by insect vectors.

With certain of these diseases the insect vectors are leaf-hoppers (family Jassidae). The recent discovery of structures resembling mycoplasma in the leaf-hoppers and in the plants which they infect raises serious doubts as to the viral nature of these particular diseases (Cousin *et al.*, 1968).

In Britain, virus diseases are common in the herbage legumes, but there is little evidence from which to assess their effect on yields. With some viruses, the herbage legumes do not show marked symptoms, whereas others cause progressive damage and yields may be seriously reduced in the second or later years of a crop. However, the infected plants are usually distributed among healthy neighbours and when infection is only slowly introduced it may have only a temporary effect on yield because healthy plants take over the sward.

Insect vectors of these diseases can be controlled using the cultural and chemical methods outlined by Gair and Roberts (1969), but the economic profitability of such measures is not known.

Viruses in white clover

Clover phyllody is transmitted by leaf-hoppers (Cousin *et al.*, 1965; Frazier & Posnette, 1957). It causes the inflorescences of clover to develop into leafy structures devoid of seed, reduces yields of seed and herbage and increases winter-killing (Table 2). In strawberries it causes the damaging green-petal disease.

The disease is common and damaging in the larger-leaved white clovers, such as the cultivar S100, but much less so in wild white clover, and is less common in white clover grown with grass than in clover grown alone, which may account for why phyllody was not found in a survey of white clover in 26 old pastures in Britain carried out by Gibbs *et al.* (1966).

The disease sometimes causes only a temporary reduction in the herbage yield of white clover, because the uninfected plants in a crop increase their contribution to the whole yield; moreover, infected plants sometimes appear to produce a healthy plant from the end of a stolon (Vanderveken, 1966).

In addition to its direct effect on the plant, phyllody appears to reduce seriously the effectiveness of *Rhizobium* in the root nodules (Joshi & Carr, 1967). Two other leaf-hopper-transmitted diseases have been frequently found in white clover seed crops in East Anglia (Carr, 1961). In a survey of 23 crops, tomato stolbur (Table 2) was observed in 22 crops and clover witches-broom in 7. However, they were rare or absent in the survey of white clover in old pastures already mentioned in connection with clover phyllody. In spaced plants of white clover, these two diseases cause much more damage than clover phyllody (Carr, 1961; Davies, 1964), but little is known of their effect in swards.

A number of other viruses are common in wild white clover and are probably present also in the larger-leaved varieties. In the survey of old pastures, 23% of randomly sampled plants contained these viruses. However, only white clover mosaic virus, which was fairly common except in a few northern sites, caused marked symptoms. This virus is thought to be spread by sap inoculation during defoliation. Two aphid-transmitted viruses were relatively more common; these were clover yellow-vein and red clover vein-mosaic viruses. Arabis mosaic virus, which has an eelworm vector (*Xiphinema diversicaudatum*) was common only in the south-west (Gibbs *et al.*, 1966).

Viruses in red clover

In a survey by Varma and Gibbs (1967) of virus diseases in red clover in permanent pastures in the southern half of Britain, no witches broom or phyllody were found. However, 5 other diseases, none of which appeared to be very damaging, were found. The two most common were aphid-transmitted; these were pea mosaic, found in 7 out of the 9 pastures, and red clover vein-mosaic, found in three of the pastures. Where either virus was found, about one out of every 10 plants was infected.

There is little or no transmission of viruses by red clover seed. No viruses were detected in seedlings grown from seed collected from a total of 89 seed crops (Varma & Gibbs, 1967) which included broad red, single-cut and late-flowering types.

Viruses in lucerne

Out of 76 lucerne crops surveyed in 1959 in Great Britain, 52 contained lucerne mosaic virus, up to 72% of the plants in a crop being infected (Gibbs, 1962). This virus, which is transmitted by aphids, appears to have little effect on yield in the early years of a crop, but it probably increases winter-kill.

In many other countries, lucerne mosaic virus is present in both red and white clovers and this increases the risk of spread to lucerne. However, it was not found in these clovers during the surveys of permanent pastures in Great Britain already mentioned (Gibbs *et al.*, 1966; Varma & Gibbs, 1967).

Viruses in sainfoin

At the Grassland Research Institute, Hurley (Unpublished), large numbers of plants in sainfoin crops have been killed by tomato bushy stunt virus.

Pests and diseases of seed crops

Organisms attacking new sowings have already been mentioned. Apart from stem eelworm and clover rot, the organisms most likely to reduce yields in established seed crops are gall midges and seed weevils and, of diseases, scorch of red clover, clover phyllody and tomato stolbur virus (Welsh Plant Breeding Station, 1967, Tables 1 and 2; Ministry of Agriculture, Fisheries & Food, 1968).

Effects on forage quality

Pests and diseases may influence the time at which forage is available. Thus some of the first spring growth and autumn growth of clovers can be destroyed by leatherjackets (*Tipula* spp.) (White & French, 1968) and, locally, by wood pigeons.

The quality of the legumes themselves can also be affected. Clover weevils, aphids and wood pigeons feed selectively upon young, highly nutritious buds, shoots and leaves. Also, certain diseases, notably clover rot and Verticillium wilt, destroy young shoots. Others, such as scorch of red clover, downy mildew, the rusts (*Uromyces* spp.) and the leaf-spot diseases caused by *Pseudopeziza* and *Stemphylium*, destroy the leaf lamina. In lucerne crops in Italy (Matta, 1962) and in the USA (Hanson, 1965), a reduction in the yield and concentration of nutrients, particularly protein, has been measured following such attacks.

Foliar diseases arising from attacks by *Pseudopeziza* spp. and *Stemphylium* spp., both of which are common in Britain, increase the concentrations of oestrogens in clover and lucerne, which may possibly lead to infertility in ewes grazing these crops (Hanson, 1965; Bickoff, 1968). Newton *et al.* (1970) showed that feeding white clover leaves infected with *Pseudopeziza trifolii*, or with another common leaf disease, *Leptosphaerulina trifolii*, to mice induced responses of the kind often attributed to oestrogens.

SEED HEALTH

Dangers arise from the fact that stem eelworm, clover rot, Verticillium wilt and bacterial wilt are all seed-borne and the disease risk is greater because a large proportion of the seed sown in Britain is imported. For example, in the seed year June 1966 to May 1967, the following figures for quantities of seed used were published by the Committee on Herbage Seed Supplies (1968):

	Grown in UK cwt	Imported cwt
Red clover	16 700	12 800
White clover	2 400	15 100
Lucerne	negligible	1 400

Government regulations have been introduced in an attempt to reduce the risk of introducing bacterial wilt or stem eelworm with imported lucerne seed. These regulations forbid the importation into Britain of lucerne seed from crops grown in areas in other countries where bacterial wilt is known to occur. Also, prior to export to Britain, all lucerne seed must be fumigated with methyl bromide to control stem eelworm (Ministry of Agriculture, Fisheries & Food, 1965b).

No other precautions to ensure seed health are compulsory, but Plant Health Inspectors are empowered to inspect any seed stock landed, or likely to be landed in Britain and, within Britain, may inspect any seed crop or seed stock in which the presence of a destructive pest or disease is suspected. These Inspectors may also order the compulsory destruction of any pest or organism found or, in the case of an imported seed stock, the re-export or destruction of the seed (Ministry of Agriculture, Fisheries & Food, 1965 a, b). However, inspection is hampered because, in a seed stock, anything but a heavy infestation or infection is difficult to detect other than by detailed examination.

In a certification scheme administered for England and Wales by the National Institute of Agricultural Botany, red clover seed crops are inspected and, if stem eelworm is found, the seed must be fumigated. Apart from this scheme, it should be noted that clover seed stocks, whether or not imported, are not required to be fumigated except on order of a Plant Health Inspector.

References

ALDRICH, D. T. A.; DOLING, D. A. 1967. Varietal resistance to clover rot in white clover. *Nature, Lond.* **214,** 946–7.

BARNES, H. F. 1946. Gall midges of economic importance. Vol. 2. Gall midges of fodder crops. London: Crosby Lockwood and Son Ltd, pp. 160.

BEHRINGER, P. 1963. [Reliability of red clover cultivation, especially by control of clover wilt (*Sclerotinia trifoliorum* Erikss.).] *Bayer. landw. Jb.* **40,** 643–70.

BICKOFF, E. M. 1968. Oestrogenic constituents of forage plants. *Review Series 1/1968 Commonw. Bur. Past. Fld crops* Farnham Royal: Commonw. Agric. Bur., pp. 39.

BINGEFORS, S. 1969. The use of nematode-resistant varieties of grasses and legumes. *Herb. Abstr.* **39,** 107–11.

BROWN, E. B. 1957a. Lucerne stem eelworm in Great Britain. *Nematologica* **2** Suppl., 369–75.

BROWN, E. B. 1957b. Lucerne stem eelworm: a serious threat to lucerne growing. *Agriculture, Lond.* **63,** 517–20.

BROWN, E. B. 1965. Cultural and biological control. *Tech. Bull. 7 Minist. Agric. Fish. Fd* London: H.M.S.O., pp. 219–32.

CAMERON, A. E. 1940. Insects and pests of 1939. *Trans. R. Highld agric. Soc. Scotl.* **52,** 91–115.

CARR, A. J. H. 1960. Clover virus diseases. *Rep. Welsh Pl. Breed. Stn 1959* pp. 61–3.

CARR, A. J. H. 1961. The significance of virus diseases in herbage crops. *Proc. 8th int. Grassld Congr., Reading, 1960* pp. 200–4.

CARR, A. J. H. 1965. Plant pathology. *Rep. Welsh Pl. Breed. Stn 1964* pp. 97–103.

CARR, A. J. H.; CATHERALL, P. L. 1964. The assessment of disease in herbage crops. *Rep. Welsh Pl. Breed. Stn 1963* pp. 94–100.

CARR, A. J. H.; LARGE, E. C. 1963. Surveys of phyllody in white clover seed crops, 1959–62. *Pl. Path.* **12,** 121–7.

CLOSE, R. C. 1967. Diseases of lucerne in New Zealand. In: Langer, R. H. M. [Ed.] The lucerne crop. Sydney: A. H. & A. W. Reed, pp. 248–56.

COMMITTEE ON HERBAGE SEED SUPPLIES [UK]. 1968. Report of the Committee on herbage seed supplies. *Cmnd. 3748* London: H.M.S.O., Appendix X.

COURTILLOT, M. 1961. Les maladies des légumineuses fourragères. *Producteur Agricole Français* No. 44 Suppl., Inst. Technique Céréales, Fourrages, Paris.

COUSIN, M.-T.; MAILLET, P.-L.; GOURRET, J.-P.; GRISON, C.; STARON, T. 1968. [Presence of particles of the 'mycoplasm' type in the phloem of plants infected by three diseases of the 'European yellows' type: stolbur of tomato, aster yellows of gladiolus, and clover phyllody. Cytological and ultrastructural studies. First trials of chemical control.] *C. r. hebd. Séanc. Acad. Agric. Fr.* **54,** 887–95.

COUSIN, M.-T.; MOREAU, J. P.; VAN LOON, L. C. 1965. [Virus diseases of clover transmitted by Cicadellids in France.] *Annls Épiphyt.* **16,** 137–48.

DAVIES, W. E. 1964. Observations on phyllody and 'red leaf' in white clover. *Pl. Path.* **13,** 40–3.

EDWARDS, C. A.; HEATH, G. W. 1964. The principles of agricultural entomology. London: Chapman and Hall Ltd.

FIDDIAN, W. E. H.; ALDRICH, D. T. A. 1964. The susceptibility of red clover varieties to clover stem eelworm. *Pl. Path.* **13,** 139–43.

FRAZIER, N. W.; POSNETTE, A. F. 1957. Transmission and host-range studies of strawberry green-petal virus. *Ann. appl. Biol.* **45,** 580–8.

FREE, J. B.; NEEDHAM, P. H.; RACEY, P. A.; STEVENSON, J. H. 1967. The effect on honeybee mortality of applying insecticides as sprays or granules to flowering field beans. *J. Sci. Fd Agric.* **18,** 133–8.

FREEMAN, B. E. 1967. The biology of the white clover seed weevil *Apion dichroum* Bedel (Col. Curculionidae). *J. appl. Ecol.* **4,** 535–52.

GAIR, R.; ROBERTS, E. T. 1969. Pest and disease control in grass and fodder crops. In: Martin, H. [Ed.] Insecticide and fungicide handbook. Oxford: Blackwell Scientific Publications, 3rd edn, chapter 8.

GIBBS, A. J. 1962. Lucerne mosaic virus in British lucerne crops. *Pl. Path.* **11,** 167–71.

GIBBS, A. J. 1964. Virus diseases of pasture and forage legumes in temperate regions. *Herb. Abstr.* **34,** 141–5.

GIBBS, A. J.; VARMA, A.; WOODS, R. D. 1966. Viruses occurring in white clover (*Trifolium repens L.*) from permanent pastures in Britain. *Ann. appl. Biol.* **58,** 231–40.

GOODEY, T. 1922. On the susceptibility of clover and some other legumes to stem-disease caused by eelworm, *Tylenchus dipsaci,* syn. *Devastatrix,* Kuhn. *J. agric. Sci., Camb.* **12,** 20–30.

GOSTICK, K. G. 1963. Control of seed-borne lucerne stem eelworm with phorate. *Ann. appl. Biol.* **51,** 503–7.

GRAM, E.; BOVIEN, P.; STAPEL, C. 1969. Recognition of diseases and pests of farm crops. London: Blandford Press, 2nd edn.

HANSON, C. H. 1965. Foliar diseases and forage quality. *Proc. 9th int. Grassld Congr., São Paulo, 1965* pp. 1209–13.

HEALE, J. B.; ISAAC, I. 1963. Wilt of lucerne caused by species of *Verticillium.* 4. Pathogenicity of *V. albo-atrum* and *V. dahliae* to lucerne and other crops; spread and survival of *V. albo-atrum* in soil and in weeds; effect upon lucerne production. *Ann. appl. Biol.* **52,** 439–51.

HESLING, J. J. 1966. Biological races of stem eelworm. *Rep. Glasshouse Crops Res. Inst. 1965* pp. 132–41.

INSTITUT NATIONAL DE LA RECHERCHE AGRONOMIQUE 1961. Maladies des légumineuses fourragères. Fiche 38, 39, 52, 53. Paris: Association de Coordination Technique Agricole.

INSTITUT TECHNIQUE DES CÉRÉALES ET DES FOURRAGES. 1964. Étude sur la nature, la fréquence et la gravité des differents parasites des plants fourragères. Paris: Inst. Tech. Céréales Fourrages.

ISAAC, I. 1957. Wilt of lucerne caused by species of *Verticillium. Ann. appl. Biol.* **45,** 550–8.

ISAAC, I. 1959. *Verticillium* wilt of lucerne. *N.A.A.S. q. Rev.* **46,** 75–9.

ISAAC, I.; HEALE, J. B. 1961. Wilt of lucerne caused by species of *Verticillium.* 3. Viability of *V. albo-atrum* with lucerne seed; effects of seed dressings and fumigants. *Ann. appl. Biol.* **49,** 675–91.

JACKSON, D. J. 1921. Bionomics of weevils of the genus *Sitones* injurious to leguminous crops in Britain. *Ann. appl. Biol.* **7,** 269–99.

JENKINS, J. R. W. 1929. Observations on the control of weevils of the genus *Apion* attacking red clover. *Welsh J. Agric.* **5,** 176–86.

JONES, F. G. W.; JONES, M. G. 1964. Pests of field crops. London: Edward Arnold Ltd.

JONES, J. M. 1950. Clover seed weevils. *Ann. appl. Biol.* **37,** 313–20.

JOSHI, H. U.; CARR, A. J. H. 1967. Effect of clover phyllody virus on nodulation of white clover (*Trifolium repens*) by *Rhizobium trifolii. J. gen. Microbiol.* **49,** 385–92.

LEGOWSKI, T. J. 1956. Clover seed weevil damage in East Anglia. *Pl. Path.* **5,** 99–105.

LEGOWSKI, T. J. 1967. Control of red clover seed weevils (*Apion aestivum* Germ. and *A. assimile* Kirb.). *Pl. Path.* **16** Suppl., 33.

LESTER, E.; LARGE, E. C. 1958. Surveys of clover rot with incidental observations on eelworm in clover: England and Wales, 1953–55. *Pl. Path.* **7,** 115–24.

LOPER, G. M.; HANSON, C. H. 1964. Influence of controlled environmental factors and two foliar pathogens on coumestrol in alfalfa. *Crop Sci.* **4,** 480–2.

MATTA, A. 1962. [Qualitative changes in lucerne plants attacked by *Sclerotinia trifoliorum.*] *Riv. Pat. veg., Pavia Ser. 3* **2,** 30–3.

METCALFE, M. E. 1933. *Dasyneura leguminicola* (Lint.) the clover seed midge. *Ann. appl. Biol.* **20,** 185–204.

MILNE, D. L. 1960. The gall midges (Diptera: *Cecidomyidae*) of clover flower heads. *Trans. R. ent. Soc. Lond.* **112,** 73–108.

MILNE, D. L. 1962. Gall midges (*Cecidomyidae*) causing seed losses in red and white clovers. *S. Afr. J. agric. Sci.* **5,** 89–93.

MINISTRY OF AGRICULTURE, FISHERIES AND FOOD. 1965a. The destructive pests and diseases of plants Order 1965. *Statutory Instrum. 216, 1965* London: H.M.S.O.

MINISTRY OF AGRICULTURE, FISHERIES AND FOOD. 1965b. The importation of plants and plant produce (health) (Great Britain) Order 1965. *Statutory Instrum. 1426, 1965* London: H.M.S.O.

MINISTRY OF AGRICULTURE, FISHERIES AND FOOD. 1966. Grass and grassland. *Bull. 154* London: H.M.S.O., 4th edn, pp. 91–7.

MINISTRY OF AGRICULTURE, FISHERIES AND FOOD. 1968. Grass and clover crops for seed. *Bull. 204.* London: H.M.S.O., pp. 35–42.

MOORE, W. C. 1959. British parasitic fungi. Cambridge: Univ. Press.

MURTON, R. K. 1965. Natural and artificial population control in the wood-pigeon. *Ann. appl Biol.* **55,** 177–92.

MURTON, R. K.; ISAACSON, A. J.; WESTWOOD, N. J. 1966. The relationships between wood-pigeons and their clover food supply and the mechanism of population control. *J. appl. Ecol.* **3,** 55–96.

NATIONAL INSTITUTE OF AGRICULTURAL BOTANY. 1968. Red clover. *Rep. natn. Inst. agric. Bot.* **48,** 21.

NEWTON, J. E.; BETTS, J. E.; DRANE, H. M.; SABA, N. 1970. The oestrogenic activity of white clover. *Occ. Symp. 6 Br. Grassld Soc.* pp. 309–14.

NOBLE, M.; RICHARDSON, M. J. 1968. An annotated list of seed-borne diseases. *Phytopath. Pap. 8* Commonwealth Mycological Institute, Kew, England, 2nd edn.

PALO, A. V. 1962. Translocation and development of stem eelworm, *Ditylenchus dipsaci* (Kühn) in lucerne, *Medicago sativa* L. *Nematologica* **7,** 122–32.

PAPE, H. 1937. [Contributions to the biology and control of clover stem rot (*Sclerotinia trifoliorum* Erikss.).] *Arb. biol. Anst.* (*Reichsanst.*) *Berl.* **22,** 159–247.

PLANT BREEDING INSTITUTE, CAMBRIDGE. 1965. Bacterial wilt of lucerne. *Rep. Pl. Breed. Inst. 1963–4*, p. 82.

PLANT BREEDING INSTITUTE, CAMBRIDGE. 1966. Lucerne. *Rep. Pl. Breed. Inst. 1964–5*, pp. 96–8.

QUANTZ, L. 1961. Investigations on virus diseases of leguminous forage crops in Germany. *Proc. 8th int. Grassld Congr., Reading, 1960* pp. 204–7.

ROBERTS, E. T. 1963. Pathogenicity of *Verticillium dahliae* to red and white clover. *Pl. Path.* **12,** 138.

ROBERTS, E. T.; LARGE, E. C. 1963. Surveys of *Verticillium* wilt in lucerne, England and Wales, 1958–60. *Pl. Path.* **12,** 47–58.

SAMPSON, K. 1928. Comparative studies of *Kabatiella caulivora* (Kirchn.) and *Colletotrichum trifolii* which causes red clover Anthracnose. *Trans. Br. mycol. Soc.* **13,** 103–42.

SAMPSON, K.; WESTERN, J. H. 1954. Diseases of British grasses and herbage legumes. Cambridge: Univ. Press.

SCHMIEDEKNECHT, M. 1967. Observations on the epidemiology of lucerne common leaf spot and trials on the diminution of losses. Cited in *Rev. appl. Mycol.* **47,** 112.

SEWELL, G. W. F. 1959. Direct observation of *Verticillium albo-atrum* in soil. *Trans. Br. mycol. Soc.* **42,** 312–21.

SEWELL, G. W. F.; WILSON, J. F. 1966. *Verticillium* wilt of the hop: the survival of *V. albo-atrum* in the soil. *Ann. appl. Biol.* **58,** 241–9.

STEPHENSON, J. W. 1968. A review of the biology and ecology of slugs of agricultural importance. *Proc. malac. Soc. Lond.* **38,** 169–76.

TOOVEY, F. W.; WATERSTON, J. M.; BROOKS, F. T. 1936. Observations on the black-stem disease of lucerne in Britain. *Ann. appl. Biol.* **23,** 705–17.

VANDERVEKEN, J. 1966. [Study of recovery of white clovers infected with clover phyllody virus.] *Meded. Rijksfac. LandbWet. Gent* **31,** 981–5.

VARMA, P.; GIBBS, A. J. 1967. Preliminary studies on sap-transmissible viruses of red clover (*Trifolium pratense* L.) in England and Wales. *Ann. appl. Biol.* **59**, 23–30.

WELSH PLANT BREEDING STATION. 1967. Principles of herbage seed production. *Tech. Bull. 1 Welsh Pl. Breed. Stn* pp. 113–16.

WHITE, J. H.; FRENCH, N. 1968. Leatherjacket damage to grassland. *J. Br. Grassld Soc.* **23**, 326–9.

See also Advisory leaflets of the Ministry of Agriculture, Fisheries and Food listed in Tables 30.1 and 30.2.

CHAPTER 31

Chemical Composition

Introduction

THE major chemical constituents of the legumes are the same as those of the grasses listed on p. 98. There are, however, consistent differences between legumes and grasses in the percentage contents of some constituents, and these are noted in the appropriate sections of this chapter. The greater tendency for livestock to develop bloat when consuming legume rather than grass herbage is presumably due to a difference in chemical composition, but it has not yet been established which constituents are involved in this disorder. Although less information is available on the composition of legumes than on that of grasses, it is clear that there are greater differences between legume species than between the cultivated grass species. One important reason for this is the morphological difference between white clover and the three other legume species considered in this book. When cut at a height of 1–1·5 in. (2–4 cm), white-clover herbage consists of leaf and petiole, together with flowers and peduncles during the flowering stage, but little or no stolon material, whereas the herbage of the other species also contains stem material, the proportion of which generally increases with advancing maturity.

The constituents to be considered in this chapter are described in the same sequence as in Chapter 13, namely structural carbohydrates, non-structural carbohydrates, lignin, N compounds, lipids, organic acids, vitamins, oestrogens and goitrogenic compounds, and the mineral elements of significance in plant and animal nutrition. As in Chapter 13, all values refer to analysis on a dry matter (DM) basis except where otherwise stated.

Structural carbohydrates

CELLULOSE

The cellulose content of legumes is generally lower than that of grasses (Hirst *et al.*, 1959), although values higher than 25% have been reported in mature plants. Increases in cellulose content with increasing maturity from the succulent to the full-bloom stage have been reported by Baumgardt and

Smith (1962) for white clover (18·4–26·6%), red clover (16·2–30·4%) and lucerne (19·6–36·2%). Lower values for red clover (9·2% at the leafy stage and 16·9% at the flowering stage) were found by Salo (1965). For lucerne, Armstrong *et al.* (1950) reported an increase from 17·7% in mid-April to 29·9% in early June.

When herbage is cut several times during the season, there is little change in cellulose content. Ferguson (1963) reported that the cellulose content of white clover and lucerne cut 4 times between June and September varied only between 13·0 and 15·6%, and from 15·3 to 19·8%, respectively. Jarrige (1963) reported two consecutive cuts of one-month-old white clover herbage to have cellulose contents of 13·5 and 14·5%. Two-month-old herbage cut in mid-July had a content of 18·9%, and herbage of the same age cut in early September, 13·6%. White clover grown in New Zealand was found to have very low cellulose contents, ranging from 1·8 to 6·5% in one-month-old regrowth (Bailey, 1964). Three cuts of lucerne taken on 10 June, 11 July and 7 September had cellulose contents of 19·3, 12·5 and 16·1%, and two cuts of sainfoin taken on 16 July and 31 August had 18·5 and 16·4%, respectively. (Kerguelen, 1960).

Differences between species are small for herbage at the early vegetative stage, but the cellulose content increases with advancing maturity to a lesser extent in white clover than in red clover, lucerne and sainfoin.

HEMICELLULOSE

Legumes generally have lower contents of hemicellulose than grasses. The data of Hirst *et al.* (1959) on lucerne, and of Salo (1965) on red clover, indicate that hemicellulose contents do not change much with increasing maturity, although the highest values usually occur in mature herbage.

In lucerne samples taken during a two-year period, the sum of the hemicellulose sugars (arabinose, xylose, galactose and mannose) in the total herbage ranged from 6·2 to 16·4% (Hirst *et al.* 1959). Salo (1965) reported contents of hemicellulose (including pectin) of 21% in red clover at the leafy stage and 22·8% at the flowering stage. According to Jarrige (1960), average hemicellulose contents are 13·5% for lucerne and 12·0% for white clover. Bailey (1964) reported that the hemicellulose content of white clover grown in New Zealand and sampled at monthly intervals during a 6-month period ranged from 3·6 to 6·6%.

PECTIN

In contrast to hemicellulose, pectin contents are generally higher in legumes than in grasses. Hirst *et al.* (1959) reported pectin contents in lucerne of 7·6 to 13·1%, the highest content being at the seed-setting stage; leaf material contained 10–12% pectin. Bailey (1964) reported pectin contents of 3·5–7·7% in white clover herbage, and Wright (1961) 1·5–5·9% in red clover herbage.

Non-structural carbohydrates

The composition of the non-structural carbohydrate fraction of clover and lucerne herbage differs from that of temperate grasses, in that sucrose and starch, rather than fructosan, are present as reserve materials (Smith *et al.*, 1964). Fructosan has not been identified in clover or lucerne (Ferguson, 1963; Salo, 1965), but there appears to be no information on this point for sainfoin. Because starch is only slightly soluble in water, extraction with water does not provide a satisfactory estimate of the total non-structural ('available') carbohydrate content in herbage legumes. Grotelueschen and Smith (1967) considered that extraction with takadiastase enzyme gave the most accurate value for non-structural carbohydrates in legume herbage.

There are considerable diurnal variations in the contents of non-structural carbohydrates. With white clover and red clover, Bailey (1958) found that starch contents were lowest after several hours of darkness and highest after several hours of bright sunshine. With lucerne, Holt and Hilst (1969) reported that water-soluble carbohydrate contents reached a peak between noon and 1800 h, but that acid-soluble polysaccharide contents were highest at 1800 h.

Comparative data for total available carbohydrates in white clover, red clover and lucerne as assessed by enzyme extraction have been reported by van Riper and Smith (1959) and for water-soluble carbohydrates by Davies *et al.* (1966). The results for primary spring growth, summarized in Table 31.1, show very close agreement. Changes with advancing maturity showed no consistent trend.

Table 31.1. Non-structural carbohydrate content of legumes—mean and range values (% of DM)

	White clover	Red clover	Lucerne	References
Total available carbohydrates	9·4–12·8	8·0–11·8	7·4–8·9	Van Riper and
	11·4	9·8	7·8	Smith (1959)
Water-soluble carbohydrates	8·1–13·1	8·5–11·0	6·6–8·0	Davies *et al.* (1966)
	11·4	9·9	7·2	

In white clover cut at monthly intervals during a 6-month period, the water-soluble carbohydrate content varied from 5·2 to 14·6% (Bailey, 1964). Values in S100 white clover, cut 4 times per year but sampled more frequently, varied from 6·3 to 10·5% (Vez, 1961). On average, sucrose constituted about half the total water-soluble carbohydrate content, and starch and reducing sugars about one quarter each. Two other cultivars of white clover had similar composition. Lucerne sampled at 6 stages of primary growth was found to have water-soluble carbohydrate contents of 8–11%, with a maximum value in early May (Terry & Tilley, 1964).

A certain amount of work carried out on the available carbohydrate content of stolons, crowns and roots of legume species has been summarized by Weinmann (1961).

Lignin

Little information is available on contents of lignin in legumes, particularly for species other than lucerne. In lucerne and red clover, the lignin content has been reported to increase from about 4–5% in young herbage at the vegetative stage in April, to 8–10% in mature herbage at the flowering stage (Armstrong *et al.*, 1950; Kivimäe, 1959; Terry & Tilley, 1964; Salo, 1965). Lucerne at the seeding stage was reported by Hirst *et al.* (1959) to contain 15·8% lignin. White clover appears to have a lower lignin content than red clover and lucerne. In herbage sampled at monthly intervals during a 3-month period, Bailey (1964) found a lignin content of 2·0–2·6%. Ladino white clover was reported by Plummer (1953) to have a lower content than red clover.

Nitrogen compounds

Total N contents are highest in young vegetative growth and decline with increasing maturity. The decline is normally less in white clover than in red clover, lucerne and sainfoin (Baker *et al.*, 1952; Fleming & Coulter, 1963; Whitehead & Jones, 1969), although Davies *et al.* (1966) found a similar decline in all 4 species. Even at a young stage of growth, white clover has a higher N content than the other species. Data on changes in N content of legumes are shown in Fig. 31.1 overleaf. Differences between white clover, red clover and lucerne are not very great, but sainfoin tends to have the lowest content (see Table 31.2).

Table 31.2. N contents of legume species—mean and range values (% of DM)

White clover	Red clover	Lucerne	Sainfoin	References
3·42–4·73	2·34–4·57	2·29–5·19	...	Van Riper and Smith (1959)
3·94	3·34	3·47		
3·9–5·3	2·6–4·7	...	...	Fleming and Coulter (1963)
4·4	3·4			
2·66–4·69	3·10–4·08	2·34–3·76	...	Davies *et al.* (1966)
3·50	2·93	2·82		
3·38–5·21	2·50–4·18	2·06–4·30	2·00–4·56	Whitehead and Jones (1969)
4·42	3·40	2·94	2·87	

With frequent defoliation, the N content remains at a relatively high level. In New Zealand, white clover sampled at monthly intervals over a 6-month period had N contents in the range 4·2–5·2% with no particular trend (Bailey, 1964). With 4 cuts per year but more frequent sampling, Vez (1961) found N contents of 3·2–4·8%.

PROTEINS

Of the total herbage N approximately 75% occurs in protein form. For

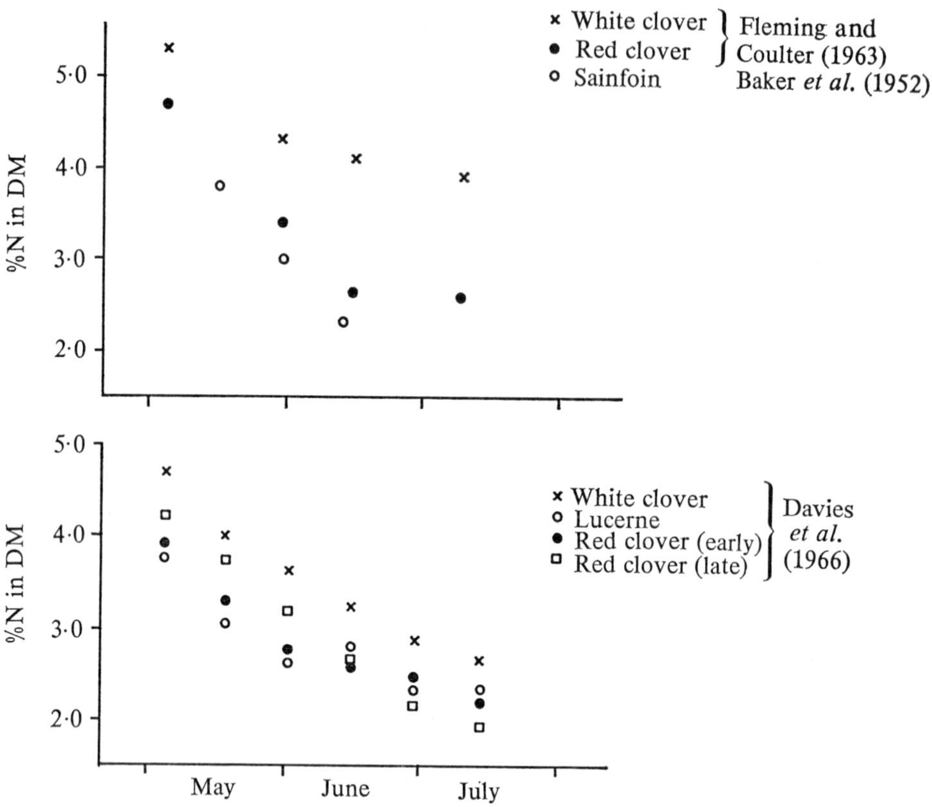

Fig. 31.1. Changes in N content of legumes with advancing maturity.

white clover, a value of 75% was reported by Ferguson and Terry (1954), but 90% by Bailey (1964) for plants grown in New Zealand. For lucerne, 71% was reported by Ferguson and Terry (1954) and 63–71% by Wilson and Tilley (1965). There was no consistent change with advancing maturity.

Protein preparations extracted from red clover, lucerne and sainfoin by Gerloff *et al.* (1965) were similar in amino-acid composition to a preparation from ryegrass. Wilson and Tilley (1965) also found that protein preparations from lucerne were similar in amino-acid composition to those from various grasses.

SOLUBLE ORGANIC NITROGEN COMPOUNDS

The investigation of Ferguson and Terry (1954) indicated that, of the non-protein N in white clover and lucerne, rather more than half was present as peptides. The remainder consisted mainly of amino acids, amides and bases such as purine and betaine.

INORGANIC NITROGEN

Contents of inorganic N in legume herbage are normally low. Ferguson and Terry (1954) found that in both white clover and lucerne, about 0·6% of the total N was present in ammonium form. Nitrate N represented 1·0% of the total N (0·040% in the DM) in white clover and 1·3% (0·037% in the DM) in lucerne. There are several other reports that white clover grown without fertilizer N contained less than 0·08% nitrate N in the DM or had less than 2% of the total N in this form (Bailey, 1964; Smith & Sund, 1965; Murphy & Smith, 1967). With lucerne, however, Smith and Sund (1965) reported contents of up to 0·14% nitrate N, equivalent to 2·7% of the total N, in plants grown without fertilizer N.

When fertilizer N is applied to legumes, relatively high levels of nitrate can accumulate in the herbage. Murphy and Smith (1967) reported 0 38% nitrate N in Ladino white clover a few weeks after the application of 336 kg N as ammonium nitrate per ha, but only 0·08% in similarly treated lucerne. Smith and Sund (1965), however, reported a content of 0·23% nitrate N in lucerne one month after application of 448 kg N as ammonium nitrate per ha.

Lipids

Legumes appear to be similar to grasses in lipid content (Van Riper & Smith, 1959; Shorland, 1961) and to have a similarly high proportion of galactolipids (Weenink, 1961; Shorland, 1961). Smith (1964) reported an average fat (ether-extract) content of 2·95% in Ladino white clover, 2·65% in red clover and 2·03% in lucerne. There was no consistent change with advancing maturity. Sainfoin sampled at various growth stages by Baker *et al.* (1952) had an average fat content of 2·5% and showed no definite change with maturity.

Organic acids

Hirst and Ramstad (1957) and Ferguson (1963) have stated that clovers and lucerne appear to contain greater quantities of organic acids than grasses, but there is little information on this point. Ferguson (1948) reported values up to 11·6% for red clover; and Vez (1961) found that the organic acid content in shoots of white clover ranged from 6 to 8% and from 3 to 5% in the stolons. From the data on ash alkalinity and cation-anion balance given by Whitehead and Jones (1969) and assuming all the organic acid to be present as malic acid, it can be calculated that the average organic acid contents are: white clover 9·5%, red clover 9·2%, lucerne 8·1% and sainfoin 5·4%.

Richardson and Hulme (1957) reported that the principal organic acids in lucerne were succinic, malic and malonic acids but gave no information on total quantities present. *Trans*-aconitic acid occurs in legume species and

Stout *et al.* (1967) reported the rather high level of 0·9% in one sample of lucerne.

Vitamins, oestrogens and goitrogenic compounds

VITAMIN A AND CAROTENE

Legumes generally contain rather more carotene than grass, contents usually being in the range 20–25 mg/100g according to Ferguson (1963).

Van Riper and Smith (1959) found the following means and ranges of contents (mg/100g) of carotene: white clover 26 (24–28), red clover 34 (28–36), lucerne 19 (11–23). The content increased until the pre-bud stage and then tended to decline. A similar decline in carotene content after the pre-bud stage has also been reported in white clover, the range of values found being 14–55 mg/100 g by Smith and Wang (1941), in red clover 7–45 mg/100 g and in lucerne 9–27 mg/100 g by Kivimäe (1959), and in sainfoin 9–25 mg/100 g by Baker *et al.* (1952).

VITAMIN B COMPLEX

The data summarized by Cowlishaw and Eyles (1957) indicate that contents of most of the vitamins of the B complex are generally higher in legumes than in grasses.

VITAMINS C AND D

See pp. 114-15.

VITAMIN E

White clover and lucerne were found by Brown (1953) and by Booth (1964) to be similar to grasses in α-tocopherol content. This content more than doubled during April-November (Booth, 1964).

OESTROGENIC COMPOUNDS

Information on the oestrogenic activity in legume herbage and on the effects of oestrogenic compounds on animal production has been reviewed by Moule *et al.* (1963), Hanson and Loper (1965) and Bickoff (1968).

In many investigations, oestrogenic activity has been assessed only by biological methods, with no chemical analysis of the herbage. In other investigations attempts have been made to isolate the oestrogenic compounds, and several of these have been identified (Bickoff *et al.*, 1965). However, Lindner (1967) has pointed out that "no consistent correlation has yet emerged between the concentration of any one of the known oestrogenic isoflavones in clover pastures and the severity of the reproductive disturbances in livestock consuming them."

Much of the early work on oestrogenic activity was carried out in Australia where subterranean clover was found to have particularly high levels of activity. Subsequent work has demonstrated appreciable levels of activity in red clover, white clover and lucerne in Britain (Newton & Betts, 1968), as well as in other countries. Sainfoin showed only slight activity (Newton & Betts, 1968).

Various environmental factors have been reported as influencing the level of oestrogenic activity in herbage. With subterranean clover, P deficiency was found to increase the concentration of oestrogenic isoflavones estimated chemically (Rossiter & Beck, 1966). The stage of maturity of the herbage and season of the year were found to influence the activity in various legume species, the highest values generally occurring in spring and the lowest in mid-summer (Kitts *et al.*, 1959). With lucerne, work by Loper and Hanson (1964) indicated that the degree of infection by pathogenic fungi was the dominant factor governing the coumestrol content as assessed by chemical analysis. The content was 0·4–2·1 ppm. in non-infected leaf herbage and 28–184 ppm. in leaves infected with leaf-spot pathogens. Differences in coumestrol content due to stage of growth, temperature and P supply were found to be not significant in this investigation.

Activity of an anti-oestrogenic type, assessed biologically, has also been detected in various legume species (Biely & Kitts, 1964) but its chemical nature has not been elucidated.

GOITROGENIC COMPOUNDS

Evidence for the occurrence of appreciable quantities of goitrogenic compounds in legume herbage has been summarized by Calderbank (1963). Cyanogenetic glucosides appear to be the main source of goitrogenic activity. They are hydrolysed in the rumen with the release of hydrocyanic acid and this, in turn, is converted into thiocyanate, which inhibits the uptake of I by the thyroid gland (Russell & Duncan, 1956; Calderbank, 1963). There appear to be no comparative data for the different legume species.

Mineral elements

As in Chapter 13, the mineral elements considered are those essential for plants or animals, as well as others which can have undesirable nutritional effects on animals.

PHOSPHORUS

Legume herbage is similar to grass herbage in P content. This content shows a fairly steady decline with advancing maturity in red clover, lucerne and sainfoin (Thomas *et al.*, 1952; Van Riper & Smith, 1959; Fleming & Coulter, 1963; Davies *et al.*, 1966; Davies *et al.*, 1968; Whitehead & Jones, 1969). Davies *et al.* (1968) found a similar decline in white clover, but the other workers who studied white clover found the decline in P content to be considerably less than in the other legume species. Comparative data for the 4 legume species are summarized in Table 31.3.

A critical deficiency value (defined on pp. 115-16) of 0·23% P for white clover (mature leaf+petiole, sampled immediately before flowering) has been reported by Andrew (1960).

There is some evidence suggesting that legumes show greater increases in P content after applications of P than do grasses (Whitehead, 1966).

Table 31.3. P content of legume species—mean and range values (% in DM)

White clover	Red clover	Lucerne	Sainfoin	References
...	...	0·24–0·66 0·40	0·22–0·55 0·39	Thomas *et al.* (1952)
0·28–0·45 0·34	0·22–0·40 0·29	0·22–0·51 0·32	...	Van Riper and Smith (1959)
0·35–0·47 0·40	0·26–0·45 0·34	...	...	Fleming and Coulter (1963)
0·19–0·32 0·25	0·14–0·24 0·19	0·14–0·22 0·16	...	Davies *et al.* (1966)
0·32–0·47 0·38	0·18–0·34 0·27	0·20–0·38 0·26	0·20–0·42 0·28	Whitehead and Jones (1969)

POTASSIUM

Contents of K are similar in grasses and legumes. In all the investigations summarized in Table 31.4, K contents were higher in young than in mature herbage. However, some fluctuations occurred, and in white clover the decline with maturity was less marked than with the other species.

A critical deficiency value of 1·8% K in whole leaves (leaflets+petioles) was found for both white clover and red clover by McNaught (1958). However, the data obtained by Andrew (1960) indicated a critical value for white clover of 1·1% K in mature leaf+petiole material. For lucerne, critical values were reported to range from 1·0–1·75% (see MacLeod, 1965). The increase in herbage K content due to the application of K was found to be greater in clover than in grass (Hemingway, 1961).

Table 31.4. K content of legume species—mean and range values (% in DM)

White clover	Red clover	Lucerne	Sainfoin	References
...	...	1·55–3·92 2·61	1·56–3·69 2·54	Thomas *et al.* (1952)
1·54–3·06 2·09	0·88–2·68 1·59	1·06–2·18 1·55	...	Van Riper and Smith (1959)
2·7–3·1 2·9	1·8–3·9 2·9	...	...	Fleming and Coulter (1963)
2·4–3·8 3·11	2·6–4·1 3·42	1·8–3·5 2·55	...	Davies *et al.* (1966)
1·56–2·82 2·26	1·40–2·68 2·07	1·28–2·10 1·65	1·18–2·52 1·73	Whitehead and Jones (1969)

CALCIUM

Ca contents in white clover, red clover and lucerne are generally about three times as high as those in grasses sampled on the same date. Sainfoin has a lower content, about half that of the other legumes; comparative data are given in Table 31.5. Differences among red clover cultivars and to a lesser extent those of white clover have been reported (Davies *et al.*, 1968). Changes in Ca content with advancing maturity are not consistent, and most investigations have shown fluctuating values.

Table 31.5. Ca content of legume species—mean and range values (% in DM)

White clover	Red clover	Lucerne	Sainfoin	References
...	...	1·86–2·57 2·15	0·89–1·31 1·03	Thomas *et al.* (1952)
1·20–1·55 1·37	1·36–1·84 1·60	1·13–1·57 1·36	...	Van Riper and Smith (1959)
1·4–1·6 1·5	1·4–1·6 1·5	...	...	Fleming and Coulter (1963)
1·22–1·51 1·36	1·19–1·45 1·25	0·90–1·20 1·05	...	Davies *et al.* (1966)
1·86–2·31 2·10	1·36–2·42 1·84	1·54–2·27 1·82	0·84–1·02 0·94	Whitehead and Jones (1969)

MAGNESIUM

Mg content is generally rather higher in legumes than in grasses, especially in spring (Whitehead, 1966). Red clover tends to have a higher content than the other 3 legume species (see Table 31.6).

Mg content fluctuates with advancing maturity (see references in Table 31.6) but in regularly cut herbage it tends to rise steadily throughout the season, often until October (Hemingway, 1961; Todd, 1961; Reith, 1963).

Table 31.6. Mg content of legume species—mean and range values (% in DM)

White clover	Red clover	Lucerne	Sainfoin	References
...	...	0·50–0·64 0·56	0·55–1·16 0·80	Thomas *et al.* (1952)
0·20–0·29 0·24	0·31–0·38 0·34	...	...	Fleming and Coulter (1963)
0·17–0·21 0·18	0·20–0·30 0·26	...	...	Jones (1963)
0·17–0·21 0·19	0·17–0·25 0·22	0·15–0·20 0·17	...	Davies *et al.* (1966)
0·15–0·21 0·18	0·18–0·24 0·21	0·11–0·17 0·15	0·14–0·20 0·17	Whitehead and Jones (1969)

SULPHUR

Legumes contain approximately the same amount of S as grasses growing under the same conditions. Comparative data for the four main legume species considered are scanty. Whitehead and Jones (1969) obtained the following mean and seasonal range values: white clover 0·29% (0·24–0·36), red clover 0·21% (0·15–0·24), lucerne 0·27% (0·20–0·32), and sainfoin 0·24% (0·20–0·34). Apart from lucerne, which showed a peak value in late June, there was a trend for S contents to decline with advancing maturity. A critical deficiency value of 0·22% S in the herbage of mature lucerne was found by Harward *et al.* (1962).

SODIUM

Na content in legumes, as in grasses, shows wide variation. Comparative data for the 4 species are summarized in Table 31.7. On average, Na content in white clover is considerably higher than in the other species. The values for sainfoin show poor agreement; Thomas *et al.* (1952) and Whitehead and Jones (1969) found very low and Kerguelen (1960) relatively high contents. Differences between red clover have been reported by Davies *et al.* (1966, 1968).

Table 31.7. Na content of legume species—mean and range values (% in DM)

White clover	Red clover	Lucerne	Sainfoin	References
...	...	0·04–0·10 0·07	0·02–0·05 0·04	Thomas *et al.* (1952)
0·41–0·48	0·14	0·18–0·63	0·20–0·32	Kerguelen (1960)*
0·30–0·46 0·41	0·12–0·20 0·17	...	...	Fleming and Coulter (1963)
0·20–0·33 0·28	0·03–0·08 0·06	0·11–0·20 0·14	...	Davies *et al.* (1966)
0·05–0·20 0·12	0·02–0·05 0·04	0·04–0·08 0·06	0·01–0·01 0·01	Whitehead and Jones (1969)

* Samples were taken irregularly and included some regrowths.

CHLORINE

The only comparative data for the Cl content of the 4 legume species appear to be those obtained recently by Whitehead and Jones (1969). Mean and seasonal range values were: white clover 0·76% (0·62–0·91), red clover 0·53% (0·32–0·62), lucerne 0·60% (0·54–0·72), and sainfoin 0·38% (0·32–0·46). Stewart and Bear (1951) reported a range of 0·34–1·56% Cl for Ladino white clover.

IRON

Contents of Fe average nearly twice as high in legumes as in grasses (Whitehead, 1966). The limited amount of data available (Table 31.8) suggest that differences between the 4 cultivated species are small. Beeson and MacDonald (1951) noted that Fe content increased somewhat with advancing maturity, but Thomas *et al.* (1952) and Loper and Smith (1961) found that it generally decreased.

Table 31.8. Fe content of legume species—mean and range values (ppm. in DM)

White clover	Red clover	Lucerne	Sainfoin	References
120–195	...	85–120	...	Beeson and MacDonald (1951)
155		102		
...	...	217–558	131–360	Thomas *et al.* (1952)
		291	231	
202–448	164–362	153–596	...	Loper and Smith (1961)
291	217	260		
102–137	74–97	78–182	73–132	Whitehead and Jones (1969)
117	85	108	93	

MANGANESE

In some cases Mn content has been found to be lower in legumes than in grasses growing on the same soils (Fleming, 1965). Differences between the legume species are small (Table 31.9). Mn contents were not found to change consistently with advancing maturity, although in several investigations highest values were found at the earliest growth stage.

Table 31.9. Mn content of legume species—mean and range values (ppm. in DM)

White clover	Red clover	Lucerne	Sainfoin	References
47–63	...	29–37	...	Beeson and MacDonald (1951)
56		31		
48–54	43–62	46–73	...	Loper and Smith (1961)
51	50	55		
87	75	...	...	Fleming (1965)
40–57	36–52	35–52	44–62	Whitehead and Jones (1969)
49	44	42	52	

ZINC

Legumes appear to have Zn contents similar to those of grasses, but information on this point is limited. Comparative data are given in Table 31.10.

Table 31.10. Zn content of legume species—mean and range values (ppm. in DM)

White clover	Red clover	Lucerne	Sainfoin	References
25–32	30–35	23–36	...	Loper and Smith (1961)
29	32	28		
22–31	21–29	20–34	20–41	Whitehead and Jones (1969)
25	24	24	28	

COPPER

Legumes show a greater variation in Cu content than do grasses. Mitchell *et al.* (1957) found that clovers growing on various sites in north-east Scotland had Cu contents of 1·7–12·3 ppm., whereas the range for ryegrass on the same sites was 2·0–4·3 ppm. The data reported in Table 31.11 suggest that red clover has a rather higher Cu content than the other legume species. In lucerne and red clover, the highest values are found in young herbage, but in white clover Cu content changes little with advancing maturity.

Table 31.11. Cu content of legume species—mean and range values (ppm. in DM)

White clover	Red clover	Lucerne	Sainfoin	References
7·5–9·7	...	8·3–9·3	...	Beeson and MacDonald (1951)
8·7		8·9		
...	...	6·8–12·1	5·0–10·4	Thomas *et al.* (1952)
		9·1	8·1	
7·4–9·5	9·2–11·6	5·8–9·0	...	Lôper and Smith (1961)
8·0	10·2	6·9		
7·8–8·0	7·5–10·8	...	...	Fleming (1965)
7·9	9·0			
5·4–8·2	5·8–9·1	6·0–8·3	5·9–8·2	Whitehead and Jones (1969)
7·3	7·4	7·0	7·0	

COBALT

Legumes generally contain considerably more Co than grasses growing under the same conditions (Mitchell, 1955; Latteur, 1962; Hayakawa, 1962; Andrews, 1966), but Thomas *et al.* (1952) found no significant difference between several grasses and lucerne and sainfoin. Kubota (1964) reported that white clover and red clover had almost identical Co contents when grown together on a range of soil types, and that lucerne generally had a lower though more variable Co content. Comparative data are given in Table 31.12.

Table 31.12. Co content of legume species—mean and range values (ppm. in DM)

White clover	Red clover	Lucerne	Sainfoin	References
0·15–0·26	...	0·09–0·18	...	Beeson and MacDonald (1951)
0·22		0·14		
...	...	0·11–0·25	0·10–0·24	Thomas *et al.* (1952)
		0·15	0·16	
0·10–0·22	0·12–0·23	0·08–0·28	...	Loper and Smith (1961)
0·14	0·15	0·13		
0·18–0·38	0·17–0·36	...	...	Andrews (1966)
0·24	0·23			
0·10–0·15	...	0·14–0·39	...	Andrews (1966)
0·13		0·23		
0·15–0·23	0·10–0·26	0·13–0·20	0·12–0·24	Whitehead and Jones (1969)
0·19	0·21	0·16	0·17	

IODINE

There appears to be no marked difference between legumes and grasses in I content. Johnson and Butler (1957) reported that the I content of white clover is about double that of red clover (0·9 *v.* 0·4 ppm.); Alderman and Jones (1967) found that the I content of single samples of ten varieties of white clover ranged from 0·14 to 0·44 ppm.

MOLYBDENUM

The results obtained by Whitehead and Jones (1969) indicated considerable differences in Mo contents between legume species grown on a calcareous soil, mean values in May–July being (ppm.): white clover 0·64, red clover 0·44, lucerne 0·18, sainfoin 0·18.

SELENIUM

There is some evidence that legumes tend to have lower contents of Se than grasses at low levels of supply, but higher contents at high levels (Davies & Watkinson, 1966).

The Se content of white clover growing on an Se-deficient soil in New Zealand ranged from 0·005 to 0·023 ppm. (Davies & Watkinson, 1966), whereas white clover growing on a seleniferous soil in Ireland contained 153 ppm. Se and red clover, 103 ppm. (Fleming, 1962).

LEAD

The data of Mitchell and Reith (1966) indicate that the Pb content of clovers is similar to that of grasses and, as with grasses, it increases markedly with senescence.

SILICA

Legumes generally contain only 5–10% as much silica as grasses (Jones & Handreck, 1967).

References

ALDERMAN, G.; JONES, D. I. H. 1967. The iodine content of pastures. *J. Sci. Fd Agric.* **18,** 197–9.

ANDREW, C. S. 1960. The effect of phosphorus, potassium and calcium on the growth, chemical composition and symptoms of deficiency of white clover in a subtropical environment. *Aust. J. agric. Res.* **11,** 149–61.

ANDREWS, E. D. 1966. Cobalt concentrations in some New Zealand fodder plants grown on cobalt-sufficient and cobalt-deficient soils. *N.Z. Jl agric. Res.* **9,** 829–38.

ARMSTRONG, D. G.; COOK, H.; THOMAS, B. 1950. The lignin and cellulose contents of certain grassland species at different stages of growth. *J. agric. Sci., Camb.* **40,** 93–9.

BAILEY, R. W. 1958. Carbohydrates in pasture species. 1. The starch content of clovers and ryegrass. *J. Sci. Fd Agric.* **9,** 743–7.

BAILEY, R. W. 1964. Pasture quality and ruminant nutrition. 1. Carbohydrate composition of ryegrass varieties grown as sheep pastures. *N.Z. Jl agric. Res.* **7,** 496–507.

BAKER, C. J. L.; HEIMBERG, M.; ALDERMAN, G.; EDEN, A. 1952. Studies on the composition of sainfoin. *J. agric. Sci., Camb.* **42,** 382–94.

BAUMGARDT, B. R.; SMITH, D. 1962. Changes in the estimated nutritive value of the herbage of alfalfa, medium red clover and bromegrass due to stage of maturity and year. *Res. Rep. 10 Wis. agric. Exp. Stn* pp. 17.

BEESON, K. C.; MACDONALD, H. A. 1951. Absorption of mineral elements by forage plants. 3. The relation of stage of growth to the micro-nutrient element content of timothy and some legumes. *Agron. J.* **43,** 589–93.

BICKOFF, E. M. 1968. Oestrogenic constituents of forage plants. *Review Series 1/1968 Commonw. Bur. Past. Fld Crops* Farnham Royal: Commonw. Agric. Bur., pp. 39.

BICKOFF, E. M.; LIVINGSTON, A. L.; GUGGOLZ, J. 1965. Isolation of phenolic compounds from ladino clover. *J. agric. Fd Chem.* **13,** 151–5.

BIELY, J.; KITTS, W. D. 1964. The anti-oestrogenic activity of certain legumes and grasses. *Can. J. Anim. Sci.* **44,** 297–302.

BOOTH, V. H. 1964. The α-tocopherol content of forage crops. *J. Sci. Fd Agric.* **15,** 342–4.

BROWN, F. 1953. The tocopherol content of farm feeding stuffs. *J. Sci. Fd Agric.* **4,** 161–5.

CALDERBANK, G. 1963. Iodine. In: Worden, A. N.; Sellers, K. C.; Tribe, D. E. [Eds] Animal health, production and pasture. London: Longmans, Green and Co. Ltd, pp. 681–736.

COWLISHAW, S. J.; EYLES, D. E. 1957. The nutritive value of herbage for poultry. *Nutr. Abstr. Rev.* **27,** 983–96.

DAVIES, E. B.; WATKINSON, J. H. 1966. Uptake of native and applied selenium by pasture species. 1. *N.Z. Jl agric. Res.* **9,** 317–27

DAVIES, W. E.; GRIFFITH, G. AP; ELLINGTON, A. 1966. The assessment of herbage legume varieties. 2. *In vitro* digestibility, water-soluble carbohydrate, crude protein and mineral content of primary growth of clover and lucerne. *J. agric. Sci., Camb.* **66,** 351–7.

DAVIES, W. E.; THOMAS, T. A.; YOUNG, N. R. 1968. The assessment of herbage legume varieties. 3. Annual variation in chemical composition of eight varieties. *J. agric. Sci., Camb.* **71,** 233–41.

FERGUSON, W. S. 1948. The analysis of plant material with particular reference to structural constituents. *Agric. Prog.* **23,** 129–37.

FERGUSON, W. S. 1963. Pasture Chemistry. In: Worden, A. N.; Sellers, K. C.; Tribe, D. E. [Eds] Animal health, production and pasture. London: Longmans, Green and Co. Ltd, pp. 92–127.

FERGUSON, W. S.; TERRY, R. A. 1954. The fractionation of the non-protein nitrogen of grassland herbage. *J. Sci. Fd Agric.* **5,** 515–24.

FLEMING, G. A. 1962. Selenium in Irish soils and plants. *Soil Sci.* **94,** 28–35.

FLEMING, G. A. 1963. Distribution of major and trace elements in some common pasture species. *J. Sci. Fd Agric.* **14,** 203–8.

FLEMING, G. A. 1965. Trace elements in plants with particular reference to pasture species. *Outl. Agric.* **4,** 270–85.

FLEMING, G. A.; COULTER, B. S. 1963. Mineral elements in pasture plants. *Proc. 1st Reg. Conf. int. Potash Inst., Wexford (Ireland) 1963* pp. 63–70.

GERLOFF, E. D.; LIMA, I. H.; STAHMANN, M. A. 1965. Amino acid composition of leaf protein concentrates. *J. agric. Fd Chem.* **13,** 139–43.

GROTELUESCHEN, R. D.; SMITH, D. 1967. Determination and identification of non-structural carbohydrates removed from grass and legume tissue by various sulfuric acid concentrations, takadiastase and water. *J. agric. Fd Chem.* **15,** 1048–51.

HANSON, C. H.; LOPER, G. M. 1965. Variation in coumestrol content of alfalfa as related to location, variety, cutting year, stage of growth and disease. *Tech. Bull. 1333 U.S. Dep. Agric.* pp. 72.

HARWARD, M. E.; CHAO, T. T.; FANG, S. C. 1962. The sulfur status and sulfur supplying power of Oregon soils. *Agron. J.* **54,** 101–6.

HAYAKAWA, T. 1962. *Natn. Inst. Anim. Hlth Q., Tokyo* **2,** 172 [Cited by Underwood, E. J. 1966. The mineral nutrition of livestock. FAO and Commonw. Agric. Bur. p. 157.]

HEMINGWAY, R. G. 1961. Magnesium, potassium, sodium and calcium contents of herbage as influenced by fertilizer treatments over a three year period. *J. Br. Grassld Soc.* **16,** 106–16.

HIRST, E. L.; MACKENZIE, D. J.; WYLAM, C. B. 1959. Analytical studies on the carbohydrates of grasses and clovers. 9. Changes in carbohydrate composition during the growth of lucerne. *J. Sci. Fd Agric.* **10,** 19–26.

HIRST, E. L.; RAMSTAD, S. 1957. Changes in organic acid content of perennial ryegrass during conservation. *J. Sci. Fd Agric.* **8,** 727–32.

HOLT, D. A.; HILST, A. R. 1969. Daily variation in carbohydrate content of selected forage crops. *Agron. J.* **61,** 239–42.

JARRIGE, R. 1960. The membrane constituents of herbage. *Proc. 8th int. Grassld Congr., Reading, 1960* pp. 628–34.

JARRIGE, R. 1963. [The membrane constituents of forage plants.] *Annls Biol. anim. Biochim. Biophys.* **3,** 143–90.

JOHNSON, J. M.; BUTLER, G. W. 1957. Iodine content of pasture plants. *Physiologia Pl.* **10,** 100–11.

JONES, L. H. P.; HANDRECK, K. A. 1967. Silica in soils, plants and animals. *Adv. Agron.* **19,** 107–49.

KERGUELEN, M. 1960. [Variations in the chemical composition of some forage crops in relation to species, vegetation stage, and soil and manurial conditions.] *Annls Amél. Pl.* **10,** 177–236.

KITTS, W. D.; SWIERSTRA, E.; BRINK, V. C.; WOOD, A. J. 1959. The estrogen-like substances in certain legumes and grasses. 2. The effect of stage of maturity and frequency of cutting on the estrogenic activity of some forages. *Can. J. Anim. Sci.* **39,** 158–63.

KIVIMÄE, A. 1959. Chemical composition and digestibility of some grassland crops. *Acta Agric. scand.* Suppl. 5, pp. 142.

KUBOTA, J. 1964. Cobalt content of New England soils in relation to cobalt levels in forages for ruminant animals. *Proc. Soil Sci. Soc. Am.* **28,** 246–51.

LATTEUR, J. P. 1962. Cobalt deficiencies and sub-deficiencies in ruminants. Brussels: Centre d'Information du Cobalt, pp. 48.

LINDNER, H. R. 1967. Study of the fate of phyto-oestrogens in the sheep by determination of isoflavones and coumestrol in the plasma and adipose tissue. *Aust. J. agric. Res.* **18,** 305–33.

LOPER, G. M.; HANSON, C. H. 1964. Influence of controlled environmental factors and two foliar pathogens on coumestrol in alfalfa. *Crop Sci.* **4,** 480–2.

LOPER, G. M.; SMITH, D. 1961. Changes in micronutrient composition of the herbage of alfalfa, medium red clover, ladino clover and bromegrass with advance in maturity. *Res. Rep. 8 Wis. agric. Exp. Stn* pp. 19.

MACLEOD, L. B. 1965. Effect of nitrogen and potassium on the yield and chemical composition of alfalfa, bromegrass, orchard grass and timothy grown as pure species. *Agron. J.* **57,** 261–6.

MCNAUGHT, K. J. 1958. Potassium deficiency in pastures. 1. Potassium content of legumes and grasses. *N.Z. Jl agric. Res.* **1,** 148–81.

MITCHELL, R. L. 1955. Trace elements and liming. *Scott. Agric.* **34,** 139–43.

MITCHELL, R. L.; REITH, J. W. S. 1966. The lead content of pasture herbage. *J. Sci. Fd Agric.* **17,** 437–40.

MITCHELL, R. L.; REITH, J. W. S.; JOHNSTON, I. M. 1957. Soil copper status and plant uptake. In: Plant analysis and fertilizer problems. I.R.H.O., Paris, pp. 249–59.

MOULE, G. R.; BRADEN, A. W. H.; LAMOND, D. R. 1963. The significance of oestrogens in pasture plants in relation to animal production. *Anim. Breed. Abstr.* **31,** 139–57.

MURPHY, L. S.; SMITH, G. E. 1967. Nitrate accumulations in forage crops. *Agron. J.* **59,** 171–4.

NEWTON, J. E.; BETTS, J. E. 1968. Seasonal oestrogenic activity of various legumes. *J. agric. Sci., Camb.* **70,** 77–82.

PLUMMER, B. E. 1953. Chemical composition of grasses and legumes in Maine. *Bull. 513 Maine agric. Exp. Stn* pp. 29.

REITH, J. W. S. 1963. The magnesium contents of soils and crops. *J. Sci. Fd Agric.* **14,** 417–26.

RICHARDSON, A.; HULME, A. C. 1957. The non-volatile organic acids of lucerne. *J. Sci. Fd Agric.* **8,** 326–30.

RIPER, G. E. VAN; SMITH, D. 1959. Changes in the chemical composition of the herbage of alfalfa, medium red clover, ladino clover and bromegrass with advance in maturity. *Res. Rep. 4 Wis. agric. Exp. Stn* pp. 25.

ROSSITER, R. C.; BECK, A. B. 1966. Physiological and ecological studies on the oestrogenic isoflavones in subterranean clover (*T. subterraneum L.*). 2. Effects of phosphate supply. *Aust. J. agric. Res.* **17**, 447–56.

RUSSELL, F. C.; DUNCAN, D. L. 1956. Minerals in pasture: deficiencies and excesses in relation to animal health. *Tech. Commun. 15, Commonw. Bur. Anim. Nutr.* Farnham Royal: Commonw. Agric. Bur., 2nd edn, pp. 170.

SALO, M. L. 1965. On the content of cell wall constituents in various plant materials. *Maataloust. Aikakausk.* **37**, 127–34.

SHORLAND, F. B. 1961. Acetone-soluble lipids of grasses and other forage plants. 2. General observations on the properties of the lipids with special reference to the yield of fatty acids. *J. Sci. Fd Agric.* **12**, 39–43.

SMITH, A. M.; WANG, T. 1941. The carotene content of certain species of grassland herbage. *J. agric. Sci., Camb.* **31**, 370–8.

SMITH, D. 1964. Chemical composition of herbage with advance in maturity of alfalfa, medium red clover, ladino clover and birdsfoot trefoil. *Res. Rep. 16 Wis. agric. Exp. Stn* pp. 10.

SMITH, D.; PAULSEN, G. M.; RAGUSE, C. A. 1964. Extraction of total available carbohydrates from grass and legume tissue. *Pl. Physiol., Lancaster* **39**, 960–2.

SMITH, D.; SUND, J. M. 1965. Nitrate in forage. Influence of stage of growth and soil nitrogen on nitrate content of herbage of alfalfa, red clover, ladino clover, trefoil and bromegrass. *J. agric. Fd Chem.* **13**, 81–4.

STEWART, I.; BEAR, F. E. 1951. Ladino clover, its mineral requirements and chemical composition. *Bull. 759 New Jersey agric. Exp. Stn* pp. 32.

STOUT, P. R.; BROWNELL, J.; BURAU, R. G. 1967. Occurrence of transaconitate in range forage species. *Agron. J.* **59**, 21–4.

TERRY, R. A.; TILLEY, J. M. A. 1964. The digestibility of the leaves and stems of perennial ryegrass, cocksfoot, timothy, tall fescue, lucerne and sainfoin as measured by an *in vitro* procedure. *J. Br. Grassld Soc.* **19**, 363–72.

THOMAS, B.; THOMPSON, A.; OYENUGA, V. A.; ARMSTRONG, R. H. 1952. The ash constituents of some herbage plants at different stages of maturity. *Emp. J. exp. Agric.* **20**, 10–22.

TODD, J. R. 1961. Magnesium in forage plants. 1. Magnesium contents of different species and strains as affected by season and soil treatment. *J. agric. Sci., Camb.* **56**, 411–15.

VEZ, A. 1961. [A study of the development of white clover in relation to the fluctuations of some organic substances.] *Ber. schweiz. bot. Ges.* **71**, 118–72.

WEENINK, R. O. 1961. Acetone-soluble lipids of grasses and other forage plants. 1. Galactolipids of red clover (*Trifolium pratense*) leaves. *J. Sci. Fd Agric.* **12**, 34–8.

WEINMANN, H. 1961. Total available carbohydrates in grasses and legumes. *Herb. Abstr.* **31**, 255–61.

WHITEHEAD, D. C. 1966. Nutrient minerals in grassland herbage. *Mimeo Publ. 1/1966 Commonw. Bur. Past. Fld Crops* Farnham Royal: Commonw. Agric. Bur., pp. 86.

WHITEHEAD, D. C.; JONES, E. C. 1969. Nutrient elements in the herbage of white clover, red clover, lucerne and sainfoin. *J. Sci. Fd Agric.* **20**, 584–91.

WILSON, R. F.; TILLEY, J. M. A. 1965. Amino acid composition of lucerne and of lucerne and grass protein preparations. *J. Sci. Fd Agric.* **16**, 173–8.

WRIGHT, D. E. 1961. Bloat in cattle. 19. The metabolism of pectin by rumen microorganisms. *N.Z. Jl agric. Res.* **4**, 203–15.

CHAPTER 32

Response to Management and Environment

Response to fertilizers

NITROGEN

Legumes in general show little response to fertilizer N, as this merely substitutes for the atmospheric N fixed by bacteria in the root nodules. If the supply via the nodules is insufficient for the requirements of the plant, as for example in acid soils in which the nodules do not flourish, or when the early nodulation of a newly sown crop is slow, then legumes will utilize N supplied in fertilizers and show an increase in yield; however, it has been shown that fertilizer N may retard nodulation.

PHOSPHORUS

In the past, P has been the principal fertilizer element applied to legumes in Britain (Cooke, 1964), but recent experiments have shown that in arable areas yields are more frequently increased by fertilizer K than by P. This probably reflects the general increase in the P status of soils through the continued use of fertilizer P.

On soils low in available P, the incorporation of 20 kg P/ha into the soil at 2–5 cm below the seed at the time of sowing is recommended in addition to a broadcast dressing of 50 kg P/ha worked into the soil during seedbed preparation.

POTASSIUM

On soils low in available K, application of K is essential for the success of legumes, except for sainfoin, which is tolerant of a low K supply. When fertilizer K is required, it is generally recommended that 50–100 kg K/ha should be ploughed in or otherwise worked deeply into the soil prior to the sowing of the crop. For perennial crops such as lucerne, an annual rate of

100–150 kg K/ha, applied as a top dressing in autumn on soils not readily leached, will maintain production. On light soils liable to leaching, the annual requirement is more effectively met by two dressings applied during the growing season.

CALCIUM

As legumes require a near-neutral soil (pH 6–7), lime should be applied as necessary to produce this pH value. In addition, lucerne and particularly sainfoin thrive on soils containing free calcium carbonate. The application of 250 kg calcium carbonate/ha drilled with the seed was shown by Williams and Back (1961) to benefit lucerne sown on a soil of pH 6–7. The calcium carbonate promoted rapid and effective nodulation.

Response to irrigation

In general, the principles suggested for the irrigation of grasses also apply to the legumes. However, there are greater differences among the various species and cultivars of legumes in susceptibility to drought than among the agriculturally important grasses.

The responses of the four major legume species to irrigation are considered in detail in chapters 33–36. At one extreme of the response range is lucerne which, under the conditions experienced in the British Isles, can withstand a soil water deficit of 150–200 mm without its growth being appreciably affected (Stiles, 1966), and at the other extreme is white clover, which appears to be more sensitive to drought than the grasses with which it is usually grown (Low & Armitage, 1959; Stiles, 1966; Garwood, 1969). Because legumes are commonly associated with grasses in swards, it is difficult to define the true response of many legumes to irrigation. In irrigation experiments it is also impracticable to maintain clover swards free from grasses for a long period (Stiles, 1966). Interactions between the relative proportions of grass and clover in a sward and the quantities of N and water applied and the frequency of irrigation may also be complicating factors in defining the response.

Differences in effective rooting depth between the legumes may account for their differences in response to irrigation. Thus, the ability of lucerne to withstand the soil water deficits quoted above is related to its deep tap root and to its independence of N supplied from the soil or from fertilizer. By contrast, the drought-susceptible white clover is a shallow-rooting species.

Response to temperature

As with the grasses, direct experimental evidence of the effect of temperature on the yield of legumes is almost non-existent. Some correlational analyses have been attempted, notably by Black (1955) who found that the growth of the subterranean clover cultivar Bacchus Marsh was independent of temperature in the range 8–25°C, but was significantly correlated with the amount of light energy received. His conclusions were, however, rejected by Morley (1961), largely on the grounds that the technique is suspect if two or more of the independent variables are closely correlated, as were light, temperature and daylength (which Black did not consider), and if growth responses are non-linear. Certainly, the few results from controlled-environment studies available indicate that north-temperate legumes are as sensitive to temperature as are the grasses. Maximum growth rates are usually recorded between 20° and 30° C, with a higher optimum at higher light intensities.

The response of legumes to temperature is complicated by the fact that the optimum for N fixation, and hence for the growth of inoculated plants, may differ slightly from that for the growth of non-inoculated plants. Thus, Meyer and Anderson (1959) reported a greater rate of dry matter (DM) production from non-inoculated clover plants at 30° than at 20° C, whether or not they were given extra additional N, but a greater rate at 20° than at 30° C with inoculated plants. They concluded that N fixation was restricted at temperatures above 25° C. Other workers have found little difference between the optimum temperatures for growth of inoculated and non-inoculated plants; for example, Gibson (1963) found both to have optima between 22° and 26° C, whether or not additional N was given.

In general, as with grass swards, it would seem likely that legume yields are severely restricted by low temperature during winter, but that in summer light is a more important determinant of yield than temperature.

Response to defoliation

The response of the legumes to defoliation can be considered from the viewpoint of response to frequency, to severity and to pattern of defoliation (for definition, see Chapter 14).

Legumes give their highest annual DM yield with infrequent defoliation (two or three cuts per year). With the exception of some white clover varieties, the loss in DM production as the frequency of defoliation increases is associated with a progressive decline in their regrowth potential.

High growth rates are associated with flowering, and to obtain maximum yield the first cut should be taken after flowering has started. Defoliation before flowering has begun causes a serious loss in yield at the early harvests and growth in the second half of the season is also greatly reduced.

From the little information available, there is no evidence to suggest that lenient cutting (to 10 cm or more above ground level) has any advantage over more severe defoliation, but differences between species exist in the optimum cutting height. There is no published work on the effect of pattern of defoliation on legumes.

Specific measurements of the gross energy or calorific value of legumes have not been carried out, but from the wide range of plants studied it is unlikely that harvesting management will cause any change in this value, or in the metabolizable or net energy values.

Legumes are frequently sown in mixtures with grasses. The defoliation management will greatly affect the proportions of grasses and legumes in the mixture and therefore the botanical composition, which in turn influences the digestibility and crude protein content of the herbage. The more vigorous the companion grass, the more important it is to ensure that the defoliation management adopted should reduce its vigour to an extent sufficient to maintain a good proportion of the often slower-growing legumes. As grasses usually tolerate defoliation better than legumes, the defoliation management chosen should be that best suited to maintaining the legume component of the sward.

References

BLACK, J. N. 1955. The interaction of light and temperature in determining the growth rate of subterranean clover (*Trifolium subterraneum* L.). *Aust. J. biol. Sci.* **8,** 330–43.

COOKE, G. W. 1964. Fertilizers and profitable farming. London: Crosby Lockwood and Son Ltd, 2nd edn, pp. 83–4.

GARWOOD, E. A. 1969. Seasonal tiller populations of grass and grass/clover swards with and without irrigation. *J. Br. Grassld Soc.* **24,** 333–44.

GIBSON, A. H. 1963. Physical environment and symbiotic nitrogen fixation. 1. The effect of root temperature on recently nodulated *Trifolium subterraneum* plants. *Aust. J. biol. Sci.* **16,** 28–42.

LOW, A. J.; ARMITAGE, E. R. 1959. Irrigation of grassland. *J. agric. Sci., Camb.* **52,** 256–72.

MEYER, D. R.; ANDERSON, A. J. 1959. Temperature and symbiotic nitrogen fixation. *Nature, Lond.* **183,** 61.

MORLEY, F. H. W. 1961. Subterranean clover. *Adv. Agron.* **13,** 57–123.

STILES, W. 1966. Ten years of irrigation experiments. *A. Rep. Grassld Res. Inst. Hurley 1965* pp. 57–66.

WILLIAMS, T. E.; BACK, H. L. 1961. The combine drilling of lime and lucerne. *Exps Prog. 13 Grassld Res. Inst. 1959–60* pp. 107–9.

CHAPTER 33

White Clover *(Trifolium repens)*

Morphology

WHITE clover (Plate 14) is a creeping, much branched, glabrous perennial. Leaves glabrous, trifoliate, borne on upright petioles; leaflets ovate to circular, conspicuously toothed. Stipules small, lanceolate, pointed, forming a tube around the stem. Flowers papilionaceous, densely crowded in globose racemes borne on long axillary peduncles; pedicels deflexed after fertilization. Pods oblong, dehiscent, 3- to 6-seeded. Root system, a tap root in young plants, with fibrous roots at the stolon nodes. The mature plant has two root systems, a tap root developed from the primary root, and adventitious roots formed at the stolon nodes. White clover is essentially a shallow-rooted plant with the greater part of the root mass in the upper 10 cm of the soil and a rooting depth of up to 60 cm, depending on soil texture and drainage. Some nodules occur on the young tap root but most are carried on the fine secondary roots. The nodules are characteristically club-shaped to ellipsoidal, small (1·5 × 3 mm) and pinkish-white in colour. Larger fan-shaped nodules occasionally occur.

The seedling develops a short primary stem with 5–8 nodes (Sanderson, 1966). Internodes on the primary stem do not elongate and the leaves become crowded, with primary stolons developing from the axils of these leaves to form a rosette. Thereafter, the plant develops from a series of secondary and tertiary branching stolons. Adventitious roots develop at all or some of the stolon nodes. Eventually, in two years or less (Sanderson 1966), the proximal ends of the stolons degenerate and individual stolons take on an independent existence. The length and number of stolons derived from a single plant or within a dense stand are variable and depend on environment.

The leaves are borne alternately on long petioles. Sanderson (1966) showed that petiole length was related to the degree of shading. Leaflet size and shape are functions of variety and position on the stolon.

Four main groups of varieties can be distinguished on the basis of leaf size. Erith (1924) gave detailed descriptions of a wide range of material, and

Hawkins (1960) used the length/width ratio of leaflets, flowering time and cyanophoric reaction as criteria for varietal identification. Wild types, as for example the cultivar S184, are small-leaved (leaflets 10–20 × 12–14 mm), with short petioles and dense, many-branched stolons forming roots at most nodes. The large-leaved types are less prostrate and have shorter, less-branched stolons, large leaflets measuring up to 50 × 40 mm, and long petioles. Nevertheless, there is much variation among varieties; also, leaf size and length of stolon may be modified by nutrient and moisture supply and by cutting frequency. Current classification (see National Institute of Agricultural Botany Farmers' Leaflet No. 23, 1966) recognizes a group of small-leaved cultivars, two medium-leaved groups and a large-leaved category.

Stolons and stolon apices are seldom severed by cutting or grazing and leaves, petioles and inflorescences are the main parts removed. This is reflected in the high digestibility of white clover herbage.

Distribution

Trifolium repens, a native species, occurs very commonly in grassy places in all vice-counties of the British Isles and at all altitudes (Clapham *et al.*, 1962; Perring & Walters, 1962).

There appear to be few climatic or plant factors that would limit its distribution in Britain. Cold-hardiness during establishment might just conceivably be a factor in a very few cases, as this appears to vary with age; e.g. after freezing plants to −10°C for 12 h there was no survival in 4-day-old seedlings. In 19-day-old seedlings there was 47·2% survival, 79·3% in 24-day-old plants, 50·7% in 36-day-old plants, and 48·3% in 44-day-old plants (Rachie & Schmid, 1955). This, however, would hardly be a common effect.

White clover occurs on all soil types, especially on clays, but usually dies if the salt concentration rises above 0·3% (Chippindale, 1954). The optimum pH range is 5·0–7·0 (Spurway, 1941).

The factors most commonly limiting the occurrence of white clover are probably soil pH, Ca or P status, the absence of *Rhizobium* and competition from grasses.

Compatibility with grasses

It is not yet practicable to grow white clover in pure stands because of ingress by weeds. White clover is normally used to provide a source of N for a sown companion grass while itself yielding herbage rich in protein and minerals and of high digestibility.

The cultivated grasses vary in compatibility with white clover. The work of Cowling and Lockyer (1965) and earlier work at the Grassland Research Institute, Hurley (1952), suggested that meadow fescue was the most compatible; perennial ryegrass was initially more competitive owing to its rapid early growth, but subsequently allowed a similar amount of white clover to grow in association. Conversely, clover established well with timothy, which was initially slow to develop, but which became more competitive in later years. Cocksfoot has been consistently highly competitive to white clover. In this work each species was represented by a cultivar bred by the Welsh Plant Breeding Station.

In other trials at Hurley, Green and Corrall (1965) reported that white clover grew better with Northern Irish commercial perennial ryegrass than with either of the bred cultivars S23 or S24. In their work, which was under grazing conditions, parity of clover content was obtained with a seed rate of 22 kg/ha for the Northern Irish material, compared with only 5 or 6 kg/ha for S23 or S24. When the seed rate of S23 and S24 was reduced from 22 to 5 kg/ha, there was only a small effect on yield of ryegrass, and this was offset by an increase in white clover. In a comparison of legumes grown with Italian ryegrass at Hurley (1958), the yield of the Italian ryegrass component was as high with white clover as with any other legume in the first harvest year, but the yield of white clover was low, so that total herbage yields were higher from Italian ryegrass with broad red clover or with lucerne.

Davies (1928), summarizing a series of experiments at the Welsh Plant Breeding Station, stated that all the larger clovers were aggressors in relation to wild white clover. Garner and Sanders (1939) reported that the presence of red clover in seeds mixtures reduced the proportion of white clover which developed. This was also the case when broad red clover was used as a nurse crop for low seed rates of perennial ryegrass with white clover at the Grassland Research Institute, Hurley (1962). In later work at Hurley with nurse crops (1963), a combination of broad red clover and commercial Westerwolds ryegrass added to timothy/white clover mixtures was found to repress white clover less than did red clover alone and also provided satisfactory weed control.

In a comparison of direct-sown perennial ryegrass/white clover with that undersown in a barley crop at the Grassland Research Institute (1965), there was little difference over two years between the two when no N was applied or when 67 kg N/ha was applied to the seedbed, but on a third occasion the establishment and development of white clover were markedly less after undersowing than when direct-sown.

Walker *et al.* (1952, 1953) showed that the amount of white clover in various mixtures with grasses was drastically reduced by applying N. In experiments reported by Cowling (1961a) on the effect of applying up to 235 kg N/ha to mixtures of S37 cocksfoot and S100 white clover, the contribution of white clover to the yield during three years when no N was applied was 40%, but only 5% when 235 kg N/ha was applied; similarly, the proportion of the combined yield which could be said to be attributable to the clover fell from 77% to less than 8%.

Seed

CHARACTERS

The seed of white clover is usually heart-shaped, with the radicle projecting to the side of the cotyledons and extending to their length; a groove or notch may define the position of the radicle (Adams, 1926; Percival, 1949). The average length of the seed is 1·5 mm (Peto, 1961), with a range of 0·9–1·8 mm (Adams, 1926), and the average width is 1·0 mm (Peto, 1961). The colour of the seed varies from yellow to brown (Percival, 1949) and the percentage of hard seeds in commercial samples has been estimated as 7–13 (Robinson, 1953). It is probable that the proportion of hard seed is higher on the plant, but that a measure of scarification occurs during threshing and cleaning (Hyde, 1950).

The 1000-seed weight for wild white clover is 0·557 g (Robinson, 1953), with a range of 0·41–0·75 g (Adams, 1926); for cultivated white clover it is 0·612 g (Robinson, 1953), with a range of at least 0·60–0·71 g (Adams, 1926).

PRODUCTION

Seed of wild white clover is normally obtained from ryegrass/white clover pastures, particularly in Kent (Haggar *et al.*, 1963). White clover grown specifically for production of seed is usually sown under a cereal cover crop early in spring, but the seed rate and amount of N given to the cereal are reduced. Companion grasses are often sown with clover; meadow fescue is preferable to ryegrass, but the highest yields of clover seed are obtained from clover sown by itself (National Institute of Agricultural Botany, 1959; Zaleski, 1964b).

The recommended seed-rate is 2–4 kg/ha (National Institute of Agricultural Botany, 1959), but subsequent seed yields are higher if successful establishment can be obtained from lower seed-rates (Zaleski, 1961, 1964a, b). The seed should be sown at a depth of 10–15 mm, either broadcast or in narrow drills (National Institute of Agricultural Botany, 1959; Zaleski, 1965). Application of 30 kg P and 60 kg K/ha in autumn is recommended (National Institute of Agricultural Botany, 1959; Zaleski, 1964b). Grass/clover pastures should be grazed or cut in spring to suppress the grass, but the clover should not be defoliated after mid-May or the clover heads will be too low to be harvested successfully. The highest yields of seed are usually obtained from pure clover stands which have not been cut or grazed (Haggar *et al.*, 1963; Zaleski, 1961, 1964b, 1965). Crops should be cut for seed when 90% of the flower heads are brown and reflexed (Haggar *et al.*, 1963). Crops should be isolated during the period of anthesis (flowering); the requirement is 180 m for crops of less than 2 ha, declining to 45 m for crops covering 7 ha or more (National Institute of Agricultural Botany, 1959). Principles and procedures for harvesting are well explained elsewhere (Ministry of Agriculture, Fisheries & Food, 1968).

The annual yield of seed/ha can vary from 8 to 525 kg, with the mean yield ranging between 25 and 215 kg, depending on cultivar and year (Lincs. Seed Growers' Association, 1965).

Yield

VARIETIES

The majority of trials with white clover varieties have been carried out in the presence of a companion grass. Under these conditions at least, the larger-leaved varieties appear to outyield the small-leaved ones of the 'wild white' type. Because of the very low growth habit of the latter, cutting trials may underestimate their productivity. From the results of a sheep-grazing trial at Wye (Davis & Cooper, 1951), it could be inferred that the amount of forage available to sheep was as great on Kent wild white as on S100 or New Zealand C.M.S. (now Grasslands Huia). However, the medium-leaved varieties in common use (S100, Grasslands Huia and Kersey) do offer a crop for harvest a week or two earlier in spring than the smaller-leaved S184 cultivar and local varieties of wild white clover. The large-leaved Ladino rarely outyields other varieties, and in most instances has failed to sustain a satisfactory yield after a second winter.

Varieties of white clover do not differ significantly in N content. There is not sufficient evidence to comment on their comparative digestibility.

YIELD OF WHITE CLOVER GROWING WITH GRASS

The following general statements concerning yield and quality are based on experiments with white clover varieties of the S100 type.

Because white clover is almost invariably grown with a companion grass or grasses, it is pertinent to assess its potential contribution in a mixture. Any random collation of data from mixed swards will show how variable the actual contribution can be. Some of the variation is due to factors such as the grasses used, seed rate, frequency of harvesting and rate of N fertilizer applied, which are discussed elsewhere.

The data assembled in Table 33.1 are all derived from plots which received no top-dressing of N and were cut every 4–6 weeks, i.e. conditions relatively favourable to the clover component. Even these data show a range from 300 to 6 400 kg/ha in the dry matter (DM) yield from clover.

Much of this variation stems from the fact that white clover is sensitive to superficial droughts, and that the effect of drought is exacerbated by competition from companion grasses. Fig. 33.1 shows how erratic the seasonal production of white clover can be in a grass/clover mixture, in the absence of irrigation.

The status of white clover in a mixed sward is greatly enhanced, and its yield more predictable, if the sward is irrigated. At the Grassland Research

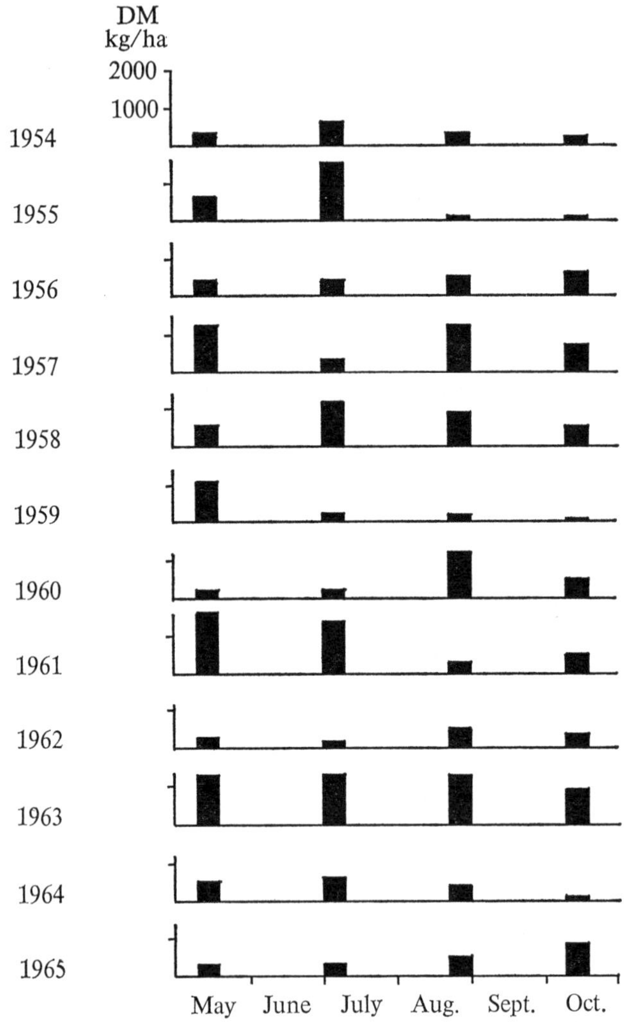

Fig. 33.1. Yield of white clover (growing with S23 perennial ryegrass) in 4 cuts. (In most years the data are averages from 1st, 2nd and 3rd harvest-year swards. No N, no irrigation. (Grassld Res. Inst., Hurley, Exp. H9, 1954–65)

Table 33.1. Annual yield of white clover growing with grass, without irrigation or N (kg DM/ha)

Location	Hurley*								Edinburgh**	
Grass cv.	S23	S37	S23	S37	Various	S24	S24	S24	S23	S143
Year										
1955	2 520								2 460	2 240
1956	2 220	1 260	1 220						2 980	1 680
1957	3 870	3 520	2 010						2 360	900
1958	3 580	3 740	3 260	2 040						
1959	1 750	670	300	600						
1960	2 450			1 460	1 060	1 460				
1961	4 280				4 800	1 500	2 570			
1962	1 620				3 000	590	1 460			
1963	6 400									
1964	1 670									
1965								4 280		
1966								2 660		

Linehan and Lowe (1960) sowed white clover with perennial ryegrass and with tall fescue in 1949. The yield of clover in 1950 was described as low. The average annual yield of clover 1951–1955 was 2670 kg/ha.

Lowe (1967) gave data for 4 successive years, unspecified, of a New Zealand perennial ryegrass/N.Z. white clover sward from which yields of DM from the white clover have been calculated. These are respectively 2300, 3500, 3200 and 1300 kg/ha.

* Data compiled by J. O. Green (Grassland Research Institute, Hurley) from literature cited in the text.

** From Herriott and Wells (1960).

Table 33.2. Annual yield of white clover growing with grass, with irrigation without N (kg DM/ha)

Grass cv.	S23		S24	S24	
Year	Irrigated	Unirrigated	Irrigated	Irrigated	Unirrigated
1956	3 340	1 220	...	...	...
1957	4 640	2 010	...	...	...
1958	3 540	3 260	...	...	...
1959	3 940	300	...	...	...
1963	...	...	4 280	...	...
1964	...	...	2 490	...	...
1965	...	...	...	5 000	4 280
1966	...	...	...	3 740	2 600

Data compiled by J. O. Green (Grassland Research Institute, Hurley) from literature cited in the text.

Institute, Hurley, white clover grown under an irrigation regime designed to maintain soil moisture near to field capacity has given comparatively consistent yields of around 4 000 kg DM/ha from year to year (see Table 33.2, p. 353; low yields in 1956 and 1964 reflect protracted low temperatures in spring).

YIELD OF WHITE CLOVER SOWN PURE

There are few published data relating to white clover sown pure, and even less for white clover that has been maintained weed-free. Those that are available indicate a higher potential yield than might be deduced from the performance of white clover growing in association with grasses. Results of experiments at the Grassland Research Institute, Hurley, and at Drayton in the west Midlands are summarized in Table 33.3 and show that without

Table 33.3. Annual yield of S100 white clover sown alone (kg DM/ha)

Year	Irrigated	Unirrigated
1946	...	7 000
1947	...	3 900
1948	...	7 300
1949	...	6 500
1950	...	6 900
1957	11 800	7 400
1958	...	8 650
1961	11 250	7 850
1965	6 510	...
1966	6 980	...

Data compiled by J. O. Green (Grassland Research Institute, Hurley) from published and unpublished information. Data for years 1946–50 derived from experiments at Drayton, Warwickshire, and for subsequent years from experiments at Hurley, Berkshire.

irrigation it can yield between 6 500 and 8 650 kg DM/ha in a normal to wet season. However, if a dry summer follows a late spring, as in 1947, it may yield less than 4 000 kg/ha. In 1949, also a dry summer, white clover produced very little after 1 July; the relatively high yield for that year can be ascribed to the fact that growth began exceptionally early in spring.

In the two experiments at Hurley in which the effect of irrigation was measured, the irrigated plots yielded more than 11 000 kg DM/ha, but a portion of this was contributed by grass weeds. In a further experiment in which the grasses were controlled by spraying with paraquat during the winter and thus contributed very little to yield, clover yields were somewhat lower, but it is possible that the white clover was slightly checked by the paraquat treatment.

Data from weed-free crops of white clover are summarized in Tables 33.4 and 33.5. Table 33.4 shows the average expectation of production under a

N

Table 33.4. Yield, digestibility and protein content of S100 white clover under low-rainfall conditions

Cutting date		DM kg/ha			DM digestibility %	D-value %	Digestible OM kg/ha			N %	N kg/ha			CP %	CP kg/ha		
1st cut	12 May	1 200			82	74	890			4·5	56			28·3	340		
	27 May		2 000		79	71		1 420		4·0		80		25·0		500	
	11 June			2 700	77	69			1 860	3·6			97	22·5			610
2nd cut	28 June	1 800			70	63	1 130			3·8	68			23·7	430		
	12 July		1 500		70	63		940		3·8		57		23·7		360	
	25 July			1 500	70	63			940	3·8			57	23·7			360
3rd cut	28 Aug.	1 600			76	68	1 090			4·0	64			25·0	400		
	12 Sept.		1 600		76	68		1 090		4·0		64		25·0		400	
	28 Sept.			1 600	76	68			1 090	4·0			64	25·0			400
4th cut	15 Oct.	400			80	72	290			4·5	18			28·3	110		
Total yield		5 000	5 100	5 800			3 400	3 450	3 890		206	201	218		1 280	1 260	1 370

Average performance in lower rainfall conditions comparable to those under which the other legume species approach their potential yields. Data compiled by J. O. Green (Grassland Research Institute, Hurley) from literature cited in the text. For conversion formulae see Chapter 5.

3- to 4-cut system, which is appropriate to the other main legume species, without irrigation. On this basis white clover compares unfavourably with the other legumes in yield; it produces forage of higher quality, but the yield of nutrients is low, although in view of the annual variation in yield of this crop such average data are of little relevance anyway.

The data summarized in Table 33.5 are derived from irrigated plots and therefore tell us more about the potential yield of white clover. Also, they relate to a monthly cutting system, which appears to be appropriate to this crop with its stoloniferous habit and relatively rapid turnover of leaves.

Table 33.5. Yield, digestibility and protein content of S100 white clover irrigated and cut monthly

Cutting date	DM kg/ha	DM digestibility %	D-value %	Digestible OM, kg/ha	N %	N kg/ha	CP %	CP kg/ha
7 May	1 600	83	75	1 200	4·5	72	28	450
3 June	2 000	82	74	1 480	4·5	90	28	560
2 July	1 800	80	72	1 300	4·0	72	25	450
29 July	1 300	82	74	960	4·5	59	28	360
27 Aug.	1 300	82	74	960	4·5	59	28	360
23 Sept.	700	82	74	520	4·5	31	28	200
Total	8 700			6 420		383		2 380

Data compiled by J. O. Green (Grassland Research Institute, Hurley) from literature cited in the text. For conversion formulae see Chapter 5.

Under these favourable conditions, white clover is capable of a DM yield approaching that of the erect legumes under optimum cutting treatment. Since this DM has comparatively high concentrations of digestible organic matter and N, the yields of these components are comparable to those provided by the erect legumes (to cut the latter at a stage which would give forage of the same quality would both reduce their yield drastically and shorten their useful life).

SEASONAL YIELD

White clover becomes productive in spring later than the majority of grasses. In association with grass, it normally contributes little before the end of April. As with the other legumes, however, a large proportion of its produce is harvested after midsummer. In a mixture (see Fig. 33.2) it often yields more after the middle of July than before that date. In the absence of a companion grass (Fig. 33.3) white clover is able to express its full potential in May and June, but still gives a large proportion of its yield later in the season. The first two trials represented in Fig. 33.3 took place in years when the calculated irrigation need was slight. The results of the third trial show once again that

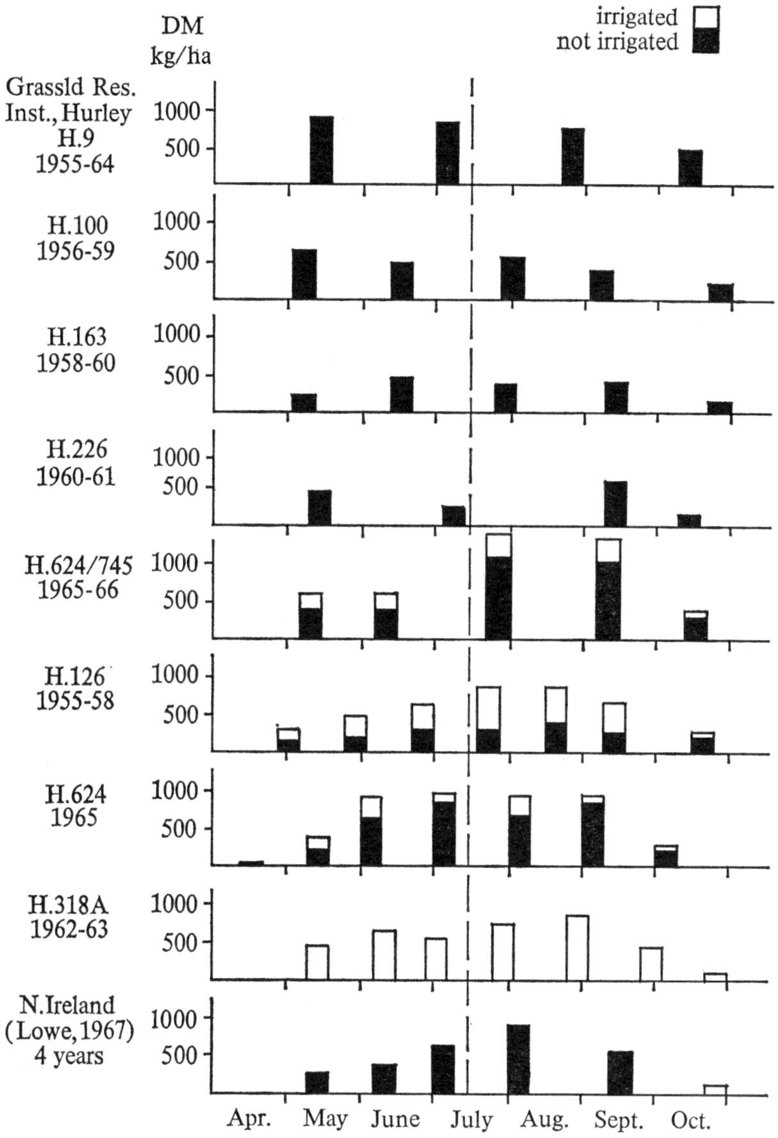

Fig. 33.2. Seasonal yield of white clover growing with grass, with no N applied.

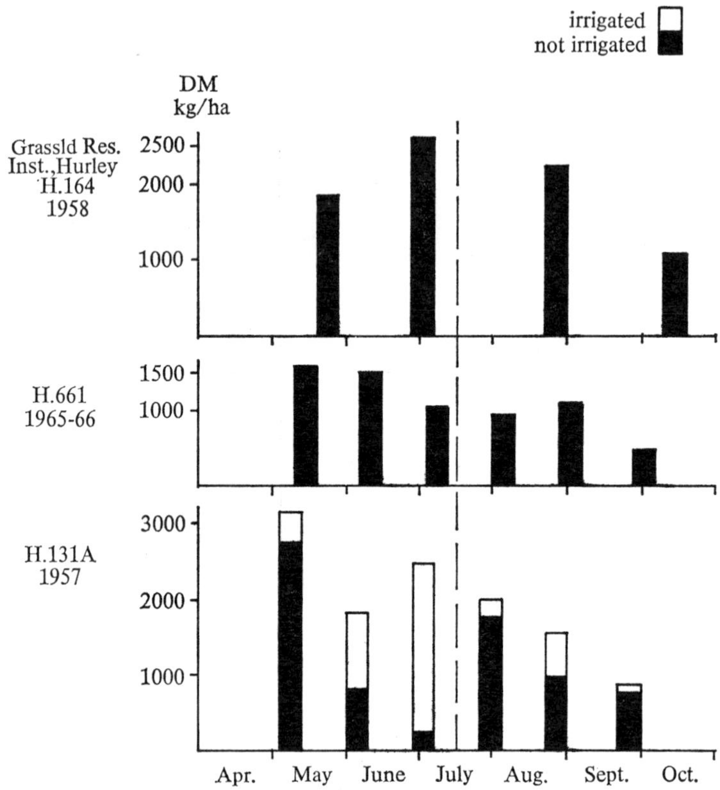

Fig. 33.3. Seasonal yield of white clover growing alone.

Table 33.6. Potential annual yields of energy and protein from white clover (DM basis)

	Unirrigated		Irrigated
	Mean	Range	Mean
DM digestibility (%)	75	74–77	82
Yield of digestible OM (kg/ha)	3 580	3 400–3 900	6 420
Yield of metabolizable energy (Mcal/ha)	14·5	13·2–15·4	25·0
Crude protein (%)	24·8	23·8–26·2	28·0
Yield of crude protein (kg/ha)	1 300	1 260–1 370	2 380
Yield of digestible crude protein (kg/ha)	1 030	982–1 080	1 960

Data compiled by J. O. Green (Grassland Research Institute, Hurley) from literature cited in the text. For conversion formulae see Chapter 5.

moisture stress, in this instance during May and June, can distort the inherent pattern of production from this crop.

YIELD OF ENERGY AND PROTEIN

A summary of examples from the literature is given in Table 33.6.

Response to management and environment

RESPONSE TO FERTILIZER

Nitrogen

White clover will utilize N applied as fertilizer. Sown with 25 kg N/ha, it gave 10% greater yield than when given no N but nodule establishment was halved (Thornton, 1956). However, this author concluded that such an application 'may be beneficial' for assisting establishment and recommended that it should be made just prior to sowing or as a top-dressing after sowing, rather than with the seed.

On an established white clover crop, Cowling (1961b) showed clearly that there was no increase in yield from the application of N at rates up to 225 kg/ha, and that the weight of nodules on clover plants given 110 kg N/ha was half that on plants given no N. Similar results were obtained by Reid (1961).

Potassium

Andrew (1960) has stated that white clover requires a concentration of 1·2% K in the DM if yield is not to be restricted by K deficiency. In experiments he carried out, the application of 30 kg K increased the concentration of K in the DM from 0·7 to 1·3% and at the same time increased DM yields from 1 500 to 2 500 kg/ha. On an annual basis, a crop yielding 10 000 kg DM/ha will remove 120 kg K. On soils which are themselves able to provide some K, the apparent requirement for applied K will be reduced; conversely, other soils may fix a proportion of any applied K in forms unavailable to the plants and the requirement for applied K will be increased accordingly.

Phosphorus

In the absence of precise guidance from trials under British conditions, Cooke (1964) recommends an annual application of 25 kg P/ha for clover. Under Australian conditions, Andrew (1960) established that a level of 0·23% P in the DM of white clover was necessary to avoid deficiency; an application of 50 kg P/ha provided this concentration in the plant.

Calcium

White clover is at its most vigorous on weakly acid soils, the optimum pH value being 6·2 (Edinburgh School of Agriculture, 1963). Lime is generally required to produce this pH. Ecological populations have been found which have become adapted to more acid or more alkaline conditions than those stated above (McNeur, 1954).

RESPONSE TO IRRIGATION

Information on the response of white clover when grown alone is limited, but Stiles (1966) obtained large responses in dry years. The mean response of 4 cultivars of white clover was 1·41 kg DM/m^3 of water applied. In the same year and with a similar level of irrigation, the mean response of grasses giving a similar maximum annual yield was 0·95 kg DM/m^3 of water. White clover therefore appears to be rather more sensitive to drought than the agriculturally important grasses. Unlike the grasses, however, there is some variation in the response of the individual cultivars; Ladino clover, originating from North Italian strains that have been developed under irrigation, appears least sensitive to drought.

The response of white clover in grass/clover swards is in accord with the greater sensitivity to drought of the clover; with increasing soil water deficit the yield and percentage contribution of clover to the total herbage yield fall more rapidly than do those of the grass component (Munro, 1958; Low & Armitage, 1959; Stiles, 1966). Clover tolerates the application of fertilizer N better under irrigated than non-irrigated conditions (Robinson & Sprague, 1947; Nelson & Robins, 1956; Munro, 1958) although the continued use of N depresses and will eventually eliminate it from the sward (Stiles, 1966).

Increased growth of clover in grass/clover swards in the year following irrigation has been reported (Garwood, 1965), but Reid and Castle (1965) obtained slightly lower subsequent yields after irrigation and also slightly reduced clover contents in grass/clover swards during the year in which they were irrigated. Appadurai and Holmes (1964) similarly reported a depression in yield of clover in mixed swards with irrigation, but amounts of water applied in their experiments appeared excessive. Excess rainfall or irrigation during flowering may have a harmful effect on seed production, according to Zaleski (1966).

RESPONSE TO TEMPERATURE

Information on the response of white clover to temperature is restricted almost entirely to data from studies in New Zealand and the U.S.A. Mitchell (1956) and Mitchell and Lucanus (1962) reported optima ranging from 19°C to 24°C for a New Zealand certified white clover cultivar grown in 2 700 f.c. (27 000 lux). The optima reported by Beinhart (1962, 1964) for plants grown in 600 f.c. of light (6 000 lux) fall mainly between 17°C and 23°C, although there are exceptions. In one experiment in which 6 clones were compared, three achieved maximum growth rates at 17°C, one did equally well at 17°C and at 23°C, one appeared little affected by temperature over the range 10°

to 30°C, and one still showed an increase in dry weight at a temperature of 30°C. This last value seems particularly high, although Sprague (1943) did record a greater yield of DM from Ladino clover in a regime of 29°/21°C than in 21°/15°C or 38°/29°C. There is general agreement that the lower the light intensity, the lower the optimum temperature for growth, that yield is affected more by day than by night temperature, and that the optimum regime is one in which the night temperature is equal to, or slightly less than, that of the day.

RESPONSE TO DEFOLIATION

As has been stated, the varieties of white clover can be divided into 4 main groups on the basis of leaf size. Each group responds differently to defoliation, varieties in the small-leaved group being adapted to withstand frequent defoliation, and those in the two intermediate-leaved groups requiring more lenient harvesting. The large-leaved white clovers are not of agricultural significance in Britain (National Institute of Agricultural Botany, 1966).

The medium- and large-leaved groups of white clovers have been shown to produce high DM yields when cut twice during the season (June and August), yielding some 70% of the total yield at the first cut (Davies, 1958). White clovers are more usually found in swards that are grazed, and nearly all the published results refer to clover/grass mixtures. However, the small-leaved varieties of white clover grow well under frequent and severe defoliation (e.g. Reid, 1966) and no harmful effect of severe defoliation on the clover fraction has been found. The ability of small-leaved white clovers to withstand close grazing distinguishes them from the other legumes and from the other groups of white clover.

In comparison with grass/clover mixtures, it has been shown that pure clover swards produce the highest crude protein yields (Hunt *et al.* 1965), so that a cutting management for a mixed sward that favours the growth of clover will also increase the crude protein concentration in the harvested material.

Quality

CHEMICAL COMPOSITION

Owing to its persistent leafiness throughout the season, white clover is generally lower in cellulose and lignin content and higher in content of available carbohydrate, N, P and S than the other forage legume species. It also appears to be generally higher than the other species in Na, Cl and Mo content. Further details of chemical composition are given in Chapter 31.

WATER CONTENT

White clovers appear to be characterized by high water content. Thus Hayward (1953) cut 8 cultivars on 20 May and found that water content varied between 89·3% for New Zealand white clover and 85·8% for Kent white clover.

The availability of water and the application of fertilizer N have a modifying effect on the water content of white clover. Dančík (1963) found that the water content in Ladino clover varied with rainfall and the amount of irrigation, and was a little higher than that of other clovers. Ivins (1952) found that the water content of S100 white clover was higher when 2 cwt of N fertilizer/ac (246 kg/ha) was applied than when no fertilizer was used. Davies *et al.* (1966) found a progressive fall in water content when white clover was cut at intervals from May to July.

DRY MATTER DIGESTIBILITY

Young leafy herbage frequently cut

The digestibility of white clover in spring is very high and values of over 80% DM digestibility have often been recorded (Davies *et al.*, 1966). Although the digestibility of immature clover at other times of the year does not seem to have been recorded, from values obtained from two-month-old aftermaths (see below) there is good reason to believe that young material has DM digestibility values in the region of 80% at all times of the year. White clover is seldom grown in pure stands, but it is clear from these figures that the proportion of white clover in a mixture can have a marked effect on its digestibility, the clover being more digestible than grass at a corresponding growth stage.

Primary growth

Fig. 33.4, top shows that the digestibility of S100 white clover is higher in spring than that of S24 perennial ryegrass. Digestibility falls more slowly, and Harkess (1963) found the rate of fall to be 0·15%/day compared with 0·5%/day for S24 perennial ryegrass. The DM yield from white clover is, however, comparatively low and the yield of digestible nutrients far below that which can be obtained from perennial ryegrass (Fig. 33.4, bottom). It is, however, clear that the presence of white clover in a mixture results in higher digestibility than that of a pure grass sward.

Regrowths

Work at the Grassland Research Institute, Hurley, and by Harkess (1963) has shown that regrowths cut 7–8 weeks after a primary cut usually have a DM digestibility of between 70 and 75%. Yields are, however, lower than in the regrowths of some other legumes such as lucerne and red clover.

The main characteristics of white clover are therefore a low yield potential but a capacity for maintaining high digestibility with increasing maturity. Age for age, white clover is more digestible than red clover, lucerne, sainfoin

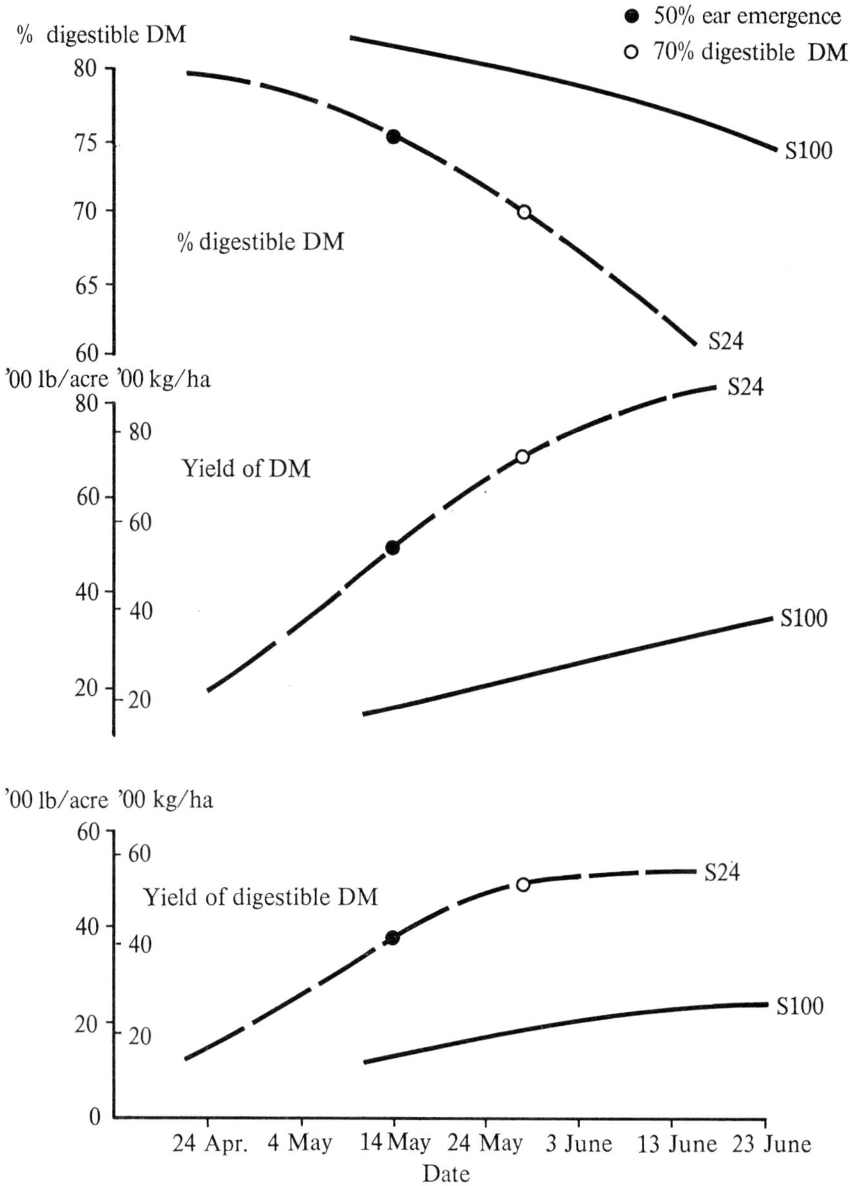

Fig. 33.4. White clover S100. Seasonal changes in DM yield and digestibility. Values for S24 perennial ryegrass shown for comparison. (From Green & Corrall, 1967)

or the grasses. It also has a high protein content which adds to its value as a high-quality component in mixtures.

ANIMAL INTAKE CHARACTERISTICS

Thomson (Private communication, 1965) demonstrated that the voluntary consumption of white clover by sheep was considerably greater than that of ryegrass of similar digestibility. Wilson (1966) has suggested that the higher milk production by cattle and liveweight gains by sheep on ryegrass/white clover swards than on pure ryegrass swards reflect the higher voluntary DM intakes on the mixed swards, although Johns *et al.* (1963) suggested that differences in the efficiency of use of the digested energy may also be a factor.

Longevity and persistency

The percentage survival of germinating seedlings of white clover is reduced at temperatures of 5°C and below (Mence, 1964). Growth patterns in S100 white clover have been studied by Sanderson (1966). In seedlings, the first organs to senesce were the cotyledons, the unifoliate leaf and the leaves at the lower nodes of the main stem. In spaced plants, the main axis lived for 4–26 months after germination and 30% of main stems of spring-sown plants died in late summer and autumn, whereas under sward conditions the main stems did not start to die until the plants were 10 months old, and only 20% had died after 12 months. Observations on plants grown in sterilized soil suggested that in a suitable environment the main stem would tend to be at least biennial; though its apical meristem declined in activity with age, it was not inherently short-lived. The proximal ends of primary stolons usually died in acropetal order on the main stem, and internodes senesced in similar order on the stolons.

Hollowell (1966) showed that severing the stolons from the seedling plants had no effect on stolon growth rate. In his opinion the apparently one-directional translocation of minerals and metabolites to meristematic tissues was the primary reason for the short life of the main axis, and the destruction of roots and stolons through infection by mildly pathogenic fungi was secondary. Leaves of white clover survived for about 40 days in summer, but much longer in winter (Brougham, 1958; Sanderson, 1966). They remained fully open for 24–26 days in summer and 63–72 days in winter (Sanderson, 1966). In Japan, leaf longevity ranged from 30–35 days in summer to 40–50 days in autumn (Hashimoto *et al.*, 1964).

Estimates of tap-root longevity vary from less than 12 months in wild white clover, longer in S100 (Vez, 1961), to 24 months in Ladino clover (Westbrooks & Tesar, 1955). Sanderson (1966) found that removal of the tap root caused temporary or permanent wilting of leaves, but the plants recovered. The tap root by itself was able to support top growth, but at a lower rate than where nodal roots were also present. Hollowell (1966) found that it lived only slightly longer than when nodal roots were allowed to function. It penetrated more deeply than did nodal roots, and in plants 23 weeks old it still played a significant part in water uptake. When it died, the ability to utilize moisture from the lower soil horizons was lost. Nodal roots developed singly at virtually all nodes on the primary stolons, 1–3 nodes behind the most apical expanded leaf (Sanderson, 1966). Under conditions of non-limiting soil moisture in the field, a node on an intact stolon invariably developed one root. Absence of water for 24 hours was usually critical for the survival of a newly-emerged root. Roots of white clover senesce at the same time as the stolons.

The plant eventually consists of branched stolons bearing a single leaf and a single root at each node; some nodes may bear a single axillary inflorescence. Senescence and death of the distal ends of stolons proceeds as growth progresses and, except in the early stages of growth from seed, no part of the plant lives for much more than one year (Hollowell, 1966; Sanderson, 1966). The plant is capable of persisting indefinitely only because of its capacity for continual vegetative renewal, and in order to survive it needs to be able to gain a progressive foothold among older, established grasses at times when there are competitive stresses for water, nutrients and light. In winter, the rate of senescence of stolons may exceed the growth rate (Sanderson, 1966); this may give an indication as to why some cultivars are more winterhardy than others. In summer, its high rate of growth may enable it to outgrow attacks by diseases, as has been shown by Vanderveken (1966) for phyllody virus.

As far as is known no records have been published of longevity in clones of cultivated white clover. Assessments of the age of wild clones have, however, been made by Harberd (1963), who estimated that most of the clones he measured were of the order of 20 years old; however, one clone may have been 60–100 years old, or even much older.

Since white clover is seldom sown as a pure species but almost invariably with a companion grass in swards which are to be grazed or cut, the persistence of clover is largely dependent upon the vigour of the accompanying grasses, as long as soil nutrients and soil moisture are favourable (Robinson & Sprague, 1947). When not defoliated for long periods, clover is suppressed by the more extensive growth of the grasses, hence the detrimental effect on clover of cutting grass/clover swards for hay. If grazing is delayed in the spring, the grasses tend to dominate the sward, since they start growing earlier (Jones, 1933a). Alternatively, if vigorous grasses (e.g. ryegrass, cocksfoot, etc.) are severely defoliated during the early period of their spring growth, a subsequent period of rest favours the later-growing legume (Jones, 1933b; Hubbard, 1954). The force of these reactions may be increased when fertilizer N is applied, for the adverse effect of N on clover appears generally to result from the greater encouragement given to vegetative growth of the grasses (Cowling & Green, 1956), which then make better use of light (Stern &

Donald, 1962) and possibly of soil moisture also. However, there are cases where N appears to suppress white clover by increasing the ability of the roots of the grasses to compete for soil nutrients (e.g. P) which may be in short supply (Heddle, 1966; Kydd, 1965).

In general, whenever it is desirable that white clover should persist in the sward, the vigour of sown grasses must be controlled. The severity of defoliation required to achieve this may vary and, under sheep grazing, if too severe, the persistence of both the legume and the sown grasses may suffer. To prevent this, it is probably easier to adjust stocking under a system of rotational as opposed to continuous grazing.

References

ADAMS, F. M. J. 1926. Some observations on white clover, and a method of distinguishing between the seeds of wild white and dutch clover. *Ann. appl. Biol.* **13,** 339–57.

ANDREW, C. S. 1960. The effect of phosphorus, potassium and calcium on the growth, chemical composition and symptoms of deficiency of white clover in a subtropical environment. *Aust. J. agric. Res.* **11,** 149–61.

APPADURAI, R. R.; HOLMES, W. 1964. The influence of stage of growth, closeness of defoliation and moisture on the growth and productivity of a ryegrass/white clover sward. 1. Effect on herbage yields. *J. agric. Sci., Camb.* **62,** 327–32.

BEINHART, G. 1962. Effects of temperature and light intensity on CO_2 uptake, respiration and growth of white clover. *Pl. Physiol., Lancaster* **37,** 709–15.

BEINHART, G. 1964. Free sugar concentrations in white clover, *Trifolium repens* L., grown at different temperatures and light intensities. *Crop Sci.* **4,** 625–31.

BROUGHAM, R. W. 1958. Leaf development in swards of white clover. *N.Z. Jl agric. Res.* **1,** 707–18.

CHIPPINDALE, H. G. 1954. The restoration of grassland flooded by sea water. *Agric. Prog.* **29,** 25–33.

CLAPHAM, A. R.; TUTIN, T. G.; WARBURG, E. F. 1962. Flora of the British Isles. Cambridge: Univ. Press, 2nd edn.

COOKE, G. W. 1964. Fertilizers and profitable farming. London: Crosby, Lockwood and Son Ltd, pp. 86.

COWLING, D. W. 1961a. The effect of white clover and nitrogenous fertilizer on the production of a sward. 1. Total annual production. *J. Br. Grassld Soc.* **16,** 281–90.

COWLING, D. W. 1961b. The effect of nitrogenous fertilizer on an established white clover sward. *J. Br. Grassld Soc.* **16,** 65–8.

COWLING, D. W.; GREEN, J. O. 1956. The influence of white clover on the sward. *Proc. Eur. Grassld Conf. 1954* pp. 169–74.

COWLING, D. W.; LOCKYER, D. R. 1965. A comparison of the reaction of different grass species to fertilizer nitrogen and to growth in association with white clover. *J. Br. Grassld Soc.* **20,** 197–204.

DANČÍK, J. 1963. [Possibility of growing the high yielding type of white clover, Ladino (*Trifolium repens* var. *giganteum*) in our conditions.] *Ved. Pr. výsk. Úst. rastl. Výroby Piešt'anoch* **2,** 219–32.

DAVIES, W. 1928. The factor of competition between one species and another in seeds mixtures. *Bull. H8 Welsh Pl. Breed. Stn.*

DAVIES, W. E. 1958. The yields of pure sown plots of eight white clover strains under cutting. *J. Br. Grassld Soc.* **13,** 34–8.

DAVIES, W. E.; GRIFFITH, G. AP; ELLINGTON, A. 1966. The assessment of herbage legume varieties. 2. *In vitro* digestibility, water, carbohydrate, crude protein and mineral content of primary growth of clover and lucerne. *J. agric. Sci., Camb.* **66,** 351–7.

DAVIS, A. G.; COOPER, M. M. 1951. White clover trial. *J. Br. Grassld Soc.* **6,** 299–43.

EDINBURGH SCHOOL OF AGRICULTURE. 1963. Grassland management for beef production in the East of Scotland. *Tech. Bull. 48 Edinb. Sch. Agric.* pp. 1–7.

ERITH, A. G. 1924. White clover. A monograph. London: Duckworth & Co., pp. 150.

GARNER, F. H.; SANDERS, H. G. 1939. Four year leys. The inclusion of red clover. *J. agric. Sci., Camb.* **29,** 164–73.

GARWOOD, E. A. 1965. Some factors which influence the root growth of herbage species. M. Sc. Thesis, Univ. Reading.

GRASSLAND RESEARCH INSTITUTE. 1952. *Exps Prog. 5 Grassld Res. Inst. 1952* pp. 23–31.

GRASSLAND RESEARCH INSTITUTE. 1958. *Exps Prog. 10 Grassld Res. Inst. 1956–7* p. 21.

GRASSLAND RESEARCH INSTITUTE. 1962. *Exps Prog. 14 Grassld Res. Inst. 1960–61* p. 16.

GRASSLAND RESEARCH INSTITUTE. 1963. *Exps Prog. 15 Grassld Res. Inst. 1961–62* p. 15.

GRASSLAND RESEARCH INSTITUTE. 1965. *Exps Prog. 17 Grassld Res. Inst. 1963–64* pp. 23–4.

GREEN, J. O.; CORRALL, A. J. 1965. The testing of grass varieties in swards with clover: the effect of grass seed-rate on comparisons of gross yield. *J. Br. Grassld Soc.* **20,** 207–11.

GREEN, J. O.; CORRALL, A. J. 1967. The Grassland Research Institute, Hurley. Unpublished data.

HAGGAR, R. J.; HOLMES, W.; INNES, P. 1963. Wild white clover seed production. 1. The effects of defoliation and fertilizer treatment on flowering and seed yields from ryegrass/white clover swards. *J. Br. Grassld Soc.* **18,** 97–103.

HARBERD, D. J. 1963. Observations on natural clones of *Trifolium repens* L. *New Phytol.* **62,** 198–204.

HARKESS, R. D. 1963. Studies in herbage digestibility. *J. Br. Grassld Soc.* **18,** 62–8.

HASHIMOTO, T.; YOSHIDA, K.; SAITO, T. 1964. Growth habit of Ladino clover during its growth period in Hokuriku district. *Bull. Hokuriku agric. Exp. Stn* **7,** 111–28.

HAWKINS, R. P. 1960. Investigations on local strains (varieties) of herbage plants. 4. White clover. *J. Br. Grassld Soc.* **15,** 28–33.

HAYWARD, P. R. 1953. *5th Rep. Trial exp. Wk Throws 1952–53* pp. 48.

HEDDLE, R. G. 1966. The after-effects of varying rates of spring defoliation on the yield and botanical composition of a grass/clover sward. *J. Br. Grassld Soc.* **21,** 251–7.

HERRIOT, J. B. D.; WELLS, D. 1960. Clover nitrogen and sward productivity. *J. Br. Grassld Soc.* **15,** 63–9.

HOLLOWELL, E. A. 1966. White clover (*Trifolium repens* L.) annual or perennial? *Proc. 10th int. Grassld Congr., Helsinki, 1966* pp. 184–7.

HUBBARD, C. E. 1954. Grasses. Harmondsworth, Middlesex: Penguin Books.

HUNT, I. V.; HARKESS, R. D.; MARTIN, T. W. 1965. Comparison of the production of four varieties of white clover. *Res. Bull. 34 W. Scotl. agric. Coll.*

HYDE, E. O. C. 1950. Studies on the development of white clover seed. *Proc. 12th Conf. N.Z. Grassld Ass.* pp. 101–7.

IVINS, J. D. 1952. The relative palatability of herbage plants. *J. Br. Grassld Soc.* **7,** 43–54.

JOHNS, A. T.; ULYATT, M. J.; GLENDAY, A. C. 1963. Pasture type in relation to live-weight gain, carcass composition, iodine nutrition and some rumen characteristics of sheep. 4. Some rumen characteristics of the sheep. *J. agric. Sci., Camb.* **61,** 201–6.

JONES, E. 1963 Studies on the magnesium content of mixed herbage and some individual grass and clover species. *J. Br. Grassld Soc.* **18,** 131–8.

JONES, M. G. 1933a. Grassland management and its influence on the sward. 2. The management of a clovery sward and its effects. *Emp. J. exp. Agric.* **1,** 122–7.

JONES, M. G. 1933b. Grassland management and its influence on the sward. 3. The management of a grassy sward and its effects. *Emp. J. exp. Agric.* **1,** 223–34.

KYDD, D. D. 1965. Fertiliser and management responses on permanent downland pasture grazed by cattle. *J. agric. Sci., Camb.* **64,** 335–42.

LINCOLNSHIRE SEED GROWERS' ASSOCIATION. 1965. Report for 1964. Agriculture House, Woodhall Spa, pp. 3–7.

LINEHAN, P. A.; LOWE, J. 1960. Yielding capacity and grass/clover ratio of herbage swards as influenced by fertilizer treatments. *Proc. 8th int. Grassld Congr., Reading, 1960* pp. 133–5.

LOW, A. J.; ARMITAGE, E. R. 1959. Irrigation of grassland. *J. agric. Sci., Camb.* **52,** 256–62.

LOWE, J. 1967. Botanical development and output of a sward seeded with perennial ryegrass and white clover under stated fertilizer treatments. *Rec. agric. Res. (Nth. Ireld)* **16,** 75–91.

MCNEUR, A. J. 1954. Effect of soil acidity on growth of pasture plants. A field trial with white clover (*Trifolium repens*). *N.Z. Jl Sci. Technol. Sect. A* **36,** 167–76.

MENCE, M. J. 1964. The differential mortality of *Trifolium repens* and *Phleum pratense* seedlings in relation to temperature. *Pl. Soil* **21,** 113–24.

MINISTRY OF AGRICULTURE, FISHERIES AND FOOD. 1968. Grass and clover crops for seed. *Bull. 204* London: H.M.S.O., pp. 42–60.

MITCHELL, K. J. 1956. Growth of pasture species under controlled environment. 1. Growth at various levels of constant temperature. *N.Z. Jl Sci. Technol. Sect. A* **38,** 203–15.

MITCHELL, K. J.; LUCANUS, R. 1962. Growth of pasture species under controlled environment. 3. Growth at various levels of constant temperature with 8 and 16 hours of uniform light per day. *N.Z. Jl agric. Res.* **5,** 135–44.

MUNRO, I. A. 1958. Irrigation of grassland. The influence of irrigation and nitrogen treatments on the yield and utilisation of a riverside meadow. *J. Br. Grassld Soc.* **13,** 213–21.

NATIONAL INSTITUTE OF AGRICULTURAL BOTANY. 1959. Growing clovers for seed. *Herbage Seed Growers' Leaflet No. 3.*

NATIONAL INSTITUTE OF AGRICULTURAL BOTANY. 1966. Varieties of white clover. *Farmers' Leaflet No. 23.*

NELSON, C. E.; ROBINS, J. S. 1956. Some effects of moisture, nitrogen fertilizer, and clipping on yield and botanical composition of Ladino clover-orchardgrass pasture under irrigation. *Agron. J.* **48,** 99–102.

PERCIVAL, J. 1949. Agricultural botany; theoretical and practical. London: Duckworth, 8th edn.

PERRING, F. H.; WALTERS, S. M. 1962. Atlas of the British flora. London: Thomas Nelson and Sons Ltd, pp. 456.

PETO, D. H. 1961. Methods and problems of cleaning herbage seeds. *Rep. Lincs. Seed Growers' Assn 1960* Agriculture House, Woodhall Spa, pp. 19–21.

RACHIE, K. O.; SCHMID, A. R. 1955. Winter hardiness in birdsfoot trefoil strains and varieties. *Agron. J.* **47,** 155–7.

REID, D. 1961. Factors influencing the role of clover in grass-clover leys fertilised with nitrogen at different rates. 1. The effects of the variety of companion grass on the yields of total herbage and of clover. *J. agric. Sci., Camb.* **56,** 143–53.

REID, D. 1966. Studies on cutting management of grass-clover swards. 4. The effects of close and lax cutting on yield of herbage from swards cut at different frequencies. *J. agric. Sci., Camb.* **66,** 101–6.

REID, D.; CASTLE, M. E. 1965. The response of grass-clover and pure grass leys to irrigation and to fertilizer nitrogen treatment. 1. Irrigation effects. *J. agric. Sci., Camb.* **64,** 185–94.

ROBINSON, D. H. 1953. Leguminous forage plants. London: Arnold and Co., 2nd edn.

ROBINSON, R.R.; SPRAGUE, V.G. 1947. The clover populations and yields of a Kentucky bluegrass sod as affected by nitrogen fertilization, clipping treatments, and irrigation. *J. Am. Soc. Agron.* **39,** 107–16.

SANDERSON, J. F. 1966. Aspects of the growth of white clover with particular reference to the effects of defoliation and shade. Ph.D. Thesis, Univ. Reading.

SPRAGUE, V. G. 1943. The effects of temperature and daylength on seedling emergence and early growth of several pasture species. *Proc. Soil Sci. Soc. Am.* **8,** 287–94.

SPURWAY, C. H. 1941. Soil reaction (pH) preferences of plants. *Spec. Bull 306 Mich. agric. Exp. Stn* pp. 36.

STERN, W. R.; DONALD, C. M. 1962. Light relationships in grass-clover swards. *Aust. J. agric. Res.* **13,** 599–614.

STILES, W. 1966. Ten years of irrigation experiments. *A. Rep. Grassld Res. Inst. Hurley 1965* pp. 57–66.

THOMSON, D. J. 1965. Grassland Research Institute, Hurley. Personal communication.

THORNTON, G. D. 1956. The effect of fertiliser nitrogen on nodulation, growth and nitrogen content of several legumes grown in sandy soils. *Proc. Soil Crop Sci. Soc. Fla* **16,** 146–51.

VANDERVEKEN, J. 1966. [Study of recovery of white clovers infected with clover phyllody virus.] *Meded. Rijksfac. LandbWet. Gent* **31**, 981–5.

VEZ, A. 1961. [Development of white clover with respect to fluctuations in some organic compounds.] *Ber. schweiz. bot. Ges.* **71**, 118–72.

WALKER, T. W.; EDWARDS, G. H. A.; CAVELL, A. J.; ROSE, T. H. 1952. The use of fertilizers on herbage cut for conservation. 1. *J. Br. Grassld Soc.* **7**, 107–30.

WALKER, T. W.; EDWARDS, G. H. A.; CAVELL, A. J.; ROSE, T. H. 1953. The use of fertilizers on herbage cut for conservation. 3. *J. Br. Grassld Soc.* **8**, 45–70.

WESTBROOKS, F. E.; TESAR, M. B. 1955. Tap root survival of Ladino clover. *Agron. J.* **47**, 403–10.

WILSON, G. F. 1966. Ryegrass varieties in relation to dairy cattle performance. 2. The influence of ryegrass varieties on intake, digestibility, and on some characteristics of rumen fermentation. *N.Z. Jl agric. Res.* **9**, 1053–63.

ZALESKI, A. 1961. White clover investigations. 1. Effect of seed rates and cutting treatments on flower formation and yield of seed. *J. agric. Sci., Camb.* **57**, 199–212.

ZALESKI, A. 1964a. Effect of density of plant population, photoperiod, temperature and light intensity on inflorescence formation in white clover. *J. Br. Grassld Soc.* **19**, 237–47.

ZALESKI, A. 1964b. Survey of methods used for white clover seed production. *J. natn. Inst. agric. Bot.* **10**, 110–15.

ZALESKI, A. 1965. White clover investigations. 2. Effect of companion grasses and methods of sowing on seed production. *J. agric. Sci., Camb.* **64**, 77–82.

ZALESKI, A. 1966. White clover investigations. 3. Effect of irrigation on seed production. *J. agric. Sci., Camb.* **67**, 249–53.

CHAPTER 34

Red Clover *(Trifolium pratense)*

Morphology

SPECIFIC CHARACTERS

Red clover (Plate 14) is an erect to decumbent, very hairy or almost entirely glabrous, biennial or perennial. Root system, a stout branched tap root. Leaves alternate, large, trifoliate, with ovate entire leaflets carried on long petioles. Flowers papilionaceous, densely crowded in ovoid racemes carried by axillary peduncles. Fruit a 1- to 2-seeded pod.

MORPHOLOGICAL CHARACTERS

The root system develops from a single seedling tap root which becomes branched in older plants; numerous fine lateral roots are formed, particularly in the upper 10 cm. Little detailed research on root distribution has been carried out. The tap roots grow to a length of 60–90 cm and penetrate less deeply than those of sainfoin and lucerne. A greater part of the root mass occurs in the top 10 cm layer of the soil. The majority of the numerous small (2–3 × 1–2 mm) ovoid, pinkish-white nodules are carried on the fine lateral roots.

During the establishment period, the internodes remain short and a rosette with axillary buds near ground level is formed. In the subsequent years the basal buds produce upright stems, on the upper part of which flowers arise. Such stems are annual and the plant overwinters as a loose rosette.

The large range of cultivars grown in Britain are grouped into three major morphological types, namely 'broad red' (double-cut, early-flowering), 'single-cut' (intermediate-flowering) and 'late-flowering'. The 'early-flowering' types are biennial, erect, will give two cuts a year, and are less well adapted to grazing than the late-flowering types. The late-flowering types are short-lived perennials, more or less decumbent and can withstand grazing. There are also 'wild' types, *T. pratense* var. *pratense*, that are smaller than the cultivated types and are strongly perennial.

Williams (1927) described a wide range of cultivars. Table 34.1 indicates some of the morphological differences between two of the cultivated types.

Table 34.1. Morphological characters at flowering in early- and late-flowering broad red clover

	Early-flowering	Late-flowering
Height, cm	60–80	40–80
Maximum leaflet size, mm	50 × 30	40 × 25
No. of internodes	6–9	14–15
No. of stems/plant (spaced plants)	15–17	19–43

After Williams (1927)

Hawkins (1953) distinguished cultivars on the basis of number of internodes.

For a species so widely cultivated in Britain, there is a surprising dearth of published data on leafiness in relation to growth stage. Some indication of leafiness is given in Table 34.2.

Table 34.2. Leaf percentage on a DM basis in red clover.

	Hay stage	Aftermath	Under grazing
Early-flowering	39	65	86
Late-flowering	53	39	85

After Williams (1927)

Distribution

Trifolium pratense is a native species occurring in all vice-counties of the British Isles where it is generally distributed in grassy places, though it is slightly less common in north-west Scotland. It is absent only at the highest altitudes (Clapham *et al.*, 1962; Perring & Walters, 1962) and usually dies if the salt concentration of the soil exceeds 0·3% (Chippindale, 1954). Its optimum pH range is 5·5–7·5 (Spurway, 1941).

Its distribution in Britain bears no relation to isotherms (Perring & Walters, 1962), but it normally grows at temperatures between 7 and 38°C. Above 38°C root and shoot development are reduced, while between 40 and 45°C plants become chlorotic and die. High temperatures are more detrimental to establishment, growth and persistency than are low temperatures (Gist & Mott, 1957). Doubt has been cast on its winterhardiness, but in the USA a variety has survived for 19 weeks at an average weekly temperature below 0°C and sometimes falling to −17·7°C (Bula & Smith, 1954). Thus temperature is not a limiting factor for red clover in Britain.

Early types can flower under daylengths of 9–12 h, but flowering is delayed by several weeks under shorter daylengths than this. Later types require a daylength of more than 12 h to flower (Tincker, 1925). Thus, in Britain, the daylength is sufficient to allow flowering throughout the summer.

Compatibility with grasses

Red clover is sown in mixtures with other species as a major constituent in short-term leys and to provide early herbage production when the companion species are slow-developing. In short-term leys, red clover is often grown with Italian ryegrass. In a comparison of 5 legume species at the Grassland Research Institute, Hurley (1958), broad red clover and lucerne gave the highest herbage yield in the first harvest year when grown with Italian ryegrass. In another experiment at the Grassland Research Institute (1959) (see Table 16.4), the mixtures sown consisted of 0·5–20 kg S22 Italian ryegrass with 20–0·5 kg S123 red clover/ha. This mixture yielded more than pure grass only when the seed rate was less than 10 kg/ha for Italian ryegrass and more than 10 kg/ha for red clover. Stapledon and Davies (1928) found that the addition of a suitable grass to red clover helped to combat weeds and gave a higher yield than a pure red clover sward. They found that, of the 'larger' grass species, timothy hampered clover development least.

In mixtures with species which establish slowly, the double-cut type of red clover is more suitable than the other types because its rapid development and short life reduce the risk of depressing the companion species. At the Grassland Research Institute, broad red clover has been used in the establishment of swards of timothy with white clover (1961) and of perennial ryegrass sown at low rates with white clover (1962) and has raised yields in the year of sowing considerably, with little or no depression of yield in the first harvest year. Red clover has certain advantages over Italian or Westerwolds ryegrass for this purpose. Crops harvested in the sowing year will have a higher protein content and the grasses will obtain some benefit from N fixed by the red clover. Timothy has been found to establish better with red clover than with Italian ryegrass or with 'improved' Westerwolds ryegrass. In this respect, red clover was similar to 'commercial' Westerwolds (Grassland Research Institute, 1961). On the other hand, red clover was less effective than Westerwolds in controlling weeds (Grassland Research Institute, 1963). It has also been found that the establishment of white clover is reduced if red clover is added to the mixture (Grassland Research Institute, 1961, 1962; Stapledon & Davies, 1928).

In an experiment at the Grassland Research Institute, Hurley (1962, 1963), commercial Westerwolds ryegrass at 6·7 kg seed/ha and broad red clover at 4·5 kg/ha were added to timothy and white clover. The mixture proved very effective in controlling weeds and gave a good herbage yield in the sowing year. It was noted that there was less harm to the timothy than from adding Westerwolds ryegrass alone, and less suppression of white clover than from adding red clover alone.

Garner and Sanders (1939) found that the inclusion of a late-flowering red clover in rather complex 4-year ley mixtures increased yield in the first two years with no reduction in the following years although the establishment of white clover was slightly reduced.

Lampeter (1960) (see Table 16.3) showed that in the first harvest year the suppressive effect of a companion grass on the legume was very similar for lucerne and red clover when grown with perennial ryegrass, Italian ryegrass, cocksfoot, timothy or meadow fescue.

Seed

CHARACTERS

The seed of red clover is oval, slightly flattened and with a distinct projection formed by the radicle (Percival, 1949; Robinson, 1953). Its length varies from about 1·5 to 2·2 mm, the width across the broadest part is 1·0–1·5 mm, and the thickness is about 1 mm (Percival, 1949; Robinson, 1953). The colour of the seed may be yellow and purple, yellow, or brown (Stapledon, 1920; Percival, 1949; Robinson, 1953). In general, seed-weight and germination capacity are highest in purplish and lowest in brown seeds (Stapledon, 1920). The percentage of hard seeds has been found to range from 3 to 10 (Stapledon, 1920; Robinson, 1953).

The 1 000-seed weight is 1·773 g on average (Robinson, 1953), but varies from 1·29 g for brown seeds and 1·52 g for normal samples of mixed colours, up to 2·52 g (Stapledon, 1920).

PRODUCTION

When grown for seed, red clover is best sown pure at 6–8 kg/ha but a companion grass, sown at not more than 5 kg/ha, may be included to increase herbage production from the stand (National Institute of Agricultural Botany, 1959; Hawkins, 1956). Red clover should be sown alone in drills 18 cm apart, or in alternate rows 60 cm apart with a companion grass, at a maximum depth of 2·5 cm (National Institute of Agricultural Botany, 1959; Hawkins, 1956, 1958). It is usually sown under a cereal cover crop, and the stands are normally cut for fodder or grazed before the seed is harvested; application of fertilizer is not required unless the land is deficient in P (Hawkins, 1959; National Institute of Agricultural Botany, 1959). The highest seed yields are obtained from crops not previously cut in the year of harvest (Hawkins, 1956), but a cut for hay or silage can be taken before the end of May for the late-flowering type and before mid-May for the single-cut type and such cuts may help to produce a standing crop at harvest (Hawkins, 1956, 1958, 1959; National Institute of Agricultural Botany, 1959). Pollination is a major problem; if bumble bees are scarce, it may be advantageous to produce seed in small fields, to cut part of large red clover areas for hay, or to introduce hives of honey bees (Hawkins, 1958, 1959). Yields of seed are correlated with the amount of rainfall during flowering and with rainfall and hours of sunshine during seed ripening (Hawkins, 1956). Crops grown for seed should be isolated during flowering, the required distance being 180 m between crops of less than 2 ha, falling to 45 m between crops covering 7 ha or more (National Institute of Agricultural Botany, 1959). Principles and procedures for harvesting are well explained elsewhere (Ministry of Agriculture, Fisheries & Food, 1968). The annual yield of seed can vary from 25 to 875 kg/ha, with mean yields ranging from 40 to 440 kg depending on cultivar and year (Lincs. Seed Growers' Association, 1965).

Yield

CULTIVARS

Red clover cultivars are classified according to flowering date into three main types: early (broad red), intermediate (single-cut), and late. The difference in flowering date between the earliest and latest varieties is about 4 weeks and there is a corresponding difference in spring growth. When the first cut is taken at the same growth stage (25% of inflorescences in bloom), the late type yields slightly more than the early type, but the latter makes up for this in subsequent production during the first harvest year. Since the earliest varieties are ready for a first cut in late May or early June and the latest varieties almost a month later, the former give two cuts of considerable proportions in July and September, whereas the latter normally give only one further cut in August, followed by a period of relatively slow growth which may provide autumn grazing.

Broad red clover is normally ploughed at the end of the first harvest year, since its productivity in the second year is unreliable. Single-cut red clover is more persistent and late red clover still more so, but their yields in the second harvest year are usually about one third less than in the first.

Recently several tetraploid cultivars of the broad red type have been bred and are now under test in field trials; so far they have outyielded diploids in the same maturity group by 15–40% and have also proved much more persistent than the diploids. Both these features are thought to be due largely to their high resistance to fungal diseases, especially clover rot (*Sclerotinia trifoliorum*).

Because red clover has been grown widely and over a long period in Britain, clover rot and clover stem eelworm (to mention but two of the more important soil-borne disease organisms) are endemic and frequently responsible for very low yields. The data on yield and seasonal productivity of broad red clover shown in Table 34.3 are based on reasonably healthy stands of the diploid local variety Essex, rather than on its average performance. Nevertheless, these data may not be indicative of the potential of broad red clover as represented by tetraploid cultivars. This potential should become apparent in the near future when sufficient trial results become available.

Table 34.3. Yield, digestibility and protein content of broad red clover, Essex type

	Cutting date	DM kg/ha			DM digestibility %	D-value %	Digestible OM kg/ha			N %	N kg/ha			Crude protein %	Crude protein kg/ha		
1st cut	12 May	2 000			75	67	1 340			4·0	80			25·0	500		
	27 May		4 500		70	63		2 830		3·6		162		22·5		1 010	
	11 June			6 000	65	59			3 540	2·8			168	17·5			1 050
2nd cut	28 July	4 000			65	58	2 320			2·8	112			17·5	700		
	12 July		3 600		68	61		2 190		3·0		108		18·7		680	
	25 July			3 000	69	62			1 860	3·2			96	20·0			600
3rd cut	28 Aug.	2 700			65	58	1 570			3·4	92			21·3	570		
	12 Sept.		2 300		66	59		1 360		3·7		85		23·1		530	
	28 Sept.			2 000	66	59			1 180	3·9			78	24·4			490
4th cut	15 Oct.	800			66	59	470			4·0	32			25·0	200		
Total yield		9 500	9 900	11 000			5 700	6 380	6 580		316	355	342		1 970	2 220	2 140

Data compiled by J. O. Green (Grassland Research Institute, Hurley) from literature cited in the text. For conversion formulae see Chapter 5.

Red clover yields in the first harvest year show considerable variation, not so much between years as between sites. This suggests that establishment conditions and the health of the stand are the major variables. A well established red clover crop is not very different from lucerne in its inherent productive capacity during the first harvest year. However, even the early type usually yields less than lucerne in the spring. As red clover is more digestible than lucerne at a given date in spring, it can be allowed to grow on until a later date before the first cut is taken. As shown in Tables 34.3 and 35.3, the annual dry matter (DM) yield from broad red clover first cut on 11 June was 11 000 kg/ha, against 13 300 kg/ha from lucerne first cut on 27 May. Both would have a digestibility (DOMD) of about 59 % (DM digestibility 65 %) on the first cut; the red clover regrowths were of similar digestibility, and the lucerne regrowths of slightly lower digestibility.

SEASONAL YIELD

The seasonal distribution of yield in broad red clover is similar to that in lucerne. As the late red clover type is much later maturing and also recovers more slowly, the proportion of its production given after midsummer is less than that of the broad red type.

ENERGY AND PROTEIN YIELDS

Data from the literature are summarized in Table 34.4.

Table 34.4. Potential annual yields of energy and protein from red clover (DM basis)

	Mean	Range
DM digestibility (%)	66	61–70
Yield of digestible OM (kg/ha)	5 840	5 700–6 090
Yield of metabolizable energy (Mcal/ha)	21·5	20·5–21·9
Crude protein (%)	21·5	17·5–25·0
Yield of crude protein (kg/ha)	1 950	1 930–1 970
Yield of digestible crude protein (kg/ha)	1 580	1 550–1 690

Data compiled by J. O. Green (Grassland Research Institute, Hurley) from literature cited in the text. For conversion formulae see Chapter 5.

Response to management and environment

RESPONSE TO FERTILIZER

Nitrogen

Application of N has been shown to reduce nodulation in red clover (Dilz & Mulder, 1962) with the result that the fertilizer N taken up by the plant does not promote growth but merely substitutes for the atmospheric N which would otherwise have been fixed by the nodule bacteria (Allos &

Bartholomew, 1955; McAuliffe *et al.*, 1958; Thornton, 1956). Exceptions to this have been found, mainly on acid soils (pH less than 5·5) where nodulation is poor. Under these conditions, Dilz and Mulder found that the application of 60 kg N/ha increased the fresh weight yield of red clover from 4 to 14 t/ha, and 240 kg N/ha increased it to 36 t/ha. At a higher pH (6·8) the corresponding yields were 28, 43 and 52 t/ha. This response of red clover to fertilizer N on near-neutral soil was attributed to the abnormally sunny, warm and dry climate of the year; no response to fertilizer N was obtained in the following year when more normal conditions of high rainfall and moderate temperatures prevailed.

Phosphorus

The concentration of P in the DM of red clover should be 0·2–0·4% if yield is not to be restricted by P deficiency. For an annual maximum yield of about 10 000 kg DM/ha, the P requirement of red clover is therefore between 20 and 40 kg P/ha per year. In a pot experiment, Robinson & Sprague (1947) obtained a four-fold increase in yield from an application equivalent to 50 kg P/ha on a soil low in available P and a two-fold increase in yield on soil of medium P status; an additional 50 kg P/ha gave only small further increases in yield. Fertilizer applied in a band 2·5 cm below the soil surface was more than twice as effective in increasing yield as the same quantity of fertilizer mixed uniformly into the top 6 cm of soil.

Potassium

The critical content of K below which red clover suffers from K deficiency has been estimated by McNaught (1958) to be 1·8% in the DM. For an annual yield of 10 000 kg DM/ha, red clover therefore requires 180 kg K/ha per year. Many soils can supply this amount of K (Arnold & Close, 1961) and on such soils no response to K fertilizer would be expected. On soils in which part of the K applied as fertilizer is fixed in a form unavailable to plants, rates higher than 180 kg K/ha would be necessary to ensure an adequate supply of K to red clover. On a fen peat soil low in available K, Zürn (1958) increased the yield of red clover hay from 1 800 to 11 100 kg/ha by the application of 135 kg K/ha.

Calcium

On acid soils, it is generally necessary to apply lime to bring the soil pH up to 6 in order to ensure satisfactory nodule formation in red clover. Dilz and Mulder (1962) reported almost no nodule formation at pH values below 5·2, moderate nodulation between pH 5·2 and 5·8, and free nodulation above pH 6. By liming a soil of pH 4·6, Sergejev (1958) obtained an increase in the hay yield of red clover of 3 700 kg/ha; on a less acid soil (pH 5·5) the response was 300 kg/ha. In addition to yield increases, Izawa and Okamoto (1957) reported increases in the N, P, K, Ca, Mg and S content of red clover as a result of the application of lime. Their findings were confirmed by Beaton *et al.* (1962) but not by Snaydon (1962).

RESPONSE TO IRRIGATION

There is a lack of information on the response of red clover to irrigation under British conditions. However, this response appears to be much the same as for white clover, although Dančík (1963) found that the increase in yield of red clover brought about by irrigation was intermediate between that of irrigated lucerne and that of irrigated white clover. Dančík (1962) and Julén (1951) reported an increase in the proportion of red clover in grass/clover swards under irrigation. In pure swards of red clover, Steinbrenner and Eich (1962) found an average increase in yield of 43% over a 5-year period and Klatt (1964) in 1963 obtained responses of 81 and 123% with 70 and 100 mm of water, respectively, applied during the year.

RESPONSES TO TEMPERATURE

The only precise data so far available on the response of red clover to temperature are derived almost entirely from work in the USA on the cultivar Kenland. Gist and Mott (1957) found little effect of temperature over the range 15·5–27°C, but a marked reduction in growth at 32°C, whether or not plants were grown under light intensities of 200, 600 or 1 200 f.c. (2 000, 6 000 or 12 000 lux) of light; if anything, maximum yields were achieved at 15·5°C and under 1 200 f.c. (12 000 lux). Kendall (1958) reported a rather higher optimum temperature for red clover growth. He grew plants in a range of day and night temperatures and found that growth was better at a day temperature of 24°C than at 18°C or 35°C, combined with a night temperature of 18°C rather than 13°C or 27°C; day temperature was a more important determinant of yield than night temperature. Finally, Mes (1959) found less N fixation, and hence less growth, at temperatures of 25–27°C than at 18–21°C, and concluded that red clover was more sensitive to temperature than many other legumes.

RESPONSE TO DEFOLIATION

Cultivars of red clover have been selected to produce their highest DM yields under infrequent cutting and are now divided into four varietal groups based on their time of maximum flowering (National Institute of Agricultural Botany, 1966). In the first harvest year the early-flowering types, cut for the first time when 25% of flowers had emerged, yielded less than the three other groups, but gave two cuts later in the season and produced higher regrowth yields. Cutting the other groups at the same growth stage allowed only one aftermath cut, which gave about 80% of the regrowth yields obtained from early types. The cv. Altaswede appeared exceptional in that its regrowth production was as little as 50% of that from early cultivars (National Institute of Agricultural Botany, 1964).

Red clover is not usually managed under a frequent defoliation system. Williams (1924), working with 5 early and 5 late cultivars, showed that cutting 8 times at monthly intervals during the growing season reduced the yield to below 50%, and cutting 4 times to below 70% of that given by a single cut for hay. Variation among cultivars in response to defoliation was high, but the average reduction in yield for late cultivars (77%) was greater than for earlier cultivars (67%) under the 8-cut management. Under the 4-cut management the situation was reversed and late cultivars yielded more than the early ones.

Quality

CHEMICAL COMPOSITION

For many chemical constituents, the composition of red clover is similar to that of lucerne and intermediate between that of white clover and sainfoin. Details are given in Chapter 31. Red clover, however, has generally a rather higher content of Mg than the other legume species.

WATER CONTENT

There is evidence that both stage of growth and N application influence the water content of red clover. Thus, Hvidsten and Pedersen (1950), who sampled red clover at weekly intervals during uninterrupted growth throughout the growing period for 5 successive years, found that water content fell with development stage. Kivimäe (1959) also found that water content fell with stage of development. The DM percentage plotted against cutting date followed a concave curve, the increase during the leaf stage being small, but becoming greater as development continued. Davies *et al.* (1966) found a decrease in water content of both early and late red clover when successive cuts were taken from May to July.

Ivins (1952) found that red clover herbage from areas that had received fertilizer N at the rate of two cwt/ac (246 kg/ha) had a higher water content than that from areas which had received no N.

DRY MATTER DIGESTIBILITY

Young leafy herbage frequently cut

Broad red clover has been shown to have a DM digestibility of about 75% in spring, and Davies *et al.* (1966) have shown that in the late-flowering red clover cultivar S123 it can be as high as 80%. Although no data are available on the digestibility of immature regrowths, it is probable that they would be similar to spring herbage in digestibility.

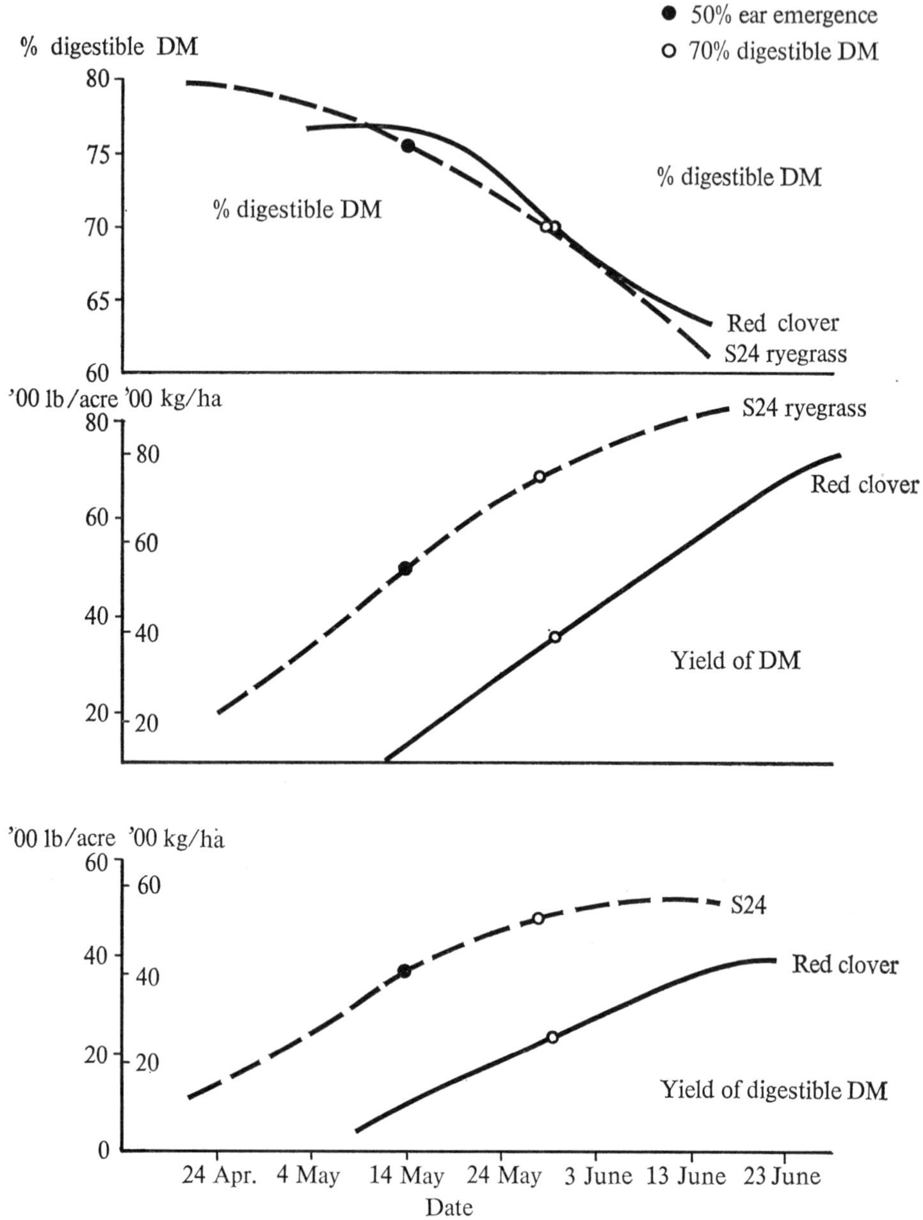

Fig. 34.1. Red clover Essex Broad Red. Seasonal changes in DM yield and digestibility. Values for S24 perennial ryegrass shown for comparison. (From Green & Corrall, 1967)

Primary growth

Red clover is mainly grown for conservation and most research workers have therefore concentrated their studies on primary growth. Fig. 34.1 shows that the main period of growth in broad red clover is about 3 weeks later than in S24 perennial ryegrass, but that the former reaches 70% DM digestibility only a day or two later than S24. A comparison shows that for any yield level, broad red clover is less digestible than ryegrass, or conversely that for any digestibility level ryegrass outyields broad red clover, although the latter has a higher protein content. These statements assume an adequate N supply for the ryegrass.

Regrowths

Experiments at the Grassland Research Institute, Hurley (Green & Corrall, 1967) have shown that after 7 weeks, red clover regrowth reaches a DM digestibility of about 68%, but Dent (1963) found 8-week-old regrowth to have values between 61 and 67%. These values are lower than those for perennial ryegrass, but red clover regrowths have a considerably higher protein content.

Red clover is traditionally grown for hay and is often cut in late June to enable its full yield potential to be realized. Broad red clover reaches 70% DM digestibility in late May or early June, depending on season, and after this time falls by about 0·5% per day, although the rate of fall later becomes less. The DM digestibility of red clover hay cut in late June must therefore be well below 70% DM digestibility. Since the regrowth is often harvested for seed, its nutritive value is of little practical significance. Late-flowering red clovers, such as S123, reach 70% DM digestibility about 10 days later than the broad red clovers. Since the crop is growing rapidly at this time, its DM yield at a DM digestibility of 70% is about 1 000 lb/ac (454 kg/ha) higher than that of the broad red clovers.

Red clover has a high protein content, which may justify its inclusion in mixtures with grasses which may be low in protein; otherwise it is difficult to justify the use of red clover in view of its digestibility characteristics.

ANIMAL INTAKE CHARACTERISTICS

The limited evidence available suggests that the intake of red clover is much higher than that of grasses and is intermediate between that of sainfoin and lucerne (Osbourn *et al.*, 1966).

Longevity and persistency

When established, red clover plants produce a rosette of leaves in the axils of which the ascending aerial stems are produced (Becker-Dillingen, 1929). These stems do not root at the nodes. Their growth is determinate, the growing points developing into inflorescences, and they survive for only one season. The length of life of individual leaves and stems of red clover does not appear to have been studied.

The course of growth and senescence below the soil surface has been described by Terekhova (1956). The diameter of the tap root at the base of the stem was greatest in late-maturing plants within a population; this was associated with higher numbers of aerial stems, better development of adventitious roots and greater longevity. Williams (1927) also found that late-maturing cultivars persisted longer. Therrien and Smith (1960) found that plants which produced only a rosette during the year of sowing were least susceptible to winter killing, whereas those producing many flowering stems and no rosette were the most susceptible. Terekhova (1956) noted that in more than 20% of plants the central vascular portion of the tap root became black and disintegrated towards the end of the year of sowing, but stated that this was not normal. Similar observations, however, have been made by Leffel and Graham (1966), which seem to indicate that the tap root is not inherently long-lived; they found a positive correlation between internal breakdown and the diameter of the tap root at 6·5 months after sowing. Terekhova (1956) noted an increase in disintegration of the tap root with age, and by the end of the third and fourth years the tap root was completely inactive in many plants and in some plants had split into several parts. First-order lateral roots grow slowly sideways and then rapidly downwards; sideways growth was greatest in the latest of the 3 cultivars tested by Terekhova and was about twice as much as in an early, non-persistent cultivar. Higher-order fibrous roots began to die off when the flowerheads turned brown.

Red clover has 4 types of adventitious roots. These are formed on the hypocotyl during aerial shoot formation and absorb nutrients from the top-soil. They are more numerous in single-cut than in double-cut types and are best developed during the sowing year. They usually die off completely in the second year, though the largest of them are difficult to distinguish from first-order lateral roots. Towards the end of the first year they become outnumbered by adventitious roots formed at the junction between the stem and the hypocotyl. Root contraction results in the sinking of stem bases into the soil. Adventitious roots develop on both nodes and internodes of these stem bases and often form the major part of the root system in the third and fourth years of life, when older roots have senesced and died.

The maximum life-span of red clover plants is not known with certainty. Becker-Dillingen (1929) quotes 2–3 years. Krylova (1965), presumably working with red clover established from seed, found no plants which survived their fifth winter. Of 26 4-year-old plants examined, all but 6 died out towards spring. The surviving plants had a suppressed central rosette, hardly any buds in the axils of dead rosette-leaves, completely degenerated tap roots, and did not flower; these plants were those which had developed least vigorously during the previous year.

Clonal material appears to have a similar life-span, according to results obtained by Taylor *et al.* (1962). Of 2 080 plants established from cuttings, the majority did not persist more than 3 years, although a few plants were alive in the fifth year. They found longevity to be associated with the production of fibrous secondary roots; seed production was not found detrimental to persistence, though results obtained by Smith (1963) indicated that it was.

Williams (1927) found that red clover was much more persistent when grown as spaced plants than when grown under competitive conditions in the field. The poor ability of red clover to compete for nutrients has been pointed out by Becker-Dillingen (1929) and others. Williams (1927) also found that frost-heaving was a major cause of winter killing, particularly in plants which had been repeatedly cut or grazed and therefore had less well developed roots. Late-maturing and wild cultivars had a more strongly developed lateral root system (see also Terekhova, 1956) and were therefore less liable to be lifted out of the soil by frost.

Factors adversely affecting longevity of red clover include cutting or grazing in late autumn, spring drought, poor sandy soils, excessive drainage and deep tillage (Becker-Dillingen, 1929). An attractive hypothesis to explain the limited persistency of red clover is that senescence of the below-ground internodes may reduce their capacity to produce adventitious roots and so decrease the ability of the plant to withstand competition and, in particular, to survive the winter, during which the majority of plant deaths occur. The apparent similarity in longevity of plants raised from cuttings and from seed provides evidence in this direction, but further investigations are needed.

Red clover as a herbage plant is generally represented by one of three main types. Broad red clover is the shortest-lived, and gives no worthwhile yield in the second harvest year. Single-cut red clover may be used for early grazing before being cut, but produces very little regrowth after cutting. The third type, late-flowering red clover, is of greater inherent persistence and is often included in long-term herbage seeds mixtures (National Institute of Agricultural Botany, 1966). This type is late to start growth in the spring but, if the crop is cut, good regrowth either for cutting or grazing can be expected. To some extent the persistence of red clover is inherent, but can be modified by grazing management. Thus in New Zealand it has been found that frequent hard grazing during the winter encouraged the later growth, but similar management during the spring reduced both yield and plant survival compared with its effect on other legumes. In a mixed sward frequent hard grazing during the summer reduced the yield of all species (Brougham, 1959). Again, in a mixed sward, red clover was found to persist under frequent or infrequent grazing provided that the intensity of grazing was not severe (Brougham, 1960).

References

ALLOS, H. F.; BARTHOLOMEW, W. V. 1955. Effect of available nitrogen on symbiotic fixation. *Proc. Soil Sci. Soc. Am.* **19,** 182–4.

ARNOLD, P. W.; CLOSE, B. M. 1961. Release of non-exchangeable potassium from some British soils cropped in the glasshouse. *J. agric. Sci., Camb.* **60,** 329–40.

BEATON, J. D.; GOUGH, N. A.; ROWLES, C. A. 1962. The influence of base saturation and fertility treatments on yield and cation composition of crops grown on two British Columbia soils high in organic matter. 2. Red clover. *Emp. J. exp. Agric.* **30,** 49–56.

BECKER-DILLINGEN, J. 1929. Handbuch des gesamten Pflanzenbaues, einschliesslich der Pflanzenzüchtung. 3. Handbuch des Hülsenfruchtbaues und Futterbaues. Berlin: Paul Parey, pp. 670.

BROUGHAM, R. W. 1959. The effect of frequency and intensity of grazing on the productivity of a pasture of short rotation ryegrass and red and white clover. *N.Z. Jl agric. Res.* **2,** 1232–48.

BROUGHAM, R. W. 1960. The effects of frequent hard grazings at different times of the year on the productivity and species yields of a grass-clover pasture. *N.Z. Jl agric. Res.* **3,** 125–36.

BULA, R. J.; SMITH, D. 1954. Cold resistance and chemical composition in overwintering alfalfa, red clover and sweet clover. *Agron. J.* **46,** 397–401.

CHIPPINDALE, H. G. 1954. The restoration of grassland flooded by sea water. *Agric. Prog.* **29,** 25–33.

CLAPHAM, A. R.; TUTIN, T. G.; WARBURG, E. F. 1962. Flora of the British Isles. Cambridge: Univ. Press, 2nd edn.

DANČÍK, J. 1962. [Contribution to the influence of irrigation on clovers and their mixtures with grasses.] *Pol'nohospodárstvo* **9,** 587–96.

DANČÍK, J. 1963. [Effect of irrigation on yield and quality of clover species and their mixtures with grasses.] *Ved. Pr. výsk. Úst. rastl. Výroby Piešt'anoch* **2,** 233–52.

DAVIES, W. E.; GRIFFITH, G. AP; ELLINGTON, A. 1966. The assessment of herbage legume varieties. 2. *In vitro* digestibility, water, carbohydrate, crude protein and mineral content of primary growth of clover and lucerne. *J. agric. Sci., Camb.* **66,** 351–7.

DENT, J. W. 1963. Applications of the two-stage *in vitro* digestibility method to variety testing. *J. Br. Grassld Soc.* **19,** 181–9.

DILZ, K.; MULDER, E. G. 1962. The effect of soil pH, stable manure and fertilizer nitrogen on the growth of red clover and red clover associations with perennial ryegrass. *Neth. J. agric. Sci.* **10,** 1–22.

GARNER, F. H.; SANDERS, H. G. 1939. Four year leys. The inclusion of red clover. *J. agric. Sci., Camb.* **29,** 164–73.

GIST, G. R.; MOTT, G. O. 1957. Some effects of light intensity, temperature and soil moisture on growth of alfalfa, red clover and birdsfoot trefoil seedlings. *Agron. J.* **49,** 33–6.

GRASSLAND RESEARCH INSTITUTE. 1958. *Exps Prog. 10 Grassld Res. Inst. 1956–7* p. 21.

GRASSLAND RESEARCH INSTITUTE. 1959. *Exps Prog. 11 Grassld Res. Inst. 1957–8* p. 23.

GRASSLAND RESEARCH INSTITUTE. 1961. *Exps Prog. 13 Grassld Res. Inst. 1959–60* p. 14.

GRASSLAND RESEARCH INSTITUTE. 1962. *Exps Prog. 14 Grassld Res. Inst. 1960–61* p. 16.

GRASSLAND RESEARCH INSTITUTE. 1963. *Exps Prog. 15 Grassld Res. Inst. 1961–62* p. 15.

GREEN, J. O.; CORRALL, A. J. 1967. Grassland Research Institute, Hurley. Unpublished data.

HAWKINS, R. P. 1953. Investigations of local strains of herbage plants. 2. Types of red clover and their identification. *J. Br. Grassld Soc.* **8,** 213–38.

HAWKINS, R. P. 1956. A preliminary survey of red clover seed production. *Ann. appl. Biol.* **44,** 657–64.

HAWKINS, R. P. 1958. A survey of late flowering and single cut red clover seed crops. *J. natn. Inst. agric. Bot.* **8,** 450–61.

HAWKINS, R. P. 1959. Problems of red clover seed production. *J. natn. Inst. agric. Bot.* **8,** 525–9.

HVIDSTEN, H.; PEDERSEN, E. 1950. Investigations on the content of dry-matter, crude protein and carotene in meadow and pasture plants. *Forsk. Fors. Landbr.* **1,** 311–45.

IVINS, J. D. 1952. The relative palatability of herbage plants. *J. Br. Grassld Soc.* **7**, 43–54.

IZAWA, G.; OKAMOTO, S. 1957. Yield and chemical composition of red clover and ladino clover as affected by phosphorus and calcium. *Soil Pl. Fd, Tokyo* **5**, 70–2.

JULÉN, G. 1951. [The effect of irrigation with river water and sewage water upon different types of temporary ley.] *Växtodling* No. 6, pp. 141.

KENDALL, W. A. 1958. The persistence of red clover and carbohydrate concentrations in the roots at various temperatures. *Agron. J.* **50**, 657–9.

KIVIMÄE, A. 1959. Chemical composition and digestibility of some grassland crops. *Acta Agric. scand.* Suppl. 5, p. 26.

KLATT, F. 1964. [Results and experience with field sprinkler irrigation, 1963.] *Dt. Landwirt.* **15**, 300.

KRYLOVA, N. P. 1965. [On the expression of the senile period in the life cycle of the clovers *Trifolium pratense* L. and *T. hybridum* L.] *Bot. Zh. SSSR* No. 50, 692–4.

LAMPETER, W. 1960. [Mutual influences of higher plants regarding growth of shoots and roots, mineral content and water consumption—studied in some economically important forage plants.] *Wiss. Z. Karl-Marx-Univ. Lpz.* **9**, 611–722.

LEFFEL, R. C.; GRAHAM, J. H. 1966. Influence of ecotype, day length, and temperature on morphological development and internal breakdown of red clover, *Trifolium pratense* L. *Proc. 10th int. Grassld Congr., Helsinki, 1966* pp. 99–103.

LINCOLNSHIRE SEED GROWERS' ASSOCIATION. 1965. Report for 1964. Agriculture House, Woodhall Spa, pp. 3–7.

MCAULIFFE, C.; CHAMBLEE, D. S.; URIBE-ARANGO, H.; WOODHOUSE, W. W. 1958. Influence of inorganic nitrogen on nitrogen fixation by legumes as revealed by N^{15}. *Agron. J.* **50**, 334–7.

MCNAUGHT, K. J. 1958. Potassium deficiency in pastures. 1. Potassium content of legumes and grasses. *N.Z. Jl agric. Res.* **1**, 148–81.

MES, M. G. 1959. Influence of temperature on the symbiotic nitrogen fixation of legumes. *Nature, Lond.* **184**, 2032–3.

MINISTRY OF AGRICULTURE, FISHERIES AND FOOD. 1968. Grass and clover crops for seed. *Bull. 204* London: H.M.S.O., pp. 42–60.

NATIONAL INSTITUTE OF AGRICULTURAL BOTANY. 1959. Growing clovers for seed. *Herbage Seed Growers' Leaflet No. 3.*

NATIONAL INSTITUTE OF AGRICULTURAL BOTANY. 1964. Grass and forage legume trials *45th Rep. natn. Inst. agric. Bot. 1964* pp. 16–18.

NATIONAL INSTITUTE OF AGRICULTURAL BOTANY. 1966. Varieties of red clover. *Fmrs' Lflt 18* pp. 4.

OSBOURN, D. F.; THOMSON, D. J.; TERRY, R. A. 1966. The relationship between voluntary intake and digestibility of forage crops, using sheep. *Proc. 10th int. Grassld Congr., Helsinki, 1966* pp. 363–7.

PERCIVAL, J. 1949. Agricultural botany; theoretical and practical. London: Duckworth, 8th edn.

PERRING, F. H.; WALTERS, S. M. 1962. Atlas of the British flora. London: Thomas Nelson and Sons Ltd, pp. 456.

ROBINSON, D. H. 1953. Leguminous forage plants. London: Arnold and Co., 2nd edn.

ROBINSON, R. R.; SPRAGUE, V. G. 1947. The clover populations and yields of a Kentucky bluegrass sod as affected by nitrogen fertilization, clipping treatments and irrigation. *Agron. J.* **39**, 107–16.

SERGEJEV, P. A. 1958. [Methods for increasing yields of red clover.] *Zemledelie, Mosk.* **6**, 66–72.

SMITH, D. 1963. Reliability of flowering as an indicator of winter survival in red clover. *Can. J. Pl. Sci.* **43**, 386–9.

SNAYDON, R. W. 1962. The response of certain *Trifolium* spp. to calcium in sand culture. *Pl. Soil* **16**, 381–8.

SPURWAY, C. H. 1941. Soil reaction (pH) preferences of plants. *Spec. Bull 306 Mich. agric. Exp. Stn* pp. 36.

STAPLEDON, R. G. 1920. Seed studies: red clover with special reference to the country of origin of the seed. *J. agric. Sci., Camb.* **10**, 90–120.

STAPLEDON, R. G.; DAVIES, W. 1928. Seeds mixture experiments with special reference to the influence of environmental factors. *Bull. H8 Welsh Pl. Breed. Stn* p. 79.

STEINBRENNER, K.; EICH, D. 1962. [Results of long-term irrigation experiments in Müche-berg.] *Ztschr. Landeskult.* **3**, 276–92.

TAYLOR, N. L.; STROUBE, W. H.; KENDALL, W. A.; FERGUS, E. N. 1962. Variation and relation of clonal persistence and seed production in red clover. *Crop Sci.* **2**, 303–5.

TEREKHOVA, A. F. 1956. [Study of the root system in red clover.] *Bot. Zh. SSSR* **41**, 553–8.

THERRIEN, H. P.; SMITH, D. 1960. The association of flowering habit with winter survival in red and alsike clover during the seedling year of growth. *Can. J. Pl. Sci.* **40**, 335–44.

THORNTON, G. D. 1956. The effect of fertiliser nitrogen on nodulation, growth and nitrogen content of several legumes grown in sandy soils. *Proc. Soil Crop Sci. Soc. Fla.* **16**, 146–51.

TINCKER, M. A. H. 1925. The effect of length of day upon the growth and reproduction of some economic plants. *Ann. Bot.* **39**, 721–54.

WILLIAMS, R. D. 1924. The productivity of different strains and nationalities of red clover under hay and pasture conditions. *Bull. H3 Welsh Pl. Breed. Stn* pp. 131–68.

WILLIAMS, R. D. 1927. Red clover investigations, 1919–1926. *Bull. H7 Welsh Pl. Breed. Stn* pp. 13.

ZÜRN, F. 1958. [Development and yield in fodder plants on low-lying heath under the influence of potassium fertilization.] *Bayer. landw. Jb.* **35**, 101–13.

CHAPTER 35

Lucerne *(Medicago sativa)*

Morphology

LUCERNE (Plate 15) is a perennial with numerous upright or decumbent stems. Root system a stout tap-root. Leaves alternate, borne on short petioles, glabrous or pubescent, trifoliate. Leaflets narrow-ovate with serrated upper margin and mucronate tips. Stipules broad, serrate, sharply pointed. Flowers papilionaceous, borne in short axillary racemes. Pods dehiscent, several-seeded, spirally twisted.

The root system consists of a stout tap-root which may have a few large lateral roots. On most soils the tap-root is 60–90 cm long, but in deep, well-drained soils rooting depths of more than 7 m have been recorded.

The young plant has a single upright stem. Buds at the base of this stem grow to form erect, leafy stems which, in turn, produce basal buds. This leads to the formation of branched crowns with buds at or near ground level.

A large number of cultivars exist, those grown in Britain being predominantly upright. Rhizomatous forms are being developed in North America and Australia. Cultivars can be classified according to flowering date. Dent and Zaleski (1954) recognized early-flowering, mid-season, semi-late and late-flowering groups. Gill and Vear (1958) described the early types as erect, with long leaflets, and the later-maturing types as shorter, with smaller, narrower leaflets. In a detailed study on the identification of lucerne cultivars, Zaleski (1954a) examined time of flowering, number of internodes, number of branches, length of stem, and the length/width ratio of the leaflets. Late types were shorter and more decumbent at the early flowering stage than early types. There were varietal differences in leaflet shape and length/width ratios, leaflets varying from long-wide to short-narrow. The proportion of leaf and stem varies with development stage, as shown by Terry and Tilley (1964) (Table 35.1).

Dent and Zaleski (1954) described early cultivars as stemmy and late cultivars as leafy, but Willey and Zaleski (1955) stated that the leaf/stem ratio calculated on a DM basis is fairly constant for all strains (= cultivars).

Table 35.1. Percentage of leaf in the DM of an early variety of lucerne during growth of the first crop

Date	Leaf %
2 Apr.	70
16 Apr.	57
1 May	50
14 May	40
4 June	36
25 June	35

After Terry and Tilley (1964)

More recent work by Dent and Zaleski (1960) showed that at the same stage of growth, early cultivars had 45% leaf on a DM basis, compared with 41% for mid-season ones and 44% for a late-season cultivar.

Distribution

Medicago sativa is an introduced species in Britain, where it occurs in 102 vice-counties (Clapham *et al.*, 1962; Perring & Walters, 1962); it is usually sown, but occurs in a more or less naturalized state on waste ground. It is fairly common in south-east England, becoming rarer towards the north and west. It is largely absent in the south-east and north-west of England, in Wales and in the west and north of Scotland, and at altitudes above 300 m (Perring & Walters, 1962). It occurs mainly in areas where the annual rainfall is less than 750 mm and only occasionally where it exceeds 1 000 mm.

Temperature does not appear to be the main factor limiting the distribution of lucerne, since in the USA this species has survived for 19 weeks at temperatures below freezing and sometimes as low as −17·7°C (Bula & Smith, 1954); Rachie and Schmid (1955) also reported that 19-day-old lucerne seedlings showed 91·3% survival when subjected to −10°C.

In Britain the distribution of lucerne tends generally, though not invariably, to be associated with the presence of calcareous soils, especially in the south-east (Perring & Walters, 1962). Although the optimum soil pH range for lucerne is slightly alkaline (6·2–7·8), the total range is 5·5–8·8 (Spurway, 1941). In soils of pH 5·3, Fox and Lipps (1955) found that root penetration was only 90 cm in the first year, nodulation was poor and the incidence of winter-kill high, whereas in soils of pH 6·3, lucerne established well and its roots penetrated to a depth of 150 cm. Lucerne roots penetrate soil made saline with up to 0·25% NaCl.

Compatibility with grasses

Lucerne is grown with a companion grass for 3 main reasons:

(i) the grass competes with weeds which invade pure stands of lucerne, particularly during the dormant winter phase;

(ii) hay and silage are easier to make from grass/legume mixtures than from pure legumes (Mansfield, 1945);
(iii) grass/legume mixtures provide a more balanced feed than pure lucerne.

CHOICE OF COMPANION GRASS

Bolton (1962) suggested that selection for competitive ability was not likely to become a major breeding objective for lucerne and that the desired result could be obtained through management practices, including the choice of companion grass. As lucerne growth is suppressed to some extent by the presence of any of the cultivated grasses (Lampeter, 1960) (Table 16.3), the choice becomes a matter of compromise. In farm trials, Williams (1950) found that lucerne was more adversely affected by being grown with cocksfoot than with ryegrass, timothy or meadow fescue; similarly, Lampeter found that timothy and meadow fescue caused less reduction in the growth of lucerne than did cocksfoot or ryegrass. As a result of experiments at the Grassland Research Institute, Hurley, Green and Davis (1955) concluded that of these 4 grasses, meadow fescue was the most suitable in that it was least competitive to lucerne, controlled weeds effectively and made a palatable addition to lucerne for summer and autumn grazing. Davies (1960) also considered meadow fescue to be an excellent companion grass for lucerne.

Rather and Harrison (1939) found that carbohydrate reserves accumulated less readily in lucerne sown with a grass than in lucerne alone. They suggested that it was important that the grass sown with lucerne should be palatable, otherwise the relatively palatable lucerne might suffer excessive defoliation. Under British conditions this would constitute an argument against the use of cocksfoot or tall fescue, because the mature herbage of these species is relatively unpalatable.

MANAGEMENT

Green and Davis (1955) stressed that the growth of the companion grass should be controlled by cutting or grazing in late autumn in order to provide optimum conditions for spring growth of the lucerne. They also reported that with lucerne drilled in rows 2 ft (61 cm) apart and with intermediate drill rows of cocksfoot, timothy, meadow fescue, perennial ryegrass or tall fescue, it was possible to maintain such a combination for several years without serious sward deterioration. The use of alternate drills was also suggested by Bolton (1962).

Seed

CHARACTERS

The seed is irregular, somewhat angular and kidney-shaped, up to 3 mm long, 1·5 mm wide and 1 mm thick (Percival, 1949; Robinson, 1953). The radicle usually projects from the seed and lies in a different plane from the cotyledons; its length is more than half that of the cotyledons (Percival, 1949; Robinson, 1953). The seed is yellow, sometimes with a buff or brown tinge (Percival, 1949). The percentage of hard seeds has been estimated at 6 (Robinson, 1953). The 1 000-seed weight has been given as 2·059 g (Robinson, 1953) and 2·61 g (Williams, 1956).

PRODUCTION

Lucerne grown for seed is usually sown under a thinly sown cereal cover crop in spring, or without a cover crop in spring to August (Zaleski, 1954a, 1956a, b; National Institute of Agricultural Botany, 1959). Recommendations are that any Ca deficiency in the soil should be rectified well in advance of sowing and, unless fertility is high, 30 kg P and 120 kg K/ha should be applied to the seedbed, and in subsequent years 20 kg P and 120 kg K/ha in early spring. In the following spring the crop should not be cut for fodder but allowed to flower so that pollination and seed ripening can take place (Zaleski, 1954b, 1956a,b, 1957, 1959; National Institute of Agricultural Botany, 1959). The aftermath can then be cut or grazed in late Oct.–Nov. with little adverse effect on the seed yield in subsequent years (Zaleski, 1956b, 1959). Lucerne is pollinated by both wild bees and honey bees, but it has been shown that honey bees are principally responsible for the amount of seed set (Zaleski, 1957, 1959; National Institute of Agricultural Botany, 1959). Lucerne seed crops should be isolated from other lucerne stands during anthesis, the required distance being 160 m for crops covering less than two ha, falling to 45 m for crops covering 7 ha or more (National Institute of Agricultural Botany, 1959).

The average annual yield of seed in Britain has been reported to be in the range of 100–200 kg/ha (FAO, 1961).

Yield

VARIETIES

Lucerne is one of the few forage crops in which one particular type clearly outyields others under British conditions. This is the Flemish or Flamande type, which has consistently yielded at least 10% more than Provence or Grimm types. This higher yield is mainly due to earlier growth in spring and later growth in autumn in comparison with other types.

The first named cultivar of the Flemish type was du Puits, introduced into Britain from France in about 1948. Several other named cultivars of the same type are now available from France and Belgium. Cv. Eynsford, selected by W. Alexander from an earlier importation of French seed, is the only named British cultivar of the Flemish type.

ANNUAL YIELD

A healthy lucerne crop gives consistent yields from year to year, largely because of its virtual immunity to dry weather. Table 35.2 gives lucerne yields in various years, including 1955 and 1959 which both had very dry summers.

Table 35.2. Annual yields of lucerne (kg DM/ha)

Lucerne with a little meadow fescue	1954 13 300	1955 13 900	1956 10 500	Reference Green (1957)
Lucerne alone	1957 6 130	1958 8 000	1959 10 000	Davies and Tyler (1962)

Lucerne of the Flemish type produces herbage from mid-March to late September or mid-October, but its use for forage is governed by considerations of crop management. In order to maintain high productivity from year to year, lucerne should not be cut or grazed before mid-May, and a period of at least 6 weeks should be allowed between successive cuts. A departure from these guide lines results in loss of vigour and yield, and eventual suppression by weeds, especially grasses. An indication as to the potential yield of lucerne is given by the results of the trials described below, in which a first cut taken between mid-May and mid-June was followed by two further cuts, one in July and one in late August or September.

Willey (1956) summarized the results from trials with lucerne varieties carried out by the National Institute of Agricultural Botany in 1937–45 at a wide range of sites from Kent to Leicestershire and Carmarthenshire, each trial lasting 3 to 5 years. In most trials, lucerne was cut 3 times per year, the first cut being taken between 13 May and 25 June, the second between 29 June and 11 August, and the third between 29 August and 22 October. The average dry matter (DM) yield was about 10 000 kg/ha per year. Willey observed that yields varied widely between sites, and though he did not quote actual figures, it is reasonable to assume that at the more favourable sites the annual yield was well in excess of 10 000 kg DM/ha. As few of the trials summarized by Willey included cultivars of the Flemish type, the average yield was heavily weighted by the yields of the cultivars of the Provence and Grimm types. In more recent trials in a similar geographical area, cultivars of the Flemish type have predominated, and total DM yields have ranged from 8 400 to 17 200 kg/ha, with an average of about 12 000 kg/ha (Aldrich, personal communication).

It is highly probable that grass weeds made a significant contribution to the gross yields quoted above, and more so in the earlier trials than in the later ones in which Flemish-type cultivars predominated. Davies and Tyler (1962) reported on a 3-year experiment at Aberystwyth in which the cultivar du Puits was sown alone. The annual average DM yield was 13 600 kg/ha, but the lucerne itself contributed only 10 200 kg, or 74%. In areas of lower rainfall, grass weeds are less prominent, though still likely to compete with the lucerne.

A trial in 1964 with the cultivar Eynsford at the Grassland Research Institute, Hurley, in which grass weeds were successfully controlled by winter treatment with paraquat, gave some indication of the yield that can be expected from a healthy, weed-free crop of Flemish-type lucerne. The results in 1965 and 1966 from this trial were very similar and average values are therefore given in Table 35.3. They indicate a potential yield of more than 13 000 kg DM/ha (5 tons/acre) with a digestibility (D-value) of 56–58 %, or approximately 7 500 kg digestible organic matter/ha.

Like red clover and sainfoin, lucerne cut in May and again at intervals of about 6 to 8 weeks has an N content of at least 3 % in the DM. The annual yield of N from lucerne can therefore exceed 400 kg/ha.

SEASONAL YIELD

The data given in Table 35.3 are typical in that the highest yield from a cut of lucerne is usually obtained from the first, but more than half of the total yield is harvested later in the season. In the trials reported by Willey (1956), the average dates of harvest and the average DM yields obtained were:

	kg/ha
4 June	5 500
20 July	3 400
25 Sept.	2 000

A crop of the cultivar Chartrainvilliers (Flemish type) grown at The Grassland Research Institute, Hurley, with a small quantity of fescue, gave the following DM yields, with very little grass in the second and third cuts (Green, 1957):

	kg DM/ha		
	1954	1955	1956
1 June	5 500	4 900	5 000
15 July	4 400	5 200	3 400*
15 Sept.	3 400	3 800	2 100*

* decline due to Verticillium wilt

A pure stand of the cultivar du Puits sown at Hurley in 1958 gave the following yields during the very dry summer of 1959:

	kg DM/ha
5 June	5 500
20 July	4 500
15 Sept.	3 200

An important characteristic of lucerne is that, after a cut has been taken in late May or early June, it recovers rapidly and gives another large yield by mid-July. Subsequent regrowth is also good, but slightly less rapid than the first regrowth. The yield in September is distinctly higher from the Flemish than from other types.

Table 35.3. Yield, digestibility and protein content of lucerne, Flamande type

Cutting date		DM kg/ha			DM digestibility %	D-value %	Digestible OM kg/ha			N %	N kg/ha			CP %	CP kg/ha		
1st cut	12 May	4 000			72	64	2 560			3·8	152			23·7	950		
	27 May		5 650		65	58		3 280		3·3		187		20·6		1 170	
	11 June			6 850	62	55			3 770	2·8			192	17·5			1 200
2nd cut	28 June	3 450			64	57	1 970			3·5	121			21·9	760		
	12 July		4 350		61	55		2 390		3·2		139		20·0		870	
	25 July			4 850	59	53			2 570	3·2			155	20·0			970
3rd cut	28 Aug.	3 700			60	54	2 000			3·0	111			18·7	690		
	12 Sept.		3 300		62	56		1 850		3·0		99		18·7		620	
	28 Sept.			2 750	64	58			1 600	3·0			82	18·7			510
4th cut	15 Oct.	1 500			60	54	810			3·0	45			18·7	270		
Total yield		13 000	13 300	14 450			7 340	7 520	7 940		429	425	429		2 680	2 660	2 680

Data compiled by J. O. Green (Grassland Research Institute, Hurley) from literature cited in the text. For conversion formulae see Chapter 5.

ENERGY AND PROTEIN YIELD

Data from the literature are summarized in the following table.

Table 35.4. Potential annual yields of energy and protein from lucerne (DM basis)

	Mean	Range
DM digestibility (%)	59	55–64
Yield of digestible OM (kg/ha)	7 500	7 300–8 000
Yield of metabolizable energy (Mcal/ha)	26·9	25·5–28·5
Crude protein (%)	20·0	17·0–24·0
Yield of crude protein (kg/ha)	2 700	2 600–2 800
Yield of digestible crude protein (kg/ha)	2 000	1 900–2 100

Data compiled by J. O. Green (Grassland Research Institute, Hurley) from literature cited in the text. For conversion formulae see Chapter 5.

Response to management and environment

RESPONSE TO FERTILIZER

Nitrogen

N fertilizer does not increase the yield of established vigorous crops of lucerne (Gerwig & Ahlgren, 1958; Manil & Nys, 1961). An increase in N content in lucerne herbage found by Carter and Foth (1960) as a result of applying N in pot trials was not confirmed in the field. Giddens (1959) considered that the only justifiable use of N for lucerne was to apply a low rate (35 kg/ha) to the seedbed in order to offset poor growth caused by ineffective early nodulation. The response to such applications varied from one season to another and ranged from nil to an increase of 500 kg DM/ha. Bergmann (1961) recommended a top-dressing of 15 kg N/ha as calcium nitrate in the year of sowing or, if the stand is thin, in the following spring.

Phosphorus

The critical level of P in the DM of lucerne, below which yield falls through P deficiency, was considered by Gerwig and Ahlgren (1958) to be less than 0·25%. Larson *et al.* (1952) reported 0·18% P as a critical value for lucerne and, on P-deficient soils, obtained yield responses of 10–40 kg DM/kg P for rates of applied P up to 125 kg/ha. The soils in this trial contained 3–8 ppm. available P as determined by the method of Bray and Kurtz (1945). For British soils low in P, Cooke (1964) recommended that 20 kg P/ha should be incorporated into the soil in a band 2–5 cm below the lucerne seed at the time of sowing, in addition to 50 kg P/ha broadcast and worked into the seedbed prior to sowing. The fertilizer placed below the seed assists in the rapid establishment of the lucerne before its roots are able to extend sufficiently to benefit from the broadcast dressing.

Potassium

The critical level of K in the DM of lucerne below which yield declines through K deficiency was considered by Gerwig and Ahlgren (1958) to vary from season to season, but to lie between 1·4 and 1·9%. In pot experiments, MacLeod (1965) found that K did not limit yield even at a concentration in the DM as low as 1%, but in field trials Kresge and Younts (1962) found that critical values for K were 1·7–2·0%. A figure of 1·8% would seem acceptable for practical purposes.

The response of lucerne to K fertilizer was found by Gerwig and Ahlgren (1958) to range from 20 to 30 kg DM/kg of K applied. The amount of K needed to maintain maximum yield increased with sward age, being 45, 90 and 180 kg K/ha in the first, second and third years, respectively; in these trials the fertilizer was applied as a single dressing each spring. Kresge and Younts (1962) found that 180 kg K/ha applied in the spring produced the same effect on the yield of lucerne as 90 kg K/ha split into two equal dressings, the first applied in the spring and the second after the first cut; split applications greatly increased the efficiency with which K was utilized and also maintained a more uniform K content in the cut herbage over the growing season.

Calcium

The optimum soil pH for lucerne is 6·2–7·8, and sufficient lime should be applied to achieve this value. According to Sutton and Hallsworth (1958), it is the high H-ion concentration associated with acid soils that adversely affects root growth, particularly at low temperature and high light intensity. Gachon (1963) considered that Ca deficiency was responsible for the poor growth of lucerne on acid soils in the Massif Central region of France.

The trials of Williams and Back (1961) showed that apart from its effect on soil pH, calcium carbonate had an intrinsic effect of its own on the growth of lucerne. On a soil already limed, 250 kg $CaCO_3$/ha drilled with the seed increased early plant vigour by 50% and also yield and N content; early establishment of large, vigorous clusters of root nodules on the roots was also observed.

RESPONSE TO IRRIGATION

Lucerne is deep rooting and its growth is not measurably affected by a soil-moisture deficit of less than 6–8 in. (Williams & Stiles, 1962; Stiles, 1966). Thus, provided the soil is sufficiently deep, that subsoil conditions are suitable for root growth (Baker & Chard, 1964), and that the water-holding capacity of the soil is adequate, it is unlikely that irrigation of lucerne will be worthwhile in Britain.

RESPONSE TO TEMPERATURE

Garza *et al.* (1965) found that the cultivars Culver and Tanverde yielded more at alternating temperatures of 30°/15°C than at constant temperatures of 30° or 15°C. The yield of Ranger lucerne was higher at 15·5° than at 27°C, and greatly reduced at 32°C (Gist & Mott, 1957). This agrees with the field observations of Feltner and Massengale (1965), who noted that during mid-summer the yields of the cultivars Moapa and Lahontan were reduced when maximum and minimum day temperatures were of the order of 35° and 21°C, respectively. Both this study and that of Smith *et al.* (1964) with Vernal lucerne suggested that high summer temperatures reduce yield, although less markedly than does light intensity.

Lucerne appears to be less adversely affected by high temperature than other forage species, although there may be wide differences in sensitivity among varieties. Ludwig and Evans (1961) found that the cultivar Hairy Peruvian was not affected by temperatures of 25°–15°C, combined with night temperatures of 20°, 15°, 10° and 5°C, except during early regrowth when low temperatures depressed yields greatly. Rambler, on the other hand, was much more sensitive to temperature, particularly in terms of leaf-area expansion, both high and low extremes of temperature restricting its growth.

RESPONSE TO DEFOLIATION

Lucerne cultivars can be classified into early, mid-season and late types. The early type includes all cultivars normally grown in Great Britain, e.g. du Puits, except the one mid-season cultivar Provence. When Provence was grown at a number of sites and cut for the first time at flowering and twice subsequently, its average yield was only 78% of that of du Puits. The early cultivars can give one cut at flowering and 3 subsequent cuts (National Institute of Agricultural Botany, 1966).

The yields of 8 cultivars of lucerne were reduced when they were cut 4 times between May and September, compared with 3 times (May, July and September); the latter gave the highest DM production (Davies, 1960). However, cutting 3 times reduced yield in the following year, so that to maintain production from year to year, not more than two cuts per season should be taken. Very frequent cutting (6–10 times during the season) resulted in complete crop failure (Klapp, 1942). Since lucerne is not damaged by severe defoliation (Ridgman, 1960), it can be maintained under a grazing management (Barker *et al.*, 1957). Ridgman (1960) found that the small benefit to lucerne regrowth from the presence of long stubbles as a result of cutting at heights of 10, 15 or 20 cm in no way counterbalanced the large loss in DM yield at each cut, compared with cutting at 5 cm, provided that extreme defoliation at ground level was avoided.

Both digestibility and N content are higher in lucerne leaves than in stems, and the digestibility of stem material and the N content of both leaves and stems falls with increasing maturity. In the leaves there is no decrease in digestibility during the season and the rate of decrease in digestibility of the whole plant is not affected by the onset of flowering (Terry & Tilley, 1964). Since the proportion of leaf in the lucerne crop also falls with increasing maturity (Dent & Zaleski, 1954), the timing of defoliation in relation to maturity stage influences the nutritive value of the harvested material.

Quality

CHEMICAL COMPOSITION

Lucerne is generally similar in chemical composition to red clover, but appears to have rather lower contents of available carbohydrates, K and Mg (see Chapter 31).

WATER CONTENT

The water content of lucerne is related to its growth stage and decreases with maturity. Kivimäe (1959) found that DM content increased throughout the growth period in accordance with the linear regression $y = 14{\cdot}5 + 0{\cdot}286\ x$, where y = DM% and x = growth stage in days. However, work by Delić *et al.* (1964) indicated that the effect of growth stage on water content is not the same for all plant parts. In his trials, lucerne was cut at the early bud stage, when 75% of the buds had been produced, at early flowering, and finally at the 50% flowering stage. At the later stages of development, a decrease in the water content of the stem was observed, but that of the leaves increased.

Miltimore *et al.* (1964) found a relationship between the water and N contents of the crop. A lucerne ley was cut daily for 111 days, and the contents of water, N and other plant nutrients determined. The percentage of water varied from 85·4 to 56·6, while that of N varied from 2·0 to 4·8 during this period. The percentage of water was found to decrease with increase in N content.

DRY MATTER DIGESTIBILITY

Young leafy herbage frequently cut

Lucerne does not tolerate frequent defoliation, hence its nutritive value at early growth stages is of little practical significance. However, as shown in Fig. 35.1, top, the digestibility of lucerne even in early spring is not high, being only about 75% at the beginning of the season.

First growth

An early cultivar such as du Puits usually flowers in late May. The seasonal trend of its digestibility is similar to that of perennial ryegrass (Fig. 35.1, top), but lucerne reaches a given digestibility value about 10 days earlier, falling to 70% DM digestibility about 10 days before flowering. This emphasizes the need to take the first cut before flowering, as recommended by Dent (1955), if the maximum seasonal yield of nutrients is to be obtained. Data from Davies *et al.* (1966) indicated that du Puits had a DM digestibility of 72% on 19 May

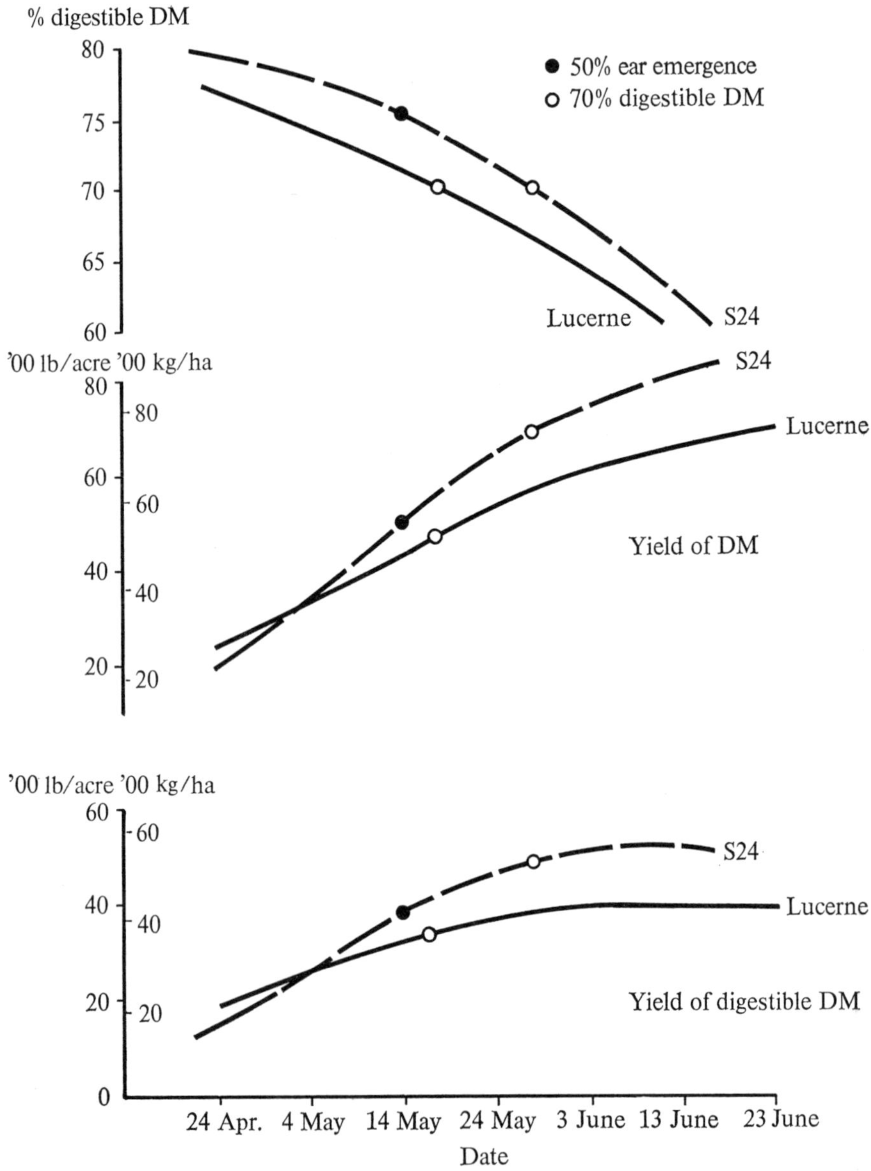

Fig. 35.1. Lucerne. Seasonal changes in DM yield and digestibility. Values for S24 perennial ryegrass shown for comparison. (From Green & Corrall, 1967)

and of 62·5% on 1 June. Terry and Tilley (1964) have shown that the young stem tips of lucerne are always highly digestible, and it is the maturing basal parts of the stem which depress overall digestibility. Therefore taking a high cut, or allowing cattle to graze only the tips of the plants, improves the apparent digestibility but also results in a low proportion of the crop being utilized. The yield of lucerne is similar to that of S24 perennial ryegrass in early spring; it falls behind in May, but lucerne yields more than broad red clover.

Regrowths

The DM digestibility of regrowths falls to 60–65% after 6–8 weeks of growth. The digestibility of lucerne is somewhat inferior to that of S24 perennial ryegrass but, provided that the first cut is taken in May and the regrowths are not allowed to become over-mature, it produces a high total annual herbage yield of reasonable digestibility.

ANIMAL INTAKE CHARACTERISTICS

The intake of lucerne of a given digestibility has invariably been greater than that of the grass species with which it has been compared. Reid *et al.* (1962), Van Soest (1965) and Osbourn *et al.* (1966) have presented evidence suggesting that the voluntary intake of lucerne is less than that of sainfoin and red clover.

Longevity and persistency

Information on leaf longevity is meagre. In the USA Pearce *et al.* (1968), who grew the cultivar Williamsburg in the field and also as cuttings under controlled-environment conditions, noted that leaves lived much longer (54–60 days) under controlled conditions than they did in dense stands in the field, where yellowed leaflets were being abscissed 19 days after unfolding and maximum duration of active photosynthesis was of the order of 30 days. Light intensity, which was much higher in the field, was thought to be the major controlling factor. The above-ground parts of this species are annual.

Lucerne produces several types of roots. The tap-root, which frequently reaches depths of 2–3 m and branches at depth (Heuser, 1931), is long-lived, but its persistence may not be essential to survival, since Becker-Dillingen (1929) stated that lucerne can be grown from cuttings and can be successfully transplanted after severing the tap-root at a depth of 10–12 cm. The descending primary lateral roots which arise from the tap-root, usually not in the top layer of the soil (Heuser, 1931), persist and bear fibrous roots, the majority of which have a limited life span. Zykov (1968) found 11·6 fourth-order lateral roots per aerial stem in the year of sowing, 14·8 in the second year, 16·3 in the third and 16·8 in the fourth year. During the third year, root weight ceased to increase and 22–30% of these small roots died and were mineralized.

Underground stems which grow from the crown of the plant produce adventitious roots in the topsoil and also bear aerial shoots. In Germany, occasional lucerne plants have survived for upwards of 100 years, but longevity of individual plants and their progeny is variable (Heuser, 1931). Lucerne stands remain productive in Australia for 4–10 years (Bolton, 1962) and in Germany for 5–12 years (Heuser, 1931). In the UK they formerly remained productive for at least 4 years, but the rapid spread of Verticillium wilt in 1950–7 caused many crops to be ploughed-in after only two or three years (Isaac & Lloyd, 1957).

Mortality of lucerne is greater at high plant populations (Jarvis, 1962). Kreuz (1969) found lucerne to be less tolerant of crowding than any other herbage species tested, and obtained reductions in total DM yield where populations exceeded $600/m^2$ in the year of sowing, of $100–200/m^2$ in the first harvest year and $60–80/m^2$ in the following year. Persistence varies with cultivar (Willey & Zaleski, 1955), and is adversely affected by the frequent defoliation of young growing shoots, which leads to weed invasion, notably by *Poa trivialis* in pure stands, and to the dominance of the grass component in mixed legume/grass stands. Successful management systems involve infrequent cutting (e.g. 3 cuts/year) or the equivalent amount of grazing (Green & Davis, 1955). Plant vigour is reduced more by grazing than by cutting, and the crop is often killed out by more than 4 grazings per year (Whitear, Hanley & Ridgman, 1962). Cutting in early May or late September is harmful (Green *et al.*, 1955).

Persistence is greatly affected by management in autumn, at which time it is important to maintain a high level of carbohydrate reserves in the roots by allowing good recovery growth immediately prior to the last grazing, and delaying the time of this grazing until growth has virtually ceased. Extensive new growth at this time of year has been found to deplete carbohydrate reserves and to render the plants particularly susceptible to winter-kill (Jones, 1955).

When all active growth has ceased, clearing the top growth by grazing during the winter has been found to improve persistence. On light land, the effect of concentrated treading during this operation may help to suppress weeds, for under this treatment almost pure stands have remained productive in the third and fourth year after sowing. On heavy land, winter grazing appears to be less efficient as a means of weed control, but does not harm the lucerne crop (Green & Davis, 1955).

References

Aldrich, D. T. A. National Institute of Agricultural Botany. Personal communication.

Baker, H. K.; Chard, J. R. A. 1964. Investigations into establishment, growth and use of grasses and legumes on Bagshot sands. *N.A.A.S. q. Rev.* **66,** 62–8.

BARKER, M. G.; HANLEY, F.; RIDGMAN, W. J. 1957. Studies on lucerne and lucerne-grass leys. 4. The effect of system of grazing management on the persistence of a lucerne-cocksfoot ley. *J. agric. Sci., Camb.* **48,** 361–5.

BECKER-DILLINGEN, J. 1929. Handbuch des gesamten Pflanzenbaues, einschliesslich der Pflanzenzüchtung. 3. Handbuch des Hülsenfruchtbaues und Futterbaues. Berlin: Paul Parey, pp. 670.

BERGMANN, W. 1961. [Fertilizing lucerne.] *Phosphorsäure* **21,** 176–87.

BOLTON, J. L. 1962. Alfalfa. Botany, cultivation and utilization. London: Leonard Hill Books Ltd, pp. 472.

BRAY, R. H.; KURTZ, L. T. 1945. Determination of total organic and available forms of phosphorus in soils. *Soil Sci.* **59,** 35–45.

BULA, R. J.; SMITH, D. 1954. Cold resistance and chemical composition in overwintering alfalfa, red clover and sweet clover. *Agron. J.* **46,** 397–401.

CARTER, C. R.; FOTH, H. D. 1960. The effect of nitrogen fertilizer on yield and protein content of alfalfa and companion crops. *Q. Bull. Mich. St. Univ. agric. Exp. Stn* **42,** 737–43.

CLAPHAM, A. R.; TUTIN, T. G.; WARBURG, E. F. 1962. Flora of the British Isles. Cambridge: Univ. Press, 2nd edn.

COOKE, G. W. 1964. Fertilizers and profitable farming. London: Crosby, Lockwood and Son Ltd, pp. 83–4.

DAVIES, W. 1960. The grass crop. London: E. and F. N. Spon Ltd, 2nd edn.

DAVIES, W. E. 1960. The relative effect of frequency and time of cutting lucerne. *J. Br. Grassld Soc.* **15,** 262–9.

DAVIES, W. E.; GRIFFITH, G. AP; ELLINGTON, A. 1966. The assessment of herbage legume varieties. 2. *In vitro* digestibility, water, carbohydrate, crude protein and mineral content of primary growth of clover and lucerne. *J. agric. Sci., Camb.* **66,** 351–7.

DAVIES, W. E.; TYLER, B. F. 1962. The yield and composition of lucerne, grass and clover under different systems of management. 4. Further studies on the effect of frequency of cutting lucerne and lucerne grown with grass. *J. Br. Grassld Soc.* **17,** 306–14.

DELIĆ, I.; PAVLOVIĆ, M.; ČELIKOVIĆ, E.; KRIVOKUĆIN, A. 1964. Effect of harvesting time and various stages of development of alfalfa on its chemical composition, carotene content and leaf/stem ratio. *Savr. poljopr.* **12,** 655–64.

DENT, J. W. 1955. Seasonal yield and composition of lucerne in relation to time of spring cutting. *J. Br. Grassld Soc.* **10,** 330–40.

DENT, J. W.; ZALESKI, A. 1954. Leafiness and chemical composition of some lucerne strains. *J. Br. Grassld Soc.* **9,** 131–40.

DENT, J. W.; ZALESKI, A. 1960. Leafiness, chemical composition and yield of some lucerne varieties. *J. Br. Grassld Soc.* **15,** 21–7.

FAO. 1961. Agricultural and horticultural seeds. *FAO agric. Stud. 55* Rome: FAO.

FELTNER, K. C.; MASSENGALE, M. A. 1965. Influence of temperature and harvest management on growth, level of carbohydrates in the roots and survival of alfalfa (*Medicago sativa* L.). *Crop Sci.* **5,** 585–8.

FOX, R. J.; LIPPS, R. C. 1955. Influence of soil profile characteristics upon distribution of roots and nodules in alfalfa and sweet clover. *Agron. J.* **47,** 361–7.

GACHON, L. 1963. [Chemical factors responsible for the stunting of lucerne on acid soils.] *Bull. Ass. franç. Ét. Sol* Nos 6 and 7, 291–354.

GARZA, T. R.; BARNES, R. F.; MOTT, G. O.; RHYKERD, C. L. 1965. Influence of light intensity, temperature and growing period on the growth, chemical composition and digestibility of Culver and Tanverde alfalfa seedlings. *Agron. J.* **57,** 417–20.

GERWIG, J. L.; AHLGREN, G. H. 1958. The effect of different fertility levels on yield, persistence and chemical composition of alfalfa. *Agron. J.* **50,** 291–4.

GIDDENS, J. 1959. Nitrogen applications to new and established stands of alfalfa. *Agron. J.* **51,** 574.

GILL, N. T.; VEAR, K. C. 1958. Agricultural Botany. London: Duckworth and Co. Ltd.

GIST, G. R.; MOTT, G. O. 1957. Some effects of light intensity, temperature and soil moisture on growth of alfalfa, red clover and birdsfoot trefoil seedlings. *Agron. J.* **49,** 33–6.

GREEN, J. O. 1957. Lucerne in Britain. *Yb. Ass. green Crop Driers 1956–57* p. 76.

GREEN, J. O.; CORRALL, A. J. 1967. The Grassland Research Institute, Hurley. Unpublished data.

GREEN, J. O.; COWLING, D. W.; ALDER, F. E. 1955. A comparison of types of lucerne. In: Grassland Research Institute. Memoir No. 1. Lucerne investigations, 1944–53, pp. 16–30.

GREEN, J. O.; DAVIS, A. G. 1955. The seasonal use of lucerne-grass mixtures for grazing. In: Grassland Research Institute. Memoir No. 1. Lucerne investigations, 1944–53, pp. 63–77.

HEUSER, O. 1931. Die Luzerne. Berlin: Paul Parey, pp. 228.

ISAAC, I.; LLOYD, A. T. E. 1957. Wilt of lucerne caused by species of *Verticillium*. 2. Seasonal cycle of disease; range of pathogenicity; host-parasite relations; effects of seed dressings. *Ann. appl. Biol.* **47**, 673–84.

JARVIS, R. H. 1962. Studies on lucerne and lucerne-grass leys. 5. Plant population studies with lucerne. *J. agric. Sci., Camb.* **59**, 281–6.

JONES, T. 1955. Management of lucerne in the critical autumn period. In: Grassland Research Institute. Memoir No. 1. Lucerne investigations, 1944–53, pp. 53–7.

KIVIMÄE, A. 1959. Chemical composition and digestibility of some grassland crops. *Acta Agric. scand.* Suppl. 5, p. 26.

KLAPP, E. 1942. Development, root formation and storage of matter in forage plants. *Pflanzenbau* 1942, **18**, 347–52 and 367–87. Reviewed by Roseveare, G. M. 1943. Stubble, roots and reserves. *Herb. Abstr.* **13**, S38-S42.

KRESGE, C. B.; YOUNTS, S. E. 1962. Effect of various rates and frequencies of potassium application on yield and chemical composition of alfalfa and alfalfa-orchardgrass. *Agron. J.* **54**, 313–16.

KREUZ, E. 1969. [On the influence of spacing on mortality and yield of grass and clover for fodder.] *Albrecht-Thaer-Arch.* **13**, 197–212.

LAMPETER, W. 1960. [Mutual influences of higher plants regarding growth of shoots and roots, mineral content and water consumption—studied in some economically important forage plants.] *Wiss. Z. Karl-Marx-Univ. Lpz.* **9**, 611–722.

LARSON, W. E.; NELSON, L. B.; HUNTER, A. S. 1952. The effects of phosphate fertilization upon the yield and composition of oats and alfalfa grown on phosphate deficient Iowa soils. *Agron. J.* **44**, 357–61.

LUDWIG, J. L.; EVANS, L. T. 1961. Effect of temperature on growth of lucerne. *Rep. Div. Pl. Ind. C.S.I.R.O. Aust. 1960–61* pp. 105–6.

MACLEOD, L. B. 1965. Effect of nitrogen and potassium on the yield and chemical composition of alfalfa, bromegrass, orchard-grass and timothy grown as pure species. *Agron. J.* **57**, 261–6.

MANIL, P.; NYS, L. 1961. [Effect of mineral-nitrogen manuring on lucerne sown on bare soil.] *Bull. Inst. agron. Gembl.* **29**, 335–8.

MANSFIELD, W. S. 1945. Lucerne and sainfoin. *Agriculture, Lond.* **52**, 255–8.

MILTIMORE, J. E.; MASON, J. L.; MCARTHUR, J. M. 1964. Bloat investigations. Relationship between daily bloat incidence and grass minimum temperatures, dry matter and certain plant nutrient contents of forages. *Can. J. Anim. Sci.* **44**, 309–16.

NATIONAL INSTITUTE OF AGRICULTURAL BOTANY. 1959. Growing clovers for seed. *Herbage Seed Growers' Leaflet No. 3.*

NATIONAL INSTITUTE OF AGRICULTURAL BOTANY. 1966. Varieties of lucerne. *Farmers' Leaflet No. 4.*

OSBOURN, D. F.; THOMSON, D. J.; TERRY, R. A. 1966. The relationship between voluntary intake and digestibility of forage crops, using sheep. *Proc. 10th int. Grassld Congr., Helsinki, 1966* pp. 363–7.

PEARCE, R. B.; BROWN, R. H.; BLASER, R. E. 1968. Photosynthesis of alfalfa leaves as influenced by age and environment. *Crop Sci.* **8**, 677–80.

PERCIVAL, J. 1949. Agricultural botany; theoretical and practical. London: Duckworth, 8th edn.

PERRING, F. H.; WALTERS, S. M. 1962. Atlas of the British flora. London: Thomas Nelson and Sons Ltd, pp. 456.

RACHIE, K. O.; SCHMID, A. R. 1955. Winter hardiness in birdsfoot trefoil strains and varieties. *Agron. J.* **47**, 155–7.

RATHER, H. C.; HARRISON, C. M. 1939. Starch reserves in the roots of pastured alfalfa when grown alone and in mixtures. *Q. Bull. Mich. St. Univ. agric. Exp. Stn* **21**, 281–91. Cited in BOLTON, J. L. (1962)

REID, R. L.; BURR, C.; WELCH, J. A.; DOSZA, L.; JUNA, G. A. 1962. Investigations of plant species and maturity stages on forage nutritive value as determined by *in vitro* digestion techniques. *Final Report U.S.D.A. Contract No. 12-14-100-4524 (24) W.Va agric. Exp. Stn.*

RIDGMAN, W. J. 1960. The effect of height of cutting on the subsequent growth and yield of lucerne. *J. Br. Grassld Soc.* **15**, 291–5.

ROBINSON, D. H. 1953. Leguminous forage plants. London: Arnold and Co., 2nd edn.

SMITH, J. G.; MOTT, G. O.; BULA, R. J. 1964. Ecological parameters of an alfalfa community under field conditions. *Crop Sci.* **4**, 577–80.

SOEST, P. J. VAN. 1965. Symposium on factors influencing the voluntary intake of herbage by ruminants: voluntary intake in relation to chemical composition and digestibility. *J. Anim. Sci.* **24**, 834–43.

SPURWAY, C. H. 1941. Soil reaction (pH) preferences of plants. *Spec. Bull. 306 Mich. agric. Exp. Stn* pp. 36.

STILES, W. 1966. Ten years of irrigation experiments. *A. Rep. Grassld Res. Inst. Hurley 1965* pp. 57–66.

SUTTON, C. D.; HALLSWORTH, E. G. 1958. Studies on the nutrition of forage legumes. 1. The toxicity of low pH and high manganese supply to lucerne, as affected by climate factors and calcium supply. *Pl. Soil* **9**, 305–17.

TERRY, R. A.; TILLEY, J. M. A. 1964. The digestibility of the leaves and stems of perennial ryegrass, cocksfoot, timothy, tall fescue, lucerne and sainfoin, as measured by an *in vitro* procedure. *J. Br. Grassld Soc.* **19**, 363–72.

WHITEAR, J. D.; HANLEY, F.; RIDGMAN, W. J. 1962. Studies on lucerne and lucerne-grass leys. Further studies on the effect of systems of grazing management on the persistence of a lucerne-cocksfoot ley. *J. agric. Sci., Camb.* **59**, 415–28.

WILLEY, L. A. 1956. Lucerne strain trials. *J. natn. Inst. agric. Bot.* **7**, 597.

WILLEY, L. A.; ZALESKI, A. 1955. Lucerne strains. In: Grassland Research Institute. Memoir No. 1. Lucerne investigations, 1944–53, pp. 9–15.

WILLIAMS, T. E.; BACK, H. L. 1961. The combine drilling of lime and lucerne. *Exps Prog. 13 Grassld Res. Inst. 1959–60* pp. 107–9.

WILLIAMS, T. E.; STILES, W. 1962. *Exps Prog. 14 Grassld Res. Inst. 1960–61* pp. 45–6.

WILLIAMS, W. 1950. Experiments on the yield of lucerne strains as pure plots and on the behaviour of lucerne when sown in a mixture with various species of grasses. *J. Br. Grassld Soc.* **5**, 113–29.

WILLIAMS, W. A. 1956. Evaluation of the emergence force exerted by seedlings of small seeded legumes using probit analysis. *Agron. J.* **48**, 273–4.

ZALESKI, A. 1954a. Lucerne investigations. 1. Identification and classification of lucerne 'varieties' and 'strains'. *J. agric. Sci., Camb.* **44**, 199–220.

ZALESKI, A. 1954b. Lucerne seed growing. *Agriculture, Lond.* **61**, 418–23.

ZALESKI, A. 1956a. Some of the factors affecting lucerne seed production. *J. Br. Grassld Soc.* **11**, 23–33.

ZALESKI, A. 1956b. Progress with lucerne seed production. *Agriculture, Lond.* **63**, 233–7.

ZALESKI, A. 1957. Lucerne investigations. 2. Pollination and seed setting in lucerne strains. *J. agric. Sci., Camb.* **48**, 236–44.

ZALESKI, A. 1959. Technique and methods of lucerne seed production. *J. Br. Grassld Soc.* **14**, 277–84.

ZYKOV, YU. D. 1968. [The development of roots and bacterial nodules in lucerne cv. Semirechenskaya.] *Vest. sel'.-khoz. Nauki, Alma-Ata* No. 10, 17–23.

CHAPTER 36

Sainfoin *(Onobrychis viciifolia)*

Morphology

SAINFOIN (Plate 15) is a slightly pubescent perennial with numerous erect stems. The leaves, borne on long petioles, are imparipinnate (5–14 pairs of leaflets and a terminal leaflet); leaflets oblong, mucronate. Stipules broad, pointed, membranous, surrounding the stem. Flowers papilionaceous in dense racemes, carried on long axillary stalks. Pods single-seeded, indehiscent, flattened, with marked reticulate ridges.

The root system consists of a stout tap-root with a few main branches and numerous fine lateral roots. The main tap-root grows to a length of up to 1·5 m. No critical studies have been undertaken on the root system of cultivated sainfoin. Most of the nodules occur on the fine lateral roots, but a few occur also on the juvenile tap-root. The nodules are large (3 × 6 mm), wedge-shaped, and orange-white in colour.

A range of types exists, but these are mainly local selections rather than bred cultivars. Two main groups are distinguished. These are Common and Giant, a division similar to that of the late-flowering and broad-red types in red clover. In Common sainfoin, stem elongation is limited during the establishment year and flowering rarely occurs until the year after. At flowering, the stems are 60–90 cm long and are decumbent at the base. The leaflets of Common sainfoin tend to be smaller than those of Giant, measuring 5–8 × 30–40 mm. Gill and Vear (1958) stated that the leaf of Common sainfoin has fewer leaflets than that of Giant. Observations on a restricted range of material at the Grassland Research Institute, Hurley, have suggested that this is not a consistent feature and that the number of leaflets in a leaf varies within a single plant with the position of the leaf on the stem. Common sainfoin produces one conservation cut per year.

Giant sainfoin is short-lived, with a useful life-span of 2–3 years. Stem elongation and flowering occur during the establishment year, and basal buds develop only after flowering or cutting. The stems are longer, less dense and more upright than those of Common sainfoin. Giant sainfoin gives two conservation cuts per year.

Both Common and Giant sainfoin produce new stems from basal buds on a branched crown near ground level. Axillary buds develop on the nodes of the flowering stem below the point of severance.

In Table 36.1 the proportions of leaf and stem in Common and Giant sainfoin are compared.

Table 36.1. Percentage of leaf in the DM of Common and Giant sainfoin during growth of the first crop and in aftermath cuts

First growth			Regrowth		
Date	Common	Giant	Date	Common	Giant
26 May	43	40	12 July	65	47
6 June	29	32			
15 June	28	38			
2 July	25	39	11 Aug.	90	57

After Fagan and Rees (1930)

Terry and Tilley (1964) reported leaf percentages (leaf plus inflorescences) of 45 in Common and 47 in Giant sainfoin when cut at the end of May at the early flowering stage.

Distribution

Onobrychis viciifolia is a native species of limited distribution in Britain, occurring in only 67 vice-counties from Somerset and Kent to Shropshire and Norfolk, and also in Glamorgan, Carmarthen and Flint. It is probably native on chalk and limestone areas, but now often occurs only as a relic of cultivation (Clapham *et al.*, 1962; Perring & Walters, 1962). It is not found at altitudes above 300 m.

There is no close relationship between the distribution of sainfoin and isotherms. Most wild sainfoin occurs within the 15°C July isotherm, but distribution is not uniform and seems confined naturally to the chalk and limestone areas of south, south-east and south-central England (Perring & Walters, 1962). Its optimum soil pH range is 6·0–7·5 (Spurway, 1941). In Britain, sainfoin is near the limit of its climatic range and it is not winter-hardy north of the river Humber (Thomson, 1938).

Compatibility with grasses

Very little has been published on the compatibility of sainfoin with grasses. The unpublished results of a single experiment at the Grassland Research Institute (1956) are summarized in Table 36.2. In this, the Common sainfoin variety Cotswold was broadcast in April 1952 at 62 kg milled seed/ha alone or with S37 cocksfoot, S48 timothy or S215 meadow fescue, each at 3 kg/ha, and two cuts were taken per year, the first in late May. At first the plots sown with sainfoin alone yielded most, but during the next 4 years they became invaded by grass weeds, so that at the end of the experiment their total yield was similar to those of plots sown with a companion grass. In general, sainfoin grew better with timothy and meadow fescue than with cocksfoot, and this was reflected in the total herbage yield.

Table 36.2. Yields of sainfoin sown alone and with 3 companion grasses ('00 kg DM/ha)

		Sainfoin alone	Sainfoin with cocksfoot	Sainfoin with timothy	Sainfoin with meadow fescue
1953	Sainfoin	80·6	56·3	61·0	56·9
	Gross yield	106·4	104·6	106·6	109·1
1954	Sainfoin	71·8	56·7	62·7	62·9
	Gross yield	95·5	88·6	96·9	95·6
1955	Sainfoin	52·8	40·1	51·7	45·6
	Gross yield	80·5	75·5	85·2	80·9
1956	Sainfoin	47·6	41·1	46·6	48·4
	Gross yield	58·7	60·0	67·6	61·6
Total	Sainfoin	252·8	194·2	222·0	213·8
4 years	gross yield	341·1	328·7	356·3	347·2

Plots sown Apr. 1952. 2 cuts per year. Gross yield includes weeds.
From Grassland Research Institute, unpublished data, 1956.

Seed

CHARACTERS

Thomson (1951b) gives the following information on sainfoin seed. The true (milled) seed is kidney-shaped, with the hilum situated at about the middle of the concave edge; it is 2·5–4·5 mm long, 2·0–3·5 mm wide and 1·5–2·0 mm thick. Seed colour ranges from dark olive to brown or black and, on average, 30% of the seeds in a sample may show brownness. Brown seeds appear to have lighter weight and lower germination capacity than other seeds. Sainfoin seed is also sold, and sown, in unmilled form; in this case the fibrous pericarp encloses the true seed and forms a one-seeded, indehiscent fruit. The unmilled seed is bilaterally symmetrical about a plane passing through the ventral and dorsal sutures; in side view, such seed is semi-circular, the ventral suture being straight and the dorsal suture curved. On the surface is a network of prominent vascular ridges, with projecting spines on about 85% of unmilled seed. The percentage of hard seeds varies from 4·0 to more than 20.

The 1 000-seed weight ranges from 13·2 to 16·8 g in true seed, and from 18·2 to 23·6 g in unmilled seed; there is no significant difference in seed weight between Common and Giant sainfoin.

PRODUCTION

Sainfoin is seldom grown specifically for seed, but a cut for seed may be taken in lieu of hay in the case of Common sainfoin (which flowers only once a year), and after a hay cut in the case of Giant sainfoin (National Institute of Agricultural Botany, 1959). The seed is sown broadcast or in drills 18 cm apart at a depth of 2·5 cm in spring under a cereal cover crop. The seed rate for pure stands is 55–65 kg/ha for milled seed and 90–100 kg/ha for unmilled seed, but the rate may be reduced and companion grasses or clovers added (National Institute of Agricultural Botany, 1959; FAO, 1961). Sainfoin should be sown in soil with adequate reserves of lime, and application of 40 kg P and 120 kg K/ha is recommended. Giant sainfoin does not generally persist for more than two years, and seed is taken either in the first or in the second year (National Institute of Agricultural Botany, 1959). Common sainfoin can persist for 7 years or more, and to maintain its persistence the seed is normally harvested during or after the third year (National Institute of Agricultural Botany, 1959; Robinson, 1953).

The average annual yield of seed in Europe, including Britain, has been reported to be 40–800 kg/ha (FAO, 1961).

Yield

DRY MATTER

Although sainfoin has a long history of cultivation in parts of southern England, the area under the crop is now almost negligible. Information on the yields of sainfoin under modern husbandry conditions is therefore very scanty. Only data obtained in regions of calcareous or well-limed soils are likely to be relevant in any discussion of potential yields. A further problem in interpreting the results of trials with sainfoin has been the presence of grass weeds, which has made it difficult to assess its inherent productivity.

The National Institute of Agricultural Botany (1968) has reported the following mean dry matter (DM) yields in variety trials: lucerne, 100 cwt/acre, red clover 90 cwt, and sainfoin 70 cwt (12 500, 11 250 and 8 750 kg/ha, respectively). The proportion of cultivars of Giant and Common sainfoin and the duration of the trials were not stated. Results from trials carried out by the Grassland Research Institute on loam overlying chalk at Hurley and on lias clay in Warwickshire showed a similar relationship between yields of sainfoin and lucerne. In the most recent comparison of the two crops at Hurley in which grass weeds were effectively controlled, Cotswold Common sainfoin and Eynsford lucerne, cut 3 times beginning when organic matter digestibility (D-value) was approximately 65%, yielded 11 000 and 12 000 kg DM/ha, respectively, in both the first and the second harvest year. Under similar conditions English Giant sainfoin outyielded Common sainfoin in the first harvest year by virtue of its more rapid regrowth; Giant sainfoin may therefore be comparable to lucerne in the first year.

Common sainfoin resembles the single-cut, late-flowering red clover in that both species are ready for the first cut in June and this first cut gives about half of the annual yield. It is, however, more persistent than red clover and capable of maintaining its performance for several years if kept free from weeds.

ENERGY AND PROTEIN

Data compiled from the literature are shown in Table 36.3.

Table 36.3. Potential annual yields of energy and protein from sainfoin (DM basis)

	Mean value	Range
DM digestibility (%)	62	57–68
Yield of digestible OM (kg/ha)	6 400	6 000–6 800
Yield of metabolizable energy (Mcal/ha)	22·4	22·1–22·8
Crude protein (%)	21·0	17·0–25·0
Yield of crude protein (kg/ha)	2 200	2 100–2 300
Yield of digestible crude protein (kg/ha)	1 710	1 650–1 870

Data compiled by J. O. Green (Grassland Research Institute, Hurley) from literature cited in the text. For conversion formulae see Chapter 5.

Response to management and environment

RESPONSE TO FERTILIZER

Baker *et al.* (1952), working with the Giant type in Britain, considered that sainfoin preferred soils with a free lime content, similar to those preferred by lucerne, but that it was less sensitive than lucerne to the low K status common with these soils.

In Switzerland, Badoux (1965) found that the yield of Giant sainfoin was not increased by N fertilizer, and that a 4% reduction resulted from the application of 90 kg N/ha per year for two successive years. Similarly, applications of 50 kg P and 250 kg K/ha per year had no appreciable effect on yield. In glasshouse experiments, Badoux found that the application of nitrates reduced root and foliage weight, as well as the number and size of root nodules. P increased root and foliage weight but K had little effect.

RESPONSE TO IRRIGATION

Sainfoin is generally considered drought-resistant. There are, however, no experimental data on its reaction to irrigation under conditions in Britain.

RESPONSE TO TEMPERATURE

Little is known of the response of sainfoin to temperature.

RESPONSE TO DEFOLIATION

All sainfoin cultivars have evolved under a hay-cutting regime and aftermath and the best management for optimum yield is to harvest the bulk of the annual yield (about 60 %) at a stage when flowering is well advanced (Evans, 1961). After the sowing year, Common sainfoin has yielded more DM than Giant sainfoin whether cut infrequently for conservation (4 cuts per year) or frequently to simulate grazing (monthly cuts for 8 months) (Thomson, 1951a). In this trial, yields of both types were greatly depressed by the 8-cut treatment but less so in Common (50 %) than in Giant sainfoin (63 %), and the regrowth potential under frequent cutting remained adequate in both. Common sainfoin also withstands the severe defoliation imposed by sheep grazing (Evans, 1961).

Evans (1961) stressed the significance of leafiness in assessing the nutritive value of the crop, and Baker *et al.* (1952) stated that leaves have a higher N content than the stems. However, Terry and Tilley (1964) found little difference in digestibility between leaves and stems in late May, the stem tops being the most digestible part of the plant.

Quality

CHEMICAL COMPOSITION

Little is known about the chemical composition of sainfoin, especially in relation to other legume species. However, there is evidence that it contains only about half as much Ca as white clover, red clover and lucerne, and that some cultivars appear to be inherently low in Na content (See Chapter 31).

WATER CONTENT

The water content of sainfoin varies with growth stage. Fagan and Rees (1930) cut both Common and Giant sainfoin at 4 growth stages during May to July. Water content decreased with delay in cutting between 26 May and 2 July, falling from 83·6 % on 26 May to 70·1 % on 2 July in Common sainfoin, and from 83·3 to 76·1 % in Giant sainfoin. Thus, the decrease in water content with increasing maturity was less marked in Giant than in Common sainfoin. Baker *et al.* (1952) found that the water content of sainfoin was influenced by season and stage of maturity. The water content of sainfoin in the hot, dry season of 1949 was lower than in the wet, cool season of 1950. There was also a fall in water content as the crop matured. The first growth cut at the pre-flowering, early-flowering and full-flowering stages contained 78·2, 77·5 and 76·1 % water in 1949, and 85·0, 82·3 and 74·6 % in 1950, respectively.

DRY MATTER DIGESTIBILITY

Young leafy herbage frequently cut

Sainfoin is seldom grazed or mown when immature, but data from the

Grassland Research Institute, Hurley, suggest that its DM digestibility is seldom higher than 75%.

Primary growth

The curve of DM yield of sainfoin (Fig. 36.1) is approximately parallel to that of S24 perennial ryegrass, sainfoin being about two weeks later. Data from Green and Corrall (1967) indicate that sainfoin maintains its digestibility well up to the end of May, reaching 70% digestibility a few days after S24 ryegrass. Thereafter, the fall in digestibility in the two species is similar. A typical DM yield of sainfoin at 70% DMD is 6 200 kg/ha (5 600 lb/acre) compared with 5 300 kg/ha (4 800 lb/acre) for lucerne, 5 500 kg/ha (5 000 lb/acre) for red clover and 7 500 kg/ha (6 800 lb/acre) for perennial ryegrass.

Regrowths

The regrowth yield of sainfoin is usually less than half that from the first cut and its DM digestibility falls to 65% after about 7 weeks. The main differences between sainfoin and lucerne are the slower rate of fall in digestibility of sainfoin in May and its lower regrowth yield.

ANIMAL INTAKE CHARACTERISTICS

Direct measurements of voluntary intake have suggested that mature sheep consume more sainfoin than either lucerne or red clover when these species are offered as hays of similar digestibility (Osbourn *et al.*, 1966). Young lambs allowed unlimited grazing have made better growth on sainfoin than on any of the other grasses or legumes tested (Grassland Research Institute, 1967).

Longevity and persistency

The longevity of individual parts of the sainfoin plant does not appear to have been studied. Badoux (1965), working in Switzerland, noted a decrease in the weight of the below-ground parts of sainfoin during the winter. Studies on root longevity in a crop with tap-roots which have been reported to penetrate the soil to a depth of 379 cm (Becker-Dillingen, 1929) would undoubtedly be difficult. Giant and Common sainfoin differ in longevity (Rees, 1928; Evans, 1961), Giant being the shorter-lived and used for leys lasting one or two years only. Common sainfoin usually lasts for 7 to 8 years, though leys 20 years old have been recorded. Becker-Dillingen (1929) quotes 10–15 years as the maximum duration of a useful crop, but states that it may persist for 20 years and that individual plants may survive much longer than this.

Sainfoin may multiply vegetatively by means of the short, brown, more or less ascending underground stems observed by Badoux (1965) in old plants; these stems are often branched and bear aerial stem buds at their tips. Thomson (1951a) found that when sainfoin was cut 4 times during the first harvest year or at monthly intervals (8 cuts), there was considerable mortality in

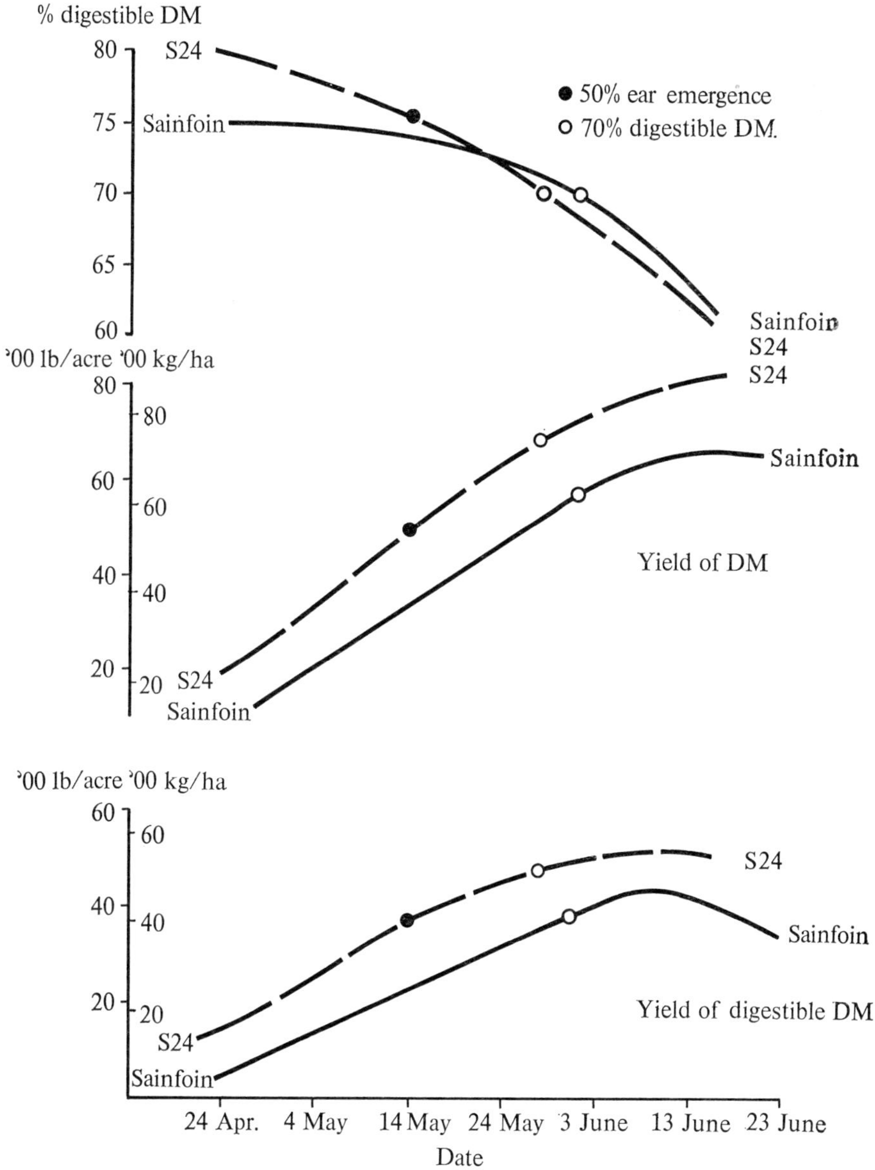

Fig. 36.1. Sainfoin. Seasonal changes in DM yield and digestibility. Values for S24 perennial ryegrass shown for comparison. (From Green & Corrall, 1967)

Giant but relatively little in Common sainfoin. Badoux (1965) also subjected sainfoin to monthly cutting throughout the season, but found that nearly all plants survived their second winter; however, drastic thinning occurred between June and September in the following year, giving 25% survival of Common, but only the isolated survival of plants of the other species and cultivars tested. Where cutting was delayed until after the start of flowering, subsequent monthly cutting had no adverse effect on longevity.

The lack of persistence of Giant has been considered to be due to the greater tendency of stems to flower, resulting in the presence of fewer leaves at soil level, destruction of a high proportion of the top growth at each cut, and to a slower rate of root development. Frequent close cutting decreases yield and eventually results in death. This would not occur in Common sainfoin, which has a winter-requirement for flowering (Badoux, 1965), produces rosettes but no stems after the first cut, and is normally used for aftermath grazing (Thomson, 1938). Sainfoin survives best on deep, porous, well-drained calcareous soils containing at least 0·3% CaO, and does not thrive on heavy clay (Becker-Dillingen, 1929). Survival is adversely affected by cutting before flowering and by hard grazing in autumn (Badoux, 1965). Sainfoin can be successfully resown if a seedling stand fails, but in Germany establishment is reputed to be difficult or impossible on fields which have carried established sainfoin during the previous 10–12 years (Becker-Dillingen, 1929).

References

Baker, C. J. L.; Heinberg, M.; Alderman, G.; Eden, A. 1952. Studies of the composition of sainfoin. *J. agric. Sci., Camb.* **42,** 382–94.

Badoux, S. 1965. [Morphological, physiological and agronomic characteristics of sainfoin populations (*Onobrychis* spp.).] *Schweiz. landw. Forsch.* **4,** 111–90.

Becker-Dillingen, J. 1929. Handbuch des gesamten Pflanzenbaues, einschliesslich der Pflanzenzüchtung. 3. Handbuch des Hülsenfruchtbaues und Futterbaues. Berlin: Paul Parey, pp. 670.

Clapham, A. R., Tutin, T. G.; Warburg, E. F. 1962. Flora of the British Isles. Cambridge: University Press, 2nd edn.

Evans, A. M. 1961. The yields of eight varieties of sainfoin cut at different growth stages. *Emp. J. exp. Agric.* **29,** 323–9.

FAO. 1961. Agricultural and horticultural seeds. *FAO Agric. Stud. 55* Rome: FAO, pp. 577.

Fagan, T. W.; Rees, J. 1930. The effect of cutting sainfoin at different stages on the yield and chemical composition of hay and aftermath. *Welsh J. Agric.* **6,** 224.

Gill, N. T.; Vear, K. C. 1958. Agricultural Botany. London: G. Duckworth and Co. Ltd.

Grassland Research Institute. 1956. Unpublished data H.26.

Grassland Research Institute. 1967. *A. Rep. Grassld Res. Inst. Hurley 1966*, pp. 48.

Green, J. O.; Corrall, A. J. 1967. Grassland Research Institute, Hurley. Unpublished data.

National Institute of Agricultural Botany. 1959. Growing clovers for seed. *Herbage Seed Growers' Leaflet No. 3.*

National Institute of Agricultural Botany. 1968. Varieties of herbage legumes. *Farmers' Leaflet No. 4.*

Osbourn, D. F.; Thomson, D. J.; Terry, R. A. 1966. The relationship between voluntary intake and digestibility of forage crops, using sheep. *Proc. 10th int. Grassld Congr., Helsinki, 1966* pp. 363–7.

PERRING, F. H.; WALTERS, S. M. 1962. Atlas of the British flora. London: Thomas Nelson and Sons Ltd, pp. 456.

REES, J. 1928. Sainfoin or French grass in South Wales. *Welsh J. Agric.* **4,** 242–50.

ROBINSON, D. H. 1953. Leguminous forage plants. London: Arnold and Co., 2nd edn.

SPURWAY, C. H. 1941. Soil reaction (pH) preferences of plants. *Spec. Bull. 306 Mich. agric. Exp. Stn* pp. 36.

TERRY, R. A.; TILLEY, J. M. A. 1964. The digestibility of leaves and stems of perennial ryegrass, cocksfoot, timothy, tall fescue, lucerne and sainfoin, as measured by an *in vitro* procedure. *J. Br. Grassld Soc.* **19,** 363–72.

THOMSON, J. R. 1938. The development of sainfoin in its seeding year. *Ann. appl. Biol.* **25,** 457–71.

THOMSON, J. R. 1951a. Sainfoin in its first harvest year. *J. Br. Grassld Soc.* **6,** 107–17.

THOMSON, J. R. 1951b. Seed studies in sainfoin. *J. Br. Grassld Soc.* **6,** 147–59.

CHAPTER 37

Other Legumes

THE preceding 4 chapters have described the legumes generally accepted as of most importance as forage crops in Britain. However, excluding trees and large shrubs, there are at least a further 70 species which would need to be considered in any attempt to list all those of actual or potential value to British agriculture. Many of these are unimproved indigenous species, others have been introduced and become naturalized, while still others are represented by species not native to Britain, but which have a sufficient degree of adaptation to British conditions to have potential value.

The present chapter deals with those species considered to be of intermediate value, together with others of lesser value but which are of sufficient interest to warrant brief mention. However, it must be stressed that the available information on many of these species is scanty and insufficient to enable any definitive assessment of their potential value to be made.

Because they do not normally have a pastoral use, such crops as peas and beans are referred to only briefly. For similar reasons, important legumes normally grown for direct human consumption are not considered.

Anthyllis vulneraria L.

Kidney vetch (Bentham, 1878; Robinson, 1949; Thomas & Davies, 1964)
Common anthyllis (Bentham, 1878)
Lady's-fingers (Bentham, 1878)
Sand clover (Robinson, 1949)
Woundswort (Whyte *et al.* 1953)

A tufted, deep-rooted legume occurring in stony places and dry pastures (Bentham, 1878) on calcareous soils (Thomas & Davies, 1964). Although the species is recorded as being perennial (Bentham, 1878; Whyte *et al.*, 1953; Thomas & Davies, 1964), it tends to behave like a biennial under cultivation (Williams, 1931; Robinson, 1949). In Britain it has been considered by some to be a valuable herbage plant (Williams, 1931). No distinct commercial varieties are available (Robinson, 1949). Potential cultivars have been selected by the Welsh Plant Breeding Station (1957), but have not yet been released.

Kidney vetch is either sown alone or under a cover crop of barley or oats, although some consider that it grows best without a cover crop (Robinson, 1949; Thomas & Davies, 1964). It has been recommended for inclusion in mixtures for permanent pastures and other long-term uses, although some authors consider that it does not last for more than several seasons (Watson & More, 1962).

Kidney vetch grows well on poor, light, dry, sandy or calcareous soils (Robinson, 1949; Whyte *et al.*, 1953; Watson & More, 1962; Thomas & Davies, 1964; Wynn Parry, 1969). It is drought-resistant and is chiefly sown in low-rainfall areas, such as East Anglia; it is also said to be cold-resistant (Robinson, 1949; Whyte *et al.*, 1953; Watson & More, 1962). It gives only one cut per year and yields much less than red clover (Robinson, 1949). There appear to be no other published data on its yield or chemical composition under British conditions. It can be grazed by sheep (Robinson, 1949; Thomas & Davies, 1964), although it is thought that livestock take some time to grow accustomed to its bitter taste.

Glycine max (L.) Merrill
=*Soya max* (L.) Piper
=*G. soja* Siebold & Zucc.

Soyabean is extensively grown in the USA for hay, pasture, silage or green fodder. Very little investigation has been carried out to determine its value in Britain (Robinson, 1949).

Lotus corniculatus L.

Birdsfoot trefoil (Whyte *et al.*, 1953)
Common lotus (Bentham, 1878)
Bird's-foot trefoil (Bentham, 1878; Thomas & Davies, 1964)
Common birdsfoot trefoil (Robinson, 1949)
There are many other common names cited by Robinson (1933).

Birdsfoot trefoil is a perennial with a well-developed tap-root. It is a very variable species within which 3 main cultivated types have been recognized (Bentham, 1878; Robinson, 1949). It is distributed extensively in Britain, being present in widely differing habitats, including both wet and dry situations (Bentham, 1878; Robinson, 1949; Whyte *et al.*, 1953). A monograph on birdsfoot trefoil has been prepared by Robinson (1933).

It can grow on land too poor, too shallow, too impermeable, too deficient in lime, too alkaline or too 'sick' to support red clover or lucerne (Robinson, 1934b). It is now seldom grown in Britain (Robinson, 1934a). It was grown in this country before the 16th century, at which time it was also recommended for use in France (Robinson, 1933), where it now has a good reputation. It has been considered a weed by some authors (Templeman, 1954).

There have been reports of poor establishment (Corrall, 1968) because of slow seedling growth (Robinson, 1933; Welsh Plant Breeding Station, 1962) and low seedling vigour (Welsh Plant Breeding Station, 1962, 1965). Even when its establishment is good, it may not persist (Thomas, 1936), survival being as short as a single season (Jones, 1967). It is generally non-aggressive (Davies, 1928) and is easily crowded out in mixtures (Robinson, 1933; Whyte *et al.*, 1953).

According to Robinson (1934a, 1934b), birdsfoot trefoil usually requires P and K fertilizer if it is to give its best yields, though heavy fertilizer applications may be deleterious. When given fertilizers, it may become abundant on poor soils (Somerville, 1918). It is usually grown in mixtures with grasses, as its stems are too weak for it to stand satisfactorily alone (Robinson, 1933; Robinson, 1934b; Thomas & Davies, 1964). In studies reported by Corrall (1968) on yields of birdsfoot trefoil/tall fescue mixtures, the total annual dry matter (DM) yield from 3 cuts in the second year was 8 200 lb/acre (9 190 kg/ha), of which 4 260 lb/acre (4 770 kg/ha) was contributed by birdsfoot trefoil. In the same trial, a tall fescue/Du Puits lucerne mixture produced 14 130 lb/acre (15 840 kg/ha), of which 12 660 lb (14 190 kg) was from lucerne. In contrast, the yield of a tall fescue/giant sainfoin mixture was only 7 130 lb/acre (7 990 kg/ha), of which 3 620 lb/acre (4 060 kg/ha) consisted of sainfoin.

In continental Europe, high yields comparable with those from lucerne are obtained from birdsfoot trefoil according to Robinson (1934b), but little experimental work has been conducted to determine its potentialities in Britain. The continental forms are more erect than the indigenous British ones, which tend to be prostrate and more suitable for grazing (Robinson, 1934a). Severe grazing may, however, be deleterious (Thomas, 1936; Jones 1967).

Birdsfoot trefoil has a crude protein content of about 17·5% (DM basis) and has been described as of low apparent nutritive value on account of its mineral and protein content (Fairbairn & Thomas, 1959). It has been regarded as a typical non-bloat-inducing legume (Macdonald, 1950), probably because of its low content of '18-S' protein, which is only about 10% of that in lucerne (McArthur & Miltimore, 1966). Seed losses due to shattering may be a problem (Whyte *et al.*, 1953; Welsh Plant Breeding Station, 1962). It can withstand severe drought (Robinson, 1933), is not affected by dodder (*Cuscuta* sp.) or broomrape (*Orobanche* sp.), and can withstand salinity better than the more important *Trifolium* species (Whyte *et al.*, 1953).

Lotus uliginosus Schkuhr (Robinson, 1949; Whyte *et al.*, 1953).

Large birdsfoot trefoil
Greater birdsfoot trefoil (Robinson, 1949)

Big trefoil (Whyte *et al.*, 1953)
Marsh birdsfoot trefoil (Robinson, 1949)
= *L. villosus* (Robinson, 1949)
= *L. major* Scop.

Although *Lotus corniculatus* and *Lotus uliginosus* have been regarded by some authorities as forms of a single species (Bentham, 1878), they differ in several morphological characters and the latter is also less winter hardy, less tolerant of salinity and more dependent on a good supply of moisture than the former (Robinson, 1949; Whyte *et al.*, 1953).

Large birdsfoot trefoil, a perennial, tends to prefer damp, shady places and can be grown with advantage on moist soils (Thomas & Davies, 1964) and on marshy or peaty soils (Robinson, 1934b), and can be grown among rushes (Robinson, 1949).

In Britain, the species has never become accepted as a crop (Robinson, 1933). In common with birdsfoot trefoil it may be difficult to establish and lacks persistency (Thomas, 1936). Compared with birdsfoot trefoil, Cowling (1954) has reported that it is less persistent and lower yielding. It is less liable to pod-shattering than birdsfoot trefoil (Welsh Plant Breeding Station, 1962).

Lupinus albus L.

White lupin is used only as a green manure crop (Robinson, 1949).

Lupinus angustifolius L. and *L. luteus* L.

Blue lupin and yellow lupin. Useful on poor, light land, especially that deficient in lime (Robinson, 1949), and for land reclamation (Watson & More, 1962). Lupins are usually grazed by sheep, although there is danger of poisoning from certain varieties (Robinson, 1949; Watson & More, 1962). In Britain, the species most commonly grown is blue lupin (Robinson, 1949).

Medicago lupulina L.

Black medick (Robinson, 1949; Whyte *et al.*, 1953; Thomas & Davies, 1964)
Trefoil (Robinson, 1949; Watson & More, 1962; Thomas & Davies, 1964)
Yellow trefoil (Williams, 1931; Robinson, 1949)
Hop clover (Whyte *et al.*, 1953; Thomas & Davies, 1964)
Hop medick (Whyte *et al.*, 1953)
Non(e)such (Bentham, 1878; Whyte *et al.*, 1953)
Black seed (Whyte *et al.*, 1953)

Black medick is an annual (Bentham, 1878; Williams, 1931; Robinson, 1949; Whyte *et al.*, 1953; Thomas & Davies, 1964) or a biennial (Robinson, 1949; Whyte *et al.*, 1953) species which occurs widely in Britain (Bentham, 1878).

It grows on a wide variety of soils, although there are indications that fairly calcareous and dry soils are preferred (Robinson, 1949; Whyte *et al.*, 1953; Watson & More, 1962; Thomas & Davies, 1964). Although natural self-seeding occurs (Robinson, 1949), a black medick population tends to die out after the first year (Watson & More, 1962).

Although black medick has been regarded by some as a valuable herbage plant (Whyte *et al.*, 1953), it has not been generally recommended except for inclusion in small quantities in seeds mixtures (Thomas & Davies, 1964). For example, it was included at 1 lb seed/acre (1·1 kg/ha) in one of the later 'Cockle Park' mixtures (Pawson, 1947), but this practice seems to have been discontinued since about 1930 (Davies, 1958). Mixtures containing a high proportion of black medick have been grown, but have not been particularly favoured (Robinson, 1949).

Black medick is reported to be of no value in permanent pastures, but is useful for establishing short-duration swards on poor, shallow, sandy soils where the better clovers will not grow (Watson & More, 1962; Thomas & Davies, 1964). However, it has been stated that it should not be sown alone on account of its weak stems (Robinson, 1949). It is sometimes sown in mixture with cereals for green manure (Thomas & Davies, 1964).

As black medick is less susceptible than red clover to clover rot (*Sclerotinia*), it can replace the latter on clover-sick land (Robinson, 1949; Watson & More, 1962). It can be grazed by sheep early in the year, but is not very acceptable to cattle (Robinson, 1949; Ivins, 1952). It has a higher Na content than other legumes (Thomas *et al.*, 1952).

Melilotus alba Desr.

Sweet clover (Robinson, 1949)
White Melilot (Robinson, 1949)

Sweet clover contains as much protein as lucerne and can be cut or grazed, but tends to become woody with age and is difficult to eradicate (Robinson, 1949). The annual cultivar Hubam can be grown on moderately acid soils (Watson & More, 1962).

Melilotus indica All.

King Island melilot (Robinson, 1949). Used occasionally for green manure (Robinson, 1949).

Melilotus officinalis Lam.

Common melilot (Robinson, 1949). Infrequently grown. It is difficult to eradicate and its seed is expensive (Robinson, 1949).

Ornithopus perpusillus L.

Common bird's foot (Bentham, 1878). Useful for green manuring on light, sandy soils (Thomas & Davies, 1964).

Ornithopus sativus Brot.

Serradella. Grown for forage. Serradella is tolerant of very acid soils (Watson & More, 1962) and capable of taking up much more P from sandy soils than red clover or lucerne. It may have a useful place in dry, *Agrostis*-dominated areas (Schofield, 1950), but its use in Britain has virtually ceased (Watson & More, 1962), probably because it does not yield well on soils of normal fertility (Cowling, 1954).

Pisum sativum L.

Field pea. Grown on light, calcareous soils, usually in mixture with beans or cereals for support. It is usually cut green for direct feeding or ensilage (Robinson, 1949).

Trifolium dubium Sibth.

Suckling clover; lesser yellow clover. Suitable for growing on unfavourable (Whyte *et al.*, 1953), poor, dry soils (Robinson, 1949; Thomas & Davies, 1964). It grows rapidly from seed (Robinson, 1949) and can be aggressive in pasture (Davies, 1928) but gives very low yields (Robinson, 1949; Thomas & Davies, 1964). A relatively vigorous variety known as 'English giant red suckling clover' has been grown on a small scale in the eastern counties of England for many years (Hawkins, 1954).

Trifolium fragiferum L.

Strawberry clover. Less vigorous than white clover (Cowling 1954) its characteristic of greatest interest is its ability to tolerate salinity and water-logged conditions. Robinson (1949) has reported that it can withstand even prolonged flooding with seawater.

Trifolium hybridum L.

Alsike clover (Bentham, 1878; Robinson, 1949; Thomas & Davies, 1964). A perennial (Williams, 1931), widespread in Britain (Bentham, 1878) and with a life-span variously reported as up to 3 (Watson & More, 1962), 3–5 (Robinson, 1949) or 6 years (Thomas & Davies, 1964). It resembles red clover in many ways, but is more adaptable (Robinson, 1949). It tolerates wet (Stapledon, 1923; Robinson, 1949; Whyte *et al.*, 1953) and acid (Robinson, 1949; Thomas & Davies, 1964) conditions, and is less susceptible to clover rot and stem eelworm (Robinson, 1949). It is valuable for soils of low fertility in high-rainfall areas (Scottish Agricultural Industries, 1967), being grown mainly in western Scotland (Laird, 1949).

Alsike clover is usually sown in mixtures with grasses (Stapledon & Davies, 1930; Dawson, 1949), combining well with timothy (Stapledon, 1923; Robinson, 1949) or cocksfoot, depending on soil type (Robinson, 1949). Up to 1909, it was recommended for inclusion in a 'Cockle Park' mixture for short-duration swards (Pawson, 1960). It is considered non-aggressive (Davies, 1928). Its yields are lower than red clover (Robinson, 1949; Watson & More, 1962) and its use in Britain has tended to decline since about 1930 (Davies, 1958).

Alsike clover herbage is high in Mn and Cu (Thomas *et al.*, 1952). It also has a tendency to cause bloat (Stapledon, 1923; Robinson, 1949) and should therefore be fed to livestock with care.

Trifolium incarnatum L.

Crimson clover. An annual, not very hardy, and grown mainly in the south of England (Robinson, 1949; Watson & More, 1962; Thomas & Davies, 1964), alone or in mixtures. It is thought that its range of cultivation could be extended (Cowling, 1954). It is grown as a catch crop for grazing or hay (Robinson, 1949; Watson & More, 1962; Thomas & Davies, 1964). The ripe, hairy fruits are dangerous to livestock because of their tendency to cause intestinal blockages (Robinson, 1949; Watson & More, 1962).

Trifolium subterraneum L.

Subterranean clover. Best suited to areas with dry summers (Whyte *et al.*, 1953), it is capable of good growth in some parts of Britain, though its true value is not known because of lack of experimental work (Robinson, 1949). According to Williams and Garrett-Jones (1949), subterranean clover is unlikely to displace white clover, even in areas with light soils and low rainfall, where lucerne would probably be a better alternative. It provides relatively more winter grazing (Robinson, 1949) than Italian ryegrass or red clover (Williams & Davies, 1924). It is important to manage it so that self-seeding occurs (Thomas & Davies, 1964).

Vicia faba L.

Field bean. Grown for soiling or silage, usually in mixtures with vetches, peas or cereals (Robinson, 1949).

Vicia sativa L.

Common vetch. Grown on most moderately calcareous soils, either alone or preferably in mixtures, as it requires a support (Paterson & Russell, 1943; Robinson, 1949; Watson & More, 1962; Thomas & Davies, 1964). It can be grazed by sheep and other livestock, cut green for direct feeding, or made into hay or silage (Watson & More, 1962; Thomas & Davies, 1964). Common vetch can cause bloat when fed green (Watson & More, 1962). Its seed yields can be seriously reduced by the vetch-leaf gall midge (Gair & Guile, 1961).

Conclusions

The available information on the legume species described in this chapter is scanty, particularly in terms of their comparative value in relation to those that are widely grown in Britain. Their value is dependent on their suitability for various climates, soil types and systems of sward management, and here, existing ecological information may be useful for predicting the value of a species in a particular agricultural situation. For example, according to Wynn Parry (1969), kidney vetch requires a high soil pH value and free Ca in the soil to make good growth, and is drought resistant only after establishment. *Lotus* species are known to be tolerant of high altitude and high rainfall, and ecotypes adapted to such conditions have developed. Suckling clover is notable for its persistence in hay meadows cut annually, and on low-fertility soils. Alsike clover thrives on peaty soils of fairly high fertility.

It is clear that many species have properties which could be exploited. Birdsfoot trefoil, suckling clover, blue lupin, yellow lupin, large birdsfoot trefoil and kidney vetch thrive to various degrees on light sandy soils or other poor dry soils. Kidney vetch and birdsfoot trefoil are drought resistant. Strawberry clover, large birdsfoot trefoil and alsike clover can withstand wet soils and, in various degrees, flooding. Strawberry clover, and probably to a lesser extent birdsfoot trefoil, also tolerate salinity. Many of these species withstand diseases which affect the more important legumes.

There is a need for a herbage legume which can make better growth than white clover on acid, dry soils of low fertility (Welsh Plant Breeding Station, 1962). Use could probably be made of serradella on very acid soils and of white melilot on moderately acid soils. These unfamiliar legumes, however, have various disadvantages; lupins, common vetch and alsike clover may cause bloat; sweet clover and common melilot are difficult to eradicate; kidney vetch and suckling clover give only low yields; seed of birdsfoot trefoil and common vetch is difficult to produce. Nevertheless, the full potentialities of these and other species such as soyabean and subterranean clover under British conditions have yet to be explored.

References

BENTHAM, G. 1878. Handbook of the British flora. London, 4th edn.

CORRALL, A. J. 1968. Grassland Research Institute, Hurley. Personal communication concerning Exps H.774 and H.375.

COWLING, D. W. 1954. Plant introduction at the Grassland Research Institute. *J. Br. Grassld Soc.* **9**, 239–43.

DAVIES, W. 1928. The factor of competition between one species and another in seeds mixtures. *Bull. H8 Welsh Pl. Breed. Stn* pp. 82–149.

DAVIES, W. 1958. Grassland developments in the United Kingdom in the last twenty-five years. *Emp. J. exp. Agric.* **26**, 195–207.

DAWSON, J. L. 1949. Grass and grazings in the North of Scotland. *J. Br. Grassld Soc.* **4**, 183–92.

FAIRBAIRN, C. B.; THOMAS, B. 1959. The potential nutritive value of some weeds common to north-eastern England. *J. Br. Grassld Soc.* **14**, 36–46.

GAIR, R.; GUILE, C. T. 1961. Life history and bionomics of the vetch-leaf gall midge. *Ann. appl. Biol.* **49**, 360–2.

HAWKINS, R. P. 1954. English giant red suckling clover. *J. Br. Grassld Soc.* **9**, 149–51.

IVINS, J. D. 1952. The relative palatability of herbage plants. *J. Br. Grassld Soc.* **7,** 43–54.
JONES, L. I. 1967. Studies on hill land in Wales. *Tech. Bull. 2 Welsh Pl. Breed. Stn* pp. 1–179.
LAIRD, R. 1949. The grassland of the West of Scotland. *J. Br. Grassld Soc.* **4,** 85–94.
MACDONALD, L. H. 1950. Birdsfoot trefoil. *Ext. Bull. 797 Cornell Univ. agric. Exp. Stn* pp. 4.
MCARTHUR, J. M.; MILTIMORE, J. E. 1966. Pasture bloat and the role of 18-S protein. *Proc. 10th int. Grassld Congr., Helsinki,* 1966 pp. 518–21.
PATERSON, W. G. R.; RUSSELL, K. N. 1943. The cultivation and utilisation of some common forage crops. *Bull. 143 W. Scotl. agric. Coll.* pp. 16.
PAWSON, C. A. 1960. Cockle Park Farm. Oxford: Univ. Press, pp. 261.
PAWSON, H. C. 1947. Fifty years' work on grassland at Cockle Park. *J. Br. Grassld Soc.* **2,** 5–10.
ROBINSON, D. H., 1933. Birdsfoot trefoil. A monograph. Ph.D. Thesis, Univ. London, pp. 174.
ROBINSON, D. H. 1934a. Birdsfoot trefoil in agriculture. *Emp. J. exp. Agric.* **2,** 274–83.
ROBINSON, D. H. 1934b. Birdsfoot trefoil. *J. Minist. Agric.* **41,** 162–7.
ROBINSON, D. H. 1949. Leguminous forage plants. London: Edward Arnold & Co., 2nd edn.
SCHOFIELD, J. L. 1950. Serradella (*Ornithopus sativus*) a legume for light arid soils. *J. Br. Grassld Soc.* **5,** 131–40.
SCOTTISH AGRICULTURAL INDUSTRIES FARM SEED CATALOGUE. 1967.
SOMERVILLE, W. 1918. 'Poverty Bottom': an experiment in increased food production. *J. Minist. Agric.* **24,** 1186–202.
STAPLEDON, R. G. 1923. Alsike clover. *J. Minist. Agric.* **30,** 303–8.
STAPLEDON, R. G.; DAVIES, W. 1930. Experiments to test the yield and other properties of various species and strains of herbage plants under different methods of management. *Bull. H10 Welsh Pl. Breed. Stn* pp. 5–41.
TEMPLEMAN, W. G. 1954. Chemical weed-control and pasture productivity. *J. Sci. Fd Agric.* **5,** 387–92.
THOMAS, B.; THOMPSON, A.; OYENUGA, V. A.; ARMSTRONG, R. H. 1952. The ash constituents of some herbage plants at different stages of maturity. *Emp. J. exp. Agric.* **20,** 10–22.
THOMAS, J. O.; DAVIES, L. J. 1964. Common British grasses and legumes. London: Longmans, Green and Co. Ltd, 4th edn, pp. 122.
THOMAS, M. T. 1936. The introduction and maintenance of nutritious and palatable species and strains. *Bull. H14 Welsh Pl. Breed. Stn* pp. 4–57.
WATSON, J. A. G.; MORE, J. A. 1962. Agriculture. The science and practice of farming. Edinburgh: Oliver and Boyd, 11th edn.
WELSH PLANT BREEDING STATION. 1957. *Rep. Welsh Pl. Breed. Stn 1950–56.*
WELSH PLANT BREEDING STATION. 1962. *Rep. Welsh Pl. Breed. Stn 1961.*
WELSH PLANT BREEDING STATION. 1965. *Rep. Welsh Pl. Breed. Stn 1964.*
WHYTE, R. O.; NILSSON-LEISSNER, G.; TRUMBLE, H. C. 1953. *FAO Agricultural Studies No. 21.* Legumes in Agriculture. Rome: FAO, pp. 367.
WILLIAMS, R. D. 1931. Fertility of various herbage legumes. *Bull. H12 Welsh Pl. Breed. Stn* pp. 221–7.
WILLIAMS, R. D.; DAVIES, W. 1924. A note on subterranean clover (Australian variety). *Bull. H1 Welsh Pl. Breed. Stn* pp. 151–8.
WILLIAMS, T. E.; GARRETT-JONES, R. 1949. Subterranean clover; field trials in East Anglia. *J. Br. Grassld Soc.* **4,** 201–8.
WYNN PARRY, D. 1969. School of Plant Biology, University College of North Wales. Personal communication.

PART IV

COMPARATIVE ASSESSMENT OF SPECIES

CHAPTER 38

Botanical Composition of the Crop

MOST cultivars of grasses and clovers have been bred from selected individual plants grown at wide spacing. Under such conditions, differences in ear emergence date, habit of growth and other morphological characters can be measured (Hawkins, 1958). In pure swards, on the other hand, plant density is high and differences in growth habit between cultivars are less well marked, though still detectable; differences in persistence and ear emergence are, however, still clear (National Institute of Agricultural Botany, 1968). The contrast between the performance of plants under wide-spaced conditions and that in a sward reflects the morphological plasticity inherent in grasses. The potential of grasses and clovers has, however, traditionally been exploited by the use of mixtures of grasses, with or without legumes, and of more than one variety within a species. In considering crop composition, questions arise as to whether such mixtures can give higher yields than those of single-species stands, and whether mixtures have advantages in feeding value. Also, it is relevant to consider whether the work of the plant breeder and the effect of seed certification measures are vitiated by the practice of using complex seeds mixtures sown at high seed rates. Such questions have been current among more enlightened farmers for at least the last 150 years, but it is pertinent to reconsider them in the light of present conditions of agriculture and the availability of new and improved varieties.

Crop composition must be related to type of farming in Britain, which ranges from the less intensive farming of the hills and of poor, sandy soils such as the Brecklands, to highly capitalized and sophisticated green-crop drying systems. Between these extremes lie various production systems involving grazing, zero-grazing and conservation, or various combinations of these. Each system involves the subjection of grasses and legumes to different intensities and frequencies of defoliation and hence to different degrees of inter-plant competition, and crop performance will be determined by the interaction between its genetic constitution and the superimposed system of management.

Dynamics of a mixture

Several theoretical studies have been made of the stresses existing between contrasting plant types within a crop stand, and useful reviews of the literature exist (e.g. de Wit, 1960; Donald, 1963; Harper, 1964). Performance of the crop may depend on the proportion of the different components within it, their density and the interactions between them. In mixtures, an aggressive component will acquire an unequal share of the limited environmental resources, a situation aggravated at high plant density (Harper, 1961). By contrast, inter-plant variation in single-species stands is masked in the assessment of total yield, and considerable plasticity appears to exist; for example, Lazenby and Rogers (1964) found no significant difference in dry matter yield between perennial ryegrass sown at 3 kg and at 40 kg/ha. Components of a mixture may either be 'complementary', in that overall yield can be predicted from the performance of the individual components grown singly, or not, in which case the overall yield cannot be predicted in this way. Theoretical aspects of these two situations have been reviewed by de Wit (1960) and numerous experiments reported by other workers have shown that both situations exist in practice.

During the establishment phase of mixed swards, if one component emerges before the others or if it has an initial advantage in speed of development, it can begin photo-assimilation earlier. This advantage is particularly marked in the case of species with large seeds. Carpenter (1960) showed that the components of a mixture which emerged early also showed a rapid increase in number of tillers per plant. The competitive advantage thus gained during the first phase of sward development was maintained at least until the spring of the second year. However, single-tiller plants survived for at least a year and were able to take over from their more aggressive neighbours if the latter died.

In swards, the influence of the amounts and proportions of the various components in the seeds mixture used and the compatibility and respective rates of early growth of the species and varieties present (dealt with in Chapters 15–22 and 33–36) is important, but the management system employed exerts an overriding effect on botanical composition. It is impossible to consider the extensive literature comprehensively and a few examples must suffice. Management implies the control of such facets as frequency and intensity of defoliation and fertilizer application, and must also take into account the spatial arrangement of crop components, e.g. whether they are grown together or in separate rows, or even the direction of the rows (Donald, 1963). The effect of management in modifying sward composition was shown in the early studies conducted at Aberystwyth, where Jones (1939) demonstrated that less intense cutting favoured certain components such as cocksfoot. Over 30 years ago it was demonstrated experimentally that a pasture could either deteriorate into a weed-infested sward or could maintain a productive balance of ryegrass and clover purely as a result of varying the timing and intensity of grazing.

If one component of the crop is in active growth when cut or grazed, it may be at a disadvantage relative to other components with different rhythms of growth. The relatively delicate balance between grass and clover in a sward has been well demonstrated by Jones (1933), who showed that heavy grazing leads to an increase in the clover component and light grazing to a decrease. Increasing the rate of applied N will progressively reduce the clover component until the point is reached where the clover has been lost—due largely to an increase in grass growth (Stern & Donald, 1962). More recently, Cowling and Lockyer (1968) have indicated that the proportions of different grass species within a mixed herbage may also be influenced by the level of applied N; for example, they showed that increases in the level of applied N caused cocksfoot to become more dominant in a mixture with timothy and perennial ryegrass. These findings, and those of Castle and Reid (1968), have practical implications in relation both to deciding the optimum level of N to apply and whether the N should be given as single or split dressings.

Practical implications

Having considered the factors most likely to affect the composition of the crop—population density, relative proportions of the components, management practices (particularly frequency and intensity of defoliation and the level of applied nutrients)—it is pertinent to relate these to agricultural practice.

Recently, there has been a trend towards increased research on crop composition (England, 1965; Cowling & Lockyer, 1968; Rhodes, 1968), though such research is limited by the time and effort required in such complex studies. That much of the research carried out in the past has tended to appear parochial has been because of the wide variety of facets to be considered within the whole spectrum of herbage crop utilization.

No full review of the factors governing the compounding of forage seeds mixtures exists for UK conditions, though the ecological and physiological factors which have to be taken into account have been reviewed by Blaser, Skrdla and Taylor (1952) for the USA. These workers concluded that uniformity of production and flexibility of use should be among the objectives pursued in the design of mixtures. Instead of designing mixtures solely for grazing or for the production of hay or silage, they should be flexible enough to permit management for more than one type of production and should be compounded for (a) grazing, (b) silage and aftermath grazing, (c) silage, hay and aftermath grazing, and (d) early, mid- and late-season grazing. The employment of several mixtures in a cropping system is logical in relation to the varied growth of forage species with season (Blaser, Skrdla & Taylor,

1952). In practice, the composition of the crop must be designed to give the maximum return of nutrients per unit area under the prevailing conditions of soil and climate but, at the same time, it must be economically integrated into the particular farming system.

Under some conditions, grass/legume associations will give more rapid establishment and better seasonal distribution of yield than grasses alone. Such an association is particularly useful in areas where terrain and other circumstances make application of fertilizer N difficult or uneconomic. At the present time, farming systems employing high inputs of N do not include a legume in the crop, but the position could change in the future. If the present programme at the Welsh Plant Breeding Station aimed at breeding a large-leaved clover tolerant of high N levels is successful, this could prove most valuable in improving liveweight gains (Miles, 1967). On the other hand, highly intensive systems of farming such as green-crop drying will utilize the potential of modern varieties as monocultures under high levels of artificial fertilizer. Under such a system the peak of quality and yield and the intensity and frequency of cutting can be defined with reasonable certainty of obtaining high returns.

How far down the scale of farming efficiency is it safe to extend the concept of monoculture? Based on experiments at the Grassland Research Institute, Hurley, Cowling and Lockyer (1968) conclude that a rational approach to grassland husbandry should be based on swards sown to a single grass variety. By using a series of pure-sown swards, it is claimed that harvest is spread over a longer period and can be integrated closely with the peaks of maximum digestibility. A slightly different opinion has been expressed by England (1968), who concludes that simple mixtures give more consistent performance under different managements than do single-species swards made up of the components of such mixtures. In his experiments the yield given by a mixture was no higher than that of the highest-yielding component of the mixture when sown pure.

The 'buffering' effect of a simple mixture has a value in many agricultural systems, but there is little good evidence to show that a complex mixture is more efficient than a simple one in exploiting the environment other than under the less efficient systems. The current situation has been considered by the Committee on Herbage Seed Supplies (1968), who concluded that official bodies and the trade should co-operate in arranging a more co-ordinated approach to the problem of seeds mixtures. At present, advice is based mainly on an official leaflet (Ministry of Agriculture, Fisheries & Food, 1970) and the experience of merchants.

References

Blaser, R. E.; Skrdla, W. H.; Taylor, T. H. 1952. Ecological and physiological factors in compounding forage seed mixtures. *Adv. Agron.* **4,** 179–219.

CARPENTER, J. A. 1960. *Mimeo Rep. Waite agric. Res. Inst., Univ. Adelaide.*
CASTLE, M.E.; REID, D. 1968. The effects of single compared with split applications of fertilizer nitrogen on the yield and seasonal production of a pure grass sward. *J. agric. Sci., Camb.* **70**, 383–90.
COMMITTEE ON HERBAGE SEED SUPPLIES. 1968. Report of the Committee on herbage seed supplies. *Cmnd. 3748* London: H.M.S.O., pp. 99.
COWLING, D. W.; LOCKYER, D. R. 1968. A comparison of the yield of three grass species at various levels of nitrogenous fertilizer sown alone or in a mixture. *J. agric. Sci., Camb.* **71**, 127–36.
DONALD, C. M. 1963. Competition among crop and pasture plants. *Adv. Agron.* **15**, 1–118.
ENGLAND, F. J. W. 1965. Interactions in mixtures of herbage grasses. *Rec. Scot. Pl. Breed. Stn 1965* pp. 125–49.
ENGLAND, F. J. W. 1968. Competition in mixtures of herbage grasses. *J. appl. Ecol.* **5**, 227–42.
HARPER, J. L. 1961. Approaches to the study of plant competition. In: Milthorpe, F. L. [Ed.] Mechanisms in biological competition. Symposia of the Society for Experimental Biology. No. 15. Cambridge: Univ. Press, pp. 1–39.
HARPER, J. L. 1964. The nature and consequence of interference amongst plants. *Proc. 11th int. Congr. Genet., 1963* pp. 465–82.
HAWKINS, R. P. 1958. The classification of strains (varieties) of herbage plants. *J. natn. Inst. agric. Bot.* **8**, 433–49.
JONES, L. I. 1939. Pasture management and its effect on the sward. *Bull. H15 Welsh Pl. Breed. Stn* p. 40.
JONES, M. G. 1933. Grassland management and its influence on the sward. 3. The management of a 'grassy' sward and its effects. *Emp. J. exp. Agric.* **1**, 223–34.
LAZENBY, A.; ROGERS, H. H. 1964. Selection criteria in grass breeding. 2. *J. agric. Sci., Camb.* **62**, 285–98.
MILES, D. G. 1967. *Rep. Welsh Pl. Breed. Stn 1966* p. 92.
MINISTRY OF AGRICULTURE, FISHERIES AND FOOD. 1970. Seeds mixtures 1970/71. *Short-term Leaflet No. 46.*
NATIONAL INSTITUTE OF AGRICULTURAL BOTANY. 1968. Grasses. *Leaflet No. 16.*
RHODES, I. 1968. Yield of contrasting ryegrass varieties in monoculture and mixed culture. *J. Br. Grassld Soc.* **23**, 156–8.
STERN, W. R.; DONALD, C. M. 1962. Light relationships in grass-clover swards. *Aust. J. agric. Res.* **13**, 599–614.
WIT, C. T. DE. 1960. On competition. *Versl. landbouwk. Onderz.* No. 66. 8, pp. 82.

CHAPTER 39

Sward Establishment

THE principles of establishment for both grasses and clovers have already been described in Chapters 8 and 26. In this chapter, establishment is taken to include the development and characteristics of the sward in the seeding year. Chippindale (1949) pointed out that the physiology of the mature plant may differ from that of the seedling. In his investigations, cocksfoot and red fescue required relatively high moisture during germination, but tolerated dry conditions when older. Timothy and meadow fescue had a low moisture requirement during germination, yet needed moist conditions later. For this and other reasons, establishment merits consideration as a separate stage in the development of the sown sward.

Though it may vary with environment, the capacity for establishment is a characteristic of each species and cultivar. In general, under British conditions cocksfoot and timothy establish less rapidly than ryegrass (Stapledon, Davies and Beddows, 1927; Milton, 1935) and both meadow fescue (Davies, 1928a) and tall fescue (Charles, 1961c) are very sensitive to competition from other grasses, particularly Italian and perennial ryegrass.

Differences between S22 Italian ryegrass and S170 tall fescue in speed of establishment were demonstrated in experiments at the Welsh Plant Breeding Station (1966) in which the average tiller numbers per plant were 19·5 and 6·1, respectively, at 7 weeks after sowing. In experiments at the Grassland Research Institute, Hurley (1967) the more rapid rate of leaf appearance and bud development in S22 Italian ryegrass resulted in this grass having 6 times as many tillers as S48 timothy 36 days after sowing. Tall fescue had only half the rate of leaf appearance of cocksfoot but about the same number of tillers 5 weeks after sowing. The rate of development of red fescue, common bent and smooth-stalked meadowgrass was slow for the first 4 weeks, but differed little from that of the other species 7 weeks after sowing. It was also noted that diploid cultivars of Italian and perennial ryegrass produced leaves faster than tetraploids.

Seed size

The importance of seed size has frequently been emphasized (Milton, 1935; Davies & Western, 1937; Hunt, 1954). Even in monospecific swards the advantage of large seeds can be considerable. This is well illustrated by Black's work (1958) on subterranean clover (*Trifolium subterraneum*) in which plants grown from small seeds comprised 25% of the total leaf area of the sward, but received only 10% of the available light energy. Elimination of seedlings developing from seed of low weight could easily occur in this situation. Seedling vigour is not, however, entirely a matter of seed weight (Davies, 1927; Heddle and Herriott, 1954), and breeding for rapid establishment ought to involve the further testing of progenies from parents of high seed weight for seedling vigour.

Seedbed preparation

The work of preparing a grass field for reseeding may commence prior to ploughing. Drainage should be assured, since destruction of the old turf and an increase in stock numbers combine to increase the danger of poaching and consequent weed invasion. Direct action against existing perennial weeds, such as mowing to weaken rushes and creeping thistle, is also helpful. The generous application of N combined with cutting for silage or hay can reduce the vigour of buttercups, and a selective herbicide can be used to aid these managements.

There is as yet no recommended treatment for the control of perennial grass weeds in newly sown swards (Weed Control Handbook, 1968). Careful cultivation during the preparation of the seedbed is therefore important, and there are advantages in using a dense, vigorous pioneer crop to suppress the remnants of the original vegetation. The use of skim coulters and a reversible plough help to give complete inversion of the old sward, and the furrows should be pressed down with a heavy roller. Nothing should subsequently be done which will bring the old sward back to the surface by such operations as discing too fast in the wrong direction or with the discs set at a wide angle.

In general, it is recommended that the tilth for a herbage seedbed should be fine and firm (Davies, 1945). There is, however, a lack of precise evidence and Williams (1922a, b, 1924) maintained that a rough surface assists the seedlings to become attached to the soil.

Minimum cultivation techniques

Large areas of Britain are too steep, stony or wet for conventional techniques of establishment to be employed. In these circumstances surface sowing methods involving only a minimum of soil cultivation offer possibilities.

The basic requirements for successful resowing are (a) adequate soil nutrient status, (b) the complete elimination of the old sward, (c) ensuring that the seed is placed in contact with the soil, and (d) appropriate management during and after establishment. The use of minimum cultivation methods is not new (Stapledon, 1935), and Scottish workers have successfully employed them over a number of years (Gardner, Hunt & Mitchell, 1954). In recent years, however, the development of new herbicides, new machinery for overdrilling or sod-sowing, the use of aircraft and improved cultivars of herbage plants have greatly extended the possibilities of minimum cultivation systems.

The existing vegetation has a major influence on the success of minimal cultivation. Areas dominated by ling (*Calluna*) are easily dealt with (Gardner, Hunt & Mitchell, 1954; Copeman & Roberts, 1960), but herbicides are a useful aid on dense, grassy swards (Elliott, 1958; Ormrod, 1960; Rowlands, 1960; Jones, 1962, Charles, 1962b; Charles & Lewis, 1962).

The number of grass plants established per unit area by minimum cultivation practices is frequently low; therefore, those species and varieties able to spread rapidly because of high tillering capacity or creeping habit are at an advantage (Copeman & Roberts, 1960). Under British conditions, perennial ryegrass is very useful, particularly such leafy pasture cultivars as S23. Cocksfoot has been easily established on drier sites in Scotland, but did not withstand hard grazing by sheep; best establishment was obtained with the cultivar S143. Timothy proved valuable on the wetter sites; Scottish and Scandinavian varieties gave good early growth, but S48 was more persistent.

Red fescue, perennial ryegrass and smaller cat's-tail (*Phleum bertolonii*) are reported to be useful for surface sowing on hill land (Hughes & Nicholson, 1961). Yorkshire fog (*Holcus lanatus*) established well but persisted only when grazed by cattle in the summer.

White clover is easier to establish by minimum cultivation than grasses, varieties recommended being New Zealand, S100, Kent and S184 (North of Scotland College of Agriculture, 1960).

Depth of sowing

As a general rule the optimum depth of sowing under British conditions ranges from 3 to 32 mm. The importance of ensuring that even the smaller-seeded species are covered with soil was clearly demonstrated by Williams (1922a, b, 1923, 1924). Drawing on experience in Wales and later in the West Midlands at the Grassland Improvement Station, Drayton, Davies (1946) emphasized the need to sow at a depth of 25 mm in districts of low rainfall, and pointed out that even in wetter districts deep sowing had the advantage of giving protection from cold spells in late spring. Drilling has advantages in the drier parts of England, as this ensures that the seed is placed in contact with moist soil. In the higher-rainfall conditions of Wales, Williams (1924) advocated broadcasting on a chain-harrowed surface and again using the chain harrow for covering the seed.

In a comparison of Italian ryegrass, cocksfoot and red clover, Williams (1923) showed that establishment of legumes was much more affected than that of the grasses by the method of covering the seed. In this investigation the complete covering of the seed was important for satisfactory establishment. In another series of trials, Williams (1922b) found that with white clover, the highest emergence of 92% was obtained with seed sown at a depth of 10 mm; at 25 mm, emergence was 66% and at 51 and 76 mm was virtually nil. The most satisfactory establishment of perennial ryegrass was obtained with sowing depths of 3 to 25 mm, but seedling numbers of cocksfoot sown 19 mm deep were higher than when sown 3 mm deep. Stapledon and Wheeler (1948) reported that timothy had better establishment from 'shallow' rather than from 'deep' sowing, and that white clover behaved similarly. These authors concluded that it would be risky to sow timothy and white clover as deeply as ryegrass, whereas cocksfoot occupied an intermediate position. It was suggested that seeds mixtures should be divided into 'deep' and 'shallow' species, the former could then be drilled and the latter broadcast and covered lightly. In sowing with a combine drill, Jones (1957) suggested mixing the heavier seeds with the fertilizer and placing the smaller seeds in the seeds hopper.

Time of sowing

In a discussion of the best time to sow herbage seeds, Davies (1946) arrived at the simple rule that sowing should take place as early in spring as soil and weather allow. Although sowing late in the year can be successful, most experimental evidence indicates the advantages of early sowing (Davies, 1927, 1928a; Chippindale, 1932, 1949; Wilks, 1945; Davis & Lambert, 1947; Stapledon & Wheeler, 1948; Martin, 1949).

In practice it may not be possible to achieve the ideal of early sowing. In the west and north of England, and in Scotland where the annual rainfall exceeds 760 mm, sowing can be deferred up to late April and early May. On marginal and hill land where annual rainfall exceeds 1 020 mm, it is normally safe to sow in May and June. In districts having less than 760 mm, with a risk of late spring and summer drought, sowing should be carried out during March to early April, or delayed until late summer.

Young seedlings of grass withstand adverse weather conditions better than those of clover (Chippindale, 1932, 1948). Clovers are much more likely to suffer from late sowing than grasses, even though legumes germinate more rapidly than grasses, particularly under adverse conditions (Stapledon & Wheeler, 1948). The latter investigators reported that the failure of clover from late sowing was almost entirely due to mortality of the young clover seedlings during winter. The advantage to grass seedlings may be accentuated under field conditions, as there are indications that clover seedlings are preferred by soil fauna (Stapledon, Davies & Beddows, 1927). On heavy clays it may be difficult to obtain a tilth for spring sowing (Davis & Lambert, 1947), and in such situations it is advisable to sow the grass seed in the autumn and the clover in the following spring.

Grass species differ in their reaction to late sowing. Stapledon and Wheeler (1948) recorded that timothy establishment was reduced by late sowing (after 8 September) and this was some two weeks before cocksfoot showed any adverse effect. Italian and perennial ryegrass could be sown successfully up to 15 October. These results are broadly similar to those obtained by earlier workers (Stapledon, Davies & Beddows, 1927; Davies, 1927). Tall fescue is also adversely affected by late sowing (Baker, 1964). In general, the critical date of sowing varies with location and from year to year. Early work at Aberystwyth led to the conclusion that the most satisfactory establishment was obtained from sowing herbage seed earlier than the end of July.

Seed rate

High seed rates are beneficial to seedling establishment and in the early stages of crop growth (Holliday, 1953). Tiller population is greatly influenced in the early stages by rate of sowing, but subsequently becomes similar for all seed rates. In the same way, Eisele and Aikman (1933) showed that seed rate had little effect on plant density in established swards. Donald (1951, 1954) also found that despite an initial yield response to increasing seed rate, plant density on all his plots reached a similar level. In Scotland, raising the seed rate of a ryegrass/cocksfoot/white clover mixture from 11·2 to 22·4 and then to 33·6 kg/ha, gave only a slight increase in dry-matter yield in the year of sowing, and even this was inconclusive owing to the high variability of the results. At the Grassland Research Institute (1959), varying the seed rate of Italian ryegrass and red clover had no marked effect (less than 5% on average) on total yield. In contrast, increasing the seed rate of red clover sown without grass from 5·6 to 11·2 kg/ha resulted in 75% higher dry matter (DM) yields in the seeding year. Greenaway (1957), however, recorded large yield increases (up to 80%) in the year of sowing as a result of increasing the seed rate of short-rotation and perennial ryegrass from 16·8 to 67·3 kg/ha. Again, rate of sowing did not greatly affect yield in the following year.

The seed rate of grasses sown in mixture with clover can have a marked effect on the establishment of the legumes. At Aberystwyth, increasing the seed rate of perennial ryegrass from 4·5 to 44·8 kg/ha reduced the mean establishment of white, Montgomery red and alsike clover from 31 to less than 15% (Davies 1928b). This effect was confirmed for both ryegrass and cocksfoot in trials at the Grassland Research Institute, Hurley (1959); in addition, the variety of grass used was shown to have a considerable effect. Thus, yield of white clover was similar when this legume was sown either with S23 perennial ryegrass at 5·6 kg/ha or with Irish perennial ryegrass at 22·4 kg/ha.

Companion crops

Herbage seeds can be sown with a cereal or other companion crop, or alone. The companion crop may aid establishment (though there is little evidence for this) or it may improve DM production in the year of sowing. Undersowing a cereal crop with herbage seed in order to reduce the period between harvesting the cereal and resowing the land to grass is an attractive proposition. Stapledon and Jones (1925), reporting work by Rhodin, noted the protection offered to young seedlings by a companion crop and the avoidance of drying-out of the soil surface. The reduction of weed growth by using a companion crop (Jones, 1933) has also been given as a reason for undersowing. However, in a survey of 417 cases of reseeding in Yorskhire, Moore (1945) found that competition from the companion crop caused 46% of the failures and, of these, 27% were caused by lodged corn. Heddle and Herriott (1955) pointed out that when competition from weeds is reduced by management, direct-sown swards show better establishment and growth than undersown swards.

Herbicides can be used to control prostrate annual weeds which are resistant to early defoliation of the sward. Chickweed (*Stellaria media*) is particularly difficult to combat, but even this is susceptible to dinoseb, care being necessary if leaf scorch of sown species is to be minimized (Weed Control Handbook, 1968). In earlier works (Stapledon & Jones, 1925; Davies, 1927; Jones, 1933; Moore, 1945), it is also stressed that the benefit to establishment of the undersown sward is outweighed by the depressive effect of undersowing on herbage growth. The open swards that result from undersown seeds invite weed ingress and Heddle and Herriott (1955) maintained that increased herbage seed rates are desirable if a companion crop is used.

Davies (1927, 1945, 1946) recommended sowing herbage at the same time as, or shortly after, the cereal because this aids herbage plant establishment. Thomas (1945) suggested that sowing in spring into the emerged cereal crop may avoid some clover kill by frost. A companion crop of Italian ryegrass, cereal or rape allows the grazing animal early access to the sward and reduces weed invasion (Davies, 1945).

Herbage species vary in their reaction to undersowing. Stapledon and Jones (1925) and Stapledon (1949) advocated a Cockle Park type mixture when a companion crop is used. This argument is based on the fact that the cereal suppresses the rapid-growing ryegrasses to a greater extent than the slower-growing cocksfoot and timothy. Tall fescue is particularly susceptible to the adverse effects of undersowing and should not be established by this method (Baker, 1964; Charles, 1965). The clover/grass balance can also be affected by the cereal during establishment (Charles, 1961b, 1963a, 1965). The ratio of clover to grass can be increased by undersowing and this effect may persist after the establishment phase (Charles, 1962a, 1963b). The balance of cultivars within a species may also be affected by undersowing (Charles, 1961a).

Fertilizer application

The rate of emergence of grass and clover seedlings from the soil is not greatly affected by fertilizer applied to the seedbed; subsequently, however, growth is increased (Davies, 1927, 1934; Stapledon, 1939; Stewart, 1952). In grass/clover mixtures the establishment of the legume may be reduced by competition from the grass following the application of phosphate (Davies, 1927, 1939; Davies & Western, 1937) and the growth of clover may be greatly reduced by N applied to the seed bed (Charles, 1961b, 1963a, 1965). Combine-drilling of N and K fertilizers with the seed may lead to reduced germination (Stewart & Reith, 1945; Stewart, 1952; Cooke, 1956).

Management of the young sward

Early grazing of the establishing sward is generally advocated and cattle are preferred to sheep (Jones, 1945). Frequent removal of herbage growth helps to control erect annual weeds and encourages the growth of white clover. However, continuous grazing should be avoided, as this can quickly lead to severe reduction of sown species, cocksfoot and timothy being particularly susceptible. The young sward is also very prone to poaching and animals should be removed when soil moisture is high.

Production in the seeding year

Production from leys in the year of sowing is not necessarily negligible, and this should be borne in mind when deciding to sow with or without a cereal cover crop. In order to make a fair estimate of the loss of grazing occasioned by the use of a cover crop, it is essential to make a comparison with swards sown early in the growing season.

Swards based on different species and varieties show a wide range of yield during establishment (Charles, 1961c). In Northern Ireland, Boyd (1959) recorded yields ranging from 1 193 kg starch equivalent/ha from meadow fescue/white clover, to 2 423 kg/ha where 22·4 kg H.1 short-rotation ryegrass seed/ha was included with timothy and clover. In other seasons H.1 at 33·5 kg/ha yielded 3 767 kg starch equivalent/ha. At another location (Chestnutt, 1959) in Ireland, Irish perennial ryegrass produced 1 569 kg starch equivalent/ha in the year of sowing. These yield figures compare favourably with a mean production of 1 256–1 883 kg starch equivalent/ha produced by established grassland in the area, and some of them approach the highest figures obtained in Northern Ireland.

On hill land, too, considerable differences in the production of some species have been recorded in the year of sowing (National Agricultural Advisory

Service, 1958). At 1100 feet (366 m) in Lancashire, S22 yielded 5 650 kg DM/ha and an S24 sward 4 771 kg/ha; with S51 timothy the yield was only 3 767 kg/ha.

Information from over 40 farms widely scattered in Britain (Clark, 1959) showed that yields were considerably affected by the mixtures used. The average number of cow-grazing days/ha obtained from swards based on Italian ryegrass was 403, but where no ryegrass was sown (swards mainly timothy/meadow fescue) the figure fell to 311. The mean number of cow-grazing days/ha from undersown Italian ryegrass was 57, and the average for all mixtures sown with a companion crop was 47; thus approximately 7 times as much grazing was obtained by sowing without a cereal.

Paterson (1956) pointed out the value of swards in their year of sowing for stimulating milk yield and maintaining it beyond early May, but observed that, in his experience, the number of cow-grazing days obtained from swards in the year of sowing is only about half that from established swards.

Varieties within a species vary in production in the year of sowing (Davies, 1934). In a comparison of 6 selections of cocksfoot during the first 3 months after sowing, Davies (1952) recorded a range from 404 kg DM/ha to 3·5 times this figure, depending on the selection sown. There appears to be considerable scope for selection within cocksfoot for high yield in the establishment phase.

References

BAKER, H. K. 1964. Tall fescue (S170) in the sward. *Agriculture, Lond.* **71,** 509–12.

BLACK, J. N. 1958. Competition between plants of different initial seed sizes in swards of subterranean clover (*Trifolium subterraneum* L.) with particular reference to leaf area and the light microclimate. *Aust. J. agric. Res.* **9,** 299–318.

BOYD, M. 1959. Personal communication.

CHARLES, A. H. 1961a. Differential survival of cultivars of *Lolium, Dactylis* and *Phleum. J. Br. Grassld Soc.* **16,** 69–75.

CHARLES, A. H. 1961b. Establishment studies. 2. The effect of method of establishment on a one year ley in the seeding year. *J. agric. Sci., Camb.* **57,** 159–72.

CHARLES, A. H. 1961c. Herbage production in Great Britain from leys in the year of sowing. *Herb. Abstr.* **31,** 1–6.

CHARLES, A. H. 1962a. Establishment studies. 3. The effect of method of establishment on a one year ley in the first harvest year. *J. agric. Sci., Camb.* **58,** 121–30.

CHARLES, A. H. 1962b. Pasture establishment by surface sowing methods. *Herb. Abstr.* **32,** 1–7.

CHARLES, A. H. 1963a. Establishment studies. 5. The effect of method of establishment on the behaviour of long term leys in the seeding year. *J. agric. Sci., Camb.* **60,** 1–10.

CHARLES, A. H. 1963b. Establishment studies. 6. The effect of method of establishment on long term leys in the first harvest year. *J. agric. Sci., Camb.* **60,** 11–17.

CHARLES, A. H. 1965. Interactions of grass clover and nurse crop in the seeding year. *J. Br. Grassld Soc.* **20,** 241–7.

CHARLES, A. H.; LEWIS, J. 1962. Observations on the use of herbicides on grassland. *Proc. 6th Br. Weed Control Conf.* pp. 75–83.

CHESTNUTT, D. M. B. 1959. Personal communication.

CHIPPINDALE, H. G. 1932. Operation of interspecific competition in causing delayed growth of grasses. *Ann. appl. Biol.* **19,** 221–42.

CHIPPINDALE, H. G. 1948. Operation of interspecific competition in causing delayed growth of grasses. *Nature, Lond.* **161,** 65.

CHIPPINDALE, H. G. 1949. Environment and germination in grass seeds. *J. Br. Grassld Soc.* **4,** 67–71.

CLARK, J. 1959. Personal communication.

COOKE, G. W. 1956. Fertilizer placement. *Outl. Agric.* **1,** 43–5.

COPEMAN, G. J. F.; ROBERTS, H. W. 1960. The development of surface seeding. *J. Br. Grassld Soc.* **15,** 163–8.

DAVIES, I. 1952. Unpublished data.

DAVIES, W. 1927. Seeds mixture problems; soil germination, seedling and plant establishment with particular reference to the effects of environmental and agronomic factors. 2. Field trials. *Bull. H6 Welsh Pl. Breed. Stn* pp. 39–63.

DAVIES, W. 1928a. The factor of competition between one species and another in seeds mixtures. *Bull. H8 Welsh Pl. Breed. Stn* pp. 82–149.

DAVIES, W. 1928b. The influence of seed rate on the establishment of perennial ryegrass, timothy and rough stalked meadow grass. *Welsh J. Agric.* **4,** 250–4.

DAVIES, W. 1934. The management and manuring of pasture plants in relation to soil establishment and productivity. *Welsh J. Agric.* **10,** 142–60.

DAVIES, W. 1939. Temporary leys: comparison of station-bred and commercial grasses used in simple mixtures. *Bull. H15 Welsh Pl. Breed. Stn* pp. 1–24.

DAVIES, W. 1945. The establishment and early management of sown pastures. *Bull. 34 Imp. Bur. Past. Forage Crops* pp. 1–21.

DAVIES, W. 1946. Reseeding and the modern ley. *Pamphlet 14 Bath and West and Southern Counties Society.*

DAVIES, W.; WESTERN, J. H. 1937. The development of certain grasses and clovers during the seeding year. *Welsh J. Agric.* **13,** 152–60.

DAVIS, A. G.; LAMBERT, J. P. 1947. Sowing grass and clover seeds in autumn. *J. Br. Grassld Soc.* **2,** 63–71.

DONALD, C. M. 1951. Competition among pasture plants. 1. Intra-specific competition among annual pasture plants. *Aust. J. agric. Res.* **2,** 355–76.

DONALD, C. M. 1954. Competition among pasture plants. 2. The influence of density on flowering and seed production in annual pasture plants. *Aust. J. agric. Res.* **5,** 585–97.

EISELE, H. F.; AIKMAN, J. M. 1933. Development and survival of species and varieties in planted pastures. *Ecology* **14,** 123–35.

ELLIOTT, J. G. 1958. The use of herbicides in surface re-seeding of old pastures. The world position and some research in the United Kingdom. *Proc. 4th Br. Weed Control Conf.* pp. 234–8.

GARDNER, A. L.; HUNT, I. V.; MITCHELL, I. W. 1954. The 'Muirfad' technique of peatland improvement. *J. Br. Grassld Soc.* **9,** 161–71.

GRASSLAND RESEARCH INSTITUTE. 1959. *Exps Prog. 11 Grassld Res. Inst. 1957–8* p 23.

GRASSLAND RESEARCH INSTITUTE. 1967. *A. Rep. Grassld Res. Inst., Hurley 1966* pp 32–3.

GREENAWAY, T. E. 1957. The book of Dunns farm seeds. pp. 11–12.

HEDDLE, R. G., HERRIOTT, J. B. D. 1954. The establishment, growth and yield of ultra-simple grass seed mixtures in south east Scotland. 1. The effects of varying grass seed rates. *J. Br. Grassld Soc.* **9,** 99–110.

HEDDLE, R. G.; HERRIOTT, J. B. D. 1955. The establishment, growth and yield of ultra-simple grass mixtures in the south east of Scotland. 3. The effects of varying seed rates of grass sown with and without a cereal cover crop. *J. Br. Grassld Soc.* **10,** 317–25.

HOLLIDAY, R. 1953. Agronomic research in grassland husbandry problems. *Agric. Prog.* **28,** 109–23.

HUGHES, R.; NICHOLSON, I. A. 1961. *Rep. 2 Hill Fmg Res. Org. 1958–61* pp. 61–5.

HUNT, I. V. 1954. Seed establishment in the west of Scotland. *J. Br. Grassld Soc.* **9,** 85–98.

JONES, D. J. C. 1957. Early stages in ley establishment. *J. Minist. Agric.* **63,** 509–13.

JONES, L. 1962. Advance notes on research. Herbicides to aid pasture renovation. *J. Br. Grassld Soc.* **17,** 85–6.

JONES, M. G. 1945. The establishment and early management of sown pastures. *Bull. 34 Imp. Bur. Past. Forage Crops* pp. 48–54.

JONES, T. E. 1933. The effect of a nurse crop on the establishment and first year yields of various grasses and clovers when sown in pure plots and in mixtures. *Welsh J. Agric.* **9,** 176–90.

MARTIN, F. M. 1949. The establishment of autumn-sown legumes. *J. Br. Grassld Soc.* **4,** 47–55.

MILTON, W. E. J. 1935. The soil establishment of pedigree and commercial strains of certain grasses. *Welsh J. Agric.* **11,** 171–81.

MOORE, H. I. 1945. The nurse crop in ley establishment. An analysis of twelve years' work. *J. Minist. Agric.* **52,** 245–8.

NATIONAL AGRICULTURAL ADVISORY SERVICE. 1958. Experimental Husbandry Farm, Great House, unpublished data.

NORTH OF SCOTLAND COLLEGE OF AGRICULTURE. 1960. *Leaflet 23 N. Scotl. Coll. Agric.*

ORMROD, J. F. 1960. The use of chemicals in the renovation of difficult swards in the East Midlands of England. *Proc. 5th Br. Weed Control Conf.* pp. 151–6.

PATERSON, R. 1956. The milk stimulating value of grass and its influence on animal health. *J. Br. Grassld Soc.* **11,** 93–9.

ROWLANDS, A. 1960. Improvement studies on *Festuca/Nardus* and *Calluna/Molinia* areas. *Proc. 5th Br. Weed Control Conf.* pp. 165–70.

STAPLEDON, R. G. 1935. The land, now and tomorrow. London: Faber and Faber.

STAPLEDON, R. G. 1939. The establishment and maintenance of temporary leys. *Bull. H15 Welsh Pl. Breed. Stn* pp. 1–24.

STAPLEDON, R. G. 1949. Seeds mixtures. Fallacies and facts. *J. Minist. Agric.* **55,** 415–19.

STAPLEDON, R. G., DAVIES, W.; BEDDOWS, A. R. 1927. Seeds mixture problems. 1. Garden trials. *Bull. H6 Welsh Pl. Breed. Stn* pp. 156–7.

STAPLEDON, R. G.; JONES, R. 1925. Seeds mixtures for temporary grass. Investigations conducted in Denmark and Sweden and observations on trials of a similar nature at Aberystwyth. *Welsh J. Agric.* **1,** 60–98.

STAPLEDON, R. G.; WHEELER, D. E. 1948. An experiment to test the influence of date and depth of sowing on four of the chief grasses and on white clover. *J. Br. Grassld Soc.* **3,** 263–71.

STEWART, A. B. 1952. Manuring for establishment and maintenance of pasture. *Proc. 6th int. Grassld Congr., Pasadena, 1952* pp. 702–4.

STEWART, A. B.; REITH, J. W. S. 1945. Comparisons of broadcast and drill applications of fertilizers. *Scott. J. Agric.* **25,** 167–71.

THOMAS, M. T. 1945. The establishment and early management of sown pastures. *Bull. 34 Imp. Bur. Past. Forage Crops* pp. 22–47.

WEED CONTROL HANDBOOK. 1968. Oxford and Edinburgh: Blackwell Scientific Publications, pp. 325.

WELSH PLANT BREEDING STATION. 1966. *Rep. Welsh Pl. Breed. Stn 1965* p. 62.

WILLIAMS, R. D. 1922a. Depth of sowing grass and clover seeds. *J. Ministr. Agric.* **29,** 53–60.

WILLIAMS, R. D. 1922b. Depth of sowing grass and clover seeds. *J. Minist. Agric.* **29,** 132–7.

WILLIAMS, R. D. 1923. Methods of covering grass and clover seeds. *J. Minist. Agric.* **29,** 1125–34.

WILLIAMS, R. D. 1924. Methods of covering grass seeds. *J. Minist. Agric.* **30,** 1134–9.

WILKS, T. N. 1945. Autumn sowing of herbage seeds. *J. Minist. Agric.* **52,** 298–300.

CHAPTER 40

Comparative Yields of Herbage Species

The grasses

THE choice of which herbage species a farmer should grow will depend largely on their expected yield and the manner in which this yield is distributed through the season; however, the performance of species or cultivar will vary with soil fertility, climate, the type of management to be imposed and other factors, all of which make it difficult to predict performance in advance.

ANNUAL YIELD UNDER HIGH-FERTILITY CONDITIONS

The highest yields of grass, in terms of dry matter (DM), are obtained on soils with abundant supplies of N and water and where only two or three cuts are taken per year. Figures available from Dent and Aldrich (1966) and reproduced in Table 40.1 show how the yields of commonly used grass species differed from each other when swards were cut either 4 or 9 times a year.

Table 40.1. Mean yields for 11 cultivars grown at 4 centres under 2 managements for the 4 years 1963–1966 (kg DM/ha). (Results are for 1st harvest years only)

		9 cuts/year	4 cuts/year
Perennial ryegrass	S24	11 009	15 303
	S23	10 696	14 901
Italian ryegrass	S22	10 696	17 198
	Prima Roskilde	10 972	17 788
Meadow fescue	S215	10 382	14 876
	S53	9 691	12 767
Timothy	S352	9 503	14 775
	S48	9 603	13 407
Cocksfoot	S37	10 319	15 252
	S143	9 754	13 043
Tall fescue	S170	10 670	15 152

From Dent and Aldrich (1966)

The results refer only to the first harvest year in each trial and are averages for 4 centres, measured in each of 4 years. Average yields of all species and cultivars were 45% higher when plots were cut 4 times a year rather than 9 times.

Comparing species only, Italian ryegrass was the most productive when all were cut 4 times only, yielding 16% more DM than perennial ryegrass or tall fescue, the two highest yielding of the perennial grasses. There was, however, little difference between perennial ryegrass (relative value 100), tall fescue (100), cocksfoot (94), timothy (93) and meadow fescue (91).

When the grasses were cut 9 times per year, perennial ryegrass yielded as much as Italian ryegrass (100); tall fescue yielded 98% as much as perennial ryegrass, cocksfoot and meadow fescue 93% and timothy 88%. The maximum difference between species under frequent cutting was therefore 12%.

These data show that yields of the commonly used grasses do not differ greatly in the first year, as was also shown to be the case for yields over a three-year period in an experiment at the Grassland Research Institute, Hurley, in which Cowling and Lockyer (1965) cut swards of various species 4 or 5 times a year and applied a total of about 360 kg N/ha to them; yields were higher from cocksfoot (100) than from timothy (92), meadow fescue (88), or the average of three varieties of perennial ryegrass (86). The maximum yield difference between species was therefore only 14%. Results published by Anslow and Green (1967) also point to a marked similarity between the annual yields of 4 species of grass grown in fertile soil and cut approximately once a month.

From results so far quoted in this chapter, it can be concluded that differences in DM yield between species are rarely great enough for this to be the sole criterion in choosing which grass to grow. The greatest difference between species in the experiments discussed was at Cambridge (Dent & Aldrich, 1966), where two cuts of cocksfoot yielded 34% less than two cuts of timothy. If this is as large a difference as might be expected, and one moreover which would be reduced under more frequent cutting, it is clear that considerations of ease of establishment, longevity, palatability and digestibility can easily outweigh those of yield alone.

ANNUAL YIELD AT LOWER LEVELS OF FERTILITY, AND UNDER GRAZING

The results and conclusions mentioned so far have been derived from situations of ample fertilizer usage. If grass crops are to be grown under conditions of periodic deficiency of nutrients, especially N, and of soil moisture, this may alter the decision as to which species to grow. However, Cowling and Lockyer (1965) found that the order of yields of the species they tested changed very little as rates of N application increased from 90 to

360 kg/ha, and comparison of the responses of grasses to irrigation at various levels of N fertilization has also emphasized the similarities rather than the differences between species (Williams, Stiles & Turner, 1959). In a later consideration of this work, it was stated that 'the production of the various varieties was very similar, both with and without irrigation and at both nitrogen levels' (Stiles, 1966).

Another aspect which must be considered when the yields of species are being compared is the reaction of each to the presence of the grazing animal. This reaction is complex, but one form of interference with yield caused by the animal, that of treading, has been examined by Edmond in New Zealand (1964), who found that the yield of grass from 9 harvests in about 10 months was reduced if plots were trodden by sheep after cutting. After being trodden by the equivalent of 32 sheep per acre (80/ha) after each harvest, the yield of perennial ryegrass was reduced to 77% of that where it was not trodden, that of short-rotation ryegrass (*Lolium perenne* × *Lolium multiflorum*) to 44%, timothy to 38% and cocksfoot to 20%. The same trends were also apparent, though less marked, at 16 sheep-equivalents/ac (40/ha). This shows that the yields of species in cutting experiments may be misleading as an index of the yields of grazed swards. Certainly, the long-term production of a species and its contribution to the yield of a pasture may be closely linked to the effects of treading; such differences between species as those which have been reported by Edmond are sufficiently large to determine which grass should be grown.

SEASONAL DISTRIBUTION OF YIELD

The annual yields of two grass species might not differ greatly, but one might be preferred to the other because of a more desirable distribution of yield during the season. If the quantity of herbage required by livestock at a critical time of the season can be supplied only by a certain species of herbage plant, it might not matter that its total annual yield is lower than that of others. Hunt (1957) examined this aspect of the growth of cultivars of perennial ryegrass, timothy, cocksfoot, meadow fescue, tall fescue and rough-stalked meadow grass when cut 5 times in each of two years. In all cultivars, one-half to three-quarters of the year's production had occurred by 8 June, only 6 weeks after growth began; this was due largely to the fact that growth rates in early-flowering varieties of perennial ryegrass and in tall fescue were highest in early May. The other grasses attained their fastest growth rate before the end of May and the writer thought that little could be done by choice of species or variety to delay this peak beyond early June.

Results which also emphasize this early date of maximum growth rate were published by Anslow and Green (1967). They found that perennial ryegrass, cocksfoot, timothy and tall fescue, cut every three or four weeks, grew most rapidly in early May in each of three consecutive years, and that at this date, perennial ryegrass and timothy showed a higher rate of DM increase (94 and

95 kg/ha per day, respectively) than either cocksfoot (80 kg) or tall fescue (84 kg).

The most marked differences between species in rates of production occur in mid-summer, even under irrigated conditions. Hunt (1957) described this interval as a summer gap, in which he found production from timothy to be negligible, and that from perennial ryegrass little better. Anslow (1965) showed that rates of production in mid-summer were only about half those recorded in early May, despite irrigation and regular applications of N; this decline was most apparent in timothy and much less evident in perennial ryegrass and cocksfoot.

After this mid-summer decline in production, there is a second, smaller peak in late summer from early July to mid-August (Hunt, 1957), in which differences between species are probably of no practical significance. In the graphs published by Anslow and Green (1967), this secondary peak occurred in late July or early August, but whether it is to be regarded as a peak or not depends on the extent of the yield decline in mid-summer. Rates of production in July/August did not exceed about one-half to three-quarters of those at the spring peak.

The decline in rates of production in autumn was shown by Anslow and Green (1967) to occur uniformly from the July/August peak to zero in late November, with no obvious differences between species and cultivars. Hunt (1957) stated that production from mid-September onwards was negligible, in which case the rate of decline would have been more rapid than that described by Anslow and Green (1967). The difference of about 5° in latitude between the sites where these experiments were carried out might explain this.

HERBAGE YIELD AND QUALITY

From the evidence brought together in the preceding paragraphs, there appears to be little to choose between bred varieties of one species and another in terms of total annual yield or, indeed, in the way in which this yield is distributed throughout the year. However, although differences in the seasonal pattern of production shown by the commonly used species and cultivars are slight, such variations as do exist can be exploited to the benefit of a particular system of animal husbandry. Thus Anslow and Green (1967) showed that when cut approximately once a month, the use of tall fescue (S170) for three years ensured an average annual period of 209 days in which production exceeded 10 kg DM/ha per day, which can be compared with a duration of only 187 days for cocksfoot (S37), which had similar values for digestibility. If the length of the growing season is as important as annual yield for a particular management system, then such a result could well be important in the selection of species.

The superiority in yield of any of the commonly sown species over any other is generally so small that it could be outweighed by relatively minor differences in palatability or digestibility. Differences between species in yield are likely to be greatest when swards are cut infrequently, especially around the time of inflorescence emergence; hence the marked changes in digestibility which occur at this stage of growth must also be borne in mind. For example, data published by the National Institute of Agricultural Botany, Cambridge (1967) demonstrate that DM yield cannot be allowed to accumulate in cocksfoot and timothy to the same extent as in ryegrass and meadow fescue if herbage of at least 70% DM digestibility is to be harvested.

YIELD OF CULTIVATED AND WEED GRASSES

Although the yields of cultivated varieties of pasture grasses differ little, it is necessary to consider whether they may differ from those of the grasses which intrude as weeds into sown swards and are common components of permanent pastures. Cowling and Lockyer (1965) found that over three years, common bent (*Agrostis tenuis*) swards yielded only 56% as much as the highest yielding species with no applied N and 51%, 65% and 75% as much as the highest yielding species at successively higher rates of applied N. In experiments conducted by Hunt (1957), the highest yield was from tall fescue, and averaging the two seasons together, rough-stalked meadow grass yielded 69% as much as tall fescue. The graphs published by Anslow and Green (1967) show that the herbage production rate throughout the season from swards of rough-stalked meadow grass was lower than from those of perennial ryegrass.

From this evidence we may conclude that although there may be only small differences in total annual yield between bred varieties of the species commonly used in sown pastures, taken as a group they give a greater harvested yield than the commonly occurring 'weed grasses'.

The legumes

The choice among the 4 legumes described in Chapters 33 to 36 is likely to be influenced less by their gross annual yield than by (a) the qualities required in the forage, (b) adaptation to soil and climate, (c) the method of utilization envisaged, (d) the duration for which the crop is required, and (e) the seasonal distribution of growth required.

White clover differs from the erect species in being adapted to a wider range of soil conditions than the other two long-term perennial legumes, lucerne and common sainfoin, especially in respect of pH value. Under favourable moisture conditions it gives a greater proportion of its production after mid-summer than do the other perennial legumes, but in regions of lower rainfall its summer production is less reliable. The major difference, however, is that white clover yields relatively well when frequently defoliated, especially in mixtures with grasses, whereas yields of red clover, lucerne and sainfoin are much reduced, and the crop may not survive if more than three harvests are taken each year. White clover will therefore be the choice for a perennial crop where the intention is to cut or graze frequently.

The erect species are at their most productive under a system in which the spring growth is cut in late May or early June. This bulky crop is normally conserved. At this stage, their potential DM yield may be lower than that of grasses well supplied with N, and their digestibility no higher than that of the grasses; but they have the advantage of a comparatively high crude protein content and, in the drier regions, their regrowth is often superior to that of either grasses or white clover.

Where yield from white clover is unreliable during July and August, one of the erect legumes may be grown primarily to supply fresh forage in that period. In this connection, it should be noted that lucerne, broad red clover and giant sainfoin are superior in rate of regrowth to late-flowering red clover and common sainfoin. When these erect legumes are used as pasture, the frequency and duration of grazing has to be more strictly controlled than would be necessary for white clover.

The choice among red clover, lucerne and sainfoin again depends largely upon factors other than yield; of these, soil type and duration of crop are important. Lucerne and common sainfoin will sustain their productivity for several years, late-flowering red clover and giant sainfoin for approximately two years, and broad red clover normally for only one full harvest year. Among the three species, lucerne would be preferred where a long crop duration combined with rapid regrowth in mid-summer was required. However, the cultivation of lucerne is more restricted by soil and climate than is that of red clover. In environments unfavourable for lucerne, it would either be necessary to forgo the advantages of a long-term crop and grow broad red clover, or to forgo the advantage of very rapid regrowth and grow a more persistent, late-flowering type of red clover. (This situation may change if the early promise of persistence in new tetraploid cultivars of broad red clover is borne out by experience in the field.) Sainfoin is even more restricted than lucerne by soil type, being rarely successful on non-calcareous soils. Where sainfoin can be grown successfully, however, it may be preferred to lucerne by virtue of its higher feeding value, even though its production is less well distributed over the growing season.

Although it has become customary to regard red clover, lucerne and sainfoin as crops primarily for conservation, and white clover as a crop primarily for grazing, this distinction should not be made too rigid. As a crop, white clover presents greater physical difficulties than the more stemmy legumes in being wilted for silage or cured for hay; also, its very high protein content would necessitate large quantities of additives for successful ensilage by conventional methods. But as the basis of a simple legume/grass mixture it is equally suitable for grazing and for conservation. So long as the mixture is cut at a reasonably early stage for conservation, white clover will continue to make a useful contribution to both yield and quality.

It is in respect of its quality, and especially of its digestibility, that white clover deserves further consideration, although its potential yield may be less than that of the other legumes even under favourable moisture conditions. Alone among the four legumes named, white clover enhances the digestibility of a legume/grass mixture, as well as its protein content. The potential value of white clover is perhaps best illustrated by reference to Table 33.5 and by comparing results from lucerne cut at 6-week intervals (Table 35.3), beginning 27 May, with irrigated white clover cut at monthly intervals. Under favourable environmental conditions and an appropriate frequency of harvesting, the yield of DM from lucerne exceeds that from white clover by 50%, but the difference in yield of digestible organic matter is only 17%, since the D-value of lucerne lies between 56 and 58% and that of white clover is between 72 and 75%. The difference in concentration of digestible organic matter and of protein may well offset the small advantage in yield of the lucerne.

References

ANSLOW, R. C. 1965. Grass growth in midsummer. *J. Br. Grassld Soc.* **20,** 19–26.

ANSLOW, R. C.; GREEN, J. O. 1967. The seasonal growth of pasture grasses. *J. agric. Sci., Camb.* **68,** 109–22.

COWLING, D. W.; LOCKYER, D. R. 1965. A comparison of the reaction of different grass species to fertilizer nitrogen and to growth in association with white clover. *J. Br. Grassld Soc.* **20,** 197–204.

DENT, J. W.; ALDRICH, D. T. A. 1966. National Institute of Agricultural Botany, unpublished data.

EDMOND, D. B. 1964. Some effects of sheep treading on the growth of ten pasture species. *N.Z. Jl agric. Res.* **7,** 1–16.

HUNT, I. V. 1957. Comparative productivity of 27 strains of grasses in the West of Scotland. Part 4. Seasonal growth. *J. Br. Grassld Soc.* **12,** 127–37.

NATIONAL INSTITUTE OF AGRICULTURAL BOTANY. 1967. Varieties of grasses. *Farmers' Leaflet No. 16* pp. 16.

STILES W. 1966. Ten years of irrigation experiments. *A. Rep. Grassld Res. Inst. Hurley, 1965* pp. 57–66.

WILLIAMS, T. E.; STILES W.; TURNER, P. J. 1959. Grassland irrigation. *Exps Prog. 11 Grassld Res. Inst. 1957–8* pp. 56–8.

CHAPTER 41

Comparative Nutritive Value of Herbage Species

THE nutritive value of different forage species can be compared either in terms of their potential nutritive value as defined in Chapter 18, or in terms of the level of animal production they are able to sustain, as described in the subsequent chapters. Whichever method of assessment is used, comparisons of nutritive value must of necessity be combined with yield data if comparisons are to be made in economic terms. In many situations the nutritive value of the forage when included in compound rations must also be established before a complete economic assessment can be undertaken (Blaxter & Wilson, 1963).

In the present chapter, comparative assessments of forage nutritive value are made in terms of the digestibility and voluntary intake of the forage and of the efficiency with which the digested nutrients consumed are utilized.

Comparative digestibility of forage species

The physical and chemical changes that occur in forage crops as they mature are accompanied by changes in digestibility which far exceed the variations in digestibility resulting from differences in growing conditions or species.

Though irrigation or the application of fertilizers may influence the yield and chemical composition of forage crops, they have little discernible effect upon dry matter (DM) digestibility. The effects of seasonal and diurnal variations in light and temperature on digestibility have so far been little studied, but it seems likely that high temperatures will accelerate maturation and senescence and thus hasten the decline in forage digestibility.

The processes of dehydrating and ensiling forage crops have only minor effects on digestibility provided they do not also cause substantial losses of DM (Harris & Raymond, 1963); however, if the losses during conservation are large, these losses generally fall on the more digestible labile constituents and so the resulting product is of lower digestibility than the original forage crop.

Changes in digestibility with advancing maturity

The changes in digestibility which occur during primary growth have been well established for grass species (Minson *et al.*, 1964). The general pattern is typified by S24 perennial ryegrass, which passes through an initial period of slow yield increase and high digestibility, with D-values (digestible organic matter in the DM) in the range 72–77% in early spring and which change little over periods of some 14 days. D-values subsequently decline by about 0·4–0·5 percentage units per day during stem elongation and earing. Typically, this decline continues from May into June, by which time digestibility may be as low as 55% D-value.

This general pattern of change in digestibility is common to all geographic locations in the British Isles, though the actual timing of the various phases varies with location and with the date when growth begins in any particular year. By its influence on the proportion of dead material (of low digestibility) in the crop, the management of a sward during autumn and winter and the prevailing weather conditions during early spring can affect the course of these changes, particularly while growth rates and yields are low (Welsh Plant Breeding Station, 1966).

Less information is available regarding the pattern of change of digestibility with age during mid- and late-season production, but the data available suggest that D-values rarely exceed 73% in practice, and that the rate of decline of digestibility with time is less than during primary growth.

Comparative patterns of digestibility in forage species during primary growth

Although all herbage species show a decline in digestibility with advancing age, not all follow the same pattern of change during primary growth. Species can most usefully be compared in terms of the range of digestibility observed, its rate of decline with time, the relative dates at which different forages reach given levels of digestibility, and their comparative DM yields at given levels of digestibility.

Observed range of forage digestibility

The range in D-value from 55 to 75% found in perennial ryegrasses is common to all the major species and varieties of sown grasses in Britain with the exception of cocksfoot, which is characterized by its relatively low digestibility in the early spring and which does not reach a D-value higher than about 71%. The erect legumes lucerne, red clover and sainfoin show a fairly wide range of digestibility, but are not normally harvested at D-values greater than 65–67%. White clover, of which only the leaflets and petioles are harvested, shows little variation in digestibility with age, rarely falling below 75% D-value in grazed swards.

Rate of decline in digestibility in forage species

Though most grasses and legumes show only a slight decline in digestibility during early spring, timothy is a major exception in that it shows no distinct period of high digestibility, but rather a continuous steady decline in digestibility from the beginning of growth onwards. As mentioned previously, the repeatability of this phase of high digestibility is variable and depends largely on the proportion of dead material present in the sward above cutting height. The rate of decline in D-value once the initial phase of high digestibility is past is fairly constant for most grass and legume species, being approximately 0·4–0·5 percentage units per day. However, white clover is an exception in showing a decline of only 0·15 percentage units per day; also, the rates of decline in timothy and Italian ryegrass are less than 0·4 percentage units per day.

The coincidence of ear emergence with the onset of rapid decline in digestibility observed in S24 perennial ryegrass and S37 cocksfoot is not common to all varieties within these species. Other species, notably timothy and S22 Italian ryegrass, may show a considerable decline in digestibility even before the flowering heads emerge.

The major significant differences between forage species are thus the relative dates at which they reach a given level of digestibility, and their relative yields at that digestibility.

Comparative yields and dates at which forages reach similar levels of digestibility

The dates on which a number of grasses and legumes reach D-values of 68, 63 and 58% during primary growth are indicated in Table 41.1, in which species are arranged in order of 'earliness' at the highest D-value. The mean yields of DM achieved with an application of 90 kg N/ha are also tabulated. While the dates and yields quoted are those found in experiments at the Grassland Research Institute, Hurley, they are likely to be broadly applicable to other sites.

It can be seen that forage of about 68% D-value can be harvested throughout the period 5 May to 3 June, and that the three cultivars of perennial ryegrass S24, S23 and Vertas Hardgrazing (now Multigrazer), span three of the four weeks during which this is possible. Herbage with D-values of 63% and 58% can be harvested between 20 May and 19 June, and 1 and 26 June, respectively. In each case, the period of availability of forage of a given digestibility is about 4 weeks. The order in time at which the different forages reach successive levels of digestibility remains roughly constant, with the major exception of S48 timothy and S22 Italian ryegrass, which reach a given D-value at progressively later dates than the other forages because of their characteristically slower rates of decline in digestibility.

Table 41.1 shows that the DM yields of different grasses at 68% D-value

Table 41.1. The dates on which selected forages achieve D-values of 68, 63 and 58% and their yields of DM (t/ha) at these digestibilities during primary growth

		DM digestibility					
		68%		63%		58%	
		Date	Yield	Date	Yield	Date	Yield
Festuca arundinacea	S170	5 May	3·11	21 May	4·82	7 June	6·58
Dactylis glomerata	S37	6 May	2·96	20 May	4·98	1 June	6·48
Lolium multiflorum	S22	10 May	4·49	30 May	6·80	25 June	8·54
L. perenne	S24	12 May	4·72	28 May	6·78	13 June	8·21
F. pratensis	S215	12 May	3·79	27 May	5·70	11 June	7·18
Phleum pratense	S48	16 May	3·54	7 June	5·65	N.R.	
P. pratense	S352	16 May	4·31	2 June	6·46	19 June	7·92
L. perenne	S23	24 May	5·20	7 June	7·08	18 June	8·31
L. perenne	Vertas Hardgrazing	3 June	6·50	19 June	8·44	N.R.	
Medicago sativa	Flamande	4 May	2·36	18 May	3·46	1 June	4·52
Onobrychis viciifolia	Cotswold Common	N.R.		27 May	4·57	8 June	5·47

N.R. = Not reported.
From Green, J. O., Grassland Research Institute. Unpublished data.
Mean values for years 1965–66, with 90 kg N/ha applied to the grasses only.

range from 2·96 to 6·50 t/ha for S37 cocksfoot and Vertas Hardgrazing (Multi-grazer) perennial ryegrass, respectively. The difference in yield is much smaller at lower D-values. Perennial ryegrass, Italian ryegrass and S352 timothy are the highest yielding grass species at all levels of digestibility. It is also noticeable that the later-maturing grasses tend to give the highest yields, though a comparison of their yields over the whole season is likely to show less marked differences. The legumes described in Table 41.1 produce markedly lower yields than the grasses and reach given levels of digestibility earlier in the season.

The differences in the range of D-values exhibited by different grass species are of little practical significance. The slow rate of decline of digestibility observed in S22 Italian ryegrass, S48 timothy and white clover are of some advantage if adverse weather conditions delay conservation. In practice, the major advantage likely to be gained from differences in seasonal patterns of digestibility is the spread of harvest that can be achieved by growing two or more forages in separate areas. Thus, by using the three cultivars of perennial ryegrass already named, it is possible to harvest material with D-values within the range 66–70% over the 5-week period from 4 May to 9 June. If only one cultivar were grown, this period would be restricted to two weeks only.

Digestibility and the grazed sward

In vitro determinations of the digestibility of different portions of forage plants have indicated the ability of animals to select a diet of higher digestibility than the average value for the herbage on offer. Terry and Tilley (1964) examined the primary growth of S24 perennial ryegrass and S37 cocksfoot and demonstrated that the DM of the leaves, leaf sheaths and stems was highly digestible at the start of the season, but thereafter declined at rates of 0·2, 0·4 and 0·8 percentage units per day (Table 41.2). The rates of decline in digestibility of the comparable morphological parts of S48 timothy and S170 tall fescue showed a similar pattern.

Table 41.2. Percentage *in vitro* DM digestibility of whole plants of immature and mature S37 cocksfoot and of some separated plant parts

	Whole plant	Leaf lamina	Leaf sheath	Stem	Inflorescence	Dead material
Immature plant	76	79	78	86	...	65
Mature plant	64	70	63	58	64	62

From Terry and Tilley (1964)

Erect legumes (Table 41.3) such as lucerne exhibit uniformly high digestibility of leaflet and petiole throughout growth, but low and rapidly declining digestibility of the stem with advancing maturity. At all stages of growth the apical 6 in. (15 cm) of the stem is much more digestible than the basal 6 in.

Table 41.3. Percentage *in vitro* DM digestibility of whole plants of lucerne and sainfoin and of some separated plant parts

	Whole plant	Leaflet and petiole	Whole stem	Stem apical 6 in.	Stem basal 6 in.
Immature lucerne	75	81	69	79	55
Mature lucerne	66	78	55	77	40
Sainfoin	70	70	72	87	58

From Terry and Tilley (1964)

The apical 6 in. of sainfoin, comprising 15% of the total DM, appears to be highly digestible.

Thus, by selecting the laminae of grasses and the leaflets, petioles and apical parts of the stems of legumes, and by avoiding inflorescences and dead material in the sward, animals are able to obtain herbage of much higher digestibility than the average for the material on offer.

Tayler and Deriaz (1963) used an *in vitro* procedure to compare the organic matter digestibility of the herbage on offer to grazing cattle with the digestibility of the diet actually consumed by the animals. The data presented in Table 41.4 demonstrate that under systems of strip grazing or set stocking, the cattle were able to select a diet 6–18 units more digestible than the average of the material on offer.

Table 41.4. Percentage *in vitro* organic matter digestibility on offer and ingested by cattle under systems of strip grazing and set stocking of a grass sward

	Strip grazed		Set stocked	
	Herbage on offer	Herbage ingested	Herbage on offer	Herbage ingested
Apr.	77	83	62	80
May	76	79	67	76
Aug.	62	70	60	74
Sept.	60	70	60	74

From Tayler and Deriaz (1963)

Comparative assessment of the voluntary intake of forage crops

The dominant influence of voluntary intake on the efficiency of food conversion by ruminants noted in Chapter 5 has been frequently emphasized (Klieber, 1936; Blaxter, 1967). The positive relationship between voluntary intake and digestibility has also been discussed in Chapter 10. Differences in voluntary intake intrinsic to particular forage species tend to be obscured by the great variation in intake resulting from changes in digestibility at different stages of development and growth of a given forage. Comparisons of the voluntary intake of different herbage species, unconfounded by level of digestibility or availability, have mainly been carried out with forages cold-stored or conserved as hay or dried grass, using mature sheep. The relative values so established have not been adequately confirmed with more productive classes of stock, although there is no evidence that relative intake levels would differ greatly.

Despite the paucity of information as yet available, some intrinsic differences between forage species in the voluntary intake of DM at given levels of digestibility have been established. Thus, in general, the intake of legumes appears to be 20–40 % higher than that of grasses of the same digestibility. This has been demonstrated for lucerne by Reid and Jung (1966) and for white clover by Joyce and Newth (1967), and similar results have been found with red clover and sainfoin (Grassland Research Institute, 1968) and birdsfoot trefoil (van Soest, 1965). There is some evidence that within the legumes, the intakes of sainfoin and birdsfoot trefoil are higher than those of lucerne of the same digestibility (Osbourn, Thomson & Terry, 1966; van Soest, 1965). Within grass species, Minson *et al.* (1964) found that the intake of S48 timothy was significantly lower than that of a number of other grass species, notably Italian ryegrass, cocksfoot and meadow fescue.

Rae *et al.* (1963) suggested that the better performance of sheep grazing Italian ryegrass swards than of those grazing swards of perennial ryegrass varieties of the same digestibility, could be the result of higher intake levels with the Italian ryegrass. Some confirmatory evidence has been obtained using sheep fed cut grass indoors (Grassland Research Institute, 1969). Demarquilly *et al.* (1966) have suggested that, at the same digestibility, the intake of meadow fescue is greater than that of perennial ryegrass, and that with all grasses the intake of herbage from the second and third harvests is greater than from the first and fourth harvests.

As with comparisons of the digestibility of forage species, scope for the immediate practical exploitation of differences in intake between forage species is limited, with the possible exception of the large differences found between grasses and legumes. This is because the effects of pasture management on the digestibility and availability of herbage to the grazing animal, and the interaction with the variable requirements of different classes and ages of livestock, are generally more important than such intrinsic differences between forage species. These aspects of the voluntary intake of forage crops will be discussed in subsequent chapters devoted to different methods of herbage utilization and to different forms of animal production.

Comparative utilization of the digested energy of forage crops

There has been little direct study of differences in the efficiency of utilization of the digested energy of different forage species. It has been suggested by Johns *et al.* (1963) and Bailey (1964) that the higher rates of gain by sheep grazing swards of Italian ryegrass compared with those grazing perennial ryegrass, and particularly the higher rates achieved on swards containing white clover, could be attributed in part to a more efficient utilization of the digested energy for productive purposes. Joyce and Newth (1967) have confirmed that not only is the intake of white clover greater than that of ryegrass of the same digestibility, but also that the efficiency of utilization of the digested nutrients is greater. Using a similar comparative slaughter technique, Thomson (1968) found no difference between Italian ryegrass and timothy in the efficiency with which the digested energy was used. Milford and Minson (1965) reported that the efficiency of utilization of digested energy by lambs grazing perennial ryegrass was higher than with S37 cocksfoot. This difference was attributed mainly to the higher digestibility of the ryegrass, and to a greater efficiency of use of digested energy associated with this higher digestibility.

It is clear from the preceding comparisons that the most marked difference in nutritive value between forages is the difference between grasses and legumes. Differences between species within these two groups are apparent, but to date have been exploited in practice only to a limited extent, as reflected by the reduced usage of cocksfoot seed in the last decade. The differences that have been established under conditions of indoor feeding in the components of nutritive value attributable to species and variety differences are often unidentifiable under grazing conditions.

The intensity of stocking, the yield of herbage available to the animals and the cumulative effects of selection by the animals are dominant factors in deciding the quantity and digestibility of the forage eaten. Differences between species and varieties in the amount animals will consume are nevertheless not inconsequential. If the causes of these differences could be defined in terms of the physical and chemical composition of the crop as determined in the laboratory, it seems probable that desirable differences could be increased by selection and combined with other desirable characteristics, such as high yield. Forages having an optimal combination of nutritive value and yield could then be utilized in systems where their superiority could be exploited to increase the yield of animal products from the limited land area in Britain available for this purpose.

References

BAILEY, R. W. 1964. Pasture quality and ruminant nutrition. 1. Carbohydrate composition of ryegrass varieties sown in sheep pasture. *N.Z. Jl agric. Res.* **7**, 496–507.

BLAXTER, K. L. 1967. The efficiency of energy transformations in ruminants. *European Ass. Anim. Prod. 4th Symp. Energy Metabolism Warsaw.*

BLAXTER, K. L.; WILSON, R. S. 1963. The assessment of a crop husbandry technique in terms of animal production. *Anim. Prod.* **5**, 27.

DEMARQUILLY, C.; BOISSAU, J. M.; CUYLLE, G. 1966. Factors affecting the voluntary intake of green forage by sheep. *Proc. 9th int. Grassld Congr., São Paulo, 1965* pp. 877–85.

GRASSLAND RESEARCH INSTITUTE. 1968. *A. Rep. Grassld Res Inst. Hurley, 1967* p. 42.

GRASSLAND RESEARCH INSTITUTE. 1969. *A. Rep. Grassld Res. Inst. Hurley, 1968.*

HARRIS, C. E.; RAYMOND, W. F. 1963. The effect of ensiling upon crop digestibility. *J. Br. Grassld Soc.* **18**, 204–13.

JOHNS, A. T.; ULYATT, M. J.; GLENDAY, A. C. 1963. Pasture type in relation to liveweight gain, carcass composition, iodine nutrition and some rumen characteristics of sheep. 4. Some rumen characteristics of sheep. *J. agric. Sci., Camb.* **61,** 201–6.

JOYCE, J. P.; NEWTH, R. P. 1967. Use of the comparative slaughter technique to estimate the nutritive value of pasture for hoggetts. *Proc. N.Z. Soc. Anim. Prod.* **27,** 166–80.

KLIEBER, N. 1936. Problems involved in breeding for efficiency of food utilization. *Amer. Soc. Anim. Prod.* **29,** 247.

MILFORD, R.; MINSON, D. J. 1965. The energy value of ryegrass and cocksfoot assessed by a slaughter technique with lambs. *Br. J. Nutr.* **19,** 373–82.

MINSON, D. J.; HARRIS, C. E.; RAYMOND, W. F.; MILFORD, R. 1964. The digestibility and voluntary intake of S22 and H.1 ryegrass, S170 tall fescue, S48 timothy, S215 meadow fescue and Germinal cocksfoot. *J. Br. Grassld Soc.* **19,** 298–305.

OSBOURN, D. F.; THOMSON, D. J.; TERRY, R. A. 1966. The relationship between voluntary intake and digestibility of forage crops, using sheep. *Proc. 10th int. Grassld Congr., Helsinki, 1966* pp. 363–6.

RAE, A. L.; BROUGHAM, R. W.; GLENDAY, A. C.; BUTLER, G. W. 1963. Pasture type in relation to liveweight gain, carcass composition, iodine nutrition and rumen characteristics of sheep. 1. Liveweight growth of sheep. *J. agric. Sci., Camb.* **61,** 187–90.

REID, R. L.; JUNG, G. A. 1966. Factors affecting the intake and palatability of forages for sheep. *Proc. 9th int. Grassld Congr., São Paulo, 1965* pp. 863–9.

SOEST, P. J. VAN. 1965. Voluntary intake in relation to chemical composition and digestibility. *J. Anim. Sci.* **24,** 834–43.

TAYLER, J. C.; DERIAZ, R. E. 1963. The use of rumen-fistulated steers in the direct determination of nutritive value of ingested herbage in grazing experiments. *J. Br. Grassld Soc.* **18,** 29–38.

TERRY, R. A.; TILLEY, J. M. A. 1964. The digestibility of leaves and stems of perennial ryegrass, cocksfoot, timothy, tall fescue, lucerne and sainfoin as measured by an *in vitro* procedure. *J. Br. Grassld Soc.* **19,** 363–72.

THOMSON, D. J. 1968. The utilisation of herbage diets by growing sheep. Ph.D. Thesis, Univ. Reading.

WELSH PLANT BREEDING STATION. 1966. *Rep. Welsh Pl. Breed. Stn 1965* p. 70.

CHAPTER 42

Comparative Suitability of Species for Cutting and Conservation

FOR many centuries it has been a common practice for grassland herbage to be cut and either fed direct to ruminants, or conserved for feeding at a later date, in an attempt to increase the efficiency of utilization over that obtained by grazing alone. In particular, conservation offers a means of utilizing surplus summer growth for winter feeding. In the present chapter the characteristics of importance in species from the viewpoint of their suitability for cutting and conservation are discussed. Distinctions are made between (i) morphological and physiological characteristics of significance for cutting and for regrowth after cutting, (ii) those characteristics of cut crops which affect their suitability for conservation by dehydration or ensilage, and (iii) the growth cycle of herbage species in relation to their use in systems involving cutting and conservation.

Cutting

For a herbage crop to be efficiently utilized by cutting, a large proportion of the aerial parts of the crop needs to be above the cutting height. The proportion of the crop which can be harvested generally increases with length of the growth period (Anslow, 1967), but differences between species and varieties in the percentage of their total yield above a height of 4 cm have also been reported by Anslow (1965). For example, in his trials this figure was always higher for S24 perennial ryegrass than for S23 perennial ryegrass throughout the growing season. The percentage of the crop which can be harvested by cutting is generally high for erect crops such as lucerne and maize.

The minimum cutting height is influenced by mechanical and topographical factors as well as by the need for rapid pasture regrowth. On relatively flat land with currently used mowing machines, the minimum cutting height is normally about 5 cm, but this will be considerably higher on uneven land.

Where the proportion of a crop which is harvested is high the rate of regrowth may be depressed. It has been shown that the regrowth of grass may be faster with lenient than with severe defoliation (Brougham, 1956), but if cutting is the only harvesting method employed, the basal part of the sward will never be utilized. Severe defoliation has been shown to maximize the harvested yield over a whole season in both grasses (Chapter 15) and lucerne (Chapter 35). The frequency of cutting also affects seasonal yield. For example, red clover, lucerne and sainfoin (Chapters 34–36) perform better with the less frequent defoliation associated with cutting than with grazing. Infrequent cutting, however, gives little advantage in yield for white clover (Chapter 33) and cocksfoot (Chapter 20).

The cost of cutting a unit dry weight of forage generally falls as the yield of a crop increases, annual forage crops which give a single cut of high quality being particularly suitable from this viewpoint. There are also differences between perennial species. The yield of white clover in primary growths (Chapter 33) is, for instance, considerably lower than that of perennial ryegrass (Chapter 15).

These factors are of importance for crops cut either for conservation or for immediate use in zero-grazing systems. The nutritive specifications for zero-grazing crops may be more exacting and, because material has to be cut and fed daily, it is important that the variation in nutritive quality from day to day should be small. This is, of course, less important in crops grown for conservation.

Conservation

The conservation of crops offers a means of separating harvest from utilization in time. The significance of conserved crops as buffers in animal production systems has been discussed by Duckham (1967). To be successful, conservation systems must preserve plant material without causing unacceptable losses of dry matter (DM) or feeding value. If the water content of herbage can be reduced to 20–25% by dehydration, the activity of microorganisms and plant enzymes is greatly inhibited. A similar degree of chemical stability can be achieved by the anaerobic fermentation of ensiled material. The factors involved differ between these two principal methods of conservation, and they are therefore discussed separately.

DEHYDRATION

Hay can normally be stored safely at an initial water content of 20–25% (Shepperson, 1959, 1965). Further drying usually occurs during storage and storage moulds are unlikely to be a problem once the water content falls below 13% (Wright, 1941). Since the water content of herbage crops at cutting is usually within the range 75–85% (Chapters 15–22, 33–36), large quantities of water have to be removed from the crop to effect conservation by dehydration. In field-curing systems, the energy used to evaporate this water is normally solar energy. In barn-drying systems, the final stages of drying are accomplished by passing warm or cool air through the forage, while in 'artificial' drying systems most of the water is evaporated by air heated to high temperatures. At present, dehydration in the field is of most importance and the remainder of this discussion is concerned primarily with the factors which render crops suitable for conservation by this method.

Photosynthesis in herbage may continue for some time after cutting (MacGregor, 1966), but losses of DM through plant and fungal respiration, leaching and the fracture and loss of plant parts (physical losses) will result in a net loss during the field-curing period. Losses of DM during this period are often above 20% and may reach 50% (Watson & Nash, 1960). The critical nature of the duration of the field-curing period has been demonstrated by Kormos and Chestnutt (1968), who found DM losses to be 1·5–2% per day in wilting periods of 4 days. The rate of loss was faster in bad than in good weather. Comparisons between herbage species in the rate of DM loss in the field have not been made. It is likely that the following factors are of importance:

(i) Crop maturity. Losses through respiration by plants and fungi will be greater in crops containing a large proportion of labile compounds.
(ii) Proportion of soluble constituents. Leaching losses will be greater for crops with a large proportion of water-soluble constituents.
(iii) Fragility. Large differences in fragility between varieties have been detected (Shepherd, 1961), so that the physical loss of plant material during harvest will be high for varieties of high fragility.

Total losses during haymaking are dominated by the duration of the field-curing period. Since there is no evidence that there are differences among crops in their safe storage water contents, the duration of the field-curing period will be determined by the initial and subsequent rates of water loss from each particular crop. The water content of different species has been discussed earlier (Chapters 15–22, 33–36). Within each species, water content is characteristically high during the early stages of growth and also when the crop is grown with large applications of N fertilizer. Diploid varieties generally have lower water contents than tetraploid varieties. White clover appears to contain more water than the other species considered, but the marked effects of environmental conditions on water content make comparisons between the results of experiments difficult.

Differences between species and varieties in the rate of water loss after cutting have not been adequately explored. Experiments in controlled conditions have shown that the rate of water loss is affected by temperature, relative humidity and the rate of air movement (Shepherd, 1965; Thaine, 1967). The resistance to water loss increases on the closure of the stomata

(Shepherd, 1964), and it is likely that the nature of the cuticle and the internal anatomy of the leaf will then be responsible for variation between crops in their resistance to water loss. Thaine *et al.* (1969) found that in similar controlled environments, detached leaves of wheat reached a relative water content of 15% in only 65% of the time required by detached leaves of cocksfoot.

The situation in the cut swath during dehydration is extremely complex. During the early stages there are large differences in water content between different parts of the swath. At a time when water loss from the top of the swath is limited by plant factors, material lower in the swath may still be fully turgid. The nature of the crop will influence the structure and micro-climate within the swath, but the effect on drying rate is not known. Mechanical treatment of the swath will have a large effect on the swath micro-climate and the nature of the mechanical treatment has a large impact on the rate of water loss.

The lack of adequate field experimental information on crop drying rates reflects the dominant effect of weather on drying, the implications of which are discussed later.

ENSILAGE

Efficient crop conservation by ensilage depends on the attainment in the ensiled material of conditions sufficiently acid to inhibit the activity of bacteria causing putrefaction. With insufficient acidity, clostridial bacteria develop and these are responsible for the secondary fermentation of lactic acid to butyric acid and for extensive de-amination and decarboxylation of amino acids (Whittenbury, McDonald & Bryan-Jones, 1967). The pH level at which clostridial activity is prevented varies with the water content of the ensiled crop; at a water content of 50%, silage may be well preserved at a pH of 5·0, whereas secondary anaerobic fermentation may occur in silage with 85% water content, even when the pH is below 4·0 (Wilkins & Wilson, 1969).

Low pH conditions may be achieved by the fermentation of crop components to acids or by adding organic or inorganic acids to the ensiled material. The major crop components contributing to acid production are the water-soluble carbohydrates, although hemicellulose, pectin and the non-volatile organic acids are often involved also (McDonald *et al.*, 1960). Some important pathways of acid production are shown in Table 42.1.

The quantity of carbohydrate required to bring about a given change in pH depends on the efficiency of its conversion to acid (in turn influenced by the proportions of the various sugars present and the nature of the bacterial flora), the particular acids produced, the buffering capacity of the crop and the change in buffering capacity which occurs during fermentation. It is evident that the water-soluble carbohydrate content of the crop is influenced by both genetic and environmental factors (Chapter 13). Legumes generally have lower contents of water-soluble carbohydrates than grasses, whereas amongst the grasses, the ryegrasses have particularly high contents. Heavy applications of N depress water-soluble carbohydrate content. In grasses and legumes, the major buffers in the pH range 4·0 to 6·0 are the organic acids (Playne & McDonald, 1966). Because their organic acid content is characteristically high (Chapter 31), it is difficult to achieve a satisfactory fermentation in legumes because of their high buffering capacity and low content of water-soluble carbohydrates. No comprehensive studies have been carried out on the organic acids in grasses (Chapter 13).

Table 42.1. Examples of the fermentation of carbohydrates by lactic-acid bacteria

Homofermentative		
1 glucose	→	2 lactic acid
1 fructose	→	2 lactic acid
1 pentose	→	1 lactic acid + 1 acetic acid
Heterofermentative		
1 glucose	→	1 lactic acid + 1 ethanol + 1 CO_2
3 fructose	→	1 lactic acid + 2 mannitol + 1 acetic acid + 1 CO_2
1 pentose	→	1 lactic acid + 1 acetic acid

From Whittenbury, McDonald and Bryan-Jones (1967)

Losses of DM during ensilage result from three main factors, effluent loss and the activity of aerobic and anaerobic bacteria.

Losses of crop DM in effluent rarely exceed 10% and are related to the density of the silage and its water content (Miller & Clifton, 1965). Differences in water content between crops at harvest have been discussed earlier in this chapter. If wilting is used to reduce water content prior to ensiling, then factors influencing the loss of both water and DM during field wilting will be important.

Aerobic microbial activity occurs when the silage surface is exposed to air; this may happen at any time during the ensiling process and can cause the loss of up to 30% of the DM ensiled (Whittenbury, 1968). These losses can be greatly reduced by adequate sealing of the silo surfaces and are not, therefore, directly influenced by crop characteristics.

Attention has already been drawn (Table 42.1) to the extent of carbon dioxide production during anaerobic fermention. During homofermentative activity, no loss of DM as carbon dioxide occurs. Heterofermentative bacteria, which produce CO_2, are normally responsible for the major part of the bacterial activity in silage and, of the losses of DM from ensiled herbage, 4–6% are generally attributable to bacterial fermentation (McDonald & Whittenbury, 1967).

This discussion has considered only the ensilage of crops by natural fermentation processes. The pattern of fermentation can, of course, be altered by chemical additives. The importance of high contents of water-soluble carbohydrate and low buffering capacity in the crop will therefore depend on the availability and cost of sugars and acids which can be used as aids in the ensiling process.

CROP GROWTH RHYTHM AND CONSERVATION

The growth cycle of the crop and the effect of this on the date at which a crop has defined nutritive characteristics will have a considerable influence on the success with which a crop can be conserved. Also, differences between varieties may be exploited to enable harvests to be taken over a long season.

This is of considerable importance for crops conserved as hay. The effect of weather on DM loss during haymaking has already been noted. In Britain, rainfall and the frequency of dry spells vary little between May, June and July. However, evaporation increases up to the middle of June and it will therefore be least difficult to dry crops in the field at that time. Thus, if grasses are compared at a particular level of digestibility of their primary growth, mean rates of evaporation at harvest will be highest for the latest-maturing varieties.

The production of feeds with a narrow digestibility range can be facilitated by the use of varieties characterized by a slow change in digestibility with time, or by the use of a series of varieties reaching a particular digestibility at different dates during primary growth. The adoption of either of these practices will reduce the susceptibility of the conservation system to weather. Differences between species in the rate at which digestibility declines during primary growth are noted in Chapter 33. The digestibility of S100 white clover falls more slowly than that of S24 perennial ryegrass, so that the D-value of a mixed grass/clover sward is likely to fall less rapidly than that of a pure grass sward. Large differences in the date at which different grass varieties reach a digestibility of 65% were noted by Cherry (1968).

The advantage of using crops with differing developmental patterns is not restricted to haymaking. Extension of the period during which conservation can take place reduces the machine capacity required to harvest a given quantity of crop. This is important with ensilage, but the significance of this factor increases as the amount of capital invested in the conservation system increases. The length of the drying season has a large impact on the total costs of conservation by high-temperature dehydration (Raymond, 1968), and this puts a premium on crops which allow the drying season to be extended.

Differences exist between crops in the ease and success with which they can be conserved. It is, however, difficult to evaluate the significance of these differences, because the relevance of specific properties varies with the conservation method used; also, non-crop inputs, such as chemical additives and heat from fuel oil, may be used to help efficient conservation.

References

ANSLOW, R. C. 1965. Grass growth in midsummer. *J. Br. Grassld Soc.* **20,** 19–26.

ANSLOW, R. C. 1967. Frequency of cutting and sward production. *J. agric. Sci., Camb.* **68,** 377–84.

BROUGHAM, R. W. 1956. Effect of intensity of defoliation on regrowth of pasture. *Aust. J. agric. Res.* **7,** 377–87.

CHERRY, M. 1968. Grass conservation handbook. London: Iliffe Books, p. 26.

DUCKHAM, A. N. 1967. Buffers in grassland systems. *Occ. Symp. 3 Br. Grassld Soc.* pp. 1–13.

KORMOS, J.; CHESTNUTT, D. M. B. 1968. Measurement of dry matter losses in grass during the wilting period. 1. The effects of the length of the wilting period. *Rec. agric. Res. (Nth. Ireld)* **16,** 145–50.

MCDONALD, P.; STIRLING, A. C.; HENDERSON, A. R.; DEWAR, W. A.; STARK, G. H. *et al.* 1960. Studies on ensilage. *Tech. Bull. 24 Edinb. Sch. Agric.*

MCDONALD, P.; WHITTENBURY, R. 1967. Losses during ensilage. *Occ. Symp. 3 Br. Grassld Soc.* pp. 76–84.

MACGREGOR, A. W. 1966. Photosynthetic activity in wilted grass. *Exp. Wk Edinb. Sch. Agric. 1966* pp. 63–4.

MILLER, W. J.; CLIFTON, C. M. 1965. Relation of dry matter content in ensiled material and other factors to nutrient losses by seepage. *J. Dairy Sci.* **48,** 917–23.

PLAYNE, M. J.; MCDONALD, P. 1966. The buffering constituents of herbage and of silage. *J. Sci. Fd Agric.* **17,** 264–8.

RAYMOND, W. F. 1968. Grass drying. *Agriculture, Lond.* **75,** 156–60.

SHEPHERD, W. 1961. The susceptibility of hay species to mechanical damage. 2. Effects of moisture content and variety. *Aust. J. agric. Res.* **12,** 783–96.

SHEPHERD, W. 1964. Paths and mechanisms of moisture movement in detached leaves of white clover (*Trifolium repens* L.). 1. Losses of petiole moisture direct from petioles and via laminae. *Ann. Bot.* **28,** 207–20.

SHEPHERD, W. 1965. Air speed effects during drying of harvested pasture material. *Aust. J. agric. Res.* **16,** 385–9.

SHEPPERSON, G. 1959. Haymaking research at N.I.A.E.—problems, progress and future trends. *Expl Husb.* No. 4, 34–40.

SHEPPERSON, G. 1965. Quick haymaking machinery and methods. *Outl. Agric.* **4,** 259–69.

THAINE, R. 1967. Plant physiology in relation to the process of conservation. *Occ. Symp. 3 Br. Grassld Soc.* pp. 42–52.

THAINE, R.; HARRIS, C. E.; LESHEM, Y. 1969. Water and nutrient loss from leaves and stems after cutting. *A. Rep. Grassld Res. Inst., Hurley 1968* pp. 93–6.

WATSON, S. J.; NASH, M. J. 1960. The conservation of grass and forage crops. Edinburgh: Oliver and Boyd.

WHITTENBURY, R. 1968. Microbiology of grass silage. *Process Biochem.* **3,** No. 2, 2–6.

WHITTENBURY, R.; MCDONALD, P.; BRYAN-JONES, D. G. 1967. A short review of some biochemical and microbiological aspects of ensilage. *J. Sci. Fd Agric.* **18,** 441–4.

WILKINS, R. J.; WILSON, R. F. 1969. Methods of conservation. *A. Rep. Grassld Res. Inst., Hurley 1968* pp. 67–9.

WRIGHT, N. C. 1941. The storage of artificially dried grass. *J. agric. Sci., Camb.* **31,** 194–211.

CHAPTER 43

Comparative Suitability of Herbage Species for Grazing

THE performance of herbage species has been discussed in this book in relation to many factors and, in this chapter, information specifically derived from the grazing situation is brought together.

To what extent can a species be described as 'suitable for grazing'? To answer such a question, the kind of grazing must necessarily be specified. The influence of grazing on the sward will depend on the environment, soil type and availability of water and soil nutrients, grazing method (e.g. whether continuous or rotational, lenient or severe), the class of animal involved and other factors, many of which are associated with the method of animal management.

It seems reasonable, however, to describe the main effects of grazing in relation to the principal ways in which it appears to differ from other methods of harvesting and to assess herbage species in these terms. It is possible that at many seasons of the year, infrequent rotational grazing of a pure stand will have an effect on sward growth similar to that of a cutting regime. This may be particularly true if the level of fertilizer input is so high that returns of dung and urine count for relatively little, if the weather and soil type rule out excessive damage by treading, and if the intensity of defoliation under grazing is similar to that under cutting.

Under other conditions, however, the effects of cutting may be quite different from those of grazing. A species may react in a different manner to grazing than to cutting when it is present in a mixture; the effects of selection by grazing animals on the botanical composition of a mixed sward can be far-reaching. Its reaction to grazing may also be altered where poaching occurs or under frequent grazing. In autumn, winter and spring when growth is slow, the sward may be more suited to grazing than cutting, and the value of different species for grazing will then presumably depend largely on their rates of growth.

In the past, most comparisons of species performance have been carried out by cutting, and there is a current need for more evaluation tests under grazing conditions. The information that is available is summarized in the following sections.

General survey of the use of species for grazing

Perennial ryegrass is used for grazing in a wide variety of situations and can respond to intensive grazing by adopting a stoloniferous habit (Kydd, 1966). Italian ryegrass can be rotationally grazed with success (Jones, 1953), though it has only limited persistence (p. 182). Westerwolds ryegrass is not a good grazing plant owing to its tendency to die out if allowed to flower (p. 189). Hybrid ryegrass (New Zealand Journal of Agriculture, 1945; Walker, 1959), timothy (Welsh Plant Breeding Station, 1961) and cocksfoot (Jones, 1939) are all suitable for grazing, the last being able to resist grazing by producing procumbent tillers (p. 227). Meadow fescue can survive only moderate severity and frequency of grazing (Jones, 1939). Tall fescue can tolerate close grazing (Jones, 1939; Matches, 1966) except during the year of establishment, but can benefit from a reasonably long period of autumn rest (p. 252).

White clover is normally grazed in mixtures with grasses and the small-leaved white clovers surpass other legumes in their ability to withstand close grazing (p. 361), which may be related to the heights of stem apices above the soil surface. Red clover may be grazed, for example, before seed is harvested (Hawkins, 1956; National Institute of Agricultural Botany, 1959), but repeated grazing may contribute to poor winter survival (Williams, 1927) and may depress yields if practised at certain times of the year (p. 383). Lucerne can be maintained under a grazing regime (Barker *et al.*, 1957); it takes little harm from severe defoliation (Ridgman, 1960), but will not tolerate frequent defoliation (p. 396; Green & Davis, 1955) and is liable to be killed by more than 4 grazings per year (Whitear, Hanley & Ridgman, 1962). Sainfoin can resist hard grazing; it is seldom grazed when immature (p. 409).

Certain fodder crops, such as the Aberystwyth varieties of perennial ryegrass (Davies, 1960), are suited to the frequent, lenient defoliation associated with grazing rather than cutting for conservation. Timothy (Jones, 1939) and lucerne (Whitear, Hanley & Ridgman, 1962) are reported to be less vigorous under grazing than under a cutting system.

Italian ryegrass is commonly used to provide grazing for dairy cows and sheep in early spring (p. 174). Sainfoin can withstand sheep grazing and in many ways is superior to lucerne or red clover for this purpose (Evans, 1961). Meadow fescue does not persist under heavy stocking by sheep, but is suited to lenient defoliation (Jones, 1939). White clover is so readily grazed by sheep in preference to grasses (e.g. ryegrass, Thomson, 1963), that there may be difficulty in maintaining it in the sward. Young lambs perform very well when given unlimited access to sainfoin (Grassland Research Institute, Hurley, 1967).

Nutritive value of different species when grazed

There may be appreciable differences in the digestibility and nutritive value of different species when subjected to the same grazing management. The digestible dry matter (DDM) content of perennial ryegrass (e.g. Table 19.6), Italian ryegrass (Table 16.7), timothy (Table 19.6), cocksfoot (Table 20.4), meadow fescue (Table 21.3), tall fescue (Table 22.3), and probably Westerwolds ryegrass also (p. 189), can be maintained at values between 70% and 80% by frequent grazing. On the other hand, the digestibility of lucerne can be improved by grazing it very leniently (p. 399).

Under grazing conditions generally, perennial ryegrass (Chapter 13), Italian ryegrass (Table 16.7), meadow fescue (Table 21.3), tall fescue (Table 22.3), timothy (Table 19.6) and cocksfoot (Table 20.4), show little difference either in crude-protein or crude-fibre content, but the last two species differ from the rest in having a lower content of soluble carbohydrates. There are reports (Greenhalgh, 1966) that the intake of perennial ryegrass by dairy cattle is greater than that of cocksfoot. It is clear (Wilson, 1966) that the presence of white clover in a grass sward increases voluntary intake and milk production in cattle and liveweight gains in sheep. Lambs may show better liveweight gains on sainfoin than on many other grasses and legumes (p. 410).

Suitability of species for grazing at different seasons

Species may vary in their suitability for grazing at different seasons of the year. Italian ryegrass is purposely grown for spring grazing (p. 174) and ought to be grazed or cut in spring even if it is grown for seed production. Grass/white clover swards grown for clover seed production should similarly be defoliated in spring in order to suppress the grass (p. 350). Red clover may provide early grazing, but lucerne should not be grazed before mid-May (p. 391). Cocksfoot shows reduced persistence if grazed in early spring (Jones, 1933; Beddows, 1959). Sainfoin (Thomson, 1938) and tall fescue (Jones & Charles, 1961) are particularly suited for aftermath grazing.

Grazing in autumn may benefit swards of perennial ryegrass if this prevents the accumulation of foggage (Baker, 1956), but severe defoliation may reduce growth early in the following year (Baker, 1957). It is thought beneficial to graze Italian ryegrass in the autumn (National Institute of Agricultural Botany, 1964); its nutritive value is very high at that time (Welsh Plant Breeding Station, 1961) and if undergrazed then, its persistence may be reduced (p. 182). In contrast, the persistence of cocksfoot (Jones, 1933; Beddows, 1959) and tall fescue (p. 241) may be lessened if they are grazed in the autumn. It may be possible to graze red clover in the autumn (p. 374), but autumn grazing will reduce the survival of sainfoin (Badoux, 1965).

Timothy (Alder, 1966), meadow fescue and tall fescue (p. 241) may provide grazing in winter. Similarly, perennial ryegrass (National Institute of Agricultural Botany, 1959), timothy (Evans, 1953, 1955; Lambert, 1966; Roberts, 1958), lucerne (p. 390), meadow fescue (Green & Evans, 1956, 1957) and cocksfoot (Green & Evans, 1956, 1957; Lambert, 1966) can be grazed in winter without reducing seed yields in the following summer. Winter grazing of cocksfoot can be successful (Alder, 1966) but may sometimes reduce subsequent growth (Green & Evans, 1956, 1957; Lambert, 1966; Jones, 1933; Beddows, 1959). Lucerne (p. 400) and tall fescue (p. 252) may provide winter foggage.

Competitive ability of species when grazed

The suitability of a species for use in a grazed sward will depend, in part, upon its ability to resist ingress by weeds if it is sown alone, and the attributes of neither dominating nor being suppressed by other sown species if it is included in a mixture. Competition under the grazing situation is therefore of importance and will be influenced by the vegetative vigour of the species and any tendency it may have to be either selected out or avoided by the grazing animal.

Perennial ryegrass will persist in grazed swards and if severely defoliated is usually only partially replaced by weeds. It requires moderately fertile soil and will be displaced by other species if fertility declines or the soil becomes poorly drained or markedly acid (p. 164). When grazed, Italian ryegrass will survive for more than the few years expected from its biennial habit, if it is not in competition with other grasses (e.g. perennial ryegrass, Charles, 1964). Also under grazing conditions, there are instances where meadow fescue has become dominant in mixtures with cocksfoot and timothy (p. 241), and tall fescue has become dominant over meadow fescue (Alder & Redford, 1958; Baker, Chard & Hughes, 1965). Meadow fescue has been reported to be less competitive than perennial ryegrass under grazing, but more so than cocksfoot (Arens, 1962).

Legumes tend to be dominated by grasses when the two are present in mixtures, partly because the legumes are selectively grazed by livestock. Mixtures of lucerne and a grass are more likely to be unstable if the grass is one of the less palatable species such as cocksfoot or tall fescue (p. 389). The proportion of cocksfoot included in mixtures with lucerne should be small (Green & Davis, 1955) and the grass must be kept in check by winter defoliation (p. 218). Perennial ryegrass, timothy or meadow fescue are more appropriate than cocksfoot for inclusion in such a mixture (Green & Davis, 1955). Mixtures of perennial ryegrass and white clover have been found to be more stable under grazing conditions if Northern Irish ryegrass is used rather than S23 or S24 (Green & Corrall, 1965). Concentrated treading and winter grazing both tend to control weeds in lucerne (p. 400).

Yield from species when grazed

The number of animals which can be *satisfactorily* fed by grazing a unit area of a sward is the ultimate criterion in judging the value of a species for grazing. The yield of animal products, which in turn is determined by the yield of herbage, has ultimately to be considered.

The yield of DM of a species under grazing depends on the type of management adopted. Perennial ryegrass does best when fertility is high and grazing is adjusted to production (p. 165). Italian ryegrass is particularly high-yielding under rotational grazing (Jones, 1953), where grazing is severe on each occasion (Stapledon *et al.*, 1927). Cocksfoot gives a satisfactory yield if the sward is not continuously grazed (Jones, 1939). The yield of tall fescue may be high when cut frequently to simulate grazing (Hunt, 1961) and common sainfoins may outyield the giant sainfoins when cut frequently (Thomson, 1951). The yield of a timothy sward has been reported to be greater than that of a mixture of meadow fescue, cocksfoot and timothy when grazed by dairy cows (Mudd & Mair, 1961).

Trampling by animals can adversely affect the dry matter yield of cocksfoot (Beddows, 1959; Edmond, 1964) and of sainfoin (Rees, 1928), but may encourage the growth of perennial ryegrass (Davies, 1938). Timothy can produce a satisfactory yield when grazed on poorly-drained land (Mudd & Mair, 1961) in conditions liable to cause poaching, whereas sainfoin cannot withstand poaching (Rees, 1928).

Conclusion

If 'profitability' is the criterion most pertinent to the agricultural situation, different species ought to be compared by using them in farm systems (or relevant parts of such systems) which are oriented towards grazing as the means of utilization. The species which either alone or in suitable mixtures gives the greatest output from the system would then be regarded as the most suitable for grazing.

Few such comparisons have yet been made and there appear to be few published records of a comprehensive examination having been made of the suitability of species for grazing systems in Britain.

Alder (1966) recognized this need and carried out 'farmlet' trials of Aberystwyth-bred and commercial strains of several herbage species. His results demonstrated a superiority of S-cultivars within a species, but do not enable detailed comparisons to be made between species.

In addition, there are many references to the reaction of species to the grazing process, but it is not known how these isolated findings could be interpreted in the context of a complete system. It is clear, therefore, that the foregoing discussion can provide only an approximate comparison of the usefulness of species for grazing and that more research is needed in order to define satisfactorily the relative suitability of species for grazing in relation to age, size and type of animal.

References

ALDER, F. E. 1966. A summary of twelve years' work on a farmlet type of experiment designed to study animal production on sown pastures. *A. Rep. Grassld Res. Inst. Hurley 1965* pp. 79–86.

ALDER, F. E.; REDFORD, R. A. 1958. Further observations on grassland management for meat production. *J. Br. Grassld Soc.* **13,** 239–46.

ARENS, R. 1962. The effect of rate of seeding on competition between species and the productivity of the first year's pasture. *Z. Acker-u. PflBau* **115,** 357–74.

BADOUX, S. 1965. Morphological, physiological and agronomic characteristics of sainfoin (*Onobrychis* spp.) populations. *Schweiz. landw. Forsch.* **4,** 111–90.

BAKER, H. K. 1956. Note on the influence of previous management on the death of perennial ryegrass during winter. *J. Br. Grassld Soc.* **11,** 235–7.

BAKER, H. K. 1957. Studies on the root development of herbage plants. 3. The influence of cutting treatments on the root, stubble and herbage production of a perennial ryegrass sward. *J. Br. Grassld Soc.* **12,** 197–208.

BAKER, H. K.; CHARD, J. R. A.; HUGHES, W. E. 1965. A comparison of cocksfoot- and tall fescue-dominant swards for out-of-season production. *J. Br. Grassld Soc.* **20,** 84–90.

BARKER, M. G.; HANLEY, F.; RIDGMAN, W. J. 1957. Studies on lucerne and lucerne-grass leys. 4. The effect of system of grazing management on the persistence of a lucerne-cocksfoot ley. *J. agric. Sci., Camb.* **48,** 361–5.

BEDDOWS, A. R. 1959. Biological flora of the British Isles. *Dactylis glomerata* L. *J. Ecol.* **47,** 223–39.

CHARLES, A. H. 1964. Differential survival of plant types in swards. *J. Br. Grassld Soc.* **19,** 198–204.

DAVIES, I. 1960. Growth stage and related factors in grassland management. *Rep. Welsh Pl. Breed. Stn 1959* pp. 104–9.

DAVIES, W. 1938. Vegetation of grass verges and other excessively trodden habitats. *J. Ecol.* **26,** 38–49.

EDMOND, D. B. 1964. Some effects of sheep treading on the growth of 10 pasture species. *N.Z. Jl agric. Res.* **7,** 1–16.

EVANS, A. M. 1961. The yields of eight varieties of sainfoin cut at different growth stages. *Emp. J. exp. Agric.* **29,** 323–9.

EVANS, T. A. 1953. Management and manuring for seed production in cocksfoot and timothy. *J. Br. Grassld Soc.* **8,** 245–59.

EVANS, T. A. 1955. Manuring and winter grazing for seed production in S48 timothy. *J. Br. Grassld Soc.* **10,** 254–62.

GRASSLAND RESEARCH INSTITUTE. 1967. Sheep ecology. *A. Rep. Grassld Res. Inst. Hurley 1966* p. 48.

GREEN, J. O.; CORRALL, A. J. 1965. The testing of grass varieties in swards with clover; the effect of grass seed-rate on comparisons of gross yield. *J. Br. Grassld Soc.* **20,** 207–11.

GREEN, J. O.; DAVIES, A. G. 1955. The seasonal use of lucerne/grass mixture for grazing. In: Grassland Research Institute, Memoir No. 1. Lucerne investigations, 1944–53, pp. 63–77.

GREEN, J. O.; EVANS, T. A. 1956. Manuring and grazing for seed production in S37 cocksfoot and S215 meadow fescue. *J. Br. Grassld Soc.* **11,** 165–73.

GREEN, J. O.; EVANS, T. A. 1957. Grazing management for seed production in leafy strains of grasses. *J. Br. Grassld Soc.* **12,** 4–9.

GREENHALGH, J. F. D. 1966. Studies of herbage consumption and milk production in grazing dairy cows. *Proc. 10th int. Grassld Congr., Helsinki, 1966* pp. 351–5.

HAWKINS, R. P. 1956. A preliminary survey of red clover seed production. *Ann. appl. Biol.* **44,** 657–64.

HUNT, I. V. 1961. Comparison of production from three varieties of tall fescue and three varieties of meadow fescue. *Res. Bull. 29 W. Scotl. agric. Coll.* pp. 1–43.

JONES, L. I. 1939. Pasture management and its effect on the sward. *Bull. H15 Welsh Pl. Breed. Stn* pp. 40–129.

JONES, L. I. 1953. Preliminary note on S22 Italian ryegrass. *J. Br. Grassld Soc.* **8,** 345–8.

JONES, L. I.; CHARLES, A. H. 1961. Observations on S170 tall fescue. *Rep. Welsh Pl. Breed. Stn 1960* pp. 55–9.

JONES, M. G. 1933. Grassland management and its influence on the sward. 3. The management of a 'grassy' sward and its effects. *Emp. J. exp. Agric.* **1,** 223–34.

KYDD, D. D. 1966. The effect of intensive sheep stocking over a five year period on the development and production of the sward. 3. Ecological discussion. *J. Br. Grassld Soc.* **21,** 298–302.

LAMBERT, D. A. 1966. Factors affecting seed yield. *Proc. Conf. Timothy Seed Production, W. Scotl. agric. Coll.* pp. 4–12.

MATCHES, A. G. 1966. Influence of intact tillers and height of stubble on growth responses of tall fescue (*Festuca arundinacea* Schreb.). *Crop Sci.* **6,** 484–7.

MUDD, C. H.; MAIR, R. B. 1961. Performance of eight seeds mixtures at Great House 1952–1958. *Expl Husb.* No. 6, 21–50.

NATIONAL INSTITUTE OF AGRICULTURAL BOTANY. 1959. Growing grasses for seed. *Herbage Seed Growers' Leaflet No. 2.*

NATIONAL INSTITUTE OF AGRICULTURAL BOTANY. 1964. Varieties of ryegrass. *Farmers' Leaflet No. 16.*

NEW ZEALAND JOURNAL OF AGRICULTURE. 1945. Short-rotation ryegrass. Its breeding and characteristics. *N.Z. Jl Agric.* **71,** 465–70.

REES, J. 1928. Sainfoin or French grass in South Wales. *Welsh J. Agric.* **4,** 242–50.

RIDGMAN, W. J. 1960. The effect of height of cutting on the subsequent growth and yield of lucerne. *J. Br. Grassld Soc.* **15,** 291–5.

ROBERTS, H. M. 1958. The effect of defoliation on the seed-producing capacity of bred strains of grasses. 1. Timothy and perennial ryegrass. *J. Br. Grassld Soc.* **13,** 255–61.

STAPLEDON, R. G.; FAGAN, T. W.; EVANS, R. E.; MILTON, W. E. J. 1927. Italian ryegrass for winter and early spring keep; the effect of methods of grazing on productivity and palatability, and on the chemical and botanical composition of the herbage. *Bull. H5 Welsh Pl. Breed. Stn* pp. 5–41.

THOMSON, D. J. 1963. Grassland Research Institute, Hurley. Personal communication.

THOMSON, J. R. 1938. The development of sainfoin in its seeding year. *Ann. appl. Biol.* **25,** 457–71.

THOMSON, J. R. 1951. Sainfoin in its first harvest year. *J. Br. Grassld Soc.* **6,** 107–17.

WALKER, M. H. 1959. H.1 ryegrass production compared with that of Italian ryegrass at the New England Experiment Farm. *Agric. Gaz. N.S.W.* **70,** 528–31.

WELSH PLANT BREEDING STATION. 1961. The Aberystwyth varieties of grasses and clovers. *Leaflet Series No. 7.*

WHITEAR, J. D.; HANLEY, F.; RIDGMAN, W. J. 1962. Studies on lucerne and lucerne-grass leys, 6. Further studies on the effect of systems of grazing management on the persistence of a lucerne-cocksfoot ley. *J. agric. Sci., Camb.* **59,** 415–28.

WILLIAMS, R. D. 1927. Red clover investigations, 1919–1926. *Bull. H7 Welsh Pl. Breed. Stn* pp. 136.

WILSON, G. F. 1966. Ryegrass varieties in relation to dairy cattle performance. 2. The influence of ryegrass varieties on intake, digestibility and some characteristics of rumen fermentation. *N.Z. Jl agric. Res.* **9,** 1053–63.

CHAPTER 44

Comparative Value of Herbage Species for Milk Production

In this chapter, the few direct comparisons that have been made of grass or legume species as feeds for milk production and those of varieties within species have been brought together. Where no results of this kind are available the suitability of species and varieties for milk production must be assessed by reference to the species chapters.

Milk production is characterized by a potential rate of production which changes during the course of a lactation in response to other physiological controls than purely nutritional ones. Should this potential rate of production not be exploited at the time, it is generally incapable of being realized at any later stage. The aim of feeding the dairy cow is usually, therefore, to ensure that she obtains sufficient nutrients to sustain the rate of milk production of which she is capable at the time. In contrast, the rate of production shown by the growing and/or fattening animal can be effectively controlled by the level of feeding during any period of the animal's life.

The most general conclusion regarding the relative suitability of herbage species for milk production has been that cocksfoot has a depressing effect on milk production per cow and per acre compared with most other cultivated herbage species.

This conclusion, which has applied to S26, S37 and S143 and other, unspecified, varieties of cocksfoot, has been deduced from practical experience by Castle (1953) and Boyd and Jennings (1960) and has been confirmed experimentally for conserved herbage (Ramage, Ogrinz, Eby & Mather, 1957; Castle, Drysdale & Watson, 1962; Apgar, Ramage & Mather, 1966), for grazed herbage (Demarquilly, 1963; Greenhalgh, 1966), and for cut herbage offered indoors to cows (MacLusky, Alexander, Leach & Black, 1963; Demarquilly, 1963). However, the results of later work by Greenhalgh and Reid (1969) have been more equivocal.

Swards dominated by cocksfoot or perennial ryegrass sown with S100 white clover were compared by Castle (1953) in an experiment designed for quite different purposes. He concluded that milk yields per cow and per acre were greater on ryegrass than on cocksfoot swards. Under practical conditions in Northern Ireland, Boyd and Jennings (1960) confirmed that in May and June, cocksfoot swards depressed the yields of dairy cows grazing them, whereas timothy/meadow fescue swards boosted yields, at least when cows first started to graze a fresh paddock. Similar results were obtained by Paterson (1959), who reported that S143 cocksfoot/white clover leys gave 116 cow-days/acre (290/ha), compared with 205 (512) from timothy/meadow fescue/white clover leys, and 152 (380) from perennial ryegrass/white clover leys; a shorter growing season was noted, however, for the timothy/meadow fescue/white clover leys.

Cocksfoot hays of different N contents were compared with lucerne hay by Ramage *et al.* (1957) for milk production when concentrates were rationed according to yield. Both hay intake and milk production were lower on cocksfoot hays than on lucerne hay.

Comparisons of artificially dried S24 perennial ryegrass and S37 cocksfoot, harvested either in the third week of May or in the second week of June, were made by Castle, Drysdale and Watson (1962). When offered to lactating cows for a 20-week experimental period at a fixed rate of 8·6 kg per day, together with concentrate at 350 g per kg milk produced, the first-cut hays showed no differences in intake or milk production by the cows. Only second-cut cocksfoot hay depressed DM intake and milk yield, and none of the hays affected milk composition.

Cocksfoot and lucerne, conserved both as silage and hay, were compared by Apgar, Ramage and Mather (1966) over three successive lactations. Small fixed quantities of silage and hay *ad lib.* were offered to the cows, together with different concentrate levels. Forage dry matter (DM) intake, fat-corrected milk (FCM) production and solids-not-fat (SNF) content of the milk were lower with cocksfoot than with lucerne; the fat contents of the milks were similar.

Murdock, Hodgson and Austenson (1959) observed that when cows grazed cocksfoot given heavy rates of fertilizer N and cocksfoot/Ladino clover mixtures at the same stocking rate, more milk per cow and per acre was produced from the mixed sward. In so far as milk production per acre was concerned, this position was reversed when the stocking rate was adjusted during the grazing season to give the same degree of herbage utilization on both swards. MacLusky *et al.* (1963) obtained a depression in herbage DM intake of some 10% and in milk yield per cow of 4–5% when cows were fed cut cocksfoot compared with ryegrass. Demarquilly (1963) found that when grazing cows were moved from ryegrass, meadow fescue or timothy to cocksfoot, tall fescue or lucerne swards, milk yields per cow fell and milk fat contents rose, whereas the reverse occurred when they were moved in the opposite direction. Both effects were also observed on herbage cut and offered fresh indoors to cows.

McDowall and McGillivray (1963) compared pure stands of New Zealand certified white clover and short-rotation ryegrass cut and fed indoors to dairy cows. Although DM intake of clover was less than that of ryegrass, yield (about 9 kg per cow per day) and composition of milk were unaffected; however, the carotene and vitamin A levels of the butterfat were lower for cows receiving ryegrass.

In grazing trials with dairy cows, in which Greenhalgh (1966) compared S24 perennial ryegrass and S37 cocksfoot, he found that perennial ryegrass sustained a higher daily milk yield than cocksfoot, but milk fat content was lower and FCM yields were not significantly different. During May and early June, Greenhalgh and Reid (1969) found that there was virtually no difference in the OM intakes and milk yields of cows grazing cocksfoot and ryegrass swards, despite a reduction of 4–5 percentage units in the digestibility of the cocksfoot. It was only in July and later in the season that milk yields on the cocksfoot swards were significantly inferior. In trials with the perennial ryegrass cultivar Grasslands Ruanui and the *Lolium perenne* × *multiflorum* cultivars Grasslands Manawa and Grasslands Ariki, Wilson and McDowall (1966) noted differences between their performances in spring and in autumn grazing. At yield levels of about 11 kg milk per cow per day in spring, Ruanui produced less milk per cow than Ariki, and Ariki less than Manawa; also, the fat content of the milk of cows grazing Manawa was significantly lower than that of those grazing Ariki and Ruanui. As a result, the butterfat yields per cow per day from these varieties did not differ significantly. At lower milk yield levels of about 7 kg per cow per day in autumn, butterfat yield per cow per day was higher on Ruanui. Wilson (1966) established that these results were due to differences in voluntary intake of individual animals, although he was unable to demonstrate consistent differences in intake among the ryegrasses.

In later experiments, Wilson and Dolby (1967) compared the perennial ryegrass cultivars Grasslands Ruanui and Grasslands Paroa and the Westerwolds ryegrass cultivar Grasslands 4707 (a tetraploid with an annual habit). Cows with yields approaching 18 kg per day in a winter experiment produced significantly less milk of higher fat content on Ruanui than on 4707. In a spring experiment in which cows were yielding about 14 kg per day, there were no differences in milk yield or fat content between the ryegrasses; however, Ruanui produced milk of lower SNF content than the other two cultivars, mainly owing to its relatively low protein content. The experiments of Wilson and McDowall (1966) and Wilson and Dolby (1967) were carried out in such a way as to produce differences in quality between the varieties; although the stocking rates were not specified, they confirmed that higher yields per cow were obtained from mixed than from single-species swards under the conditions of their experiments.

The quality and energy yield of individual species have been summarized and the species listed in order in Table 45.1. This table makes no reference to Italian ryegrass, which has the highest potential yield in its first year but declines rapidly thereafter, its main advantage being its good early-season growth. Lucerne ranks first in order of energy yield, but will not tolerate more than three defoliations in the growing season. Tall fescue becomes very coarse in mid-season, at which time Baker, Chard and Hughes (1965) observed that cocksfoot was preferred to it; the same authors also reported that S170 tall fescue began growth as early in spring as Italian ryegrass and yielded as well at that time. On the basis of quality and yield of total energy, perennial ryegrass and meadow fescue therefore appear to be the most suitable species for grazing. For conservation, lucerne appears preferable in terms of yield, although its digestibility is never very high. However, survey data collected by Baker and Baker (1965) have suggested that the management imposed by the farmer is a more important factor than herbage species in determining the total yield of a sward.

Reference to Chapters 13 and 31 and to the feeding standards proposed in the UK by the Agricultural Research Council (1965) suggests that most grasses contain insufficient magnesium to preclude the possibility that hypomagnesaemia may occur and that the P content of both grasses and legumes may be too low for high-yielding cows, particularly if N fertilizer inputs are high and available P supplies low. Timothy, meadow fescue and sainfoin are generally low in Na content for milk production. Most grass and legume species may contain less than the recommended dietary concentration of Cu for cattle, especially with advancing maturity; also, the I contents of both grasses and legumes are often insufficient for the needs of lactating cows.

Apart from the normal criteria of yield and quality, herbage species for dairy cows should be adapted to severe defoliation at fairly frequent intervals, should not depress the DM intake of the cow, should not adversely influence milk yield or composition, and should not be toxic. They should show a high response to fertilizer N and should be capable of being conserved to give a product of high quality. The levels at which each of these criteria are set will be modified by many managerial considerations, especially the seasonal calving pattern, the yield level of the cow and the supply of supplementary feeds to the animal.

References

AGRICULTURAL RESEARCH COUNCIL. 1965. The nutrient requirements of farm livestock. No. 2. Ruminants. London: H.M.S.O.

APGAR, W. P.; RAMAGE, C. H.; MATHER, R. E. 1966. Nitrogen-fertilized orchardgrass compared with alfalfa at different levels of concentrate feeding for dairy cows. *J. Dairy Sci.* **49**, 1033–7.

BAKER, H. K.; CHARD, J. R. A.; HUGHES, W. E. 1965. A comparison of cocksfoot- and tall fescue-dominant swards for out-of-season production. *J. Br. Grassld Soc.* **20,** 84–90.

BAKER, R. D.; BAKER, H. K. 1965. The output of swards on commercial farms in relation to fertilizer and other management factors. *J. Br. Grassld Soc.* **20,** 182–7.

BOYD, M.; JENNINGS, J. W. 1960. A comparison of the effects of different swards on the milk yields of dairy cows at Greenmount Agricultural College. *Res. exp. Rec. Minist. Agric. Nth. Ire.* **9,** 135–46.

CASTLE, M. E. 1953. Grassland production and its measurement using the dairy cow. *J. Br. Grassld Soc.* **8,** 195–211.

CASTLE, M. E.; DRYSDALE, A. D.; WATSON, J. N. 1962. A comparison between ryegrass and cocksfoot hay for milk production. *J. Dairy Res.* **29,** 199–206.

DEMARQUILLY, C. 1963. [The effect of different leys on the yield and composition of milk of the cow.] *Annls Zootech.* **12,** 69–104.

GREENHALGH, J. F. D. 1966. Studies of herbage consumption and milk production in grazing dairy cows. *Proc. 10th int. Grassld Congr., Helsinki, 1966* pp. 351–5.

GREENHALGH, J. F. D.; REID, G. W. 1969. The herbage consumption and milk production of cows grazing S24 ryegrass and S37 cocksfoot. *J. Br. Grassld Soc.* **24,** 98–103.

McDOWALL, F. H.; McGILLIVRAY, W. A. 1963. Studies on the properties of New Zealand butterfat. 6. Comparison of the properties and vitamin A potencies of butterfats produced by clover-fed and ryegrass-fed dairy cows. *J. Dairy Res.* **30,** 47–57.

MacLUSKY, D. S.; ALEXANDER, J. R. B.; LEACH, K. W.; BLACK, W. J. M. 1963. The effects of pure swards of cocksfoot and ryegrass on dairy cow dry matter intake, milk production and composition. *A. Rep. Univ. Durham Sch. Agric. 1963* p. 19.

MURDOCK, F. R.; HODGSON, A. S.; AUSTENSON, H. M. 1959. A comparison of orchardgrass-ladino clover and orchardgrass as pasture for milking dairy cows. *J. Dairy Sci.* **42,** 1675–85.

PATERSON, R. 1959. Sixteen years' experience of grassland recording on a farm scale. In: Ivins, J. D. [Ed.] The measurement of grassland productivity. London: Butterworths Scientific Publications.

RAMAGE, C. H.; OGRINZ, E. F.; EBY, C.; MATHER, R. E. 1957. The nutritive value of nitrogen-fertilized orchardgrass hay as compared to alfalfa hay using milk production as a measure. *J. Anim. Sci.* **16,** 1057.

WILSON, G. F. 1966. Ryegrass varieties in relation to dairy cattle performance. 2. The influence of ryegrass varieties on intake, digestibility, and on some characteristics of rumen fermentation. *N.Z. Jl agric. Res.* **9,** 1053–63.

WILSON, G. F.; DOLBY, R. M. 1967. Ryegrass varieties in relation to dairy cattle performance. 3. A comparison of the milk yield and composition from a tetraploid and two diploid ryegrass varieties. *N.Z. Jl agric. Res.* **10,** 415–24.

WILSON, G. F.; McDOWALL, F. H. 1966. Ryegrass varieties in relation to dairy cattle performance. 1. The influence of ryegrass varieties on milk yield and composition. *N.Z. Jl agric. Res.* **9,** 1042–52.

CHAPTER 45

Comparative Value of Herbage Species for Beef Production

THE two most important indexes of quality in herbage for ruminants are the quantity consumed and its digestibility; compared with mixed diets containing cereal grains, the efficiency of utilization of digested nutrients in herbage is of relatively little significance. The relative suitability of herbage species for beef production will therefore be discussed in terms of intake and digestibility and of their influence on the growth of cattle for meat production. No work has been done on the suitability of the cultivated grasses and legumes for beef production from suckler cows. The available evidence refers to the post-natal growth of artificially reared calves and to the liveweight gains of steers of various ages.

Calculations based on the measured consumption of grass by cattle of different ages, weights and rates of gain, show that rapid weight gains up to slaughter weight are associated with improved efficiency of food conversion. This stresses the importance of a high daily nutrient intake to maximize the proportion of food used for carcass growth. A high nutrient intake requires that the herbage should be both highly digestible and available in ample daily quantities. It is necessary in any critical examination of experiments on the comparative value of grazed swards to demonstrate that the quantities of herbage available per head to the cattle were equal for the different species. The relevant information will be considered for 4 main situations: grazing, fresh-crop feeding by mechanical harvesting and transport to penned cattle (zero grazing), the feeding of ensiled herbage, and the feeding of dehydrated grass and legume crops.

Grazing

The information given in earlier chapters on the average digestibility and potential energy yield of the major species under frequent defoliation is summarized below (Table 45.1). The species are listed separately in order

Table 45.1. Herbage species grouped in order of average digestibility and potential yield

Species	Order of average digesti-bility	DM digesti-bility (%)	Species	Order of maximum potential yield	Yield of digestible OM (kg/ha)
Perennial ryegrass	1	75–78	Lucerne	1	7 100–8 000
White clover	2		Tall fescue	2	
Meadow fescue	3		Perennial ryegrass	3	
Timothy	4		Meadow fescue	4	
Cocksfoot	5	73–74	Sainfoin	5	5 660–6 800
Tall fescue	6		Timothy	6	
Red clover	7	55–70	Red clover	7	
Sainfoin	8		Cocksfoot	8	
Lucerne	9		White clover	9	3 900

of digestibility and yield, as a guide to their potential yield when grazed. In this setting, the evidence in terms of the growth of cattle can be examined.

Several comparisons have been made of the liveweight increase obtainable from the species listed above. Alder and Cooper (1967) reported an experiment in which a single-species sward of perennial ryegrass (mixture of S24 and S23) was compared with cocksfoot (mixture of S37 and S143) over three seasons of grazing by March-born calves in their first summer.

Calves of the same age consumed more ryegrass than cocksfoot, but at equal liveweights their intake of dry matter (DM) of the two species was the same. Ryegrass was consistently more digestible than cocksfoot by an average of 4 to 7 digestibility units throughout the season, with the result that the calves grazing ryegrass consumed more digestible organic matter (OM) per unit of liveweight than those on cocksfoot. They also showed a significantly more rapid rate of liveweight gain, by 13%, and produced heavier carcasses than those grazing cocksfoot.

In a further experiment, Alder (1970a) compared single-species swards of perennial ryegrass (mixture of S24 and S23), meadow fescue (mixture of S53 and S215) and timothy (mixture of S48 and S51). The swards were given 361 kg N/ha each year and were grazed at intervals of 7–15 days in April, increasing to 21–25 days in June to October in each year. An additional long-rest treatment in which the rest period was 35 days, increasing to 55 days, was included in the second season. Spring-born steer calves were grazed on these swards from March to October in two consecutive years.

There were no significant differences between species in herbage consumption in either year. In both years the average liveweight gains of the calves over the whole season were not significantly different between species, but in one year the calves gained significantly less on meadow fescue than on the other species during the period mid-April to mid-June.

No comparisons between grasses and white clover each sown alone appear to have been made for beef production. One known difficulty in conducting such experiments is to maintain pure swards of white clover; however, an indication of the comparative effects of white clover on the performance of beef cattle can be judged from experiments with mixed swards of grass and clover in which different levels of N application have resulted in some swards high in the grass component and others high in clover content. Alder, Cowlishaw, Newton and Chambers (1967) studied beef production from an irrigated sward of perennial ryegrass (S24) and white clover (New Zealand and S184) over three consecutive seasons. They concluded that when clover exceeded 20–40% of the herbage DM in the sward (it reached 70–80% for prolonged periods in their experiment), the steers did not graze readily, herbage consumption and liveweight gain tended to fall and death from bloat (ruminal tympany) was a constant hazard. There were indications, however, that when the average daily amount of herbage available was equal, herbage intake and liveweight gain were greater in the swards containing moderate quantities of clover than in those consisting almost entirely of ryegrass.

Alder and Minson (1960) measured beef production over two years on swards grazed three times per season by young steers. A pure crop of lucerne was compared with a cocksfoot/white clover sward in which there was no more than 5% clover in the DM on 87% of the dates of sampling (Alder & Minson, 1963). The steers grazing lucerne showed 14% lower rates of liveweight gain than those on cocksfoot. Bloat occurred in the cattle grazing young regrowths of lucerne in September and October.

This series of experiments supports the conclusion indicated by Table 45.1 above, that perennial ryegrass is outstanding in digestibility combined with yield, and therefore for liveweight gain of cattle from grazed swards. The hazards of bloat with the legumes tested, and the low yield of white clover, reduce their suitability for inclusion in high proportions in the diet of fattening cattle. Meadow fescue and timothy equal ryegrass in their potential to produce growth in beef cattle, but they commonly yield less than ryegrass. The grazing of tall fescue has been associated with ill health of unknown etiology, in which lameness and emaciation develop which are not associated with the presence of ergot (*Claviceps purpurea*) (e.g. see Jacobson & Miller, 1961).

There is no information on the productivity of red clover and sainfoin for beef production, but the persistence of these crops under grazing is unlikely to equal that of perennial ryegrass.

Fresh-crop feeding

When fresh grass or legume crops are mechanically harvested and fed to penned or housed cattle, conditions are different in many respects from when the same crops are grazed. Less selection is possible from cut than from standing herbage. Laceration and soil contamination by the mechanical harvesting process, and heating resulting from respiration, may differ in degree between species, and many other differences could be elaborated.

As part of the experimental comparisons of perennial ryegrass and cocksfoot reported by Alder and Cooper (1967), a parallel experiment was conducted in which the same species of grass were cut daily and fed indoors. Perennial ryegrass was consumed in greater quantities daily by calves of the same age than was cocksfoot, but DM intake was equal per unit of liveweight. As a result of a higher average digestibility of the mown ryegrass, consumption of digestible OM by the calves was higher on ryegrass than on cocksfoot and liveweight gains were also higher by 22%.

When perennial ryegrass, meadow fescue and timothy were compared by feeding indoors in parallel with a grazing experiment, Alder (1969) found no significant differences between species in intake of OM or DM in either year. In only one period (early April to early July) was there a difference in liveweight increase, when calves fed on meadow fescue gained less than on the other two species. The mean liveweight gains over the whole season did not differ between species. Sonneveld (1965), however, found that cattle fed perennial ryegrass gained liveweight more rapidly than those fed timothy.

Feeding of ensiled crops

There is no evidence on the comparative value for beef production of silage made from different pure species of grasses and legumes.

Feeding of dehydrated crops

The swards of perennial ryegrass, meadow fescue and timothy which were compared by Alder (1969, 1970a) under grazing and cutting conditions were also cut on the same date in two seasons, made into hay by barn-drying and fed to 6-month-old or yearling steers (Alder, 1970b). There were no significant differences in daily consumption of hay or in liveweight gain.

Barn-dried crops of the first growth of single species of perennial ryegrass (S24), timothy (S51) and cocksfoot (S37), were fed to steers in both long and milled-and-pelleted form by Miles, Walters and Evans (1969). Ryegrass was the most digestible, timothy was intermediate and S37 the least digestible. With both forms of feeding, liveweight gain and voluntary intake varied in relation to digestibility.

When the same species of grasses were artificially dried at equal digestibility and fed in the long form, voluntary intake of DM and liveweight gain were lower in steers fed timothy than in those fed ryegrass or cocksfoot. Low gains tended to be associated with low crude-protein content in the timothy.

Dried regrowths of the three species gave equal levels of performance in cattle.

References

ALDER, F. E. 1969. Comparative studies of perennial ryegrass, timothy and meadow fescue. 1. Herbage consumption and performance of calves fed on cut grass. *J. Br. Grassld Soc.* **24,** 308–16.

ALDER, F. E. 1970a. Comparative studies of perennial ryegrass, timothy and meadow fescue. 2. Results obtained in grazing experiments. *J. Br. Grassld Soc.* **25,** 53–64.

ALDER, F. E. 1970b. Comparative studies of perennial ryegrass, timothy and meadow fescue. 3. Results obtained from feeding hay, and animal production per unit area. *J. Br. Grassld Soc.* **25,** 162–6.

ALDER, F. E.; COOPER, E. M. 1967. Comparative studies of perennial ryegrass and cocksfoot as food for the calf. *J. agric. Sci., Camb.* **68,** 331–446.

ALDER, F. E.; COWLISHAW, S. J.; NEWTON, J. E.; CHAMBERS, D. T. 1967. The effects of level of nitrogen fertilizer on beef production from grazed perennial ryegrass/white clover pastures. Pt. 1. An irrigation experiment. *J. Br. Grassld Soc.* **22,** 194–203.

ALDER, F. E.; MINSON, D. J. 1960. Lucerne as a grazing crop. *Agriculture, Lond.* **67,** 448–51.

ALDER, F. E.; MINSON, D. J. 1963. The herbage intake of cattle grazing lucerne and cocksfoot pastures. *J. agric. Sci., Camb.* **60,** 359–69.

JACOBSON, D. R.; MILLER, W. M. 1961. Fescue toxicity. *J. Anim. Sci.* **20,** 960–1.

MILES, D. G.; WALTERS, R. J. K.; EVANS, E. M. 1969. Dry matter intake and live-weight gain of cattle and sheep offered different grass varieties with and without clover. *Anim. Prod.* **11,** 19–28.

SONNEVELD, A. 1965. [Observations on intake of fresh herbage by cattle.] *Veeteelt Zuivelber.* **8,** 409–20. [Cited in *Herb. Abstr.* **36,** 1267]

CHAPTER 46

Comparative Value of Herbage Species for Sheep Production

THE grass and legume species considered in detail in Chapters 15 to 22 and 33 to 36 may not be equally suitable for different sheep in different phases of production and managed in different ways. There are three main categories in which the suitability of these crops for sheep production can be discussed: (1) in relation to their survival under sheep grazing, (2) their value as food for different sheep production processes, and (3) in terms of the annual yield of nutrients useful to sheep.

Suitability for sheep grazing

This has been discussed in Chapter 43. The discussion under (2) and (3) will be confined therefore to those species which will survive under sheep grazing or which have some other value of overriding importance.

Value of herbage species as food for sheep

Herbage species differ in value according to the physiological state of the sheep: they will therefore be discussed in relation to the main phases of the production cycle.

DRY SHEEP

Mature ewes that are neither pregnant nor lactating have a very low food requirement if the need is simply to maintain body weight. If they are intended to gain weight because growth is incomplete (i.e. they are not, in fact, mature) or because they are in a thinner condition than is desired, then better-quality herbage is needed. For maintenance, however, any of the major herbage species listed is capable of supplying the nutrient requirements of the dry ewe, provided that it is harvested at a reasonable level of digestibility. It is not known precisely how low a digestibility is tolerable for this purpose, but hay with an apparent digestible dry matter (DDM) value of 45% would be expected to give little more than maintenance and could be associated with substantial crop yields (Blaxter & Wilson, 1963). In fact, even with low-quality herbage some restriction of intake is required to prevent dry ewes gaining weight.

There is no certainty that this applies to all breeds of sheep or to all well-conserved versions of the main herbage species, but there is no reason to think that it does not.

The best herbage species for maintenance purposes is therefore that which gives the greatest yield of nutrients, even if this involves a lower digestibility, since a greater number of sheep can then be maintained.

REPRODUCTION

The only evidence that specifically relates to reproduction in sheep is that concerning the oestrogenic activity of certain legumes (reviewed by Bickoff, 1968). There is little direct evidence from Britain, but Newton and Betts (1968) have reported oestrogenic activity in S123 red clover from April to October at the Grassland Research Institute, Hurley. Activity was also observed, but to a lesser extent, in Eynsford lucerne and S100 white clover from June to October. Sainfoin showed negligible activity.

The implications are that there may be some risk of lowered reproductive performance if ewes are grazed on certain legumes, notably S123 red clover, during or shortly before mating. There is no information for rams under British conditions and the evidence for ewes is as yet very limited. A great deal of work on this subject has been carried out in Australia, however, and the reproductive performance of ewes grazing oestrogenic legumes has been found to decline progressively to an extremely low level (Bennetts, Underwood & Shier, 1946; Barrett, George & Lamond, 1965).

LACTATION

There has been little critical experimentation on the values of different species for lactating ewes. From the evidence available on sheep lactation, it is difficult to distinguish when the result is to be attributed to the species used from when it is in fact associated with the quantity of that species available, or, indeed, with any of the many other factors that affect lactation. The sheer quantity present may well be the dominant attribute of the herbage in particular experiments. For spring-lambing flocks, herbage yield and earliness of growth may be the most important determinants of suitability, since herbage species are generally high in quality in the spring, but the quantity available varies.

If available in abundance and at a dry matter digestibility of over about 65%, all the major species would be adequate for a low milk yield. It is necessary, therefore, to specify the level of yield (rarely known in sheep production) and the stage of lactation. The peak milk yield in sheep is of relatively short duration (approximately weeks 2–6 after lambing) and it could be argued that a species which is satisfactory at this time will also be adequate at any other stage of lactation. Little is known for certain about this, however. There may be special factors that apply to the first days after lambing. It is unlikely that nutritional requirements will be critical in the second half of lactation unless a substantial weight loss has been incurred earlier; it may not be possible to prevent some weight loss during the peak of a heavy lactation.

Results of grazing experiments with ewes and lambs are often difficult to interpret because both the quantity and quality of herbage may influence performance and because it is not possible to dissociate direct effects on lamb growth from those which affect the milk supply.

Hodgson (1960) compared the performance of ewes suckling lambs on three different sward mixtures and found evidence of better lamb growth on clovers than on grass. The ewes also benefited, except early in the season when the production of clover was inadequate. Timothy/meadow fescue/clover mixtures gave better results than tall fescue.

Large and Tayler (1956) found higher lamb production per acre from a perennial ryegrass/white clover mixture than from one based on meadow fescue/timothy/cocksfoot/white clover, whether the sheep were grazed rotationally or folded, and at two different stocking rates. There was, however, a much higher proportion of white clover in the perennial ryegrass mixture.

EWES AND LAMBS

It is clear that where ewes and lambs are grazed together, somewhat different needs may have to be met at the same time.

If the ewe is before or at the peak of lactation, however, the solid food intake of the lambs is generally so small that the needs of the ewe must take precedence. Later, the situation is reversed, although the needs of the lamb are less important in terms of pasture species if the milk supply continues to be high (as in lowland ewes with single lambs) or the lambs are creep-fed.

Where a ewe flock suckling lambs of various ages has to be grazed at high stocking rates, a mixed perennial ryegrass/white clover sward would be expected to give the best results.

LAMB GROWTH

There is growing evidence that lamb growth is generally faster on legumes than on grasses (Rae, Brougham, Glenday & Butler, 1963; Rae, Brougham & Barton, 1964; Spedding, Large & Kydd, 1966). The reasons for this are thought to be associated with greater intakes and higher digestibility, but it has also been suggested (Johns, Ulyatt & Glenday, 1963) that diets which include white clover give rise to a higher volatile fatty-acid content in the rumen and higher proportions of propionic acid, with consequent improvement in performance.

The relationship between herbage species and lamb growth rate may, however, depend a good deal on the season (Large & Spedding, 1965) and on how dependent the lamb is on the forage for its dietary intake. If lambs are receiving a very high proportion of their intake as milk or concentrated foods (e.g. cereals), there is little likelihood that the herbage species will matter greatly, assuming that the herbage is in any case of high digestibility and present in adequate amounts.

For lambs receiving a high proportion of their total diet from herbage, the situation may be quite different, especially if they are very young (e.g. early-weaned at 6 weeks of age).

The only evidence available for young, weaned lambs and for lambs receiving a ration of milk substitute, suggests that lamb growth rate was faster on legumes than on grasses, that sainfoin was the best of the legumes tested (sainfoin, lucerne and white clover) and that perennial ryegrass was a relatively poor feed for lambs less than about 7 weeks of age (Treacher, 1966; Spedding 1970). It is difficult to be sure in such experiments, however, that the herbage species are adequately represented by the particular swards used.

Ryegrass has generally proved a better feed for lambs than cocksfoot (Milford & Minson, 1965, 1966) but not consistently so (Spedding, Large & Kydd, 1966).

The best current (1968) estimate of the rate of growth of early-weaned lambs at Hurley is shown below:

Herbage species compared	Growth rate lb/head per day	Growth rate g/head per day
Lucerne (Eynsford) *v.* cocksfoot (S37)	0·44 *v.* 0·35	200 *v.* 159
Perennial ryegrass (S23) *v.* cocksfoot (S37)	0·43 *v.* 0·38	195 *v.* 173
White clover (S100) *v.* perennial ryegrass (S23)	0·60 *v.* 0·53	272 *v.* 241

These results were obtained with low stocking rates, however, so that the quantity grown had little influence on performance. At present it must be accepted that satisfactory growth of early-weaned lambs requires stocking rates of less than 30 lambs/acre (75 lambs/ha), even where herbage production is at a high level. A marked decline in lamb growth rate may accompany any increase in stocking rate above this level (Spedding, 1967).

Little is known about the possibility that one herbage species may be better than another in terms of infestation with worm parasites or effects on carcass composition, quality or flavour. There is evidence from New Zealand that feeding on white clover may affect the flavour of fat (Cramer, Barton, Shorland & Czochanska, 1967), and that the inclusion of white clover in grass pastures gives rise to fats with higher iodine values but lower melting points (Shorland *et al.* 1967).

Annual yield of useful nutrients

The significance of the annual yield has to be related to the production process or, more usually, to the proportions of the total food required for each part of the process (i.e. for lambs, for dry ewes, etc.). Account must also be taken of the form in which the herbage is to be used, the proportion to be conserved and in what manner, and the pattern of yield, as well as the total for the year. A legume such as lucerne may give a high yield of high quality herbage entirely suitable for all parts of the sheep production process, provided that it is all dried. Even if the crop is made into mature hay it will be perfectly satisfactory for maintaining dry ewes.

To assess the best herbage species thus really requires that the entire system in which it is to be used be defined: to assess the most profitable choice requires an even more thorough examination. It is worth noting that the results of recent work in Australia (Axelsen & Morley, 1968) with ewes and lambs grazing 8 different grass species, suggested that in some circumstances species may be a more important determinant of profit than stocking rate, especially at stocking rates near an economic optimum.

In short, as for the other production processes referred to in preceding chapters, the information provided here is needed to help make such choices, but it is not sufficient by itself.

References

AXELSEN, A.; MORLEY, F. H. W. 1968. Evaluation of eight pastures by animal production. *Proc. Aust. Soc. Anim. Prod.* **7**, 92–8.

BARRETT, J. F.; GEORGE, J. M.; LAMOND, D. R. 1965. Reproductive performance of Merino-ewes grazing red clover (*Trifolium pratense* L.), improved pasture, or native pasture. *Aust. J. agric. Res.* **16**, 189–200.

BENNETTS, H. W.; UNDERWOOD, E. J.; SHIER, F. L. 1946. A specific breeding problem of sheep on subterranean clover pastures in Western Australia. *Aust. vet. J.* **22**, 2–12.

BICKOFF, E. M. 1968. Oestrogenic constituents of forage plants. *Review Series 1/1968 Commonw. Bur. Past. Fld Crops* Farnham Royal: Commonw. Agric. Bur., pp. 39.

BLAXTER, K. L.; WILSON, R. S. 1963. The assessment of a crop husbandry technique in terms of animal production. *Anim. Prod.* **5**, 27–42.

CRAMER, D. A.; BARTON, R. A.; SHORLAND, F. B.; CZOCHANSKA, Z. 1967. A comparison of the effects of white clover (*Trifolium repens*) and of perennial ryegrass (*Lolium perenne*) on fat composition and flavour of lamb. *J. agric. Sci., Camb.* **69**, 367–73.

HODGSON, J. 1960. The growth of lambs on three different swards under intensive grazing management. Ph.D. Thesis, Univ. Leeds.

JOHNS, A. T.; ULYATT, M. J.; GLENDAY, A. C. 1963. Pasture type in relation to liveweight gain, carcass composition, iodine nutrition and some rumen characteristics of the sheep. 4. Some rumen characteristics of sheep. *J. agric. Sci., Camb.* **61**, 201–7.

LARGE, R. V.; SPEDDING, C. R. W. 1965. The growth of lambs at pasture. 4. *J. Br. Grassld Soc.* **20**, 123–8.

LARGE, R. V.; TAYLER, J. C. 1956. Lamb production from two simple mixtures under two methods of management. *J. Br. Grassld Soc.* **11**, 216–23.

MILFORD, R.; MINSON, D. J. 1965. The energy value of ryegrass and cocksfoot assessed by a slaughter technique with lambs. *Br. J. Nutr.* **19,** 373–82.

MILFORD, R.; MINSON, D. J. 1966. The energy retention of lambs grazing S37 cocksfoot, S23 ryegrass and S24 ryegrass at low grazing pressure. *J. Br. Grassld Soc.* **21,** 7–13.

NEWTON, J. E.; BETTS, J. E. 1968. Seasonal variation in oestrogenic activity of various legumes. *J. agric. Sci., Camb.* **70,** 77–82.

RAE, A. L.; BROUGHAM, R. W.; GLENDAY, A. C.; BUTLER, G. W. 1963. Pasture type in relation to liveweight gain, carcass composition, iodine nutrition and some rumen characteristics of sheep. 1. Liveweight growth of the sheep. *J. agric. Sci., Camb.* **61,** 187.

RAE, A. L.; BROUGHAM, R. W.; BARTON, R. A. 1964. A note on liveweight gains of sheep grazing different ryegrass pastures. *N.Z. Jl agric. Res.* **7,** 491.

SHORLAND, F. B.; CZOCHANSKA, Z.; BARTON, R. A.; RAE, A. L. 1967. A comparison between the effects of ryegrass/white clover on the iodine values and melting points of some depot fats of sheep. *J. agric. Sci., Camb.* **68,** 221–5.

SPEDDING, C. R. W. 1967. The grazing sheep in relation to its environment. *Proc. 9th int. Congr. Anim. Prod.* Edinburgh: Oliver & Boyd, pp. 174–87.

SPEDDING, C. R. W. 1970. Sheep production and grazing management. London: Baillière, Tindall and Cassell, 2nd edn.

SPEDDING, C. R. W.; LARGE, R. V.; KYDD, D. D. 1966. The evaluation of herbage species by grazing animals. *Proc. 10th int. Grassld Congr., Helsinki, 1966* pp. 479–83.

TREACHER, T. T. 1966. Grassland Research Institute, Hurley. Unpublished data.

WILSON, A. D. 1966. The growth rate of weaner sheep on irrigated pastures. *Proc. Aust. Soc. Anim. Prod.* **6,** 179–81.

Subject Index

1. Numbers against headings refer to pages. Bold numbers indicate main references.
2. Chemical elements are spelled out in full when used as main headings, but referred to by symbol when used as subordinate headings.
3. 'See also individual species' at a subject heading indicates that only references of general application are shown under that subject heading, and that references applicable to one species only will be found under the appropriate sub-heading under that species.